Fundamentals of Advanced Omics Technologies: From Genes to Metabolites

Comprehensive Analytical Chemistry

Volume 63

Fundamentals of Advanced Omics Technologies: From Genes to Metabolites

Comprehensive Analytical Chemistry

Volume 63

Edited by

Carolina Simó
Laboratory of Foodomics, Institute of Food Science Research (CIAL), CSIC. Nicolás Cabrera 9, Madrid, Spain

Alejandro Cifuentes
Laboratory of Foodomics, Institute of Food Science Research (CIAL), CSIC. Nicolás Cabrera 9, Madrid, Spain

Virginia García-Cañas
Laboratory of Foodomics, Institute of Food Science Research (CIAL), CSIC. Nicolás Cabrera 9, Madrid, Spain

ELSEVIER

AMSTERDAM • BOSTON • HEIDELBERG • LONDON • NEW YORK • OXFORD
PARIS • SAN DIEGO • SAN FRANCISCO • SINGAPORE • SYDNEY • TOKYO

Elsevier
The Boulevard, Langford Lane, Kidlington, Oxford OX5 1GB, UK
Radarweg 29, PO Box 211, 1000 AE Amsterdam, The Netherlands

British Library Cataloguing in Publication Data
A catalog record for this book is available from the British Library

Library of Congress Cataloging-in-Publication Data
A catalogue record for this book is available from the Library of Congress

ISBN: 978-0-444-62651-6
ISSN: 0166-526X

Contents

Contributors to Volume 63

Sara Aibar, Bioinformatics and Functional Genomics Group, Cancer Research Center (CiC-IBMCC, CSIC/USAL), Salamanca, Spain

Benjamin Balluff, Center for Proteomics and Metabolomics, Leiden University Medical Center, Leiden, The Netherlands

Salvatore Cappadona, Proteomics Unit, Centre for Genomic Regulation (CRG) and Universitat Pompeu Fabra (UPF), Barcelona, Spain

Ricardo J. Carreira, Center for Proteomics and Metabolomics, Leiden University Medical Center, Leiden, The Netherlands

Richard D. Cummings, Department of Biochemistry and the Glycomics Center, Emory University School of Medicine, Atlanta, Georgia, USA

Sayani Dasgupta, Department of Molecular Pharmacology, Albert Einstein College of Medicine, Bronx, New York, USA

Noelia Dasilva, Centro de Investigación del Cáncer/IBMCC (USAL/CSIC), IBSAL, Departamento de Medicina Unidad de Proteomica & Servicio General de Citometría, University of Salamanca, Salamanca, Spain

Javier De Las Rivas, Bioinformatics and Functional Genomics Group, Cancer Research Center (CiC-IBMCC, CSIC/USAL), Salamanca, Spain

Rubén Díaz, Hospital Sant Joan de Deu, Endocrinology, Fundacio per la Recerca Sant Joan de Deu, Barcelona, Spain

Paula Diez, Centro de Investigación del Cáncer/IBMCC (USAL/CSIC), IBSAL, Departamento de Medicina Unidad de Proteomica & Servicio General de Citometría, University of Salamanca, Salamanca, Spain

Vahid Farrokhi, Department of Chemistry, University of Connecticut, Storrs, Connecticut, USA

Lloyd D. Fricker, Department of Molecular Pharmacology, Albert Einstein College of Medicine, Bronx, New York, USA

Manuel Fuentes, Centro de Investigación del Cáncer/IBMCC (USAL/CSIC), IBSAL, Departamento de Medicina Unidad de Proteomica & Servicio General de Citometría, University of Salamanca, Salamanca, Spain

Virginia García-Cañas, Laboratory of Foodomics, Institute of Food Science Research (CIAL), CSIC. Nicolás Cabrera 9, Madrid, Spain

Helen G. Gika, Department of Chemical Engineering, Aristotle University Thessaloniki, Thessaloniki, Greece

I. Gobernado, Servicio de Psiquiatría, Hospital Ramón y Cajal, Madrid, Spain

Maria Gonzalez-Gonzalez, Centro de Investigación del Cáncer/IBMCC (USAL/CSIC), IBSAL, Departamento de Medicina Unidad de Proteomica & Servicio General de Citometría, University of Salamanca, Salamanca, Spain

Clara Ibáñez, Laboratory of Foodomics, Institute of Food Science Research (CIAL), CSIC. Nicolás Cabrera 9, Madrid, Spain

Ricardo Jara-Acevedo, ImmunoStep, Edificio Centro de Investigación del Cáncer, Avda. Coimbra s/n, Campus Miguel de Unamuno, Salamanca, Spain

A. Jimenez-Escrig, Servicio de Neurología, Hospital Ramón y Cajal, Madrid, Spain

Josep C. Jiménez-Chillarón, Hospital Sant Joan de Deu, Endocrinology, Fundacio per la Recerca Sant Joan de Deu, Barcelona, Spain

Paul C.H. Li, Department of Chemistry, Simon Fraser University, Burnaby, British Columbia, Canada

Song Li, Department of Chemistry, University of Connecticut, Storrs, Connecticut, USA

Lucia Lourido, Instituto de Investigación Biomedica da Coruña (INIBIC), Hospital Universitario A Coruña, A Coruña, Spain

Jacob W. Malcom, Department of Molecular and Cell Biology, University of Connecticut, Storrs, Connecticut, USA

John H. Malone, Department of Molecular and Cell Biology, University of Connecticut, Storrs, Connecticut, USA

Francesco M. Mancuso, Proteomics Unit, Centre for Genomic Regulation (CRG) and Universitat Pompeu Fabra (UPF), Barcelona, Spain

Liam A. McDonnell, Center for Proteomics and Metabolomics, Leiden University Medical Center, Leiden, The Netherlands

Adam J. McShane, Department of Chemistry, University of Connecticut, Storrs, Connecticut, USA

Tyler J. Moss, Department of Systems Biology, UT MD Anderson Cancer Center, Houston, Texas, USA

G.A. Nagana Gowda, Department of Anesthesiology and Pain Medicine, Northwest Metabolomics Research Center, University of Washington, Seattle, Washington, USA

Reza Nemati, Department of Chemistry, University of Connecticut, Storrs, Connecticut, USA

Alberto Orfao, Centro de Investigación del Cáncer/IBMCC (USAL/CSIC), IBSAL, Departamento de Medicina Unidad de Proteomica & Servicio General de Citometría, University of Salamanca, Salamanca, Spain

Daniel Raftery, Department of Anesthesiology and Pain Medicine, Northwest Metabolomics Research Center, University of Washington, and Public Health Sciences Division, Fred Hutchinson Cancer Research Center, Seattle, Washington, USA

Prahlad T. Ram, Department of Systems Biology, UT MD Anderson Cancer Center, Houston, Texas, USA

Marta Ramón-Krauel, Hospital Sant Joan de Deu, Endocrinology, Fundacio per la Recerca Sant Joan de Deu, Barcelona, Spain

Beatriz Roson, Bioinformatics and Functional Genomics Group, Cancer Research Center (CiC-IBMCC, CSIC/USAL), Salamanca, Spain

Eduard Sabidó, Proteomics Unit, Centre for Genomic Regulation (CRG) and Universitat Pompeu Fabra (UPF), Barcelona, Spain

A. Sanchez-Herranz, Servicio de Neurobiología-Investigación, Unidad Central de Genómica Translacional, Hospital Ramón y Cajal, Madrid, Spain

Alex Sánchez-Pla, Statistics Department, Facultat de Biologia, University of Barcelona, and Statistics and Bioinformatics Unit, Vall d'Hebron Institut de Recerca (VHIR), Barcelona, Spain

Ram Sasisekharan, Department of Biological Engineering, Koch Institute for Integrative Cancer Research, Massachusetts Institute of Technology, Cambridge, Massachusetts, USA

J.M. Sayagues, Centro de Investigación del Cáncer/IBMCC (USAL/CSIC), IBSAL, Departamento de Medicina Unidad de Proteomica & Servicio General de Citometría, University of Salamanca, Salamanca, Spain

Philippe Schmitt-Kopplin, Research Unit Analytical BioGeoChemistry, Helmholtz Zentrum München, German Research Center for Environmental Health, Neuherberg, and Chair of Analytical Food Chemistry, Technische Universität München, Freising-Weihenstephan, Germany

Abootaleb Sedighi, Department of Chemistry, Simon Fraser University, Burnaby, British Columbia, Canada

Vasudha Sehgal, Department of Systems Biology, UT MD Anderson Cancer Center, Houston, Texas, USA

Carolina Simó, Laboratory of Foodomics, Institute of Food Science Research (CIAL), CSIC. Nicolás Cabrera 9, Madrid, Spain

David F. Smith, Department of Biochemistry and the Glycomics Center, Emory University School of Medicine, Atlanta, Georgia, USA

Xuezheng Song, Department of Biochemistry and the Glycomics Center, Emory University School of Medicine, Atlanta, Georgia, USA

Nathan W. Stebbins, Department of Biological Engineering, Koch Institute for Integrative Cancer Research, Massachusetts Institute of Technology, Cambridge, Massachusetts, USA

Georgios A. Theodoridis, Department of Chemistry, Aristotle University Thessaloniki, Thessaloniki, Greece

Alberto Valdés, Laboratory of Foodomics, Institute of Food Science Research (CIAL), CSIC. Nicolás Cabrera 9, Madrid, Spain

Ian D. Wilson, Department of Surgery and Cancer, Faculty of Medicine, Imperial College, South Kensington, London, United Kingdom

Michael Witting, Research Unit Analytical BioGeoChemistry, Helmholtz Zentrum München, German Research Center for Environmental Health, Neuherberg, Germany

Xudong Yao, Department of Chemistry, University of Connecticut, Storrs, Connecticut, USA

Ying Yu, Department of Biochemistry and the Glycomics Center, Emory University School of Medicine, Atlanta, Georgia, USA

Series Editor's Preface

Over the past years we have included a number of books in the CAC series that have been written or edited by Spanish scientists. This is a direct consequence of the leadership of Prof. Dr. M. Valcarcel, a real pioneer in Analytical Chemistry in Spain, and of continued investment in the field with new chairs being created across the country, which has led to Spain becoming one of the highest ranking countries in the world for publications in analytical chemistry.

This volume, edited by C. Simó, A. Cifuentes and V. García-Cañas, is a direct consequence of the leading-edge anaytical chemistry research carried out in Spain during the last 20 years. When my colleague at CSIC in Madrid and old friend Alejandro Cifuentes told me about his project, I immediately proposed that we include a book on it in the CAC series. Why? We did not have many titles on Omics in our series. In this respect we should consider this volume a follow up to volumes 46 and 52 on Proteomics and Peptidomics, and on Protein Mass Spectrometry, respectively. This new CAC volume on the fundamentals of advanced omics technolgies is an excellent addition to these two previous books and a further volume on advanced applications will follow later in the year.

As mentioned by the editors in their introduction, this field was developed in tandem with the instrumental and methodological developments achieved at the end of the 20th Century. Powerful bioinformatic tools were also integrated in the systems, allowing us to understand the relevant mechanisms of protein synthesis. Omics tools are used nowadays for high throughput assessment of changes at genome (genomics), epigenome (epigenomics), transcript (transcriptomics), protein (proteomics) and metabolite (metabolomics) levels.

The book contains 18 chapters covering a broad range of topics. The book contains a balanced cocktail of 3-4 chapters each of fundamentals and applications of genomics such as transcriptomics, epigenetics, proteomics, metabolomics and other omics approaches. Data treatment and bioinformatics are presented in the last 5 chapters, showing the importance of these tools for Omics technologies. All chapters are written by recognized experts in the field. It will be useful to establsihed scientists in the field as well as to newcomers who want to know more about Omics.

Finally, I would like to thank the editors and all the authors for their contributions to this needed state of the art book in the CAC series. I am sure it will enjoy success among the scientific community and I expect to see many citations in the literature in the next coming years.

Prof. Dr. D. Barceló
Barcelona, Spain, December 4, 2013

Preface

Omics aims at the collective characterization and quantification of pools of biological molecules (mainly nucleic acids, proteins, and metabolites) that translate into the structure, function, and dynamics of an organism or organisms. The related suffix, *-ome*, is used to refer to the study of the complete set of these molecules (i.e., the genome, proteome, or metabolome, respectively). The instrumental and methodological developments achieved at the end of the twentieth century, driven by the Human Genome Project, now have made it possible to study these complex pools of molecules. Moreover, the advances in Omics tools observed at the beginning of the twenty-first century have made feasible analytical instruments and methodologies that were unthinkable a few years ago, including powerful bioinformatic tools to integrate and interrogate multiple Omics datasets. Interestingly, the Omics field is evolving rapidly due to the huge efforts in identifying genes and pathways, detecting metabolic bottlenecks, and understanding the relevant mechanism of protein synthesis. These efforts are generating faster, more powerful, and more sensitive Omics tools for a thorough high-throughput assessment of changes at the genome (genomics), epigenome (epigenomics), transcript (transcriptomics), protein (proteomics), and metabolite (metabolomics) levels.

The interest of the scientific community in developing new Omics technologies and the different trends in this hot area of research are well documented in the 18 chapters of this book, *Fundamentals of Advanced Omics Technologies*: *From Genes to Metabolites*. This volume provides a global perspective on the advances of the different Omics technologies for investigating the genome, epigenome, transcriptome, proteome, and metabolome, including the development and use of different bioinformatic strategies for data treatment, integration, and interrogation. This volume on the fundamentals of advanced Omics technologies will be complemented by a second volume devoted to applications of these techniques, which will also be published in this series.

Chapters 1 and 2 provide an overview and discuss the current status of DNA microarray technology, including its current challenges and future trends. Chapter 3 describes next-generation sequencing tools, while Chapter 4 introduces advanced epigenomics tools used for the wide analysis of methylation and histone-genome modifications. The following three chapters are devoted to proteomics, covering an overview of quantitative proteomic approaches (Chapter 5), the use of emerging nanotechniques in proteomics (Chapter 6), and the application of mass spectrometry imaging in proteomics and metabolomics (Chapter 7). The next three chapters are dedicated to metabolomics, namely, they describe recent advances in NMR-based metabolomics (Chapter 8), discuss the role of mass spectrometry in nontargeted metabolomics (Chapter 9), and introduce the use of direct mass spectrometry-based approaches in metabolomics (Chapter 10). The book also covers other, more specific Omics approaches; thus, Chapter 11 describes the methodologies and challenges for a global glycomics analysis, Chapter 12 presents applications of glycan microarrays for functional glycomics, and Chapter 13 describes high-resolution analytical tools for quantitative peptidomics. The last five chapters are devoted to the fundamental description of several tools used for treatment and/or integration of Omic datasets. Namely, the final chapters describe tools for gene expression analysis and profiling of microarrays data and RNA-sequencing data (Chapter 14), the insights and challenges to carry out the analysis of deep-sequencing data (Chapter 15), the description of advanced bioinformatic approaches to increase proteome coverage (Chapter 16), the technical requirements for successful transcriptome and metabolome data integration and visualization (Chapter 17), and the computational approaches now available for visualization and integration of Omics data (Chapter 18).

As the editors of this book, we would like to thank all the authors for their valuable contributions, Damiá Barceló for inviting us to prepare this piece of work, Derek Coleman and Susan Dennis for their help and support, and to those on the Elsevier team whose efforts contributed to the preparation of this volume. *Gracias*!

Carolina Simó, Alejandro Cifuentes, and Virginia García-Cañas
Laboratory of Foodomics, CIAL, Madrid
National Research Council of Spain (CSIC)

Chapter 1

DNA Microarrays Technology: Overview and Current Status

Alex Sánchez-Pla*,†
*Statistics Department, Facultat de Biologia, University of Barcelona, Barcelona, Spain
†Statistics and Bioinformatics Unit, Vall d'Hebron Institut de Recerca (VHIR), Barcelona, Spain

Chapter Outline

1 INTRODUCTION AND OVERVIEW

Since the early days of molecular biology, life scientists have been interested in being able to measure gene expression or to quantify gene variation, either at the nucleotide level or by the number of copies of a gene. Techniques to measure the expression level of a gene, such as the *Northern blot (1)*, or to quantify DNA variation, such as restriction fragment length polymorphism *(2)*, have been in use for decades and have become part of the biologist's standard toolbox. These techniques, which were considered revolutionary when they were introduced, became much less attractive when the

Comprehensive Analytical Chemistry, Vol. 63. http://dx.doi.org/10.1016/B978-0-444-62651-6.00001-5

information on genomic sequences increased exponentially as a result of sequencing projects. However, with the availability of sequence information also came the solution. The development of microarray technology allowed to enter a new era of *high-throughput* data generation and analysis where gene expression or gene variation could be measured simultaneously at hundreds or thousands of genes (for gene expression) or even millions of loci (for gene variation).

Microarrays are a group of technologies designed to perform high-throughput screening by exposing amounts of biological material to a slide of plastic or glass where known DNA sequences or proteins that act as detectors have been previously deposited. DNA sequences from the biological sample, usually known as *targets*, are transformed into a stable DNA form, such as complementary DNA (cDNA), and in the process they are labeled with a fluorescent dye. Labeled target molecules are exposed to all the sequences on the slide, usually known as *probes*. It is expected that during this exposure, target molecules have the chance to hybridize with their complementary sequences on the slide. The detection of target sequences that have been hybridized to their respective probes is done by stimulating the microarrays with the appropriate laser light, scanning the resulting images, and, by means of image analysis, identifying and quantifying the signals emitted by those target molecules that have hybridized to the probes (Figure 1) *(3)*.

Research "on" microarrays, especially on techniques for microarray data analysis, and "with" microarrays, that is, applications that use microarrays, has been very active during the first decade of the century *(4,5)*, and thousands of papers on their use, applications, and analysis have been published, as can be seen by searching PubMed for references with the term "microarray" in their title (see Figure 2) *(6)*. In this chapter the basic

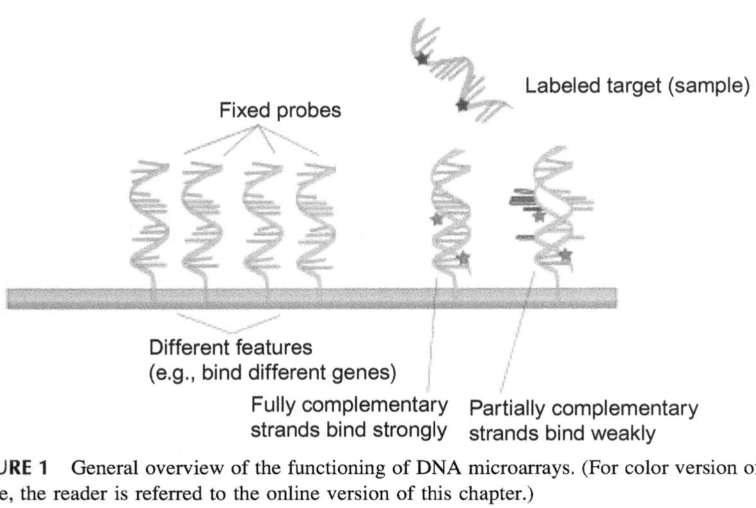

FIGURE 1 General overview of the functioning of DNA microarrays. (For color version of this figure, the reader is referred to the online version of this chapter.)

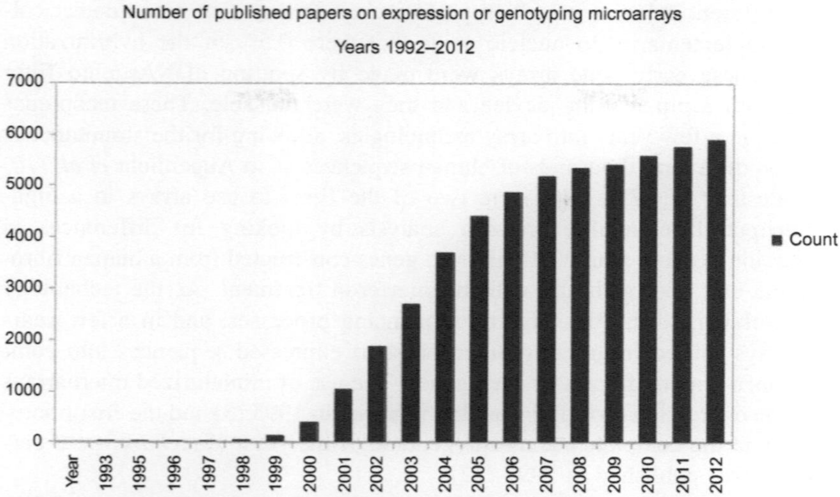

FIGURE 2 Evolution of the number of published microarray articles. The number of papers published on microarrays increased exponentially during the decade 1999–2008 and is now relatively stable. *The data from the image has been obtained from a PubMed search performed with the following terms: "microarray" OR "gene expression array" OR "(microarrays AND gene)" OR "(microarray AND gene)."* (For color version of this figure, the reader is referred to the online version of this chapter.)

principles underlying the technology of microarrays are described. In the following sections, the main different types of arrays and their applications are considered, and an overview of the current state and of future perspectives is presented.

1.1 A Brief History of Microarrays

The approach of cloning multiple sequences in one process has been used in biology for a long time. It can be traced to well-established techniques such as the Southern blot *(7)*. In this technique fragmented DNA is bound to a substrate, either of nitrocellulose or a nylon membrane. The DNA is denatured, dried, and then exposed to a labeled hybridization probe in an appropriate buffer. The blot is then extensively washed and analyzed by X-ray film, autoradiography, or membrane chromogen detection, depending on the type of probe label employed. Although in recent years this technique has been replaced by improved approaches, it can still be an option, especially in cases where the length of the expanded DNA is greater than the usual amplification ability of polymerase chain reaction (PCR). A different approach, the Grunstein and Hogness method *(8)*, was also introduced in 1975 to identify plasmid clones by colony hybridization. The method was quickly extended by Gergen *et al. (9)* to create small 144-well microplates able to produce

1728 different colonies on which perform experiments that could detect colony complementarity to nucleic acids that were 0.5% of the hybridization probe. These early gene arrays were made by spotting cDNAs onto filter paper with a pin-spotting device and they were reusable. These techniques evolved in a few years into array technologies, allowing for the simultaneous analysis of several thousands of human sequences as in Augenlicht *et al.* *(10)* or Kulesh *et al.* *(11)*, who were two of the firsts to use arrays in a high-throughput differential expression analysis by looking for differences in expression of more than 2000 different genes constructed from a human fibrosarcoma cell line, with and without interferon treatment. As the technology was evolving, so did the ongoing sequencing processes, and in a few years the arrays shifted from being based on short expressed sequences into complete open reading frames or even genes. The use of miniaturized microarrays for gene expression profiling was first reported in 1995 *(3)* and the first microarray with the complete eukaryotic genome of the yeast (*Saccharomyces cerevisiae*) was published in 1997 *(12)*.

2 TYPES OF DNA MICROARRAYS

After almost 20 years of developments, DNA microarray technology is mature, and along the years of its existence, it has evolved and experienced a high diversification so that one can find many types of microarrays. There are many criteria to distinguish between types of microarrays. Some of these criteria are introduced here, and they will be the basis for the description of the main microarray types: (i) Depending on how they have been manufactured, microarrays can be classified as *spotted*, *synthesized*, and *self-assembled* microarrays; (ii) depending on the application that microarrays are designed for, one can distinguish expression microarrays, single-nucleotide polymorphism (SNP) arrays, tissue microarrays, microRNA arrays, protein microarrays, etc.; (iii) in gene expression arrays one can distinguish between *two-channel* or *one-channel* microarrays. Section 2 is devoted to describe different types of arrays depending on how they are produced. A detailed description of DNA arrays and its characteristics based on their field of application is provided in section 3. Other types of microarrays are discussed in other chapters of the book.

2.1 Spotted or Printed Microarrays

Spotted DNA microarrays were the first to enter the market in the mid-1990s *(6)* and, in spite of improvements and competitors, they have continued in use today. The name of this type of arrays comes from the fact that the probes, which are the sequences on the array that will act as templates for the sequences from the sample target molecules, are first synthesized and then printed, or "spotted" on a glass slide *(13)*. Sequences for the probes are obtained from public information stored in sequence databases such as Genbank *(14)* and they are specifically selected to characterize as uniquely as possible the biological element (gene, exon, transcript, etc.) they are intended to represent *(15)*.

Once the sequences for the probes have been selected they are synthesized *in vitro*. Probes can be either long polynucleotides or shorter oligonucleotides. If long polynucleotides are used they are usually the products of selective amplification of cloned cDNAs by PCR *(16)*. If oligonucleotides are used they are usually commercially presynthesized sequences *(17)*. In the microarray literature the term "oligonucleotide" is often used synonymously for "*in situ* synthesized" microarrays, probably due to the fact that, due to technical limitations, this type of array can only contain short oligonucleotide probes. It is important, however, to avoid confusion because, although it is common that spotted microarrays are made with long cDNA probes, these can also consist of sort oligonucleotides *(17)*. In spotted microarrays, synthesized probes are printed on the slide usually in an automated process that uses a robotic system with several pins (Figure 3), which deposits (spots) the probes on the glass using blocks (a block corresponds to all the pins working together). Spots are organized in a rectangular grid due to the way they have been deposited by the robots and this arrangement has originated the term "microarray."

Spotted arrays have some characteristics that made them an option of choice even if new and potentially more accurate technologies have been developed. Their main advantage is probably the length of the probes they are made of, which provides a high sensitivity. This is, however, not free of problems because it makes cross-hybridization easier than with shorter sequences determining a relatively low specificity *(18)*.

2.2 *In Situ* Synthesized Microarrays

In spite of the flexibility that spotted microarrays allowed, they showed relatively low reproducibility. Besides this and independently of their quality, these arrays were known to easily present cross-hybridization problems due to the length of the probes they contained *(19)*. As an alternative, the company Affymetrix (www.affymetrix.com) developed a new type of microarrays which intended to be more precise than traditional spotted arrays, thanks to an industrial approach for their production where automatization was expected to avoid small mistakes that appeared in microarrays existing at the time *(9)*. Affymetrix probes are synthesized directly on the support using a process called photolithography *(20)* (Figure 4). This technique is based on the sequential addition of nucleotides (A, C, G, T) to each probe using masks or templates, which, in a similar way to the processes used to print electronic circuits, guide the addition or the omission of a nucleotide at each position of the sequences on a cyclic process that ends when all the nucleotides have been added. Probes synthesized using this approach could not be so long as PCR or

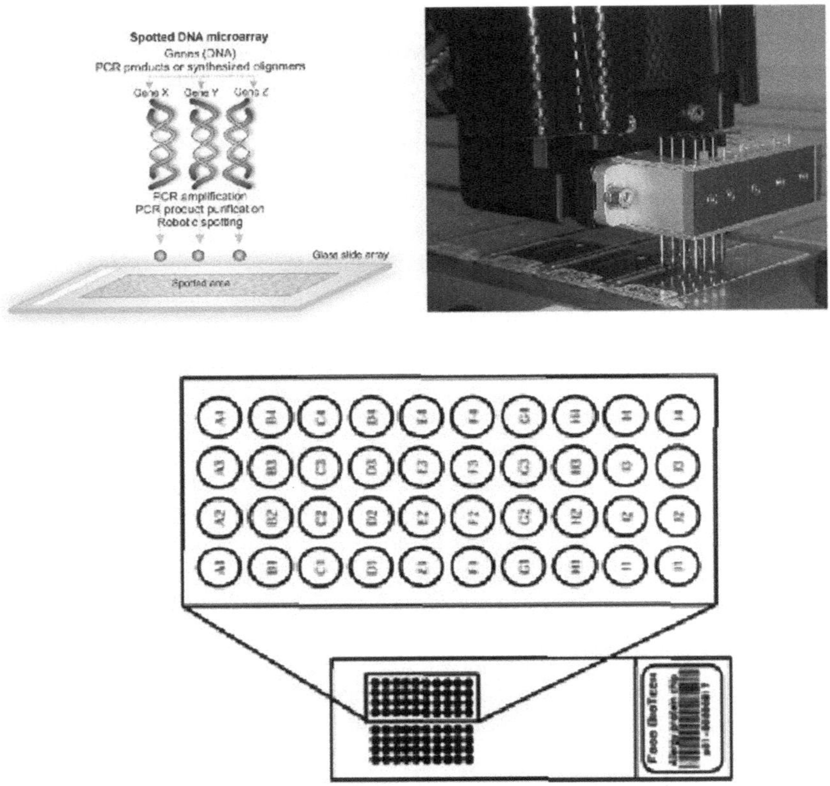

FIGURE 3 Spotted microarrays production. Spotted microarrays (schematically represented in the lower part of the image) are produced by a robotic spotter (represented in the right upper part of the image) that spots previously synthesized sequences such as PCR products, cDNAs, clone libraries, or long oligonucleotides onto coated glass slides (upper part of the image, left). Each spot on the array represents a particular gene fragment, that is, 40–70 nucleotides for oligonucleotide arrays, or several hundred nucleotides for PCR products. (For color version of this figure, the reader is referred to the online version of this chapter.)

cDNA products used in spotted arrays because, for technical reasons, they could reach 25-mer long at most. However, the higher specificity of the sequences jointly with their high density (high number of copies of each probes) determined that this type of array has been very successful in many fields, especially human biology and biomedicine.

Following the introduction of *in situ* synthesized microarrays by Affymetrix, other companies such as Roche NimbleGen (www.nimblegen.com) and Agilent Technologies (www.agilent.com) introduced their own synthesized arrays. Both platforms used longer oligonucleotide probes (60–100 bp) than Affymetrix arrays but they differed in the technology used for probe synthesis. NimbleGen uses maskless photo-mediated synthesis *(21)*, while Agilent

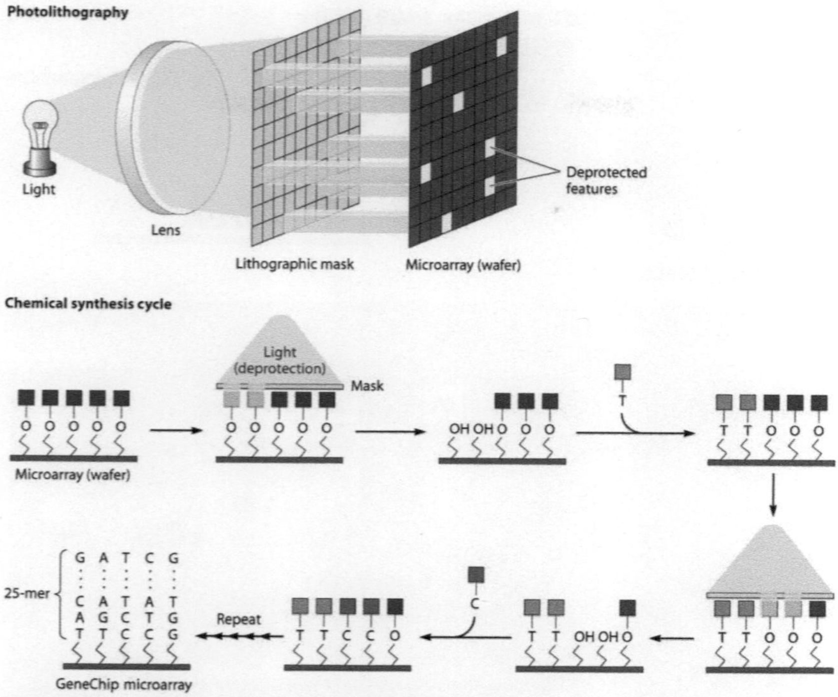

FIGURE 4 Affymetrix oligonucleotide microarrays production by photolithography. Outline of the photolithographic synthesis of oligonucleotides in Affymetrix microarrays. UV light is passed through a lithographic mask that acts as a filter to either transmit or block the light from the chemically protected microarray surface (wafer). The sequential application of specific lithographic masks determines the order of sequence synthesis on the wafer surface. (Bottom) chemical synthesis cycle: UV light removes the protecting groups (squares) from the array surface, allowing the addition of a single protected nucleotide as it is washed over the microarray. Sequential rounds of light deprotection, changes in the filtering patterns of the masks, and single-nucleotide additions form microarray features with specific 25-bp probes. *Image courtesy of Affymetrix.* (For color version of this figure, the reader is referred to the online version of this chapter.)

relies on inkjet technology for the synthesis of the probes *(22)*. The Roche NimbleGen approach to *in situ* synthesis is similar to that of Affymetrix just described, but photolithographic masks are replaced by "virtual" or digital masks (see Figure 5). Maskless array synthesizer technology uses an array of programmable micromirrors to create digital masks that reflect the desired pattern of UV light to deprotect the features where the next nucleotide will be coupled. Each NimbleGen microarray can contain more than 1 million features. On the other hand, Agilent technology rely on inkjet printing on glass slides, which eliminates the need for either lithographic or digital masks (Figure 6). The *in situ* synthesis of 60-mer oligonucleotides is achieved using five-"ink" (four bases plus catalyst) printing of nucleotide precursors

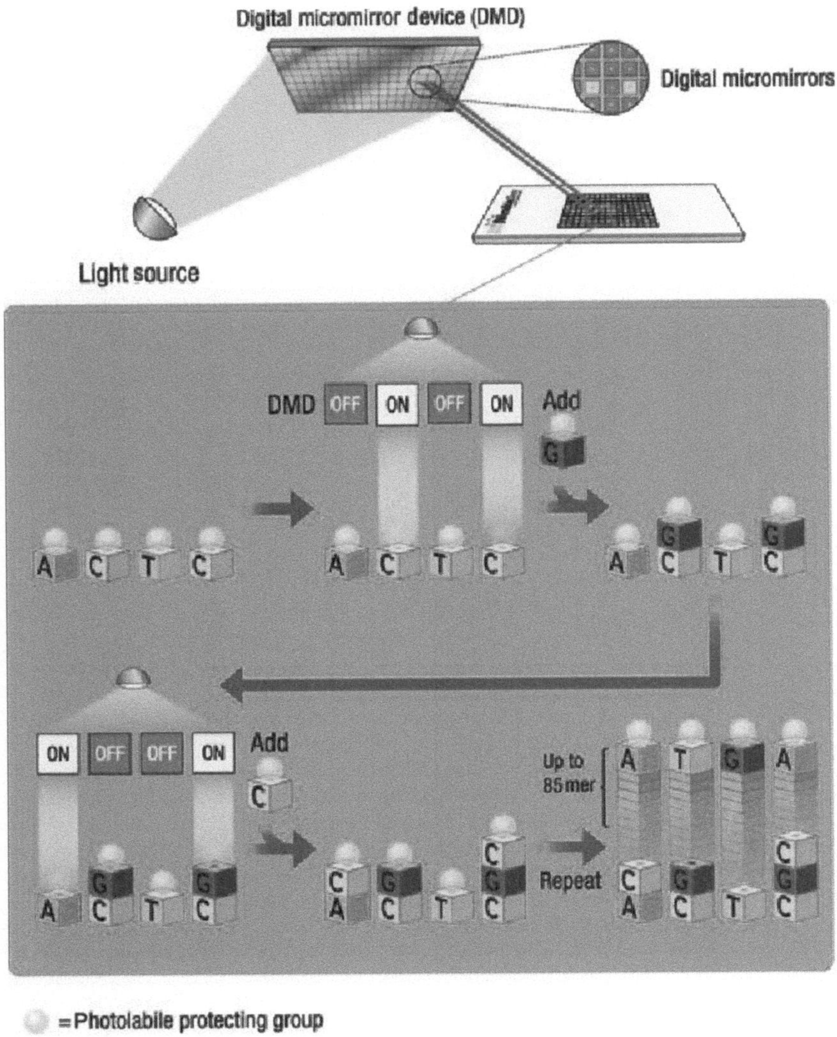

FIGURE 5 Microarray synthesis using Roche NimbleGen MAS technology. The synthesis of microarrays using NimbleGen MAS technology is very similar to traditional oligonucleotide synthesis, with some important exceptions. Unlike conventional oligonucleotide synthesis, arrays are synthesized on glass slides rather than controlled pore glass supports. Another key difference is that the deprotection steps are performed by photodeprotection rather than by acid deprotection. The illustration here depicts digital micromirrors reflecting a pattern of UV light, which deprotects the nascent oligonucleotide and allows addition of the next base. *Image courtesy of Roche NimbleGen.* (For color version of this figure, the reader is referred to the online version of this chapter.)

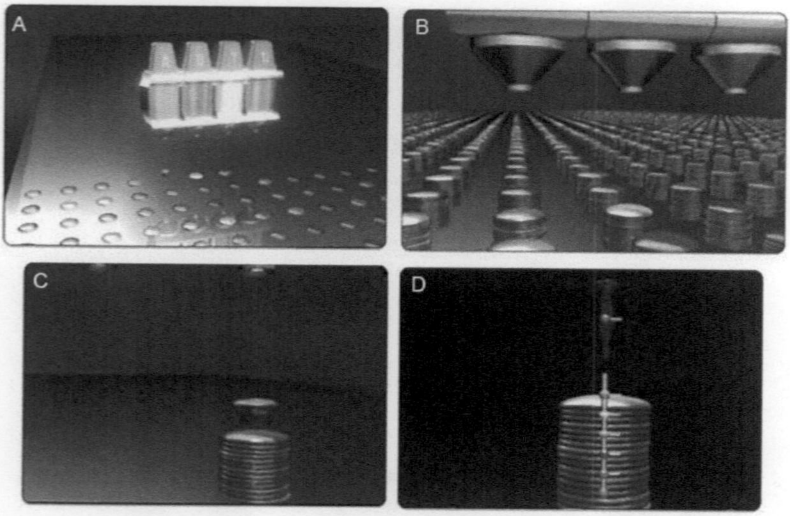

FIGURE 6 Microarray synthesis using Agilent technology. Agilent oligonucleotide microarray synthesis via inkjet printing. (A) The first layer of nucleotides is deposited on the activated microarray surface by means of noncontact printing. (B) Growth of the oligonucleotides is obtained by multiple rounds of base-by-base printing. (C) Closeup of one oligonucleotide as a new base is being added to the chain. (D) Final product is an array with thousands of copies of each probe, which consists of long (60-mer) oligonucleotides. *Images courtesy of Agilent Technologies.* (For color version of this figure, the reader is referred to the online version of this chapter.)

combined with coupling and deprotection steps. Agilent microarrays contain around 250,000 million features, although this may change depending on the evolution of their technology.

2.3 High-Density Bead Arrays

The arrays described in the previous sections are all produced by spotting or synthesizing probes onto two-dimensional substrates at known locations *(3,23)*. An alternative approach that has turned out to be very successful is the use of self-assembled arrays, also known as *bead arrays*, which are based on the random self-assembly of a bead pool onto a patterned substrate. These arrays were developed by David Walt at Tufts University *(24)* and then licensed to the company Illumina (http://www.illumina.com). The production process of these arrays starts by synthesizing DNA on microscopic (3–5 μm) silicon beads that are then deposited on the end of fiber-optic array where the ends of the fiber are etched to provide wells slightly larger than the beads. Synthesizing different DNA types on different beads and then applying the resulting mixture of beads to the fiber-optic cable produces a randomly assembled array (see Figure 7). Each bead is represented on average by around 20 copies within the array. The identity of this bead, that is, to which

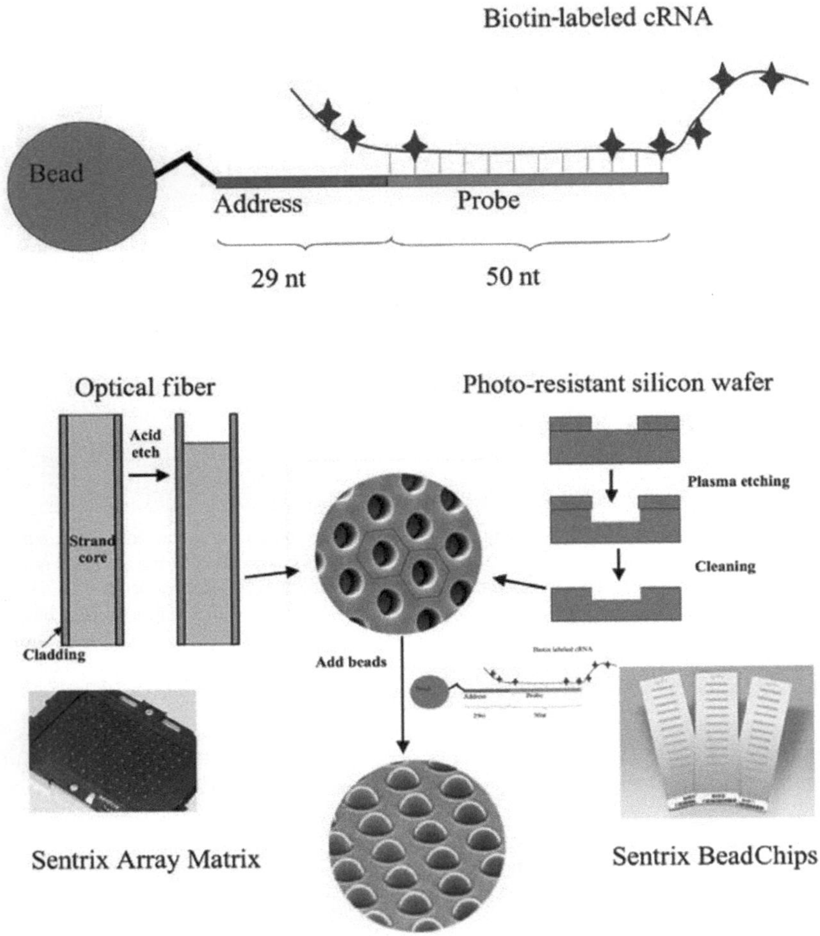

FIGURE 7 Illumina BeadArrays. The Sentrix Array Matrix contains 96 1.4-mm fiber-optic bundles (bottom left). Each bundle is an individual array consisting of 50,000 5-μm fiber-optic strands, each of which is chemically etched to create a microwell for a single bead (top left). The Sentrix BeadChips can assay 1–16 samples at a time on a silicon slide (bottom right) that has been processed to provide microwells for individual beads (top right). Both BeadArray platforms rely on 3-μm silica beads that randomly self-assemble (center and top). (For color version of this figure, the reader is referred to the online version of this chapter.)

sequence it is associated, is determined by decoding an address sequence, which is also attached to the bead. As a result of the decoding process, a unique array layout file is obtained for each array that can be used to decode the data during the scanning process *(25)*. An interesting capability of bead arrays technologies is the possibility of multiplexing, that is, of analyzing multiple samples in the same array *(10)*.

3 APPLICATIONS OF MICROARRAYS

Microarrays can be also classified on the basis of their intended use or application. The most common application of DNA microarray technology is the analysis of gene expression *(26)*, but there are many more *(5,27,28)*. In this section the main types of applications are reviewed.

3.1 Microarrays for Gene Expression Analysis

Gene expression microarrays are used to measure gene expression levels by indirectly identifying and quantifying the mRNA transcripts present in the cells in a given experimental condition. The application of these arrays is based on the assumption that the expression of a gene is, somehow, proportional to the number of mRNA molecules coming from the transcription of this gene *(29)*. That is, the higher the quantity of mRNA, the higher the expression of the gene. Although this assumption is far from being universally accepted *(30)*, there has been enough consensus on its validity that thousands of studies using gene expression microarrays have been performed and published.

3.1.1 One- and Two-Channel Microarrays

Indistinctly of how gene expression microarrays have been produced, the goal of this type of microarray is to capture the expressed mRNA in a given sample, to rely on the identity of the expressed genes (exon, transcript, etc.), and to indirectly quantify their expression. To do this, two main approaches have been used: two-channel and one-channel microarrays.

Two-channel microarrays are based on the competitive hybridization of two samples, each of which has been labeled with one different fluorescent dye (e.g., Cy3, fluorescent in the green region of the spectrum, or Cy5, fluorescent in the red region), for the probes in the microarray. After hybridization the array is exposed to red and green laser beams, which causes fluorescence signals that are proportional to the hybridized DNA. The scanned image of the microarray is then subjected to some corrections, providing expression values for each probe that represent the gene expression of one sample relative to the other *(31,32)*. Relative gene expression is a reasonable measure of gene expression if the desired goal is to establish how many times the gene is more or less expressed in one condition than in the other *(33)*. The "fold-change," the quotient between the expressions in the two conditions, has become the most frequent measure of gene expression *(34)*. Whereas this is very intuitive and simple to use in experiments with only two conditions, it becomes more complicated when there are more than two, not to say when there are multiple experimental factors, each of them with multiple conditions *(35,36)*.

An alternative approach to relative gene expression estimation is to rely on an absolute expression measure that is based on the direct quantification of the total quantity of mRNA present in the sample. Although this can be done using two-color microarrays *(37,38)*, the most common approach is to use one-color microarrays instead of the previously described, two-color microarrays.

In one-color, or one-channel, microarrays each RNA sample is analyzed in a separate microarray without relying on competitive hybridization. This approach provides absolute expression measures for each gene in each condition and has the advantage of being more flexible, especially for complicated experimental designs *(39)*. Absolute expression values are not free from criticism because, given that they are indirect measures (all microarrays measure gene expression indirectly), and that they have to go through diverse preprocessing and normalization steps, the final values are difficult, if not impossible, to give a biological interpretation *(40)*.

There are many companies providing all types of catalogue and custom arrays, and even an attempt to describe the broad variety available exceeds the purpose of this chapter. Although slightly outdated, the "Technology Features" section of *Nature (41)* provides a table of suppliers (http://www.nature.com/nature/journal/v442/n7106/pdf/4421071a.pdf) that can be consulted to have an overview of the companies and types of available microarrays.

3.1.2 Gene Expression Microarray Experiments

An important aspect in the application of DNA microarrays is how they are used to perform experimental studies. In this section a typical gene expression microarray experiment is described. Figure 8, adapted from Staal *et al. (42)*, summarizes the different steps undertaken in such experiments, highlighting common aspects and differences between one- and two-channel microarrays. There are of course as many variations of each step as different technologies are available, but, for simplicity, the description is kept generic without going into differences other than one- and two-color microarrays.

A microarray experiment, as any other experiment, starts with an experimental design where the allocation of samples to experimental conditions, the sample size, and all the usual important aspects of an experiment are performed *(36,43)*. In the following paragraph one can assume that this has been done for some experimental conditions and that one or two samples are available to be hybridized on two- or one-color microarrays. This is done following the steps described below and outlined in Figure 8.

1. The first step consists of RNA extraction and purification. Different criteria can be applied to decide which RNA fraction is to be analyzed, but it is very common to rely on total RNA. Before proceeding to the next step, the yield and quality of the purified RNA have to be determined *(44)*. RNA quantity may be determined, for example, with the Picogreen assay kit *(45)* or the Agilent Bioanalyzer *(46)*. RNA quality may also be determined using Agilent's Bioanalyzer. If quantity and quality of RNA are within acceptable limits, one can proceed to the next step.

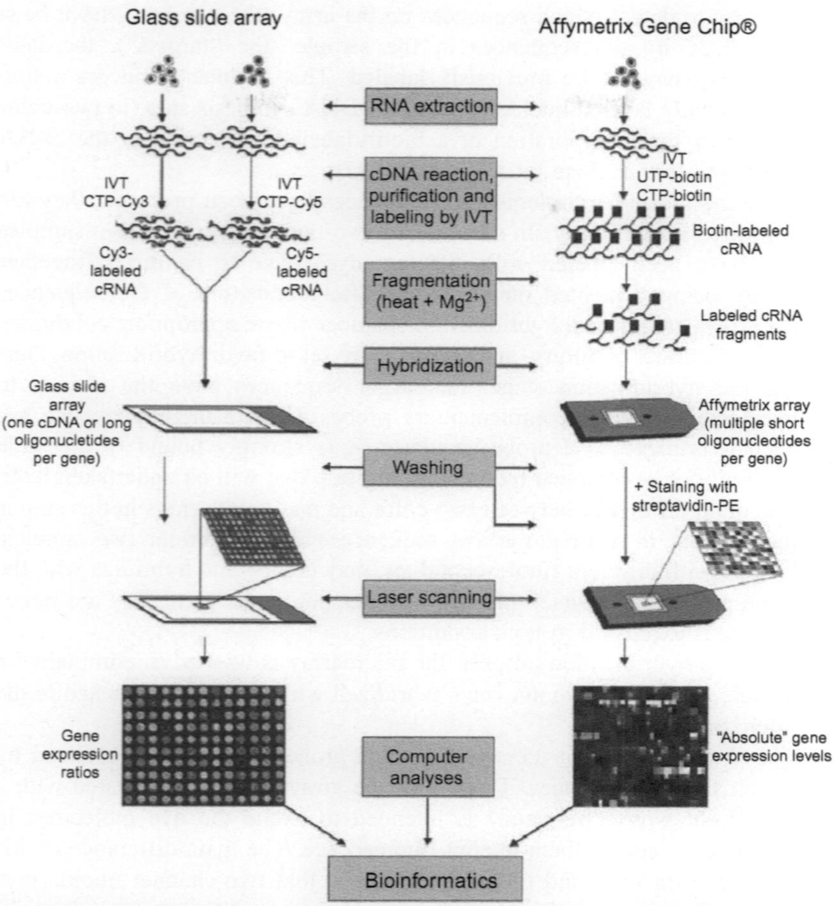

FIGURE 8 Expression array experiments. An outline of an expression microarray experiment performed with two- or one-color microarrays. For two-color microarrays analysis, RNA from the samples from two conditions are obtained, RNA is extracted, and cDNA is synthesized *in vitro* in presence of Cy3 (green) or Cy5 (red) labeled nucleotides. The two labeled cDNA samples are mixed and hybridized on a solid support. After hybridization, the microarray is scanned using two laser beams (green and red). The images obtained from each channel are digitally superposed to form the image of the microarray. For one-channel microarray analysis (e.g., with Affymetrix microarrays), only one sample is hybridized in each microarray device. Total RNA fraction is extracted and processed to be transcribed into cDNA. The cDNA is then used in an *in vitro* transcription IVT reaction to generate biotinylated cRNA. The cRNA sample is fragmented and then hybridized onto the microarrays. After incubation, the microarray is washed, stained with streptavidin conjugates of pephycoerythrin-conjugated streptidine, and then scanned on a laser scanner. (For interpretation of the references to color in this figure legend, the reader is referred to the online version of this chapter.)

2. In order to detect which sequences on the array (the "probes") have been hybridized by the sequences in the sample (the "targets"), the later sequences have to be previously labeled. This is done by incorporating fluorescently labeled nucleotides in the cDNA synthesis step (in two-color arrays) or the incorporation of a biotin-labeled nucleotide in the cRNA synthesis step, as done by Affymetrix *(47)*.
3. Once the labeled complementary sequences have been prepared, they are deposited on the array. In the case of two-color arrays, the two samples that have been labeled with different dyes have to be mixed together before being deposited on the slide. After deposition of the sequences the arrays are put in a hybridization chamber where appropriate conditions of temperature, humidity, and agitation are set to favor hybridization. During the hybridization stage, the target sequences have the chance to hybridize with their complementary probes. Once a target sequence has been hybridized to a probe sequence, it is strongly bound to the slide and will not be removed by the washing step that will be undertaken later. The main difference between two-color and one-color arrays in this step is the fact that, in two-color arrays, sequences coming from the two samples, labeled with different fluorescent dyes, may compete to hybridize with the same probes. Of course, this will only happen if the gene they are representing is expressed in both conditions.
4. After the hybridization process, the microarray is washed to eliminate the nucleic acids that have not been hybridized with the probes attached to the microarray.
5. The next step aims at identifying which probes have been hybridized by which sample sequences. To do this, the array will be illuminated with a laser light whose frequency is intended to excite the dye molecules in the targets, causing them to emit fluorescence. The main difference in this step between two- and one-color arrays is that two-channel microarrays have to be illuminated by laser beams of those two frequencies that will excite Cy5- and Cy3-labeled sequences, causing them to emit red and green lights, respectively. The two images generated, one per each different light, will be later merged using image analysis algorithms.
6. Although there is extensive variation in gene expression among individuals *(44)*, it is generally accepted that the amount of fluorescence signal, generated by the described analytical procedure, can be considered to be proportional to the amount of mRNAs present in the sample, which can be considered to be proportional to the gene expression *(45)*. The fluorescence signals will be transformed using the appropriate software and turned into numeric values that will be the basis for the subsequent bioinformatic analysis.

3.1.3 Measuring Gene Expression

In gene expression experiments, gene expression quantification is performed through the analysis of the intensity of fluorescence signals emitted after exciting labeled targets that have been hybridized to probes. Essentially, intensity values are transformed into numerical values that will be submitted to bioinformatics processing and analysis. This process, which is known as *image quantitation*, is described below and differences between two-color and one-color microarrays are briefly explained.

3.1.3.1 Relative Expression with Two-Channel Microarray

As mentioned previously, one of the main differences between one- and two-channel arrays is that the latter are based on the competitive hybridization of two RNA samples with the probes on the array. In practice, this implies that two-color arrays are intended to quantify how much each gene is expressed in one sample relatively to the other.

When the image obtained from a two-color microarray is scanned, the software produces a series of numerical values. Basically these values consist of:

1. Intensity values for each channel, usually named as "red" (R) or "green" (G) values.
2. Measures of noise or background, usually named as R_b and G_b. These are intended to provide a measure of cross-hybridization that represents the proportion of the fluorescence signal that can be attributed to causes other than hybridization.
3. Other measures intended to determine the reliability of the fluorescent signal and background values.

Altogether these values can be combined to provide different estimates of relative gene expression. For example, a natural measure of relative gene expression is the *expression ratio*, defined as:

$$M=\frac{R}{G}$$

In order to adjust this measure for background effect, the *background-corrected expression ratio* can be computed as:

$$M=\frac{(R-R_b)}{(G-G_b)}$$

It is very common not to use raw expression ratios, but instead to use the base 2 logarithm of these ratios. This has two main advantages: (i) the data on logarithmic scale tend to be more similar to a Gaussian (normal) distribution, which facilitates the statistical analysis of the expression measures, and (ii) base 2 logarithm values provide a more intuitive approach to relative expression measurement. Figure 9 outlines the process of converting intensity values into relative expression measures.

3.1.3.2 Absolute Expression with One-Channel Microarray

The analysis of gene expression in one-channel microarray is quite different than in the two-channel case. This is mainly due to expression estimates that

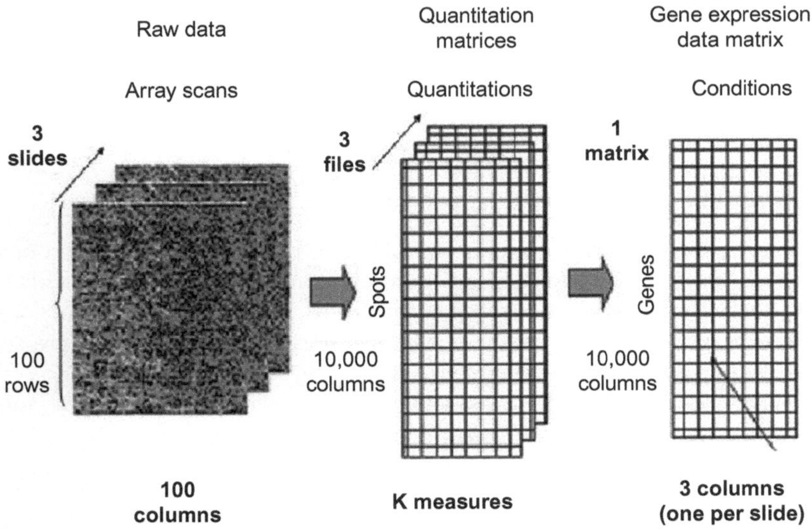

FIGURE 9 Outline of the process to quantify expression from images. On the left side are represented the raw data, representing three arrays of 100 rows and 100 columns, that is, 10,000 spots. The central image outlines the structure of a scanned image file. For each spot, data informing various aspects, ranging from intensity and background values to their quality, are generated. These measures are used to obtain an estimate of relative expression, that is, stored in the expression matrix (right) that is the basis for all subsequent analyses. (For color version of this figure, the reader is referred to the online version of this chapter.)

correspond to absolute intensities from one sample (experimental condition) not being compared with anything else. Besides this, depending on the technology used, especially if the arrays are based on short oligonucleotides such as Affymetrix, there are many single measures for each gene that have to be combined into a single gene expression measure *(46)*. This yields to the problem of combining adjustments such as background correction with other processes including the summarization of all single values. There have been developed many algorithms for doing this preprocessing microarray data, such as the proprietary MAS5 *(47)* or PLIER *(48)* and the well-known and probably most used RMA *(49)* and its extensions such as GCRMA *(50)*. There have been extensive discussions on the advantages and problems of each method *(51,52)*. Although no clear winner can be found, and some authors suggest using the appropriate method for each problem type *(53)*, the fact is that robust methods such as RMA have become the standard "*de facto*" for many common applications.

3.2 SNP Arrays for Variation Analysis and Genotyping

SNP arrays are a type of DNA array that can be used for *genotyping*, that is, for detecting SNPs within populations *(54)*. These arrays rely on the same principles as expression arrays, but the main difference is that the probes have to be able to distinguish between different allelic variations *(55)*. That is, for each polymorphism to be detected the array contains the different possible variations at the specific site.

The specific way that SNPs can be identified varies between different brands. The most common ones are *allele discrimination by hybridization* in Affymetrix arrays *(56)* and *allele-specific extension and ligation* to a bar-code oligonucleotide hybridized to a universal array, as is done by the Illumina GoldenGate BeadArray Assay *(57)*.

3.3 CGH Arrays for Comparative Genomic Hybridization

Comparative genomic hybridization (CGH) provides a way to perform genome-wide screening for copy number variations *(58)*. First developed to detect copy number changes in solid tumors, CGH uses two genomes, a test and a control, that are differentially labeled and competitively hybridized to metaphase chromosomes. The fluorescent signal intensity of the labeled test DNA relative to that of the reference DNA can then be linearly plotted across each chromosome, allowing the identification of copy number changes *(59)*. Since its introduction, CGH has proved to be a powerful technique that can be used to quickly scan an entire genome for imbalances. However, for most clinical applications the resolution of CGH was limited to alterations of approximately 5–10 Mb *(58)*.

As an alternative to overcome the aforementioned limitations, the CGH array technology of genomic BAC, P1, cosmid, or cDNA clones are used for hybridization instead of metaphase chromosomes as in conventional CGH *(60)*. After processing the array using a procedure very similar to expression microarrays (labeling/hybridization/washing/exciting), fluorescence ratios obtained from arrayed DNA elements provide a locus-by-locus measure of DNA copy number variation, which represents a means of achieving increased mapping resolution.

3.4 ChIP-on-Chip Arrays for Transcription Factor Binding Analysis

ChIP-on-chip *(61)* is a technique where DNA arrays (chip) can be used in combination with chromatin immunoprecipitation techniques (ChIP) to study interactions between proteins and DNA sequences. The most common application of this type of arrays is the study of transcription factors *(62)*, but they are also used for the analysis of dynamical transcriptional mechanisms *(63)*, or the study of replication-related proteins such as ORC or histones *(64)*. ChIP-on-chip microarray is based on the isolation of DNA sequences bound to particular proteins by immunoprecipitation (ChIP). Then, isolated DNA sequences can be hybridized on a microarray, such as a tiling array, allowing the determination of protein binding site occupancy throughout the genome.

3.5 Arrays for the Analysis of Alternative Splicing and Related Issues

There exist several types of arrays available to study different types of splicing, either to assess the effect of splicing events on transcript structure in a given sample or to compare the changes in transcript composition between two or more conditions (differential alternative splicing) *(65)*. The three main types of arrays that can be used to study splicing—exon arrays, tiling arrays, and splicing junction microarrays—are described briefly next.

Exon arrays have probes designed to identify all the known or predicted exons that are expressed in a given cell or tissue sample *(66)*. They work similarly to gene expression microarrays except by the number and distribution of probes along all exons. These arrays have been extensively used in recent years not only to study splicing events, but also as a substitute for gene expression arrays *(67)*, given that they also allow to measure global expression using all probes *(68)*.

Tiling arrays are designed to detect exon usage by setting contiguous probes matching at fixed distances across the genomes *(69)*. These arrays have different resolutions depending on the spacing between the probes and the genome for which they are designed. Based on information from the ENCODE project *(70)*, they have been used extensively to scan specific areas of the human genome and identify previously reported and novel exon usage *(71)*.

Splicing junction microarrays are designed to measure connectivity of exons through the use of probes that span known exon junctions *(72)*. These arrays are usually custom-designed with the goal of identifying splicing events but require previous knowledge of splice junction positions.

4 MICROARRAY BIOINFORMATICS

The growth in the use of DNA microarrays experienced in the past decade has been paralleled by the necessary developments in methodology, including new methods to model and analyze the data and new tools to implement these methods *(16)*.

A complementary aspect of the intensive use of microarrays is the fact that thousands, even millions of datasets have been generated along the years they have been in use. From the early times of microarrays it was noted that, with such huge datasets, it would be difficult to reproduce published results in order to obtain independent evaluations *(73)*. This led to two types of important developments:

- The Microarray Gene Expression Data Society created the MIAME (Minimum Information About a Microarray Experiment) standards for the description of microarray experiments and for the exchange of microarray data.
- Several public consortia, such as the European Bioinformatics Institute (EBI), the National Center for Biotechnical Information (NCBI), and many other organizations created databanks where data could be publicly submitted and described in a MIAME-compliant form.

Those two efforts together contributed to a number of scientific journals that require MIAME-compliant data as a condition for publishing microarray-based articles. This implies appropriate description of the data and submission of the raw data to one of the existing public repositories, which has made the microarray field much more transparent than others where, for instance, data still are kept by the authors of the studies.

4.1 The MIAME Standard

The MIAME standard was created by the Functional Genomics Data Society, formerly known as the Microarray Gene Expression Data Society (http://www.mged.org), as an effort to provide standards to specify all the information necessary to describe and interpret unambiguously the results of a microarray experiment *(74)*. The standard defines the contents required for compliance reports but it does not specify the format in which this data should be presented. As a consequence there are a number of different file formats for representing this data, and each public and subscription database has adopted its own format.

4.2 Microarray Databases

Microarray databases are repositories containing experimental microarray data, mainly microarray gene expression data. Microarray databases are used to store the results of finished experiments, and to make the data available to other users and applications, either directly or via user download. Microarray databases can fall into two distinct classes:

1. Public repositories that adhere to academic or industry standards and are designed to be used by many analysis applications and groups. A good example of this is the Gene Expression Omnibus from the NCBI (http://www.ncbi.nlm.nih.gov) *(75)* or Array Express from the EBI (http://www.ebi.ac.uk) *(76)*.
2. Specialized repositories associated primarily with the brand of a particular entity (lab, company, university, consortium, or group), an application suite, a topic, or an analysis method, whether it is commercial, nonprofit, or academic.

There are many different microarray databases and their description exceeds the goals of this chapter. Gardner *et al.* *(77)* provide a comparison between the different databases available in 2001. A more recent review can be found in Koschmieder *et al.* *(78)*.

5 DISCUSSION AND CONCLUDING REMARKS

This chapter has presented the technology of DNA arrays describing the different types of microarrays and the applications of each type of microarrays with some emphasis on gene expression microarray.

Microarrays have a short history of a little more than 15 years. However, microarrays have not been free of criticism *(79)* and by the middle of the first decade there appeared several papers criticizing their low reproducibility *(80)*. However, fast technological developments have allowed that, after a startup period where this technology was expensive and not very reliable, microarrays have become more affordable and more precise *(81,82)*. Today, DNA microarrays are the tool of choice for many types of studies that would have been unconceivable only 20 years ago *(26)*.

In response to the many criticisms against this technology, a high-scale multicentric study on the quality of microarrays, called “MicroArray Quality Control I”, was promoted by the Food and Drug Administration *(83)*. The goal of this study was to establish whether reproducibility and quality issues regarding microarray analysis were attributable to problems in the technique or to an incorrect application of the methodology. The main conclusion of these studies was that, although microarrays have their limitations, they can be considered a reliable and well-reproducible technique when adequately used *(84)*. In spite of their reliability microarrays, however, have some limitations *(85)*:

1. They provide indirect measures of gene expression based on fluorescence signals, which is known to be linear only in a given range of concentrations, but not for small or high values, where signal can be undetectable (low) or saturated (high) *(40)*.
2. It is very difficult to avoid some degree of cross-hybridization *(86)*.
3. DNA microarray are often closed platforms, thus the analysis is limited to the sequences present on the microarray.

In recent years, new reports suggest that novel next-generation sequencing (NGS) technology has the potential to quickly supersede microarrays in a wide range of applications *(80,87)*. This seems reasonable because in principle NGS can overcome the limitations (1–3) cited above *(88)*. Although it may seem a reasonable evolution, especially when NGS becomes more robust and affordable, the fact is that in 2013 microarrays are still in use and, indeed, microarrays are still the option of choice for many studies where the flexibility of NGS may not be necessary. Whether they will disappear or will stay as “one more technique” is something that is yet to be seen.

REFERENCES

1. Sambrook, J.; Russell, D. W., Eds.; *Molecular Cloning: A Laboratory Manual*; Cold Spring Harbor Laboratory Press: New York, NY, 2001.
2. Saiki, R. K.; Scharf, S.; Faloona, F.; Mullis, K. B.; Horn, G. T.; Erlich, H. A.; Arnheim, N. *Science* **1985**, *230*, 1350–1354.
3. Schena, M.; Shalon, D.; Davis, R. W.; Brown, P. O. *Science* **1995**, *270*, 467–470.
4. Kumar, R.; Sharma, A.; Tiwari, R. K. *J. Pharm. Bioallied Sci.* **2012**, *4*, 21–26.
5. Stoughton, R. B. *Annu. Rev. Biochem.* **2005**, *74*, 53–82.
6. Yauk, C. L.; Berndt, M. L. *Environ. Mol. Mutagen.* **2007**, *48*, 380–394.
7. Southern, E. M. *J. Mol. Biol.* **1975**, *98*, 503–517.
8. Grunstein, M.; Hogness, D. S. *Proc. Natl. Acad. Sci. U.S.A.* **1975**, *72*, 3961–3965.
9. Gergen, J. P.; Stern, R. H.; Wensink, P. C. *Nucleic Acids Res.* **1979**, *7*, 2115–2136.
10. Augenlicht, L. H.; Taylor, J.; Anderson, L.; Lipkin, M. *Proc. Natl. Acad. Sci. U.S.A.* **1991**, *88*, 3286–3289.
11. Kulesh, D. A.; Clive, D. R.; Zarlenga, D. S.; Greene, J. J. *Proc. Natl. Acad. Sci. U.S.A.* **1987**, *84*, 8453–8457.
12. Lashkari, D. A.; DeRisi, J. L.; McCusker, J. H.; Namath, A. F.; Gentile, C.; Hwang, S. Y.; Brown, P. O.; Davis, R. W. *Proc. Natl. Acad. Sci. U.S.A.* **1997**, *94*, 13057–13062.
13. Cheung, V. G.; Morley, M.; Aguilar, F.; Massimi, A.; Kucherlapati, R.; Childs, G. *Nat. Genet.* **1999**, *21*, 15–19.
14. Benson, D. A.; Cavanaugh, M.; Clark, K.; Karsch-Mizrachi, I.; Lipman, D. J.; Ostell, J.; Sayers, E. W. *Nucleic Acids Res.* **2013**, *41*, D36–D42.
15. Tomiuk, S.; Hofmann, K. *Brief. Bioinform.* **2001**, *2*, 329–340.
16. Stekel, D., Ed.; *Microarray Bioinformatics*; Cambridge University Press: Cambridge, 2003.
17. Lee, M.; Xiang, C. C.; Trent, J. M.; Bittner, M. L. *Anal. Biochem.* **2007**, *368*, 70–78.
18. Halperin, A.; Buhot, A.; Zhulina, E. B. *Biophys. J.* **2004**, *86*, 718–730.
19. Woo, Y.; Affourtit, J.; Daigle, S.; Viale, A.; Johnson, K.; Naggert, J.; Churchill, G. *J. Biomol. Tech.* **2004**, *15*, 276–284.
20. Dalma-Weiszhausz, D. D.; Warrington, J.; Tanimoto, E. Y.; Miyada, C. G. *Methods Enzymol.* **2006**, *410*, 3–28.
21. Singh-Gasson, S.; Green, R. D.; Yue, Y.; Nelson, C.; Blattner, F.; Sussman, M. R.; Cerrina, F. *Nat. Biotechnol.* **1999**, *17*, 974–978.
22. Carter, M. G.; Hamatani, T.; Sharov, A. A.; Carmack, C. E.; Qian, Y.; Aiba, K.; Ko, N. T.; Dudekula, D. B.; Brzoska, P. M.; Hwang, S. S.; Ko, M. S. H. *Genome Res.* **2003**, *13*, 1011–1021.
23. Holloway, A. J.; van Laar, R. K.; Tothill, R. W.; Bowtell, D. D. L. *Nat. Genet.* **2002**, *32*, 481–489.
24. Walt, D. R. *Science* **2000**, *287*, 451–452.
25. Gunderson, K. L.; Kruglyak, S.; Graige, M. S.; Garcia, F.; Kermani, B. G.; Zhao, C.; Che, D.; Dickinson, T.; Wickham, E.; Bierle, J.; Doucet, D.; Milewski, M.; Yang, R.; Siegmund, C.; Haas, J.; Zhou, L.; Oliphant, A.; Fan, J. B.; Barnard, S.; Chee, M. S. *Genome Res.* **2004**, *14*, 870–877.
26. Trevino, V.; Falciani, F.; Barrera-Saldana, H. A. *Mol. Med.* **2007**, *13*, 527–541.
27. Jaluria, P.; Konstantopoulos, K.; Betenbaugh, M.; Shiloach, J. *Microb. Cell Fact.* **2007**, *6*, 4.
28. Mao, X.; Young, B. D.; Lu, Y.-J. *Curr. Genomics* **2007**, *8*, 219–228.
29. Bast, R. C.; Kufe, D. W.; Pollock, R. E., Eds.; *Holland–Frei Cancer Medicine*; BC Decker Inc.: Hamilton, ON, 2000.

30. Vogel, C.; Marcotte, E. M. *Nat. Rev. Genet.* **2012**, *13*, 227–232.
31. Shalon, D.; Smith, S. J.; Brown, P. O. *Genome Res.* **1996**, *6*, 639–645.
32. Duggan, D. J.; Bittner, M.; Chen, Y.; Meltzer, P.; Trent, J. M. *Nat. Genet.* **1999**, *21*, 10–14.
33. Tusher, V.; Tibshirani, R.; Chu, C. *Proc. Natl. Acad. Sci. U.S.A.* **2001**, *98*, 5116–5121.
34. McCarthy, D. J.; Smyth, G. K. *Bioinformatics* **2009**, *25*, 765–771.
35. Chai, F.-S.; Liao, C.-T.; Tsai, S.-F. *Biom. J.* **2007**, *49*, 259–271.
36. Yang, Y. H.; Speed, T. *Nat. Rev. Genet.* **2002**, *3*, 579–588.
37. Shimokawa, K.; Kodzius, R.; Matsumura, Y.; Hayashizaki, Y. *Cold Spring Harb. Protoc.* **2008**, *3*, 1–3.
38. Engelen, K.; Naudts, B.; Moor, B. D.; Marchal, K. *Bioinformatics* **2006**, *22*, 1251–1258.
39. Tan, P. K.; Downey, T. J.; Spitznagel, E. L.; Xu, P.; Fu, D.; Dimitrov, D. S.; Lempicki, R. A.; Raaka, B. M.; Cam, M. C. *Nucleic Acids Res.* **2003**, *31*, 5676–5684.
40. Macgregor, P. F.; Squire, J. A. *Clin. Chem.* **2002**, *48*, 1170–1177.
41. *Nature* **2006**, *442*, 957.
42. Staal, F. J. T.; van der Burg, M.; Wessels, L. F. A.; Barendregt, B. H.; Baert, M. R. M.; van den Burg, C. M. M.; Van Huffel, C.; Langerak, A. W.; van der Velden, V. H. J.; Reinders, M. J. T.; van Dongen, J. J. M. *Leukemia* **2003**, *17*, 1324–1332.
43. Owzar, K.; Barry, W. T.; Jung, S.-H. *Clin. Transl. Sci.* **2011**, *4*, 466–477.
44. Scott, C. P.; VanWye, J.; McDonald, M. D.; Crawford, D. L. *PLoS One* **2009**, *4*, e4486.
45. Polyak, K.; Meyerson, M. *Holland–Frei Cancer Medicine;* BC Decker Inc.: Hamilton, ON, 2003.
46. Irizarry, R. A.; Hobbs, B.; Collin, F.; Beazer-Barclay, Y. D.; Antonellis, K. J.; Scherf, U.; Speed, T. P. *Biostatistics* **2003**, *4*, 249–264.
47. Pepper, S. D.; Saunders, E. K.; Edwards, L. E.; Wilson, C. L.; Miller, C. J. *BMC Bioinformatics* **2007**, *8*, 273.
48. *Guide to Probe Logarithmic Intensity Error (PLIER) Estimation*; Affymetrix, 2005. http://media.affymetrix.com/support/technical/technotes/plier_technote.pdf.
49. Irizarry, R. A.; Bolstad, B. M.; Collin, F.; Cope, L. M.; Hobbs, B.; Speed, T. P. *Nucleic Acids Res.* **2003**, *31*, e15.
50. Wu, Z.; Irizarry, R. A. *Nat. Biotechnol.* **2004**, *22*, 656–658.
51. Bolstad, B. M.; Irizarry, R. A.; Åstrand, M.; Speed, T. P. *Bioinformatics* **2003**, *19*, 185–193.
52. Autio, R.; Kilpinen, S.; Saarela, M.; Kallioniemi, O.; Hautaniemi, S.; Astola, J. *BMC Bioinformatics* **2009**, *10*, S24.
53. Harr, B.; Schlötterer, C. *Nucleic Acids Res.* **2006**, *34*, e8.
54. LaFramboise, T. *Nucleic Acids Res.* **2009**, *37*, 4181–4193.
55. Bjornsson, H. T.; Albert, T. J.; Ladd-Acosta, C. M.; Green, R. D.; Rongione, M. A.; Middle, C. M.; Irizarry, R. A.; Broman, K. W.; Feinberg, A. P. *Genome Res.* **2008**, *18*, 771–779.
56. Lamy, P.; Andersen, C. L.; Wikman, F. P.; Wiuf, C. *Nucleic Acids Res.* **2006**, *34*, e100.
57. Cunningham, J. M.; Sellers, T. A.; Schildkraut, J. M.; Fredericksen, Z. S.; Vierkant, R. A.; Kelemen, L. E.; Gadre, M.; Phelan, C. M.; Huang, Y.; Meyer, J. G.; Pankratz, V. S.; Goode, E. L. *Cancer Epidemiol. Biomarkers Prev.* **2008**, *17*, 1781–1789.
58. Weiss, M. M.; Hermsen, M. A.; Meijer, G. A.; van Grieken, N. C.; Baak, J. P.; Kuipers, E. J.; van Diest, P. J. *Mol. Pathol.* **1999**, *52*, 243–251.
59. Kallioniemi, A.; Kallioniemi, O. P.; Sudar, D.; Rutovitz, D.; Gray, J. W.; Waldman, F.; Pinkel, D. *Science* **1992**, *258*, 818–821.
60. Davies, J. J.; Wilson, I. M.; Lam, W. L. *Chromosome Res.* **2005**, *13*, 237–248.
61. Pillai, S.; Chellappan, S. P. *Methods Mol. Biol.* **2009**, *523*, 341–366.
62. Sandmann, T.; Jakobsen, J. S.; Furlong, E. E. M. *Nat. Protoc.* **2006**, *1*, 2839–2855.

63. van der Deen, M.; Hassan, M. Q.; Pratap, J.; Teplyuk, N. M.; Young, D. W.; Javed, A.; Zaidi, S. K.; Lian, J. B.; Montecino, M.; Stein, J. L.; Stein, G. S.; van Wijnen, A. J. *Methods Mol. Biol.* **2008**, *455*, 165–176.
64. Vengrova, S.; Dalgaard, J. Z., Eds.; *DNA Replication*; Humana Press: New York, NY, 2009.
65. Sánchez-Pla, A.; Reverter, F.; Ruíz de Villa, M. C.; Comabella, M. *J. Neuroimmunol.* **2012**, *248*, 23–31.
66. Gillett, A.; Maratou, K.; Fewings, C.; Harris, R. A.; Jagodic, M.; Aitman, T.; Olsson, T. *PLoS One* **2009**, *4*, e7773.
67. Lapuk, A.; Marr, H.; Jakkula, L.; Pedro, H.; Bhattacharya, S.; Purdom, E.; Hu, Z.; Simpson, K.; Pachter, L.; Durinck, S.; Wang, N.; Parvin, B.; Fontenay, G.; Speed, T.; Garbe, J.; Stampfer, M.; Bayandorian, H.; Dorton, S.; Clark, T. A.; Schweitzer, A.; Wyrobek, A.; Feiler, H.; Spellman, P.; Conboy, J.; Gray, J. W. *Mol. Cancer Res.* **2010**, *8*, 961–974.
68. Ha, K. C.; Coulombe-Huntington, J.; Majewski, J. *BMC Genomics* **2009**, *10*, 519.
69. Mockler, T. C.; Chan, S.; Sundaresan, A.; Chen, H.; Jacobsen, S. E.; Ecker, J. R. *Genomics* **2005**, *85*, 1–15.
70. The ENCODE Project Consortium. *Nature* **2012**, *489*, 57–74.
71. Bertone, P.; Stolc, V.; Royce, T. E.; Rozowsky, J. S.; Urban, A. E.; Zhu, X.; Rinn, J. L.; Tongprasit, W.; Samanta, M.; Weissman, S.; Gerstein, M.; Snyder, M. *Science* **2004**, *306*, 2242–2246.
72. Johnson, J. M.; Castle, J.; Garrett-Engele, P.; Kan, Z.; Loerch, P. M.; Armour, C. D.; Santos, R.; Schadt, E. E.; Stoughton, R.; Shoemaker, D. D. *Science* **2003**, *302*, 2141–2144.
73. Stoeckert, C. J.; Causton, H. C.; Ball, C. A. *Nat. Genet.* **2002**, *32S*, 469–473.
74. Brazma, A.; Hingamp, P.; Quackenbush, J.; Sherlock, G.; Spellman, P.; Stoeckert, C.; Aach, J.; Ansorge, W.; Ball, C. A.; Causton, H. C.; Gaasterland, T.; Glenisson, P.; Holstege, F. C.; Kim, I. F.; Markowitz, V.; Matese, J. C.; Parkinson, H.; Robinson, A.; Sarkans, U.; Schulze-Kremer, S.; Stewart, J.; Taylor, R.; Vilo, J.; Vingron, M. *Nat. Genet.* **2001**, *29*, 365–371.
75. Edgar, R.; Domrachev, M.; Lash, A. E. *Nucleic Acids Res.* **2002**, *30*, 207–210.
76. Brazma, A.; Parkinson, H.; Sarkans, U.; Shojatalab, M.; Vilo, J.; Abeygunawardena, N.; Holloway, E.; Kapushesky, M.; Kemmeren, P.; Lara, G. G.; Oezcimen, A.; Rocca-Serra, P.; Sansone, S.-A. *Nucleic Acids Res.* **2003**, *31*(1), 68–71.
77. Gardiner-Garden, M.; Littlejohn, T. *Brief. Bioinform.* **2001**, *2*, 143–158.
78. Koschmieder, A.; Zimmermann, K.; Trissl, S.; Stoltmann, T.; Leser, U. Tools for Managing and Analyzing Microarray Data. *Brief. Bioinform.* **2012**, *13*(1), 46–60. http://dx.doi.org/10.1093/bib/bbr010.
79. Frantz, S. *Nat. Rev. Drug Discov.* **2005**, *4*, 362–363.
80. Ledford, H. *Nat. News* **2008**, *455*, 847.
81. Ewis, A. A.; Zhelev, Z.; Bakalova, R.; Fukuoka, S.; Shinohara, Y.; Ishikawa, M.; Baba, Y. *Expert Rev. Mol. Diagn.* **2005**, *5*, 315–328.
82. Draghici, S.; Khatri, P.; Eklund, A. C.; Szallasi, Z. *Trends Genet.* **2006**, *22*, 101–109.
83. MAQC Consortium. *Nat. Biotechnol.* **2006**, *24*, 1151–1161.
84. Canales, R. D.; Luo, Y.; Willey, J. C.; Austermiller, B.; Barbacioru, C. C.; Boysen, C.; Hunkapiller, K.; Jensen, R. V.; Knight, C. R.; Lee, K. Y.; Ma, Y.; Maqsodi, B.; Papallo, A.; Peters, E. H.; Poulter, K.; Ruppel, P. L.; Samaha, R. R.; Shi, L.; Yang, W.; Zhang, L.; Goodsaid, F. M. *Nat. Biotechnol.* **2006**, *24*, 1115–1122.
85. Forster, T.; Roy, D.; Ghazal, P. *J. Endocrinol.* **2003**, *178*, 195–204.
86. Koltai, H.; Weingarten-Baror, C. *Nucleic Acids Res.* **2008**, *36*, 2395–2405.
87. Mardis, E. R. *Annu. Rev. Genomics Hum. Genet.* **2008**, *9*, 387–402.
88. Wang, Z.; Gerstein, M.; Snyder, M. *Nat. Rev. Genet.* **2009**, *10*, 57–63.

Chapter 2

Challenges and Future Trends in DNA Microarray Analysis

Abootaleb Sedighi and Paul C.H. Li
Department of Chemistry, Simon Fraser University, Burnaby, British Columbia, Canada

Chapter Outline

1 INTRODUCTION

Since the introduction of the first spotted DNA microarray platform in 1995, a vast development has been achieved both in the applications and in the technology *(1)*. This first microarray was created by Schena *et al.* who spotted or printed various complementary DNAs (cDNAs) on a glass microscope slide via a robotic printer and the microarray was used to monitor the differential expression of many genes in parallel. First, most of the microarray platforms have been used for obtaining the expression profiles of the genes. These expression profiling studies aim to obtain clinically relevant information from the gene expression levels *(2)*. For example, the microarray data were used to differentiate between the cancer subtypes, to provide prognostic information (e.g., likelihood of recurrence or metastasis), and, in some cases, to provide predictive information (e.g., the efficacy of chemotherapy). Second, the genotyping array has been developed to characterize the DNA (and sometimes RNA) in order to characterize viral pathogens or detect human gene mutations. While simple genotyping arrays consist of hundreds of features or spots, complex genotyping arrays utilize thousands

Comprehensive Analytical Chemistry, Vol. 63. http://dx.doi.org/10.1016/B978-0-444-62651-6.00002-7

of features to investigate mutations in many genes or to characterize multiple sequences in pathogens. Third, array-based comparative genomic hybridization (array-CGH) provides a high-resolution tool for screening of copy number variations in the whole genome and offers several advantages over the classical karyotyping techniques *(3)*. In addition to the various novel applications, as the commercial vendors took on the developments, the microarray platforms experienced many technical advances. For instance, the probes to be immobilized on the microarrays have shifted from cDNA to short oligonucleotides, and they were either presynthesized or synthesized *in situ*. These oligonucleotides demonstrated a higher specificity than cDNA probes. Glass is still the predominant substrate used in the microarray platform, but materials such as silicon and polymeric materials have also been used as the substrate. In terms of signal transduction to generate the microarray data, other aspects of the platform such as label-free detection techniques have been also developed *(4)*.

While having a remarkably fast growing pace, especially in the research area, the microarray technology has not experienced a smooth path of development for clinical applications. There are more problems in biostatistics than on technical issues. One challenge is that the microarray data have notoriously been considered as being "noisy" *(5)*. The reproducibility of the data and the validity of the data interpretation reported by prominent microarray studies have been criticized by several review articles in terms of no appropriate standardization, inadequate quality control (QC) measures, and unreliable data processing *(6)*. The uncertainty about the validity of the microarray data hindered the regulatory approval process for the development of array-based clinical tests as well as subsequent adoption of the tests by clinicians. The concerns about the validity of the clinical interpretation are more serious in applications such as expression profiling when new biomarkers are being introduced, rather than in other applications such as genotyping arrays, which are dealing with preexisting known biomarkers.

Harsh competition of microarray-based tests with PCR-based and sequencing-based tests is another challenge. For instance, simple microarray tests, when only a few genes are being monitored or a limited number of mutations are being interrogated, have to compete with the well-known PCR-based tests. On the other hand, complex microarray tests, which provide higher amounts of information out of reach of PCR-based techniques, are facing strong competition from the newly emerged next-generation sequencing (NGS) techniques. They provide detailed information about the whole genome, with the prices that have been lowered tremendously over the past few years *(7)*.

In order to use the microarray technology for point-of-care (POC) diagnostics, there are various hurdles. For instance, in the conventional format, the sample solution has to go through several steps of sample preparation, usually on bulky benchtop instruments, prior to introduction onto the DNA microarray slide. Integration of various steps of the microarray assay in a miniaturized, portable, and standalone lab-on-a-chip (LOC) device, as previously reviewed *(8)*, is a crucial requirement for using the microarray in POC diagnostics. Liquid handling and manipulation via microfluidic/nanofluidic technology plays an important role in these LOC devices. In terms of detection, recent developments in miniaturized nanoarrays, which eliminate the need for bulky fluorescent scanners, or even in label-free detection techniques are important components for successful LOC devices.

This chapter discusses the major challenges that the microarray technology has faced in the pathway of its growth over a period of less than two decades. We also highlight the significant progress that has been achieved by many researchers worldwide.

2 TOWARD MICROARRAY POC DEVICES

POC devices can bring medical laboratory tests on-site for patient care. Thanks to the innovations in POC devices over the past few decades, a variety of diagnostic tests are nowadays routinely performed in physicians' offices or even in patients' homes *(9)*. This migration from the centralized laboratory to the near-patient settings helps physicians and even patients to make informed decisions about diagnosis and treatment options. POC devices are expected to play a growing role in predictive and personalized medicine in the future *(10)*. The diagnostic tests implemented by POC devices must be robust and fast. All the steps of the tests should also be performed in the portable and standalone POC device *(9)*. On the roadmap to become a POC device, microarray technology needs to overcome several challenges. The first and probably the most significant challenge is the integration of the whole assay into a single device. Current microarray technologies use separate instruments for sample preparation, DNA hybridization, signal visualization, and data interpretation. Moreover, some of these components such as the fluorescent scanners used for signal visualization are bulky instruments that are only available in well-equipped laboratories. Another challenge comes from the long reaction time needed in DNA hybridizations, which require ~12–17 h *(11)*. All of the above-mentioned issues must be addressed in order for the microarray test to become a proper candidate for POC diagnostics. Fortunately, developments in some relevant technologies accelerate this process. For instance, some of the preparation steps for sample labeling can be avoided by using label-free detection approaches. Miniaturization of the spots also alleviates the need for bulky fluorescent scanners. More importantly with the aid of a microfluidic channel network, all steps of the microarray test can be integrated in a single miniaturized device. In the following section, we present a perspective of the challenges in developing the required technologies for microarray tests to become a suitable candidate for POC diagnostics.

2.1 Microfluidic Microarrays

Evolved after developments in microlithography techniques, microfluidics is at the heart of LOC devices *(12,13)*. Coupling liquid handling operations with microarray assays potentially adds precious advantages to them. One obvious benefit is reduced sample and reagent consumption due to small micrometer-sized channels. More importantly, highly efficient and controllable liquid and delivery in these channels allows for integration of different steps of microarray assays essential to implement the portable LOC devices *(14)*. In addition, the target molecules are delivered to the probe spots by convective flow in addition to diffusion in conventional microarrays, thereby reducing the hybridization times from hours to minutes. Last but not the least, the benefit of microfluidic microarray is its potential of high sample throughput, in addition to its sufficient probe density. Conventional microarray experiments usually allow one sample to be applied on one glass chip *(15)*; however, as is discussed in the following sections, one of the big challenges of DNA microarrays is the inevitable variations among the samples, which can lead to false-positive recognition of biomarkers in the early studies on microarrays *(6)*. These variations make replicate analysis of several samples necessary and therefore the multisample analysis capability of microfluidic microarray chips is highly valuable.

The microfluidic flow is used to deliver sample solutions over the probe spots in DNA hybridization. There are two ways to achieve this: either the microfluidic chambers or the microfluidic channels have been used to enhance the DNA hybridization of the target solutions to the probe arrays that have been conventionally pin-spotted on the surface *(8,16–17)*. First, large microfluidic chambers are used to cover the area with arrayed probes, where sample DNA solutions are hybridized with the probe molecules *(16)*. These chambers are compatible with both low-density and high-density microarrays, but it is always a challenge to design how the liquid will flow over the large chamber in such a way to achieve an equally distributed liquid movement across the arrays. Second, microfluidic channels provided better flow control of target solutions over probe arrays *(17)*. Various microfluidic chips containing straight and serpentine microchannels have been designed, mainly for low-density microarray experiments *(8)*. In these cases, the pin-spotted probe regions are usually contained along the channel length of the microchips.

The microfluidic flow is used not only to deliver the sample solutions for hybridization, but also to print probe solutions on the surface. The performance of the hybridization assay is heavily influenced by the quality of printed probe-spot morphology. Since, the spotting solutions are exposed to air in the pin-spotting methods; the solutions are subject to problems of splashing, uneven evaporation, and cross contamination *(18)*. Moreover, during the blocking and washing procedures after probe-spotting on the glass surface, the remaining unreacted probe molecules could diffuse away and smear the chip to form comet-like spots *(13)*. Furthermore, when the microchannel is used later to enclose the spotted bioarray, steel clamps must be used to ensure that the entire hybridization microchannel is well aligned to the probe rows *(13)*. By using the microchannel network as a microprinting technique, the probe spots of a high homogeneity can be obtained *(19–22)*. For instance, Wang *et al.* used networks of microchannels first for the probe printing and then for DNA hybridization to create 2D arrays, called the intersection approach (Figure 1) *(21)*. In this method, probe solutions, confined in polydimethylsiloxane (PDMS) microchannels, were first used to create an array of horizontal probe lines on a glass surface. Then, the target solutions flow on the surface along the vertical channels of another PDMS chip and the target molecules hybridize to the spotted probes at the intersection to the horizontal probe lines. The 2D microfluidic microarray format is well suited for parallel sample hybridizations. Unlike pin-spotted low-density DNA microarray, the use of long and narrow probe line arrays alleviate the need of alignment between hybridization channels and probe spots.

The 2D microfluidic microarray design is compatible with low-density DNA microarrays, which have their own diagnostic applications. Since in many gene diagnostic applications once a relatively small number of genes are identified using high-density DNA bioarrays, low-density arrays can be designed to screen these genes across many patients or to detect single-nucleotide polymorphisms (SNPs) *(13,21)*. This low-density approach has been demonstrated to be reliable, cost-effective, and fast in data analysis and interpretation *(23–26)*. Nevertheless, in order to create a global picture of cellular function in gene expression profiling, many thousands of genes are to be simultaneously monitored and, therefore, high-density bioarrays are needed.

The parallel sample hybridizations achieved in the 2D microfluidic microarray technique triggered the motivation for finding an effective way in

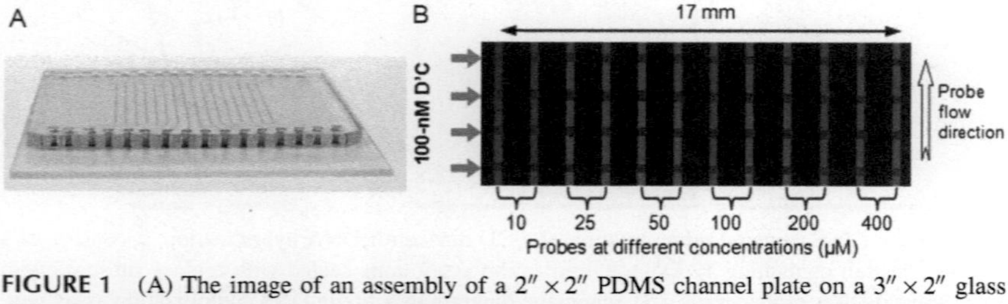

FIGURE 1 (A) The image of an assembly of a 2″ × 2″ PDMS channel plate on a 3″ × 2″ glass slide. The 16 channels filled with blue-dye solutions. (B) Dual-channel fluorescent images of DNA hybridization results with the 2D microfluidic microarray method. The overlaid images from the same glass slide show both printed probe lines (vertical green lines) and square hybridization patches (red) at intersections. *Used with permission from Dupuy and Simon* (5). (For interpretation of the references to color in this figure legend, the reader is referred to the online version of this chapter.)

simultaneous delivery of the sample solutions. In the conventional pressure-driven flow, all microchannels have to be connected to pump tubings and they have to be synchronized in order to ensure parallel flow. The arrangement becomes very complicated when many channels are to be used simultaneously. Moreover, in the pumping method a high pressure is required to transfer the liquids in long and narrow microchannels, and this in turn requires very tight sealing between the microchannel plate and the substrate *(27)*. An alternative to the pressure-driven flow is to achieve pumping by centrifugal forces. Centrifugal pumping has several advantages such as easy implementation and not being sensitive to the physiochemical properties of the liquid. Using centrifugal force the liquids can be transferred in a parallel manner in multiple channels in a wide range of sizes. Furthermore, this format is compatible with the compact disc (CD) technology and its related industries, which have been well developed over the past decades. In most biological and environmental applications, including the nucleic acid analysis applications, which utilize the centrifugal platform, only the radial channels are used for liquid handling and delivery. Li *et al.* reported a CD-like device capable of generating the reciprocating flow of DNA samples within the microchannels (Figure 2A and B). In their device, the centrifugal force was used to drive the sample solution to flow through the hybridization channel into a

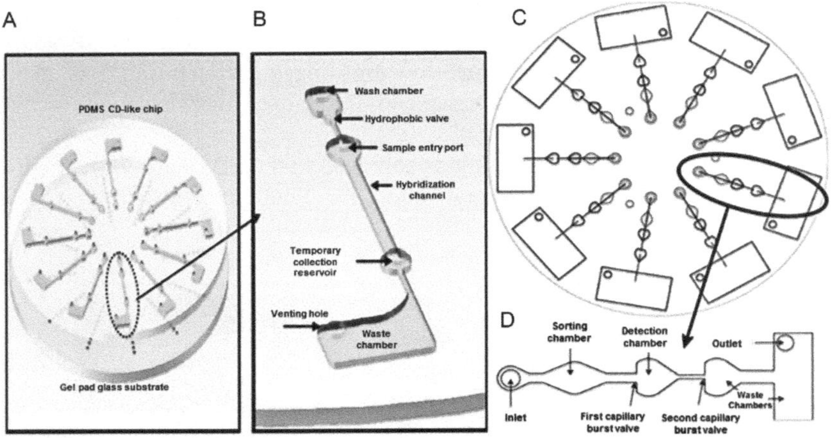

FIGURE 2 (A) Schematic representation of a CD device for DNA hybridization. It consists of a PDMS CD slab containing 12 DNA hybridization assay units sealed with a glass substrate with immobilized DNA probe arrays. (B) Schematic diagram of a single DNA hybridization assay unit. (C) Schematic of the CD nanofluidic device. (D) The magnified diagram shows the sorting chamber, the detection chamber, and a waste reservoir separated by capillary burst valves with different widths. There are 24 channels, representing the 24 different chromosomes in the human karyotype, and each channel can detect one specific translocation. *Used with permission from Li* et al. (28) *and Brøgger* et al. (29). (For color version of this figure, the reader is referred to the online version of this chapter.)

temporary collection reservoir while the capillary force pulled the solution back into the hybridization channel during the stopping period. The sample hybridization time was reduced to 90 s and the sample volume was as low as 350 nL *(28)*. Brøgger and coworkers have also developed a CD-like microarray device for the detection of chromosome translocation *(29)*. As shown in Figure 2C and D, they employed a series of capillary burst microvalves to control the stepwise fluid flow from the center toward the periphery of the CD. Using this technique they performed chromosome translocation experiments in two hybridization steps in the separate sorting and detection chambers. The radial-only format for the liquid handling and delivery limits the design of the CD because there is not enough space to accommodate the fluid structures in the radial format *(8)*. For example, if a centrifugal platform is built on a 120-mm regular CD with a 15-mm center spindle hole, the maximum limit of the length of a microchannel is 53 mm. In addition, for such a short microchannel, the capillary effect may dominate and the flow velocity cannot be easily controlled. Furthermore, by utilizing centrifugal pumping only once in the radial direction, the intersection approach cannot be applied to generate the microfluidic microarray.

In order to address these issues, Wang *et al.* exploited the centrifugal force twice based on specially designed intersecting channels in order to create a 2D microarray *(30,31)*. As shown in Figure 3, in addition to the radial microchannels, which were used for probe printing, spiral microchannels were designed and target hybridization was implemented by the intersection approach. In their design, a PDMS chip containing radial microchannels is first sealed against the glass wafer for printing the radial probe line arrays. The DNA hybridization is performed using a second PDMS chip, containing the spiral

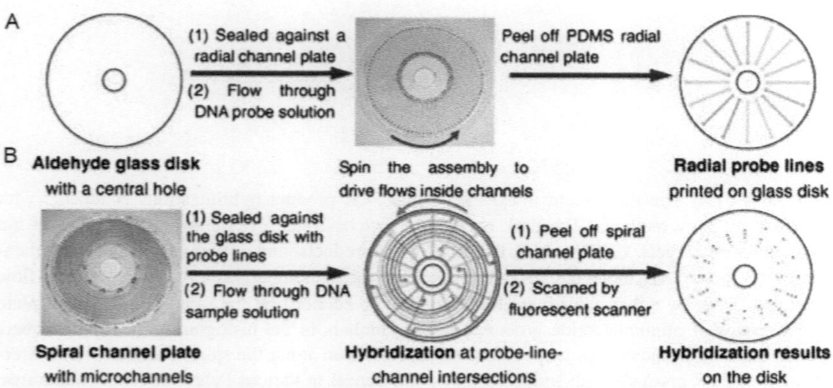

FIGURE 3 Intersection approach procedure for 2D microfluidic microarray analysis. (A) Probe line printing with the radial channel plate. (B) Hybridization procedure with the spiral channel plate. Hybridization occurring at the intersections of the spiral channels and radial probe lines, shown as colored patches in the right-most disc. *Used with permission from Wang* et al. (30). (For color version of this figure, the reader is referred to the online version of this chapter.)

microchannels, sealed against the preprinted wafer. The target molecules flow in the spiral microchannels and hybridize to their complementary probes at the intersection with the radial probe lines. Dynamic target delivery facilitated by the spiral microchannel can be conveniently controlled and synchronized *(32,33)*. 2D microarray using CD microfluidics also demonstrated a high sensitivity and specificity for DNA analysis *(30,31)*. Shown in Figure 4A is a fluorescent image comparing the hybridization results under the conditions of continuous-flow and stop-flow. Hybridization intensities from the 3-min continuous-flow are very close to those from the 2-h stop-flow. Moreover,

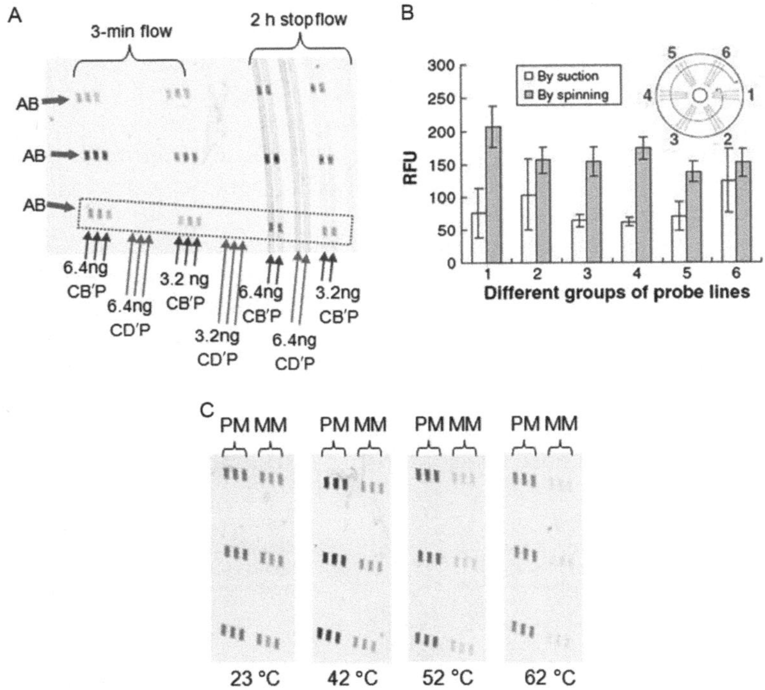

FIGURE 4 (A) The fluorescent images show the PCR product hybridizations conducted with two different flow methods. The dark rectangular patches represent the specific binding of the complementary targets. CD′P and CB′P are two PCR products, and only the latter is complementary to the probe AB molecule. (B) Hybridization signal comparison of two continuous-flow methods (driven by either centrifugal force or vacuum suction) on the same microchannel plate on six groups of oligonucleotide probes. The bar numbers in the histogram match the numbers shown in the inset, indicating the positions of hybridization along the spiral channels. (C) Differentiation of PCR products with single-base-pair differences at various hybridization temperatures. Each image was obtained from the hybridization of sample solutions in three spiral channels intersecting with three probe lines at the specified temperature. Two PCR products, perfectly matched (PM) and mismatched (MM), were tested in the experiment. *Used with permission from Wang and Li* (31). (For color version of this figure, the reader is referred to the online version of this chapter.)

nonspecific binding is negligible in the continuous-flow method because, in contrast to stop-flow, the sample flow from centrifugal pumping continually removes unhybridized DNA molecules to prevent them from accumulating and being adsorbed onto the glass surface. Centrifugal pumping, which produces liquid flows with higher consistencies in comparison with the vacuum suction method, therefore results in more consistent signals (Figure 4B). As shown in Figure 4C, adequate discrimination between single base-pair PCR products was achieved in only 3 min *(30)*. In addition to the radial-spiral approach in which the radial channels were used for probe printing and the spiral channels for target hybridization, Chen *et al.* also developed the double-spiral format in which the clockwise-spiral channels intersected with the anti-clockwise-spiral probe lines *(34)*. In this manner, a higher spot density (384×384 vs. 96×96) was achieved for the high-throughput microarray analysis on a 92-mm circular glass disc, which was of the same size as the one used previously in the radial-spiral format *(34)*.

2.2 Label-Free Detection

In the current DNA microarrays, target molecules are usually labeled with fluorescent dyes in order to achieve detection. Thanks to the discovery of efficient fluorescent dyes as well as advancements in the labeling techniques over the years, the sensitivity and stability of fluorescent-based microarrays have vastly been improved *(35)*. However, such a high sensitivity and resolution achieved in fluorescent detection can only be achieved via bulky fluorescent scanners. This limitation is one of the main obstacles to the miniaturization required for POC device development. In addition, labeling of the target molecules adds complications and cost to the assays. The efficiency of the labeling process and fluorescence quenching of the dye also affects the reproducibility of the results *(36)*. In the past decade, several approaches have been developed in order to alleviate target labeling *(36)*.

Molecular beacon (MB) arrays are one of the novel approaches that exploit both the sensitivity of fluorescence detection and the convenience of label-free target detection, though the probes are still labeled *(37)*. MB probes are single-stranded nucleic acids that retain a stem-and-loop structure and keep a pair of fluorophore-quencher at both ends of the strand in close proximity and thus quench the fluorescence emission while target molecules are absent. In the presence of nonlabeled target molecules, the loop region of MB strands hybridize to the target and consequently opens the MB structures up and enhances the fluorescence by removing quenching *(37)*.

In addition to the fluorescence platform, several other novel detection techniques, including electrochemical and optical approaches, have been developed in order to alleviate the need for target labeling *(36–44)*. Table 1 summarizes some of these approaches and their characteristics are compared with conventional fluorescence detection. Surface plasmon resonance imaging

TABLE 1 Some of the Label-Free Detection Techniques Developed for the Microarray Platforms

Detection Technique	Probe	Target	LOD (mol/L)	References
Fluorescence	Molecular beacons	DNA	$<10^{-9}$	*(37)*
SPRi	Locked nucleic acid	Micro-RNA	10^{-16}	*(38)*
Impedimetric	ssDNA	DNA	10^{-15}	*(40)*
Multiwalled carbon nanotubes	ssDNA	PCR amplicons	$\sim10^{-9}$	*(41)*
Near-field scanning microwave microscopy	Oligonucleotide	DNA and RNA	–	*(42)*
Spectral reflectance imaging biosensor	Oligonucleotide	DNA	–	*(43)*
Kelvin probe force microscopy	Oligonucleotide	DNA	$<50 \times 10^{-9}$	*(44)*

(SPRi) is an attractive detection approach for microarrays because none of the targets or probes require labeling; this is a true label-free detection. In addition, SPRi enables real-time detection and kinetic analysis *(38)*.

In 2003, Koehne *et al.* reported the detection of unlabeled PCR amplicon with electrochemical techniques using multiwalled carbon nanotubes nanoelectrode arrays *(41)*. In their technique, with the aid of $Ru(bpy)_3^{2+}$, the inherent guanine bases in the DNA amplicon target serve as the signal mediators, providing an amplified anodic current associated with the oxidation of guanine groups at the nanoelectrode surface. The abundance of signal moieties (guanine bases) in the target strands has led to a low detection limit achieved in this technique (i.e. less than ~1000 target amplicons on a microspot are detectable).

Near-field scanning microwave microscopy (NSMM) is another system that has been used for label-free detection of both DNA and RNA molecules *(42)*. NSMM monitors the microwave reflectance, a factor that depends on the dielectric permittivity profile across the microarray surface *(42)*. This parameter is, in turn, dependent on the length and surface coverage of the strands, as well as on the hybridization state of the molecules (e.g., unhybridized single-stranded probe vs. hybridized duplex). NSMM technique demonstrated an acceptable resolution (potentially less than 50 μm) and comparable sensitivity to the fluorescent detection *(42)*.

Furthermore, Ozkumur *et al.* developed an optical interferometric detection system, called a spectral reflectance imaging biosensor, for high-throughput analysis of SNPs and expression profiles on a glass microarray chip *(43)*. A comparison between the microarray images taken before and after hybridization was used in order to detect the hybridization of 20 base-pair target oligonucleotides with the surface immobilized probes. Single-nucleotide mismatch target oligonucleotides were also distinguished from perfectly matched duplexes, through dynamic data acquisition during denaturation by washing the sample with a low ionic strength buffer *(43)*.

2.3 Miniaturized Nanoarray Platforms

Fluorescence detectors utilize high numerical-aperture objectives with great magnifications in order to detect the fairly weak fluorescence signals emitted from the microarray spots. The large magnification causes the field of view of the detector to cover only a few microarray spots, which in turn necessitates the scanning of the whole array area. The need for large-format scanning fluorescent detectors is a major hindrance in developing portable microarray devices. Miniaturization of the microarray features in such a way that the whole array would be visible in the field of view of the objectives would render the scanning unnecessary. Other than alleviating the need for bulky scanning, miniaturized arrays favor faster mass and heat transports and therefore reduce the assay time *(45)*.

In order to create the arrays with submicrometer features, very accurate printing techniques with extremely high resolutions have to be developed. Among the printing techniques, scanning probe microscopy (SPM)-based techniques, such as atomic force microscopy (AFM), have received a higher level of attention in this regard. SPM-based techniques hold a high positioning precision and also their operational condition is compatible with biomolecules *(46)*. As an example of these techniques, Mirkin *et al.* used dip-pen nanolithography (DPN) to directly immobilize hexanethiol-modified oligonucleotides on the gold surface, and oligonucleotides modified with acrylamide groups on silica substrates *(47)*. They accomplished this by reducing the size of the features to 100 nm, which allows an array containing ~100,000 features to be generated in an area of 100 μm × 100 μm compatible with the size of a typical AFM scanner *(47)*. Despite the extremely high resolution of DPN, this technique is intrinsically serial and, thus, a significant amount of time must be allocated in order to generate multicomponent arrays *(48)*. Nanoimprint lithography (NIL) is another approach employed for the reduction of feature dimensions in microarrays. In the NIL printing techniques, probe oligonucleotides are either synthesized *in situ* on the chemically modified nanostructures created on a polymer surface *(49)*, or physically tethered to the surface of nanostructures before they were delivered (or stamped) on the substrate *(48)*. With this technique, two groups independently developed a soft-lithography method that was able to repeatedly replicate the whole DNA array in a single step *(50,51)*. In their technique a master substrate made of oligonucleotide probe features was immersed in a solution containing the complementary target DNAs. Following the hybridization, a secondary substrate was brought into contact with the master in order to adsorb the target DNAs (replicating the features on the second substrate). These effective nanostamping methods were able to reproduce DNA arrays with features as small as 14 nm, with the spacing of 77 nm *(52)*. Advances in miniaturization of the microarray features, to the nanometer scale, are significant steps toward portable microarray platforms. It also makes the microarray test cheaper by simplifying the detector and reducing the amount of biomolecules required.

2.4 Integrated LOC Devices

Not too long after the introduction of microarray technology researchers noticed that integration of all steps of the DNA microarray into a single device is highly beneficial. Many microdevices have been developed in order to perform all the individual steps of DNA the microarray assay. Also, great progresses in the various support systems, considered as prerequisites of a miniaturized integrated device, have been achieved during the past two decades; however, integration is not as simple as attaching these subsystems together. All aspects of the integrated system such as liquid handling techniques; thermal and pressure control; substrate material; and signal transduction unit must be compatible with every subsystem harmonized in the whole system. Despite these challenges, great effort has been made by researchers to develop the LOC devices to be able to perform as many steps of the microarray assay as possible *(53–59)*. One of the early LOC devices was reported by Anderson *et al.* in 2000 *(53)*. Their monolithic device was capable of extracting and concentrating nucleic acids from milliliter aqueous samples and performing microliter chemical amplification, serial enzymatic reactions, metering, and mixing. A commercialized Affymetrix GeneChip array chip was positioned in the center of the device to perform nucleic acid hybridization (Figure 5). A fluid barrier or hydrophobic membrane in conjunction with a pneumatically controlled diaphragm valve and hydrophobic vent was used for fluidic manipulations along a microfluidic network. The performance of the device was tested for the detection of mutations in the HIV genome *(53)*. Yeung *et al.* also developed an integrated biochip based on a multichamber design with microfluidic control for multiplex pathogen identification *(57)*. In their silicon–glass microchip, an array of indium tin oxide (ITO) electrodes was constructed within the microfluidic chamber and the oligonucleotide probes were individually positioned at each ITO electrode. Several steps of the microarray identification of the target pathogens including a thermal lysis step, magnetic particle-based isolation of the target genomes, asymmetric PCR, and electrochemical sequence-specific detection using silver-enhanced gold nanoparticles were performed in the integrated device *(57)*.

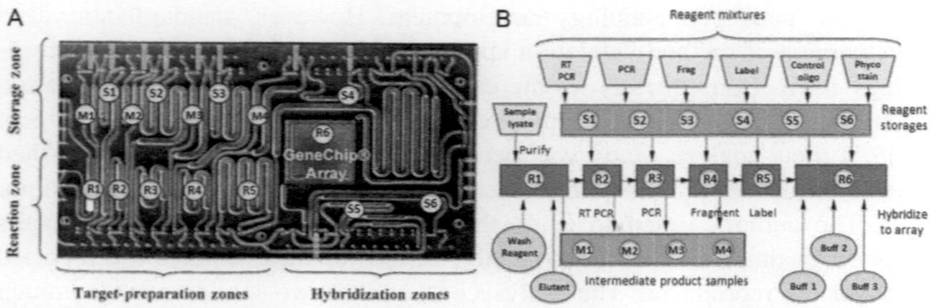

FIGURE 5 (A) An integrated device that automatically performs a multistep HIV genotyping assay. At the beginning of the process reagent mixtures are loaded into storage chambers S1–S6. The device automatically carries out RNA purification from a serum lysate (extraction), RT-PCR, nested PCR, DNase fragmentation and dephosphorylation, terminal transferase labeling, dilution and hybridization, washing, phycoerythrin staining, and washing (in chambers R1, R2, R3, R4, R5, and R6, respectively). Intermediate product samples are stored in chambers M1–M4 for later analysis. Three independently controlled temperature zones are indicated as storage zone, reaction zone, and hybridization zone on this polycarbonate device with dimensions of $8 \times 40 \times 70$ mm. (B) Process flow schematic of the automated HIV assay. The introduction of bulk reagents (washes and buffers) from externally connected vials is controlled using cartridge-based valves. Frag refers to the DNase fragmentation and dephosphorylation reaction; phyco stain refers to the streptavidin–phycoerythrin conjugate; and Buff1, Buff2, and Buff3 to buffers for hybridization. *Used with permission from Anderson* et al. (53). (For color version of this figure, the reader is referred to the online version of this chapter.)

In 2006, Liu *et al.* also integrated a DNA microarray platform, containing 12,000 features, with a microfluidic cartridge in order to automate the fluidic handling steps required for gene expression profiling assay *(58)*. The integrated electrochemical micropumps were used to transfer the liquids from the reagent chambers to the DNA microarray silicon chip. Microarray hybridization and subsequent washing and labeling steps were all performed at the self-contained device. The detection step, however, was performed on an external fluorescent scanner *(58)*. In a recent study, a miniaturized fluorescent scanner, instead of a conventional bulky one, was used for hybridization detection *(59)*. Choi *et al.* integrated an allele-specific PCR system with a disposable DNA microarray chip for multiplex SNP detection. Convective flows, created by pneumatic micropumps, were used to accelerate the hybridization process, resulting in the whole on-site assay completed in 100 min *(59)*.

Even though substantial progress has been made in creating the LOC devices, they are still not standalone (sample-in–answer-out) devices, which are required for POC diagnostics.

3 VALIDITY OF MICROARRAY DATA

The idea of DNA microarray was greatly embraced by the research community from the beginning, and by the commercial companies soon after, leading to its fast pace of technology development. However, standardization and development of method validation approaches lagged behind that of the technology itself. Consequently, in the early years a lot of concerns were raised regarding the reliability and reproducibility of the microarray data *(6,60)*. In 2005, a meta-analysis study was performed on seven of the most-published studies on cancer prognosis that used microarray-based expression profiling *(60)*. The authors aimed to study the reproducibility of microarray results reported in those studies. Surprisingly, in five of these studies the reported data was not reproduced. The analysis of the other two studies provided much weaker prognostic information than given by the original data *(60)*. This lack of reproducibility in the early microarrays originated from several sources of variability involved in the assay procedure, and they include array fabrication, biological sample matrices, DNA hybridization, signal detection, and data analysis. Table 2 summarizes some sources of variation that affect the quality of each step and consequently the final result. Following the awareness about

TABLE 2 Sources of Variation in Microarray Experiments and Currently Available Controlling Measures and Tools

Stages of the Microarray Assay	Source of Variations	Controlling Measures and Tools
Array fabrication	Spot morphology, probe solutions (quality and quantity), array design, probing approach	Replicate features, reference standards, batch quality control, DNA binding dyes
Sample preparation (extraction, purification, and labeling)	Target quality, sample matrices, extraction method, labeling technology and uniformity of dye incorporation	Target quality assessments, universal reference standards, spike-in controls
Probe-target hybridization	Hybridization buffer composition, temperature control, sequence-dependent kinetics and thermodynamics, washing process	Strict control of temperature and buffer composition, replicate samples, careful washing, positive controls
Signal detection	Sensitivity and resolution of the scanner, photobleaching	Scanner evaluation and calibration slides, scanner validation slides
Data analysis (image processing, normalization, and interpretation)	Format of produced data, normalization approach	Integrated software packages, supervised data interpretation

the shortage of method validation measures, the scientific community put much effort in preparing appropriate standards and controlling approaches and tools *(61,62)*. Aiming to provide a basis for reporting the microarray results, the MIAME (Minimum Information About a Microarray Experiment) standard was proposed in 2001 *(61)*. MIAME describes the minimum information required to ensure that microarray data can be easily interpreted and that results derived from its analysis can be independently verified. Commercial vendors of the microarray platforms improved their technologies over the years, and also set up a series of QC measures to enhance the reproducibility and accuracy of the data produced by their products. Together with regulatory agencies, they started the MAQC (Microarray Quality Control) project in 2006 *(62)*. The MAQC project intended to establish thresholds and metrics for interplatform comparison of microarray data. Several QC samples and tools are also currently available for the commercial array users as well as those who use in-house microarray platforms (Table 2).

Statistical analysis and biological interpretation is a critical issue in microarray analysis, especially in gene expression profiling where a massive amount of information is commonly produced. In order to decide if a gene is outcome-related, in gene expression studies the expression level of the gene in the patient sample is usually compared with the one from a normal sample and a fold-change and clustering analysis are used to draw the biological interpretations (e.g. class comparison or prediction of disease in cancer patients) *(6)*. However, expression levels of the genes naturally vary between different individuals and between different samples from the same individual *(60)*. Hence, a simple fold-change statistic does not account for variability across specimens and commonly creates false-positive outcomes. For instance, in 2007, Dupuy and Simon reviewed 23 studies published in 2004 that revealed the results of outcome-related gene-finding analyses, reporting false-positives in 9 of them *(5)*. Thanks to the recent advances in bioinformatics, more valid data analyses are currently available. The so-called supervised data interpretation methods are able to make distinctions among the specimens based on predefined information and create more valid information for clinical decision-making *(6)*.

4 CLINICAL ADOPTION

A quick search in the PubMed database shows that the number of scientific publications in the microarray area grew exponentially since its introduction in 1995 and remained at 7000 in recent years. Despite its revolutionary growth in the research area, the microarray technique has been slow in penetration in the molecular diagnostic market, as it accounted for only 10% of the whole molecular diagnostic market in 2010 *(7)*. The rather serious concerns about the reproducibility and robustness of microarray data, discussed previously, deterred receiving regulatory approval as well as obtaining success in the market. The poor reproducibility of the microarray data due to technical limitations or natural variation between different samples caused microarray tests to have insufficient robustness required for a diagnostic test. Gaining regulatory approval and clinical acceptance is more an issue in gene expression profiling, where the test is commonly based on the new research studies that correlates the level of expression of a set of several genes to a clinical outcome; therefore, they are not predefined and known biomarkers. In order to receive clinical acceptance and even regulatory approval, expression profiling tests are thus required to prove the correlation of their result to the clinical outcome in addition to their reliability *(63)*. Furthermore, they face the question of how the data they provide improves the outcome of the patient (clinical utility), for example, indicate a suitable type of treatment or prevent overtreatment in chemotherapy *(63)*. The other challenge of microarrays in the market is strong competition that exists at both ends of multiplicity *(7)*. PCR-based diagnostic tests are well known and flexible in processing regular sample matrices; therefore, they can compete with microarray tests if the amount of information required is within their territory. MammaPrint (Agendia, The Netherlands) was the first expression profiling test that received US Food and Drug Administration (FDA) approval in 2007. Based on research by van de Vijver *et al.* in 2002 *(64)*, MammaPrint assesses the expression profile of a set of 70 genes for prognosis application in breast cancer. The test, however, suffers from strong competition from the PCR-based Oncotype test (Genomic Health, USA), primarily because the latter is able to analyze the widely used FFPE (formalin-fixed paraffin-embedded) samples. A newer microarray-based test (Tissue of Origin; Pathwork Diagnostics, USA) with the capability to be performed on FFPE samples has recently entered the market. On the other hand, NGS techniques are well developed and their price is no longer prohibitive *(65)*. More reliable, informative (provides the sequence without prior knowledge), and convenient (digital instead of analog) data offered by sequencing techniques (e.g., RNA-Seq) can profoundly hinder the high-throughput microarray tests entering the clinical market.

Genotyping arrays, on the other hand, have fewer obstacles to overcome in gaining regulatory approval as well as clinical acceptance. Because they are usually aiming to characterize well-known and previously established sequence variation among the genome, it does not have to prove clinical correlation and utility. Nevertheless, these arrays are still required to be technically reliable and competitive in price. Low-throughput genotyping arrays use a fairly small number of features (up to a few hundred on a chip) for characterization of a pathogen or investigation of several SNP sites in a particular gene at a relatively inexpensive price. The PapilloCheck test provided by Greiner Bio-One Company (Frickenhausen, Germany) is a fairly successful low-throughput genotyping test, which obtained US FDA approval in 2009. The PapilloCheck system utilizes an array of 140 oligonucleotides to determine the HPV (human papillomavirus) subtypes in cervical smear samples. High-throughput genotyping arrays contain significantly higher numbers of features to investigate mutations in several genes or mixtures of pathogens, thus creating a huge amount of information. As an example, AmpliChip CYP450 obtained FDA approval in 2004 (the first microarray-based clinical test). The test uses 15,000 features on an Affymetrix platform to assess several types of variations in two genes involved in the metabolism of many psychoactive drugs. Table 3 summarizes some of the major microarray-based diagnostic tests that have already entered the molecular diagnosis market.

5 FUTURE TRENDS OF MICROARRAY

Some powerful LOC devices have been developed in recent years that have integrated several steps of the assay on their miniaturized platform; however, they are not yet capable of performing an assay without requiring any outside instrument. Particularly, they still need to use the conventional methods for signal detection *(53)*. Developing novel and compatible supporting technologies that allow their miniaturization is the key factor in developing LOC devices for POC. The construction of the nanoarrays help by replacing the current bulky fluorescent scanner with miniaturized fluorescence detectors in which the whole array fits in the field of view of their objective. Other than fluorescence detection, the label-free detection techniques, provided that more simple and powerful technologies will be invented in the future, will also simplify future LOC devices. Different steps of microarray assays are integrated in standalone LOC devices in a more harmonized way, which make them capable of performing sample-in–answer-out assays. These LOC devices exploit the microfluidic networks not only to connect their different compartments but also to help the device become faster to use, smaller, and conveniently controlled. Future microarray LOC devices are standalone, faster, smaller, and more efficient. These LOC devices are suitable candidates for POC diagnostics and play a major role in personalized medicine in the future.

As the microarray technology approaches its maturity, many of the issues of the early microarrays have been already overcome and consequently the microarrays of today are adequately robust and reliable. In addition, with the completion of ongoing large-scale studies that are aiming to investigate the clinical correlation and utility of the microarray data, similar to MINIDACT (Microarray in Node-Negative Disease May Avoid Chemotherapy) for the MammaPrint, gaining the regulatory approval and the acceptance by the clinicians are expected to become easier in future years *(7)*.

While the fierce competition from NGS at the high end of throughput and from PCR-based techniques at the low end has hindered the DNA microarrays in the molecular diagnostic market; however, microarrays still have advantages over these competing techniques. The price of NGS assays, although not prohibitive anymore, are still more expensive than the ones offered by

TABLE 3 Major Microarray-Based Tests in the Molecular Diagnostic Market

Category	Name	Provider	Application	Platform	Approval	Sample Matrix
Expression profiling	MammaPrint	Agendia (the Netherlands)	Prognostics in breast cancer	3000 oligo on Agilent array	CE 2005, FDA 2007	Fresh tissue
	TOO (Tissue of Origin)	Pathwork Diagnostics (USA)	Primary tumor site suggestion in metastatic cancer	2000 gene array	FDA 2008, 2010 (for FFPE)	Fresh tissue and FFPE
	AML profiler	Skyline Diagnostics (the Netherlands)	Determination of AML subtype	Affymetrix GeneChip	CE 2011, FDA (premarket approval)	Fresh bone marrow
Genotyping arrays	AmpliChip CYP450	Roche Scientific	Drug metabolism	Affymetrix GeneChip	FDA 2004	Blood
	eSensor® Cystic Fibrosis (CF) Genotyping Test	GeneMark Dx (USA)	Detection of CF mutation	Oligo array	FDA 2006	Blood
	PapilloCheck	Greiner Bio-One (Germany)	Determination of HPV subtype	140-oligo array	CE 2009	Cervical smear
	Genome-Wide Human SNP Array 6.0	Affymetrix (USA)	Whole-genome interrogation of SNP sites and CNV	1.8 million Affymetrix probe array	None, research use only	Blood and saliva
	Illumina 1M	Illumina (USA)	Whole-genome interrogation of SNP sites and CNV	~1 million probe arrays	None, research use only	Blood and saliva
Array-CGH	Human Genomic CGH microarray	Agilent Technologies (USA)	CNV detection	1 million probe arrays	None, research use only	Blood

microarray competitors. Some other unresolved difficulties in the sequencing techniques, like the necessity of multiplexing in RNA sequencing as well as of extensive sample preparation and data interpretation *(66)*, will be in favor of microarrays in the competition. On the other hand, as the microarray techniques become more familiar for clinicians and more flexible in the sample matrix they can process, they are moving to a better situation in competition with PCR-based techniques. Because of the wealth of knowledge offered by the human genome, the size of the molecular diagnostic market has been largely expanding in the past decade and is assumed to be doubled in 2017 *(67)*. In the presence of NGS *(68)*, DNA microarrays has continued to exist in the fast-growing molecular diagnostic market, and more microarray-based diagnostic tests are currently gaining regulatory approval and entering the market. According to a recent report, the microarray-based tests share the largest portion of the molecular diagnostics market with PCR-based tests *(67)* and, according to another report *(68)*, the market of microarray-based diagnostic tests is expected to grow by ~15% in the coming years. Generally speaking, microarray technology seems to gradually revive in the clinical market, even though it is not yet dominating it as was expected in the early days of its introduction.

6 CONCLUSION

Technical issues and inadequate QC measures in the early microarray studies brought about serious concerns and criticisms about the reproducibility of the microarray data provided as well as their biological interpretations inferred. These shortcomings in the microarray techniques themselves, as well as the harsh competition from other molecular diagnostic techniques, deter success in the clinical market as they were in the research counterparts.

In the pathway of maturity, DNA microarray vendors and researchers managed to resolve many of those issues, which led to regaining product confidence in the molecular diagnostic market. This is well projected in the growing number of microarray-based tests receiving regulatory approval and entering the market as well as positive foresights provided by business reports. Meanwhile, much advancement has been achieved in prerequisite technologies for developing portable and standalone LOC devices. Upon further development that allows for full integration of the assay, these devices can potentially be used for POC diagnostics.

Future DNA microarrays are expected to be faster to use, smaller, and more accurate. Supporting sub-systems, well matched with the microarray platforms, will be developed and integrated into the complex LOC devices, suitable for POC diagnostics. These LOC devices will play a central role in personalized medicine. Technical issues observed in the early microarray platforms are mostly resolved in the current platforms. New discoveries about the human genome, thereby increasing the depth of our knowledge, will initiate new clinical utilities for DNA microarrays. Enjoying all of these developments, the DNA microarray keeps a reasonable share in the fast-growing clinical diagnostic market.

REFERENCES

1. Schena, M.; Shalon, D.; Davis, R. W.; Brown, P. O. *Science* **1995**, *270*, 467–470.
2. Alizadeh, A. A.; Eisen, M. B.; Davis, R. E.; Ma, C.; Lossos, I. S. *Nature* **2000**, *403*, 503–511.
3. Shaw-Smith, C.; Redon, R.; Rickman, L.; Rio, M.; Willatt, L.; Fiegler, H.; Firth, H. *J. Med. Genet.* **2004**, *41*, 241–248.
4. Storhoff, J. J.; Marla, S. S.; Bao, P.; Hagenow, S.; Mehta, H.; Lucas, A. *Biosens. Bioelectron.* **2004**, *19*, 875–883.
5. Dupuy, A.; Simon, R. M. *J. Natl. Cancer Inst.* **2007**, *99*, 147–157.
6. Simon, R.; Radmacher, M. D.; Dobbin, K.; McShane, L. M. *J. Natl. Cancer Inst.* **2003**, *95*, 14–18.
7. Jordan, B. R. *Expert Rev. Mol. Diagn.* **2010**, *10*, 875–882.
8. Wang, L.; Li, P. C. H. *Anal. Chim. Acta* **2011**, *687*, 12–27.
9. Chan, C. P. Y.; Mak, W. C.; Cheung, K. Y.; Sin, K. K.; Yu, C. M.; Rainer, T. H.; Renneberg, R. *Annu. Rev. Anal. Chem. (Palo Alto, Calif.)* **2013**, *6*, 191–211.
10. Gubala, V.; Harris, L. F.; Ricco, A. J.; Tan, M. X.; Williams, D. E. *Anal. Chem.* **2012**, *84*, 487–515.
11. Bynum, M. A.; Gordon, G. B. *Anal. Chem.* **2004**, *76*, 7039–7044.
12. Stone, H. A.; Stroock, A. D.; Ajdari, A. *Annu. Rev. Fluid Mech.* **2004**, *36*, 381–411.
13. Sedighi, A.; Wang, L.; Li, P. C. H. In: *Nanopatterning and Nanoscale Devices for Biological Applications;* Iniewski, K.; Selimovic, S., Eds.; Taylor & Fransis, CRC Press: Boca Raton, Florida, 2013.
14. Hong, J.; Edel, J. B.; deMello, A. J. *Drug Discov. Today* **2009**, *14*, 134–146.
15. Bowtell, D.; Sambrook, J. *DNA Microarrays: A Molecular Cloning Manual;* Cold Spring Harbor Laboratory Press: New York, 2003.
16. Liu, J.; Williams, B. A.; Gwirtz, R. M. *Angew. Chem. Int. Ed.* **2006**, *45*, 3618–3623.
17. Peytavi, R.; Raymond, F. R.; Gagné, D.; Picard, F. J.; Jia, G.; Zoval, J.; Madou, M. *Clin. Chem.* **2005**, *51*, 1836–1844.
18. Campas, M.; Katakis, I. *Trend Anal. Chem.* **2004**, *23*, 49.
19. Lee, H. J.; Goodrich, T. T.; Corn, R. M. *Anal. Chem.* **2001**, *73*, 5525.
20. Situma, C.; Wang, Y.; Hupert, M.; Barany, F.; McCarley, R. L.; Soper, S. A. *Anal. Biochem.* **2005**, *340*, 123.
21. Wang, L.; Li, P. C. H. *J. Agric. Food Chem.* **2007**, *55*, 10509–10516.
22. Sedighi, A.; Li, P. C. H. *Int. J. Comput. Mater. Sci. Eng.* **2013**, *1*, 45–49.
23. Bouchie, A. *Nat. Biotechnol.* **2002**, *20*, 8.
24. Gillet, J. P.; Molina, T. J.; Jamart, J. *Leuk. Lymphoma* **2009**, *50*, 410–418.
25. Meneses-Lorente, G.; de Longueville, F.; Santos-Mendes, S. D. *Chem. Res. Toxicol.* **2003**, *16*, 1070–1077.
26. Sedighi, A.; Li, P. C. H. *Anal. Biochem.* **2014**, *448*, 58–64.
27. Wei, C. W.; Cheng, J. Y.; Huang, C. T.; Yen, M. H.; Young, T. H. *Nucleic Acids Res.* **2005**, *33*, e78.
28. Li, C. Y.; Dong, X. L.; Qin, J. H. *Anal. Chim. Acta* **2009**, *640*, 93–99.
29. Brøgger, A. L.; Kwasny, D.; Bosco, F. G. *Lab Chip* **2012**, *12*, 4628–4634.

30. Wang, L.; Li, P. C. H.; Yu, H. Z. *Anal. Chim. Acta* **2008**, *610*, 97–104.
31. Wang, L.; Li, P. C. H. *Anal. Biochem.* **2010**, *400*, 282–288.
32. Wang, L.; Kropinski, M.-C.; Li, P. C. H. *Lab Chip* **2011**, *11*, 2097–2108.
33. Peng, X. Y.; Li, P. C. H.; Yu, H. Z. *Sens. Actuators B Chem.* **2007**, *128*, 64–69.
34. Chen, H.; Wang, L.; Li, P. C. H. *Lab Chip* **2008**, *8*, 826–829.
35. Epstein, J. R.; Biran, I.; Walt, D. R. *Anal. Chim. Acta* **2002**, *469*, 3–36.
36. Sassolas, A.; Leca-Bouvier, B. D.; Blum, L. J. *Chem. Rev.* **2008**, *108*, 109–139.
37. Fang, X.; Liu, X.; Schuster, S.; Tan, W. *J. Am. Chem. Soc.* **1999**, *121*, 2921–2922.
38. Nelson, N. B. P.; Grimsrud, T. E.; Liles, M. R.; Goodman, R. M.; Corn, R. M. *Anal. Chem.* **2001**, *73*, 1–7.
39. Fang, S.; Lee, H. J.; Wark, A. W.; Corn, R. M. *J. Am. Chem. Soc.* **2006**, *128*, 14044–14046.
40. Baur, J.; Gondran, C.; Holzinger, M.; Defrancq, E.; Perrot, H.; Cosnier, S. *Anal. Chem.* **2010**, *82*, 1066–1072.
41. Koehne, J.; Chen, H.; Li, J.; Cassell, A. M.; Ye, Q.; Ng, H. T.; Han, J.; Meyyappan, M. *Nanotechnology* **2003**, *14*, 1239–1245.
42. Lee, K.; Babajanyan, A.; Melikyan, H.; Kim, C.; Kim, S.; Kim, J.; Lee, J.-H.; Friedman, B.; Levicky, R.; Kalachikov, S. *Biosens. Bioelectron.* **2012**, *42*, 326–331.
43. Ozkumur, E.; Ahn, S.; Yalçin, A.; Lopez, C. A.; Cevik, E.; Irani, R. J.; DeLisi, C.; Chiari, M.; Unlü, M. S. *Biosens. Bioelectron.* **2010**, *25*, 1789–1795.
44. Sinensky, A. K.; Belcher, A. M. *Nat. Nanotechnol.* **2007**, *2*, 653–659.
45. Tsarfati-Barad, I.; Sauer, U.; Preininger, C.; Gheber, L. A. *Biosens. Bioelectron.* **2011**, *26*, 3774–3781.
46. Xu, S.; Miller, S.; Laibinis, P. E.; Liu, G.-Y. *Langmuir* **1999**, *15*, 7244–7251.
47. Demers, L. M.; Ginger, D. S.; Park, S.-J.; Li, Z.; Chung, S.-W.; Mirkin, C. A. *Science* **2002**, *296*, 1836–1838.
48. Truskett, V. N.; Watts, M. P. C. *Trends Biotechnol.* **2006**, *24*, 312–317.
49. Moorcroft, M. J.; Meuleman, W. R.; Latham, S. G.; Nicholls, T. J.; Egeland, R. D.; Southern, E. M. *Nucleic Acids Res.* **2005**, *33*, e75.
50. Yu, A. A.; Savas, T. A.; Taylor, G. C.; Guiseppe-Elie, A.; Smith, H. I.; Stellacci, F. *Nano Lett.* **2005**, *5*, 1061–1064.
51. Lin, H.; Sun, L.; Crooks, R. M. *J. Am. Chem. Soc.* **2005**, *127*, 11210–11211.
52. Akbulut, O.; Jung, J.-M.; Bennett, R. D.; Hu, Y.; Jung, H.-T.; Cohen, R. E. *Nano Lett.* **2007**, *7*, 3493–3498.
53. Anderson, R. C.; Su, X.; Bogdan, G. J.; Fenton, J. *Nucleic Acids Res.* **2000**, *28*, e60.
54. Liu, R. H.; Yang, J.; Lenigk, R.; Bonanno, J.; Grodzinski, P. *Anal. Chem.* **2004**, *76*, 1824–1831.
55. Trau, D.; Lee, T. M. H.; Lao, A. I. K.; Lenigk, R.; Hsing, I.-M.; Ip, N. Y.; Carles, M. C.; Sucher, N. J. *Anal. Chem.* **2002**, *74*, 3168–3173.
56. Lee, T. M.-H.; Carles, M. C.; Hsing, I.-M. *Lab Chip* **2003**, *3*, 100–105.
57. Yeung, S.-W.; Lee, T. M.-H.; Cai, H.; Hsing, I.-M. *Nucleic Acids Res.* **2006**, *34*, e118.
58. Liu, R. H.; Nguyen, T.; Schwarzkopf, K.; Fuji, H. S.; Petrova, A.; Siuda, T.; Peyvan, K.; Bizak, M.; Danley, D.; McShea, A. *Anal. Chem.* **2006**, *78*, 1980–1986.
59. Choi, J. Y.; Kim, Y. T.; Byun, J. Y.; Ahn, J.; Chung, S.; Gweon, D. G.; Kim, M. G.; Seo, T. S. *Lab Chip* **2012**, *12*, 5146–5154.
60. Roussy, I. G.; Michiels, S.; Koscielny, S.; Hill, C. *Lancet* **2005**, *365*, 488–492.
61. Brazma, A.; Hingamp, P.; Quackenbush, J.; Sherlock, G.; Spellman, P.; Stoeckert, C. *Nat. Genet.* **2001**, *29*, 365–371.

62. Chen, J. J.; Hsueh, H.-M.; Delongchamp, R. R.; Lin, C.-J.; Tsai, C.-A. *BMC Bioinformatics* **2007**, *8*, 412.
63. Simon, R. *Eur. J. Cancer* **2008**, *44*, 2707–2713.
64. van de Vijver, M. J.; He, Y. D.; van't Veer, L. J.; Dai, H. *N. Engl. J. Med.* **2002**, *347*, 1999–2009.
65. Metzker, M. L. *Nat. Rev. Genet.* **2009**, *11*, 31–46.
66. Morozova, O.; Hirst, M.; Marra, M. A. *Annu. Rev. Genomics Hum. Genet.* **2009**, *10*, 135–151.
67. ReportLinker. Molecular Diagnostics Market & Forecast (By Application, Technology, Countries, Companies & Clinical Trials) to 2017: Global Analysis, April 2013; Retrieved from: http://www.reportlinker.com/p01158111-summary/Molecular-Diagnostics-Market-Forecast-By-Application-Technology-Countries-Companies-Clinical-Trials-to-Global-Analysis.html.
68. Ledford, H. *Nature* **2008**, *455*, 7512.

Chapter 3

Next-Generation Sequencing: New Tools to Solve Old Challenges

I. Gobernado*, A. Sanchez-Herranz† and A. Jimenez-Escrig‡
**Servicio de Psiquiatría, Hospital Ramón y Cajal, Madrid, Spain*
†Servicio de Neurobiología-Investigación, Unidad Central de Genómica Translacional, Hospital Ramón y Cajal, Madrid, Spain
‡Servicio de Neurología, Hospital Ramón y Cajal, Madrid, Spain

Chapter Outline

1 INTRODUCTION

As a sequel of the Human Genome Project, new technologies have been developed that have changed the way in which research and clinical medicine can be approached. A new generation of sequence technologies called next-generation sequencing (NGS) has transformed the means by which scientists extract genetic information from biological systems, revealing vast insight about the genome, transcriptome, and epigenome of any species by allowing DNA/RNA to be assayed in a rapid and cost-effective way. This capability has catalyzed a number of important breakthroughs in human disease research

Comprehensive Analytical Chemistry, Vol. 63. http://dx.doi.org/10.1016/B978-0-444-62651-6.00003-9

and clinical settings, moving to personalized medicine *(1–3)*. The rapidity with which this change has occurred has left many researchers and clinicians unaware of this new technology and of the opportunities that it provides *(4,5)*. In addition, NGS is going to replace many aspects of the current methods for assessing gene expression, protein binding, and other biological information such as microarray technology *(6)*. A brief glossary of NGS terms is provided in Table 1.

The use of NGS has allowed sequencing of hundreds of genomes of different species, even DNA from long-extinct species, and has spawned the finding of novel genes implicated in various human diseases. Nevertheless, NGS applications exceed by far the mere study of the aligned sequence of nucleotides in a genome. Detecting the molecular causes of human genetic diseases relied in the past on candidate gene studies, genome-wide association studies (GWAS) of single-nucleotide polymorphisms (SNPs), or copy number variants (CNVs) analyzed with microarrays. These techniques are being progressively replaced by NGS methods *(7)*. In addition to sequencing large regions or even a whole genome, NGS can provide very high sensitivity for quantitative applications, such as gene expression analysis. With NGS, researchers

TABLE 1 NGS Glossary

Term	Definition
Adapter	A short oligonucleotide that is attached to the DNA providing a ligation or priming site
Assembly	A reconstruction of the original sequence of DNA from shorter derived sequences
Contig	A set of overlapping contiguous DNA segments that can be used to deduce the original DNA sequence
Coverage depth	The average number of times that nucleotides, from sequencing reads, are mapped to a given position of reference genome
Exome	The part of the genome formed by exons
Homopolymer	Uninterrupted stretch of a single nucleotide type (e.g., TTTTTTT)
Indel	A single or very few nucleotides either deleted or inserted
Library	A collection of DNA fragments with adapters ligated to each end
Multiplexing	Pooling different DNA samples into a single sequencing run by using diverse adapters
Reference sequence/genome	A fully assembled version of a genome that can be used for mapping short DNA sequence reads

can quantify transcriptional activity at much higher resolution than traditional microarray-based methods *(8)*, essential for capturing subtle gene expression changes associated with biological processes. This powerful and flexible nature of NGS has made it the prevalent technique in any project that involves the study of the genome, transcriptome, and epigenome of any organism *(6,9)*.

Since it was introduced, NGS data output has increased at a rate that outpaces Moore's law, the observation made in 1965 by Gordon Moore, cofounder of Intel, that the number of transistors per square inch on integrated circuits had doubled every year since the integrated circuit was invented *(10)*. In the Human Genome Project, using Sanger technology, the first human genome required about 10 years to be sequenced and an additional 3 years for finishing the analysis. The completed human genome draft was published in 2001 with a cost nearing $3 billion *(11)*. In 2007, a single NGS sequencing run could produce a maximum of roughly one gigabase (Gb) of data. By 2011, that rate has practically reached a terabase (Tb) of data in a single sequencing run—nearly a 1000-fold increase in 4 years. It is possible now to sequence several human genomes in a single run, producing data in 1 week, for a reagent cost of less than $5000 per genome.

Even though NGS instruments are designed for massive data output, this technology is scalable and can be used for lower-output volumes for targeted studies or smaller genomes. Thus, for sequencing small bacterial/viral genomes, targeted regions, exomes, or a panel of disease-related genes, a researcher can choose to use a lower-output instrument and process a smaller number of samples per run, or can opt to process a large number of samples by multiplexing on a high-throughput instrument. Multiplexing enables simultaneous sequencing of several samples during a single experiment by tagging the DNA library generated from an individual genome with a "barcode" sequence added to each sample so they can be differentiated during the data analysis *(12)*.

In NGS, coverage depth refers to the average number of sequencing reads that align to each base within the sample DNA. For example, a whole genome sequenced at 50 × coverage means that, on average, each base in the genome was covered by 50 sequencing reads *(13)*. A sequencing run can be tailored to produce more or less data, zooming in with high resolution on particular regions of the genome, or providing a more expansive view with lower resolution. To adjust the level of resolution, a researcher can choose the coverage generated for a particular type of experiment. The required level of coverage and resolution depends on the experimental design. For instance, in cancer research, somatic mutations may only exist within a small proportion of cells in a given tissue sample. Using a mixed-cell sample of the tumor, the region of DNA harboring the mutation must be sequenced at very high level of coverage, upward of 200 ×, to detect with good resolution mutations present in a small percentage of tumor cells *(14)*. This type of analysis is not possible with Sanger sequencing because of the cost and error rate incurred with each

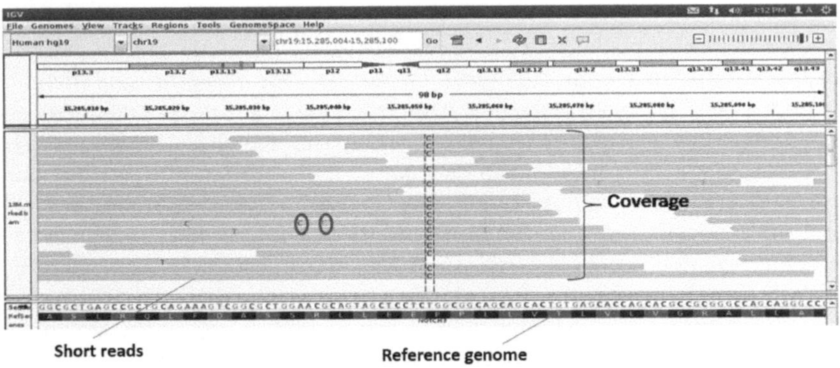

FIGURE 1 Illustration of the coverage and short reads aligned to a reference sequence in a typical NGS analysis. (For color version of this figure, the reader is referred to the online version of this chapter.)

additional read, so experiments requiring high read depths can become prohibitively expensive with Sanger sequencing. In contrast, an application like genome-wide variant discovery such as studying haplotypes requires a much lower coverage level, in a 5–10 × range. In this case scenario, it makes more sense to sequence at lower resolution but process a larger number of samples to achieve greater statistical power within a given population (Figure 1).

2 BASIS FOR NGS

NGS can be defined as sequencing techniques not based on Sanger sequencing. From a practical point of view, NGS started in the year 2005 when the first commercial sequencers were launched. At that time a run of advances began with the aim of sequencing a human genome for the price of $1000 in the shortest possible time. Sequencing applications are largely influenced by the procedures used to prepare the libraries and the way the data are analyzed, with the actual sequencing stage remaining fundamentally unchanged. There are a number of standard library preparation kits that offer protocols for sequencing whole genomes, targeted regions such as whole exomes, custom-selected regions, mRNA, protein-binding regions, and more. Protocols for NGS are generally more rapid and straightforward than those for the former capillary electrophoresis-based Sanger sequencing, since NGS starts directly from a genomic DNA or cDNA library. Usually, the DNA is sheared into shorter fragments, amplified and ligated to platform-specific oligonucleotide adapters needed to perform the sequencing, and the whole process preparation is completed in 90–240 min.

In principle, the concept behind NGS technology is similar to classic Sanger sequencing: the bases of a fragment of DNA are sequentially identified from signals emitted as each fragment is resynthesized from a DNA template strand. However, NGS extends this process across millions of reactions in a massively parallel fashion, hence the name sometimes given to this technique of "parallel sequencing" *(15)*. Thus, NGS enables rapid sequencing of short fragments of DNA, but spanning over a large part of the genome or even full genomes in the same run. The latest NGS instruments are capable of producing hundreds of Gb of data in a single sequencing run. In essence, the sequencing procedure involves an initial shearing of the DNA of the sample into a library of small fragments that can be uniformly and accurately sequenced in millions of parallel reactions. The newly identified sequences of bases, called short (~40–400 bases) reads, are then reassembled using a known reference genome as a scaffold (resequencing) or in the absence of a reference genome (*de novo* sequencing) *(16)*.

There are several NGS platforms currently available that are based on different techniques. An overview of the technologies and platforms used in NGS including their potential and limitations are provided next.

3 SAMPLE PREPARATION FOR NGS

The chosen method for RNA or DNA sample preparation previous to the sequencing step will depend on the NGS platform employed *(17)*. However, most of them follow common procedures, and are aimed to get a set of DNA short fragments with adapters ligated in their 5′- and 3′-ends. This set of DNA fragments and adapters is known as a library.

3.1 Clonal Amplification

Most detection systems require a high number of DNA template copies to get sufficient signal threshold for detection. Therefore, it is necessary to amplify every template in several orders of magnitude from the library. This clonal amplification is performed using either emulsion PCR (emPCR) or solid-phase amplification.

3.1.1 Emulsion PCR

In emPCR, every DNA template in the library is attached to a single bead and amplified in 26 μm water droplets containing the PCR reagents immersed in oil (Figure 2) *(18,19)*. In brief, DNA fragments are ligated to the adaptors, denatured to single strands, captured onto beads, and finally, amplified in a water-in-oil emulsion, one DNA fragment per droplet. Beads can be subsequently immobilized in a polyacrylamide gel on a microscope slide (Polonator), chemically cross-linked to an amino-coated glass surface (Life/APG; Polonator) or deposited into individual PicoTiterPlate wells (Roche 454) where sequencing reactions will take place *(20,21)*. This technique allows for complex genomes to be amplified with common PCR primers.

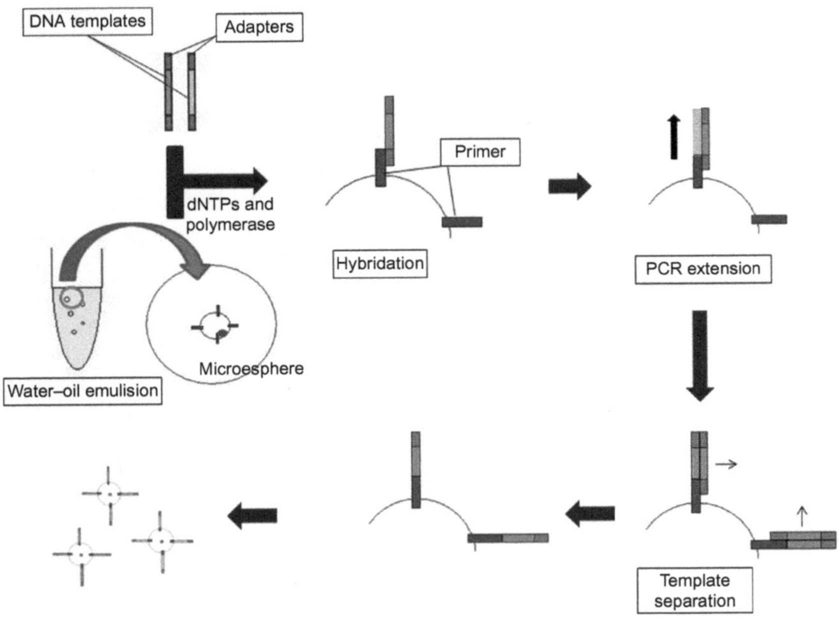

FIGURE 2 Scheme of emulsion PCR. (For color version of this figure, the reader is referred to the online version of this chapter.)

3.1.2 Solid-Phase Amplification

In solid-phase amplification, forward and reverse primers are covalently attached to a slide where DNA fragments are ligated at both ends to adapters, denatured, and immobilized, creating a densely DNA-coated surface (Figure 3) *(22)*. Each single-stranded DNA fragment, immobilized at one end on the surface, bends over to form a "bridge" structure by hybridizing by its free end to the complementary adapter on the support. In the mixture containing the amplification reagents, the adapters on the surface act as primers for the subsequent amplification cycle. After several cycles, random clusters of about 1000 copies of single-stranded DNA fragments (termed DNA "polonies," resembling cell colonies after polymerase amplification) are created on a surface *(23)*. This technique is used in Illumina sequencing platforms (formerly Solexa).

3.2 Single-Molecule Sequencing

Single-molecule sequencing represents a step forward in NGS, so it is sometimes called third-generation sequencing or next-NGS. Using single molecules for sequencing allows needing less DNA, less reagents, getting longer reads, and it is also faster and cheaper *(24)*. Several single-molecule sequencing methods are under development but currently only single-molecule

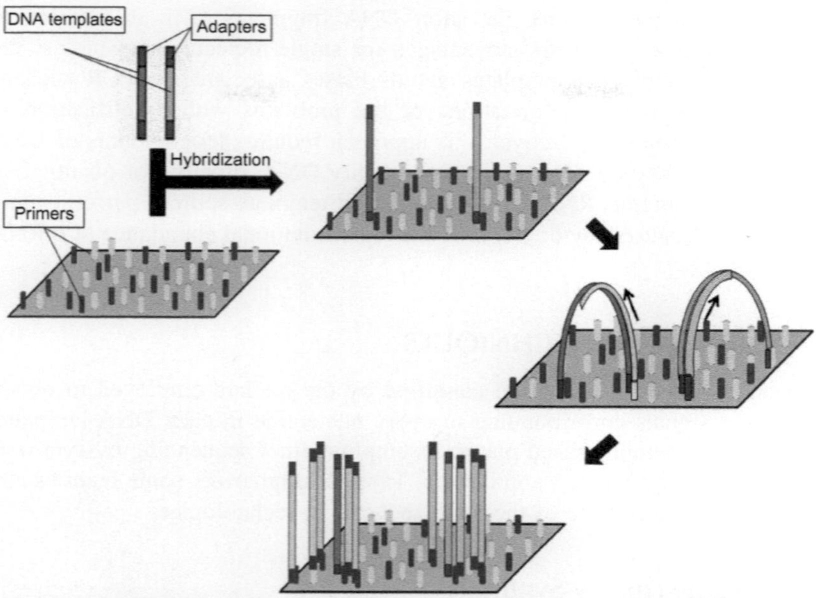

FIGURE 3 Scheme of solid-phase PCR. (For color version of this figure, the reader is referred to the online version of this chapter.)

sequencing by cyclic synthesis is advanced enough to be in use. This sequencing technology relies on the single DNA fragment library arrayed on a flat substrate by adapters ligated to primers or directly when DNA fragments covalently anchor to a solid surface. DNA polymerase and one of four fluorescently labeled nucleotides are flowed into the system at each sequencing cycle. DNA strands in the array that undergo template-directed base extension are lightened up by fluorescent label and recorded with a charge-coupled device (CCD) camera. The individual DNA strands are separated by a few microns and can be monitored as independent entities. The fluorescent signal of each incorporated labeled nucleotide is then sequentially detected using fluorescent microscopy. Because each DNA molecule is sequenced separately, there is no need for synchronization between different molecules *(25)*. Tens of millions of molecules can be sequenced in parallel in single small-reaction volume, and thus this method produces high-throughput sequencing at a very low cost. This technique is used by Helicos Biosciences platforms. Another method, used by Pacific Biosciences and reported in Life/VisiGen patents, is to anchor DNA polymerase molecules to the surface of the array and then DNA templates are sequenced as the enzymes synthesize new strands *(26)*.

Currently this technique produces short reading lengths, which make it indicated for resequencing applications in which a reference sequence is provided as a template for the short DNA fragments *(27)*. In addition to what we outlined about the advantages for single-molecule sequencing, the use of single-molecule templates avoids biases associated to PCR such as the introduction of new mutations or the problems with amplification in AT/GC-rich regions. Moreover, this approach requires less amounts of DNA (<1 μg), and they do not require high-quality DNA. Finally, for quantitative applications, such as RNA-seq, nonamplified template sources perform more effectively because they do not alter the representational abundance of mRNA molecules *(28)*.

4 SEQUENCING TECHNIQUES

Sequencing techniques can be classified by the method employed to obtain detectable signals corresponding to every nucleotide in each DNA template. Thus, most commonly used platforms employ either sequencing-by-synthesis or sequencing-by-ligation approaches. Table 2 summarizes some features and the current performance of the main sequencing technologies.

4.1 Sequencing-by-Synthesis

The NGS techniques relying on sequencing-by-synthesis require the use of a DNA polymerase during the sequencing run. Sequencing-by-synthesis can be divided into two major techniques: single-nucleotide addition and cyclic-reversible termination.

4.1.1 Single-Nucleotide Addition

In single-nucleotide addition techniques, limited amounts of an individual dNTP are introduced to the reaction media to synthesize DNA and, after washing, sequencing is resumed with the addition of another nucleotide. Limiting the amount of a given dNTP is required to minimize misincorporation effects observed at higher concentrations. A major drawback with this approach is the incomplete extension through homopolymer repeats. Pyrosequencing technology and the more recently developed proton-based personal genome machine (PGM) platforms use single-nucleotide addition.

4.1.1.1 Pyrosequencing Technology (454 Life Science/Roche)

Pyrosequencing was designed by Hyman in 1988 and developed by Ronaghi *et al.*, from the Stockholm Technological Institute, and it is one of the most widely NGS techniques used *(29,30)*. It employs a library generated by emPCR as a DNA template for sequencing using nucleotides that are supplied in a predetermined order. When the introduced nucleotide is complementary to the template, the DNA polymerase is activated, the nucleotide is incorporated to the DNA chain and a pyrophosphate molecule is released.

TABLE 2 Current Next-Generation-Sequencing Platforms

Platform	Library Generation/ Sequencing Principle	Read Length (Bases)	Runtime (Days)/Gb per Run	Pros	Cons	Biological Applications
Roche 454	Emulsion PCR/ pyrosequencing	400	0.4/0.5	Longer reads improve mapping in repetitive regions, fast runtime	High reagent cost, high error rates in homopolymer repeats	Bacterial and insect genome *de novo* assemblies, medium scale (<3 Mb) exome capture, 16S in metagenomics
Illumina HiSeq 2000	Solid-phase amplification (polony)/ sequencing by synthesis	120	10/600G	Most widely used, low cost per base, low error rate	Low multiplexing capability of samples	Variant discovery by whole-genome resequencing or whole-exome capture, gene discovery in metagenomics
Life Technologies SOLiD4/5500	Fragment/ emulsion PCR sequencing by ligation	50–60	10/600G	Low cost per base, low error rate by error correction by two-base encoding	Long runtimes/short reads increase cost and difficulty of data analysis and genome assembly	Variant discovery by whole-genome resequencing or whole exome capture, provides inherent gene discovery in metagenomics
PGM (Ion Torrent)	Emulsion PCR/ pyrosequencing, change in pH	200–400	0.5/200 Mb on 314 chip; 1 G on 318 chip	Long reads, fast runtime	High reagent cost, high error rates in homopolymer repeats	Bacterial and insect genome *de novo* assemblies, medium-scale (<3 Mb) exome capture, 16S in metagenomics

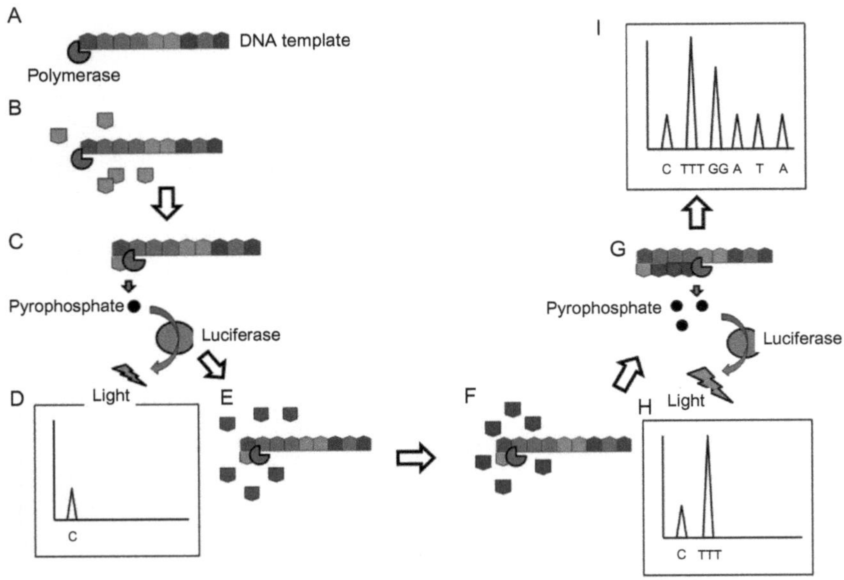

FIGURE 4 Scheme of pyrophosphate sequencing. (For color version of this figure, the reader is referred to the online version of this chapter.)

The released pyrophosphate is converted to a light signal by sulfurylase and luciferase enzymatic activities (Figure 4). The emitted light signal is detected and quantified, revealing the template DNA sequence. The light signals are collected in different peaks called pyrograms that correspond to the nucleotide that is complementary to the template DNA.

4.1.1.2 Semiconductor Technology (Ion Torrent/Ion Proton PGM/Life Technologies)

Semiconductor sequencing makes use of a highly dense microwell array in which each well acts as an individual DNA polymerization reaction chamber containing the sequencing fragment and DNA polymerase molecules. These sequencers exploit semiconductor technology, usually employing emPCR for template amplification prior to sequencing and are capable of detecting changes in the pH as nucleotides are incorporated during single nucleotide addition synthesis. Every time a nucleotide is incorporated in the complementary template, H^+ is released. Underneath the wells where sequencing takes part there is an ion-sensitive layer followed by a proprietary ion sensor, which detects changes in pH and converts the change in pH to a recordable voltage change *(31)*.

The advantage of single-nucleotide addition techniques is that they give readings quite similar to Sanger chromatograms, but they have two important caveats. First, it is done in a library of PCR-amplified fragments so it is possible to have erroneous readings because of asynchrony changes, which increase noise and limit the reading length. Second, sequencing homopolymer regions (DNA segments containing AAA, GGG, etc.) are prone to errors because there is no terminating moiety preventing multiple consecutive incorporations at a given cycle, so sequencing relies on the magnitude of light-emitted/pH changes to determine the number of repetitive bases. Therefore, the detector enables the detection of two equal and consecutive nucleotides (e.g., AA) in the template because the signal will be double with respect to the signal obtained from a single nucleotide (A). However, after several repeats the difference decreases (e.g., for 10 repeats, the difference in the signal between the 9th nucleotide and the 10th one is only 1/10th) *(32)*. As a consequence, the dominant errors for these platforms are insertions and deletions (indels) rather than substitutions. The cost per sequenced base using these platforms is higher than that of other NGS platforms (e.g., SOLiD and Illumina); therefore, they are unsuitable for sequencing large genomes. On the other hand, Roche and Life Technologies platforms have the advantage of providing longer reads (up to 400–1000 bp), facilitating *de novo* genome assembly as well as alignment of repetitive regions.

4.1.2 Cyclic-Reversible Termination

Cyclic-reversible termination techniques are based on the use of reversible terminators, which are nucleotides with a removable fluorescent dye that does not allow the incorporation of the following nucleotide to the chain until the fluorescent tag is cleaved off *(33)*. This is a cyclic procedure that involves (i) flowing of nucleotides to the sequencing cell or reaction chamber, (ii) washing-out of the unbound nucleotides, and (iii) fluorescence signals acquisition and nucleotide deprotection to allow the incorporation of the next reversible terminator. The main advantages of this technique are that it limits the possibility of erroneous readings since the fluorescent groups are eliminated. In addition, the accuracy of homopolymer readings is higher than that with pyrosequencing. There are two platforms that use this approach: Illumina (formerly Solexa) and Helicos Bioscience sequencers.

4.1.2.1 Illumina Technology

The Illumina sequencing platform, which currently is the most widely used NGS platform, was commercialized in 2006, with Illumina acquiring Solexa in early 2007 *(34)*. It is based on sequencing-by-synthesis chemistry, with reversible terminator nucleotides for the four bases, each labeled with a different fluorescent dye, and a DNA polymerase enzyme to incorporate them *(35)*. This reaction mixture for the sequencing reactions is supplied onto the "polonies" created on a surface by solid-phase amplification. After each terminator nucleotide is incorporated into the DNA strand, a fluorescent signal is emitted. A CCD camera detects the fluorescent signal and identifies the nucleotide and its position in the support. By deprotection of blocking groups and washing, a new synthesis cycle starts (Figure 4). Early Illumina sequencers provided read lengths of about 35 nucleotides. With this technology, the sequence of at least 40 million polonies can be simultaneously determined in parallel, resulting in a very high sequence throughput, on the order of gigabases per run. Further upgrades in Illumina technology allowed the development of the Genome Analyzer II and later, the HiSeq series with improved capabilities. A paired-end module for the sequencer was introduced. In the paired-end approach, a single long read is sequenced at both ends, leaving an intermediate region not sequenced. Paired-end sequences allow better detection of structural variants in the genome and increased specificity when aligning back to the genome compared to single-end fragments. They are also useful for *de novo* transcriptome assembly. HiSeq instruments bear optics and camera components that allow the system to image DNA clusters more efficiently over larger areas so that they triple the output from 1 to 3 Gb, with reads of up to 120 bp *(3)*. The system generates at least 1.5 Gb of single-read data per run, at least 3 Gb of data in a paired-end run, recording data from more than 50 million reads per flow cell. The runtime for a 36-cycle run was decreased to 2 days for a single-read run, and 4 days for a paired-end run. In contrast, the main limitations of this technology are (1) the need for library amplification, which in this case introduces errors, mainly underrepresentation of AT-rich and GC-rich regions; and (2) it suffers from biases due to DNA extension failure. Since the observed reading is a consensus of all the clonal templates, misincorporations increase the noise progressively, limiting the read length.

4.1.2.2 Helicos Bioscience Technology

In 2007, Helicos introduced the first commercial single-molecule DNA sequencing system (HeliScope sequencer). This system is based on a single-color fluorescence marker. The HeliScope principle is to scan thousands or millions of single molecules of DNA anchored to a glass cover slip flow cell. DNA is simply sheared, tailed with poly(A), and hybridized to a flow cell surface containing oligo(dT) for sequencing-by-synthesis in parallel with a primer, a modified polymerase, and analogs of labeled nucleotides named virtual terminator nucleotides. The system is able to analyze millions of single DNA fragments simultaneously, resulting in sequence throughput in the gigabase range. Although still in its first years of operation, the HeliScope sequencer has been tested and validated in several applications with promising results, for example, in the prenatal trisomy-21 (Down syndrome) test using a maternal blood sample, potentially replacing the standard test, which is associated with risks to the fetus *(36)*. Helicos technology does not require sample amplification, avoiding problems and bias associated with PCR that simplifies the DNA sample preparation process and tolerates degraded samples. On the other hand it generates reads only 32 nucleotides long, complicating *de novo* genome assembly. Also, in homopolar regions, multiple fluorophore incorporations could decrease emissions, sometimes below the level of detection.

4.2 Sequencing-by-Ligation

Sequencing-by-ligation involves the use of a DNA ligase instead of a DNA polymerase to create a complementary DNA strand *(37)*. This mode of sequencing employs several fluorescently labeled DNA probes that hybridize with the template *(38)*. This strategy is implemented in the ABI SOLiD sequencing system. Briefly, DNA fragments attached onto beads are amplified by emPCR. Then, DNA is denatured and the beads are deposited onto a glass support surface. The next step involves the hybridization of a (sequencing) primer to the adapter region of the template. After that, sequencing starts by the addition of a mixture containing four fluorescent oligonucleotides (8-mer, octamers) that compete for ligation. In each octamer, the dye color is defined by the first pair of the eight bases. After the detection of the fluorescence from the dye, bases 1 and 2 in the sequence can thus be determined. The ligated octamers are cleaved off after the fifth base, removing the fluorescent label. Then, another hybridization–ligation cycle is initiated. This last cycle allows the determination of the sixth and seventh bases in the sequence. In the subsequent cycle, the 11th and 12th bases are determined. Progressive rounds of octamer ligation enable sequencing of every five bases. After 10 cycles of ligation, the extended primer is removed from the template and the process is repeated with a universal primer that is shifted one base from the adaptor-fragment position. Shifting the universal primer in five rounds of 10 cycles enables the entire fragment to be sequenced and provides an error-correction scheme because each base position is interrogated twice. Color calls from the five ligation rounds are then ordered into a linear sequence and decoded in a process called “two-base encoding.” At present, this technology provides 35 bases read length with a reduced error rate. Sequences can be determined in parallel for more than 50 million bead clusters, resulting in a very high throughput of the order of Gb per run. Applied Biosystems produced an updated version in 2008, the SOLiD 2.0 platform, which increases the output of the instrument from 3 to 10 Gb per run. This improvement reduces the overall runtime from 8.5 to 4.5 days.

4.3 Other Sequencing Technologies

A new generation of sequencers, based on single-molecule and real-time sequencing, is rapidly emerging. An outstanding advantage of these novel technologies is that they do not require routine PCR amplification prior to sequencing, thereby avoiding systematic amplification bias occurring in current technologies. In addition to their capability for sequencing RNA directly with the corresponding savings in reagents and manpower, novel technologies can also sequence molecules in real time, decreasing the time of analysis and allowing longer read lengths. These novel platforms include Pacific Biosciences' nanopore-sequencing technologies and direct imaging of individual DNA molecules using advanced microscopy techniques.

4.3.1 Single-Molecule Real-Time Sequencing

The single-molecule real-time (SMRT) sequencer, developed by Pacific Biosciences, measures the incorporation of fluorescently labeled nucleotides into a complementary DNA template by means of a DNA polymerase using an array of densely packed zero-mode waveguide (ZMW) nanostructures that allow for optical interrogation of single-fluorescent molecules. The device allows limiting the observation field enough to capture exclusively the fluorescence emitted by a single nucleotide. Individual DNA polymerase molecules are anchored to the inner surface of ZMW wells of 30 nm diameter where the DNA synthesis can be initiated using nucleotides labeled with four different fluorescent dyes. The fluorescence pulse emitted by every incorporated nucleotide is acquired in a parallel fashion in thousands of ZMW wells *(39,40)*. The SMRT sequencer has an error rate of more than 5%, mainly at indels, due to the limitations of the technique. Owing to the high throughput of the technique, multiple reads of the same template can be obtained, which improve the accuracy up to 99.999%. Very long (up to 10,000 bp) reads facilitate the *de novo* assembly process. In addition, since it gives kinetic information of the DNA polymerase, it can be used to determine methylation patterns that can be useful for epigenomic research *(26)*.

4.3.2 Sequencing by Fluorescence Resonance Energy Transfer of Single Pairs

A new sequencer has recently been developed to sequence single molecules in real-time fashion using a sequencing-by-synthesis method *(41)*. This approach is based on fluorescence resonance energy transfer (FRET) technology and the use of a surface of 10 nm quantum dot nanocrystal. FRET is a mechanism of energy transfer between two fluorophores. Sequencing by FRET involves the use of a modified DNA polymerase with a donor fluorescent dye and four nucleotides each with a different acceptor dye. During the synthesis, when the correct nucleotide is incorporated in the DNA strand, the donor dye label in the polymerase comes into close proximity with the acceptor dye on the nucleotides and energy is transferred from donor to acceptor dye giving rise to a FRET light signal with a different frequency depending on the label incorporated in the nucleotides.

4.3.3 Nanopore Sequencing

Nanopore sequencing is a technique that determines the sequence of intact DNA molecules passing through a protein nanopore, a hollow structure slightly larger than the width of a double-stranded DNA molecule (4 nm) anchored to a membrane. Nanopore DNA sequencing is free of nucleotide labeling. Sequencing is achieved in real time as the DNA translocates the pore by recognition of the different DNA bases by the optic signal or by the electrical changes caused by the DNA chain as it goes through the nanopore.

Oxford Nanopore has implemented the nanopore technology in a commercial product (GridION system) based on two separate methods for DNA sequencing: strand sequencing and exonuclease sequencing. In the strand sequencing method, intact strands of single-stranded DNA are fed through a nanopore. Different bases will change the current in characteristic ways, allowing for the readout of individual bases. A major challenge of this technique is the fact that approximately 10–15 bases can occupy the interior of the nanopore at any one time, making the signal acquisition for a single base more challenging. In contrast, in the exonuclease sequencing method there is an exonuclease anchored to the external surface to the nanopore, which cleaves off the DNA chain into individual nucleotides (Figure 5). As each base is cleaved, it flows through the nanopore and is measured. A nucleotide passes through the nanopore every 20 ms, and the four nucleotides produce different current disruption signals, allowing the determination of DNA sequence. This method has the advantage of potentially solving the rate problem (as the speed is controlled by the exonuclease). Theoretically, 100,000 pores could sequence a human genome with 30× coverage in 30 min. It has an error rate of up to 4% in a single read, but this error rate lowers to 0.15% with 15× coverage. A further development that lowers the error rate is a single-stranded DNA passing through the pore without fragmentation and optic reading with a CCD camera that detects the light emitted by the different bases that form the DNA chain as it goes through the pore *(42,43)*.

4.3.4 Other Third-Generation Sequencers

A different technology used by third-generation sequencers is the one developed by companies such as Halcyon or ZS Genetics that uses a rapid-scan tunneling electron microscope to read the DNA chain directly *(44)*. DNA bases have to be labeled with heavy atoms to make them visible under the electron microscopy. Each nucleotide is tagged with a characteristic heavy label so that they can be distinguished in the transmission electron micrograph.

A new nanopore sequencing approach is currently under development by IBM based on nanometer artificial nanopores fabricated by e-beam drilling in 10-nm-thick membranes made of titanium nitride, separated by insulating layers of silica *(3)*. As the DNA strand is driven through the nanopore, the

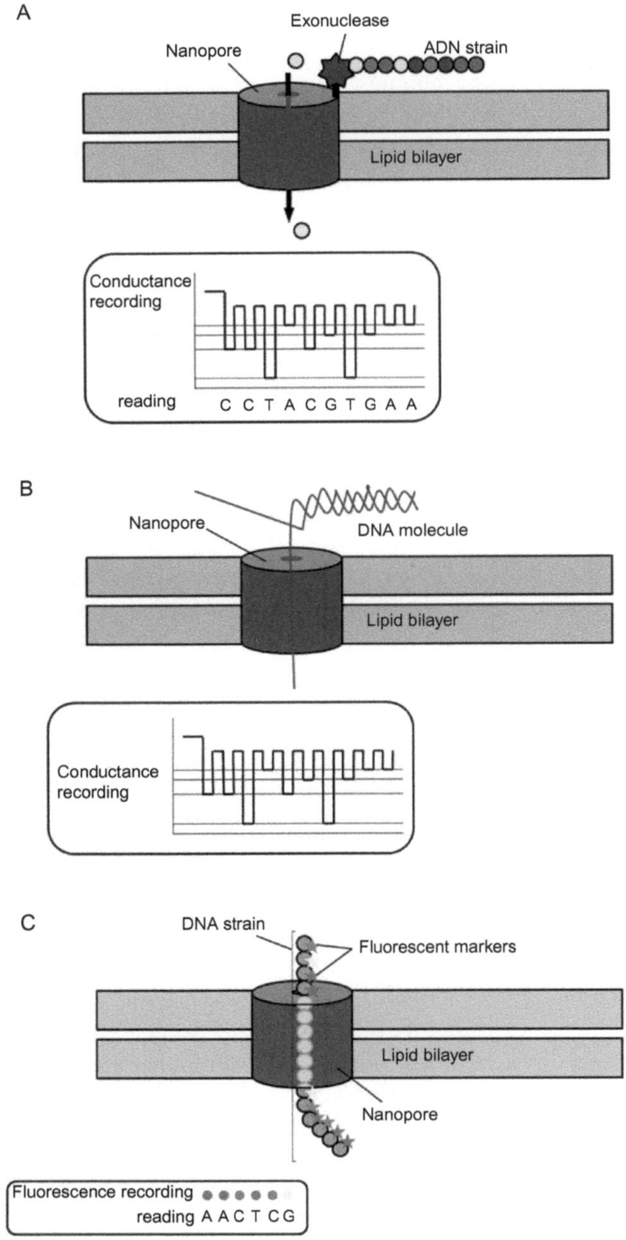

FIGURE 5 Nanopore sequencing by electric detection (A), translocation (B), and optic reading (C). (For color version of this figure, the reader is referred to the online version of this chapter.)

electrical field across the metal layers can be flipped, controlling the motion of the DNA strand and detecting changes in the ion current. This approach is similar to Oxford Nanopore detection, except that the DNA strand remains intact and it needs to be trapped for interrogation only for a millisecond. Although this technology has not been commercialized yet, it offers good potential for easy sample preparation and inexpensive sequence readouts *(45)*.

5 NGS DATA ANALYSIS

Along with sample preparation, the analysis of data generated by NGS technologies is an important aspect to consider in sequencing applications and one of the main bottlenecks for the diffusion of these techniques. Generation, analysis, and storage of NGS data demand a large computing infrastructure (at a minimum 8–32 Gb of RAM and 3–12 Tb of storage memory). Skilled IT and bioinformatics staff are needed to set up, maintain, and possibly run NGS data analysis tools. A number of data analysis algorithms have been developed to perform specific tasks related to a given application *(45)*. While some of these are commercially available from software vendors, many are freely available open-source algorithms from academic institutions. The usual analysis pipeline for NGS technology can be divided into three steps: (i) primary analysis that involves base calling at the sequencing platform—converting the detected signals to nucleotide base calls; (ii) secondary analysis that involves alignment reading and variant calling—mapping the short DNA reads to the reference genome and calling the variants present in the sequence file; and (iii) tertiary analysis that focuses on the analysis of the variant data with respect to the genetic experiment in question. Some applications, such as *de novo* sequencing, require specialized tools for assembling sequencing reads *(46)*. Other applications, such as RNA-seq, require algorithms that quantify read counts to provide information about gene expression levels *(47)*. A number of software packages exist to address the needs of each application *(46)*. An example of the pipelines and the software required for the main NGS processes are outlined in Figure 6.

6 MAIN APPLICATIONS OF NGS

NGS can be used for standard sequencing applications, such as genome sequencing and resequencing, and for novel applications in functional genomics research. Some examples of NGS applications currently working and in progress are shown schematically in Table 3.

6.1 Whole-Genome Sequencing

Before the development of NGS, sequencing an entire genome was a major challenge. Even for fairly small genomes, such as virus or mitochondrial

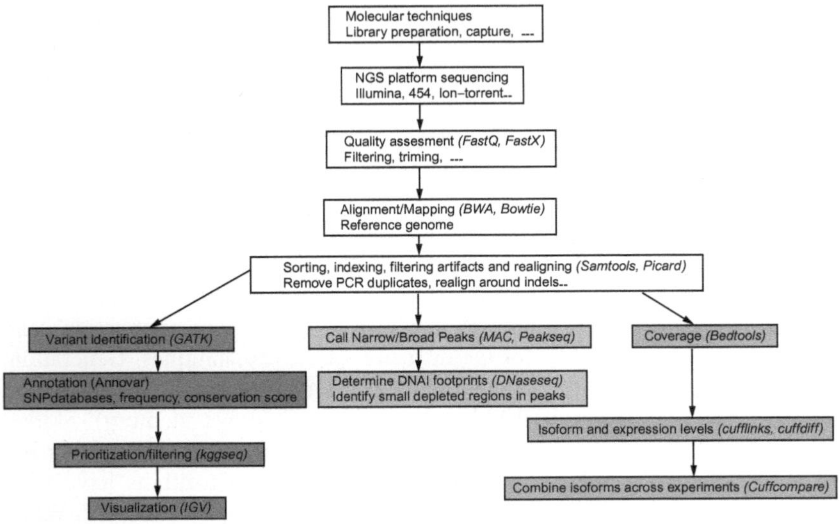

FIGURE 6 Analysis pipeline of NGS processes. Blue, targeted genome/exome/region; orange, Chip-seq; green, RNA-seq; and *cursive*, required software. (For interpretation of the references to color in this figure legend, the reader is referred to the online version of this chapter.)

genomes, whole-genome sequencing using Sanger-based technology required a significant amount of time and resources. The ability of NGS platforms to produce a large volume of data in a short period of time makes this technology a powerful tool for whole-genome sequencing, and it has completed initiatives such as the 1000 Genomes Project, a collaborative project with the goal of establishing a detailed catalog of human genetic variation *(47)*.

De novo genome sequencing refers to the generation of the genome sequence of an organism whose genome has not yet been fully sequenced. Sequencing a small genome of a microorganism like the vaccinia virus—a DNA virus of ~190 kb—using Sanger-based methods would involve roughly 2500 sequencing reactions (assuming 5× coverage and 500 bp read lengths). With NGS technology, the same sequencing project, including library preparation, can be completed in a few days, with one sequencing run at 30× coverage or greater *(77)*. Such advantages have prompted the sequencing of the genome of thousands of uncommon species, usually using a hybrid Sanger/NGS approach to resolve complex regions, but for the first time a draft sequence of the giant panda genome has been successfully generated and assembled based on NGS technology alone *(78)*.

The main challenge associated with *de novo* sequencing is the lack of a reference genome. This implies that the NGS reads have to be aligned and assembled without a reference guide. The coverage quality in terms of completeness, contiguity, connectivity, and accuracy of a *de novo* sequencing dataset depends on the quality of the contigs, which can be defined as a set

TABLE 3 Applications of NGS Technologies

Category	Applications	References
Genomics	*De novo* genome sequencing	*(48–50)*
	Whole-genome resequencing: 1000 Genomes Project, 6500 exomes (Exome Sequencing Project)	*(51–53)*
	Targeted resequencing: targeted polymorphism and mutation discovery for selected diseases or pathways	*(54–56)*
	Single-cell sequencing: find novel variants or validate the candidate variants in the target regions	*(57)*
Transcriptomics	Differential gene expression, quantification of gene expression and alternative splicing	*(58)*
	De novo transcriptome sequencing	*(59)*
	Transcriptome profiling: discovery of transcribed SNPs or somatic mutations	*(60)*
	Small RNA sequencing	*(61,62)*
	Noncoding RNA sequencing	*(63)*
Epigenomics	Transcription factor with its direct targets	*(64,65)*
	Genome-wide profiles of histone modifications	*(66–68)*
	CpG methylation	*(69–71)*
	Genomic profiles of nucleosome positions	*(72)*
Metagenomics	Environmental and human microbiome	*(73–76)*

of overlapped and aligned sequencing reads that obtain a consensus sequence for a genome region. The size and continuity of the contigs will affect the number of gaps present in the sequenced genome. An additional problem for *de novo* sequencing by NGS is that these shorter reads are also less accurate than Sanger sequences. For both reasons, these data are more difficult to assemble into a long contiguous and connected sequence, leading to a higher number of gaps in regions where no reads align. These gaps are especially frequent in genome regions containing repetitive sequence elements. To overcome this problem, some NGS platforms offer longer reads, while other sequencers can be used in combination with paired-end sequencing protocols, where both ends of a DNA fragment are sequenced, as opposed to single-read sequencing where only one end is sequenced. Paired-end data contain long-range positional information, which in turn give rise to more accurate assembly with more coverage. Also due to bidirectional sequencing, sequencing errors can be easily detected *(79)*.

In parallel with NGS technological improvements, many algorithmic advances have been made in *de novo* sequence assemblers using short-read data *(80)*. In many instances, existing computer resources in the laboratory are enough to perform *de novo* assemblies. The *Escherichia coli* genome can be assembled in as little as 15 min using a 64-bit Windows desktop computer with 32 Gb of RAM *(81)*.

While NGS is commonly associated with sequencing large genomes, especially human genomes, the scalability of the technology makes it also useful for sequencing small viral and bacterial genomes. This has been demonstrated during an *E. coli* outbreak in Germany *(82)*. Using NGS systems, researchers were able to quickly generate a high-quality whole-genome sequence of the bacterial strain in 1 week, providing new insight into the genetic mutations that conferred increased virulence.

6.2 Targeted Region Resequencing

Targeted region resequencing refers to the sequencing of a subset of genes or defined regions in a genome *(83)*. Target resequencing allows focusing on the regions of the genome of interest, saving time, cost, and data storage. In targeted resequencing applications, a region of interested is enriched for sequencing by either amplicon sequencing or capture method.

By using the amplicon sequencing method, thousands of amplicons from a large number of samples can be simultaneously generated and sequenced in a few hours. Amplicon sequencing can be used to sequence many genomes to discover, screen, or validate genetic variation within a population. It is possible to pool samples and with high sequence coverage during a single run identify rarer variants that are missed or too expensive to identify using Sanger-based sequencing approaches. With targeted sequencing it is feasible to check out a set of genes involved in a particular process, as, for example, to sequence the 86 genes involved in the hereditary ataxias *(84)*. Another common amplicon application is sequencing the bacterial 16S rRNA gene across a number of species, a widely used method for studying phylogeny and taxonomy, particularly in diverse metagenomic samples. Analysis of microbial 16S sequence data has been applied to evaluate bacterial diversity in a number of environments, allowing researchers to characterize microbiomes from samples that are otherwise difficult to study. In the past, a large number of microorganisms has evaded cultivation, but sequencing-based metagenomic analyses are finally making it possible to investigate their ecological, medical, and industrial relevance *(85)*.

An alternative to amplicon sequencing is target enrichment by capture where selected regions or genes are enriched in the library from genomic DNA using biotinylated probes and magnetic beds *(86)*. The difference with amplicon sequencing is that target enrichment allows for larger DNA insert sizes and enables a greater amount of total DNA to be sequenced per sample.

For example, rather than sequencing a few hundred exons, it is possible to sequence the entire exome to call functional SNPs for an individual *(86)*. The steps involved in preparing a NGS library for target enrichment are virtually identical to the steps involved in library generation for whole-genome sequencing, with the addition of a target enrichment step. Genomic DNA is isolated, fragmented, and purified to ensure uniform and appropriate fragment size. Their specific regions of interest, such as exons or specific genes, are captured by immobilizing DNA on a substrate such as magnetic beads. After binding the library and washing away unbound DNA, a simple elution step allows recovery of the enriched library, which can then be amplified by PCR in preparation for sequencing the selected region of DNA.

Numerous protocols and commercial kits are available for target enrichment. These can be used with custom-designed probes or with probes predesigned for widely used applications, such as exome enrichment. With custom-designed probes, researchers can target regions of the genome relevant to their research interests. This is useful for examining genes in specific pathways, or for follow-up analyses to GWAS. Regions of the genome identified in a GWAS as harboring causative variants can be enriched and sequenced to identify additional, possibly rarer, variants in the associated region. This strategy of NGS-GWAS combines NGS and genotyping to uncover novel causative genetic variants of complex diseases. Compared with traditional GWAS, it can provide more detailed information, including not only common SNPs but also rare variants *(87)*.

The application of targeted sequencing can expand over the genome with the sequencing of the ~180,000 exons of the human genome, what is called the exome *(88)*. Provided that 85% of genetic diseases are caused by mutations in the coding region, exome sequencing is currently one of the most used and powerful tools to evaluate patients with unknown gene mutations and has recently found numerous new causative genes. Different analysis strategies can be used in this regard, such as the search for common mutations in patients with the same phenotype or in affected cases in small pedigrees, or even the detection of new mutations in trios of parents and affected children *(84)*. In addition, sequencing human exomes from many individuals enables rare disease-associated alleles to be identified within a population *(89)*.

Resequencing allows identifying SNPs, indels, copy number and structural variations, in multiple individuals or strains, or a population-based sampling of a species. In the near future, genome-scale sequencing tests will become routine for the diagnosis of major medical problems for which there is a large number of possible Mendelian or high-penetrance genes. However, the current costs of NGS disease panels, such as familial cancer risk panel, cardiomyopathies, and hereditary dementias with 5–50 potentially causative genes, are lower than the costs of whole-exome tests and the interpretation of the results is far more efficient *(90)*. Thus, until whole-exome (and subsequently whole-genome) tests become more reliable, gene panels are likely to become the more consistent way for assessing Mendelian or high-penetrance disease-associated mutations in known causing genes.

Analysis of SNPs, the point mutations that arise on average every 1000 bp and occur in more than 1% of the population, and the analysis of large insertion or deletions, called CNVs, have been extraordinarily powerful when examining static germline genomes. However, there are situations in which a genome cannot reflect the pathological condition, and resequencing data are uniquely able to detect the causative alteration. In tumor cells, structural variation, or preexisting oncogene mutations may predispose a cell to uncontrolled growth and metastasis, but frequently it is *de novo* somatic mutations and rearrangements that have the main effect on tumor growth and disease progression *(91)*. In addition to *de novo* mutations, DNA translocations are also important in the etiology of many cancers and discovery of new translocations benefits from a genome-wide approach. Genomic alterations may vary even within different parts of the tumor because of variations in growth and selective pressure, so that the tumor genome can be a complex mixture of many genomes. NGS allows the discovery of a large number of mutations, many of them affecting pathways involved in tumor development and progression *(92)* both in somatic and tumoral cells. These findings have provided important therapeutic insights predicting sensitivity and resistance of a variety of therapeutic regimens and are the foundation of a personalized medicine *(93)*. A recent example of this is the 39-year-old lady with acute leukemia reported by Welch *et al. (92)*, whose metaphase cytogenetic revealed a complex pattern of deletions and rearrangements that are associated with poor long-term survival, thus making allogenic transplantation her therapeutic choice. Whole-genome sequencing in this patient identified a large region of Chr. 15 inserted into the second intron of the retinoic acid receptor alpha gene, resulting in a fusion oncogen and giving the patient a favorable prognosis, being treated with all-trans retinoic acid consolidation therapy and avoiding transplantation.

6.3 Metagenomics

NGS provides the capability to characterize a large number of different DNA sequences in a reasonable time and cost. Thus, NGS is well suited for metagenomics applications. Metagenomics is the study of the genetic material recovered from environmental sources (metagenomes). Thus, it intends to identify and quantify the microorganisms present in a particular sample *(94)*. The ability to generate NGS data without the need for culturing organisms has opened another window on biology *(95)*. Sequencing the genome of nonculturable organisms using the Sanger technique provided insight into bacteria that had been isolated as single cells. However, it is obvious that higher-throughput sequencing can provide a more detailed characterization of complex systems. In this research area, much attention is being directed to a variety of ecosystems present in the human body *(96)*. The human microbiome project (http://nihroadmap.nih.gov/hmp/) is aimed at characterizing and understanding the diversity of microbes that inhabit different parts of the human body and how they may affect health and disease *(97)*. Such studies will require NGS of organisms' genomes to fully appreciate the complexity of interactions among microbes, as well as between them and their hosts. Advances in metagenomics will provide new insight into the role that infectious agents play in disease development and progression, as well as in physiological states, for example, the influence of intestinal bacteria profile, the gut microbiome, in food absorption and obesity *(98)*.

Similarly, the genomes of some microorganisms can change rapidly, such as the HIV virus. NGS can be applied to analyze the samples isolated from multiple patients, allowing treatment protocols to be adapted to the evolving disease profile and to learn how this profile influences the pathogenicity or resistance to the therapy *(99,100)*. In addition to the direct approach just described, high-throughput sequencing can also identify viruses, bacteria, and other organisms present in a complex biological sample by identifying their genomic signatures. When an infection, tumor, or other undesirable outcome is caused by an unknown organism, sequencing all DNA and subtracting out what should be there from the total sequence profile leaves the signature of the contaminating or infecting organism *(101)*. Such an approach has been used to identify viral sequences in tumor samples, thus implicating a virus in tumor growth. In the same fashion, the viral DNA profiles from past or present infections can be obtained by testing whether a patient's genome or exome sequences align to specific viral genomes *(102,103)*.

6.4 RNA-Sequencing

The transcriptome is the complete set of transcripts in a cell and their quantity for a particular developmental stage or physiologic condition. Transcriptome analysis encompasses (i) the catalog of all transcripts, mRNAs, non-coding RNAs, and small RNAs; (ii) the transcriptional structure of genes, in terms of their start sites, 5′- and 3′-ends, splicing patterns, and other post-transcriptional modifications; and (iii) the expression levels of each transcript during development or under different physiologic and pathological conditions *(104)*.

The NGS methodologies genome-wide high-throughput transcriptomics are termed RNA-seq techniques. For RNA-seq, RNA (total or fractionated, such as polyA) is converted to a library of cDNA fragments with adaptors attached to one or both ends and each molecule is then sequenced by NGS. The resulting reads are either aligned to a reference genome or reference transcripts, or *de novo* assembled without the genomic sequence to produce a genome-scale transcription map that consists of both the transcriptional structure and/or level of expression for each gene. RNA-seq technologies provide great specificity and throughput in transcriptomic analysis, have high levels of reproducibility and do not have an upper limit for quantification *(105)*. Previous tools for studying transcription, quantitative PCR (qPCR), DNA microarrays, and serial analysis of gene expression have provided tremendous insight into many biological processes; however, each technique suffers from biases and has important limitations compared to RNA-seq. For instance, qPCR lacks high-throughput capability. On the contrary, DNA microarray offers high-throughput, but at the expense of lack of sensitivity for genes expressed either at low or very high levels of dynamic range. NGS methods offer some key advantages over existing microarrays for studying transcription. RNA-seq provides more complete information about transcriptome because it allows the direct characterization of transcript sequences, provides opportunities to discover novel classes of RNA, detects point mutations in expressed transcripts, and identifies fusion transcripts and unknown splice variants *(105)*.

Although RNA-seq is still an evolving tool, it is now preferable in most instances to microarrays since it gives more information, is more accurate, and can be more cost-effective. Therefore, it has been applied to explore the genome-wide expression profile pattern of diseases, such as neoplasms *(106)*, complex genetic traits like behavior *(107)*, or physiological states *(108)*. In addition, RNA-seq enables the investigation of noncoding RNAs, such as small RNAs, including microRNAs (miRNAs). These molecules have been poorly characterized in the past and are now attracting much attention due to their role in development processes, tissue differentiation, and cancer *(109)*. Some studies have demonstrated that specific miRNAs are directly implicated in cancer pathways, calling attention to the importance of including noncoding transcript profiling in disease research *(110)*. Indeed, the small size and high degree of similarity across different classes of miRNA molecules make them particularly well suited to new sequencing approaches. In addition to miRNAs, the expression of larger noncoding transcripts has also been found to be altered in cancer and in patients at high cardiac disease and diabetes risk *(111,112)*. The roles of these new, non-protein-coding transcripts are not yet understood, but they are believed to play fundamental roles in the regulation of gene expression and disease progression.

6.5 Other NGS Applications

6.5.1 *Epigenetics*

Epigenetics studies changes in gene expression and phenotype not depending on a DNA sequence variation, thus epigenomics is the study of these epigenetic changes on a genome-wide scale. Increasingly, we are coming to recognize that the message encoded within the DNA sequence is regulated in a variety of complex ways, aside from variations in the genome sequence. DNA methylation and histone modifications can change over time and these changes are associated with variations in gene expression. Sodium bisulfite treatment of genomic DNA, that transforms cytosines but not methylcytosines to uracil, has been widely used to determine the methylation state of the cytosines, as methylcytosines are sequenced as cytosine, and unmethylated cytosines as thymine. Sequencing of bisulfite-treated DNA has been laborious and costly, so it has been reserved for the analysis of candidate genes and thus has provided a very limited view of the genome, while NGS can overcome this issue *(113,114)*. Other methylation genome-wide techniques are MeDIP and reduced representation bisulfite sequencing (RRBS) *(69)*. MeDIP or mDIP enrich methylated DNA sequences by isolating methylated DNA fragments via an antibody raised against 5-methylcytosine. RRBS combines restriction enzymes and bisulfite sequencing in order to enrich the areas of the genome that have a high CpG content. RRBS provides an efficient way to generate absolute quantification of methylation of more than 1 million CpG sites at single base-pair resolution *(69)*.

DNA methylation variation can cause a number of human diseases such as Prader–Willi and Angelman syndromes and can inactivate tumor suppressor genes in cancer. DNA methylation can be influenced by environmental agents. Methylation differences may explain discrepancies in disease status between monozygotic twins *(115)*. In several cancer samples, variations in methylation have been identified through whole-genome bisulphite sequencing of tumor and adjacent normal tissue, suggesting that tumor progression may rely not only on the introduction of somatic mutations but also on changes in gene methylation *(116,117)*.

6.5.2 Interactomics

Interactome refers to biological interactions among genes. It is usually represented as a graphic network since interactomes may be described as biological networks (Figure 7) *(118)*. DNA-binding proteins such as histones, that bind nonspecifically to DNA and help to compact it, as well as transcription factors, that bind to specific sequence motifs and activate or repress transcription, play a role in DNA expression and repair. Although histones bind in a largely nonspecific manner, they are modified in specific ways that tag regions of the genome for selective action by transcription factors and polymerases *(119)*. Chromatin immunoprecipitation (ChIP) assay was developed to identify protein-binding sites across the genome. ChIP allows the identification of protein-binding sites within living cells by determining whether specific proteins are associated with specific genomic regions. The development of microarray technology and now advances on NGS technology enable genome-wide analysis of ChIP assay. This method allows researchers to compare protein–DNA binding profiles between tissue types, developmental stages, or differentially treated cells. Microarrays or ChIP-on-chips have the benefit of low cost of array technology, but have been limited by incomplete

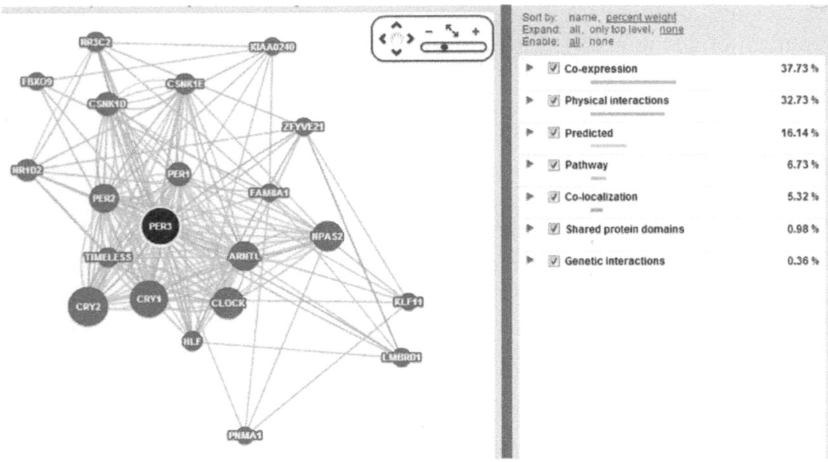

FIGURE 7 Interactome of the gene PER3 with other circadian-rhythm genes according to GENEMANIA (http://www.genemania.org). (For color version of this figure, the reader is referred to the online version of this chapter.)

coverage of the genome, difficulties in distinguishing closely related sequences, and poor resolution of sequence boundaries *(120)*. ChIP followed by sequencing (ChIP-seq) approaches are far more sensitive and allow complete coverage of the entire genome, providing more accurate pictures of the complex regulatory landscape within the genome *(121)*. ChIP-seq has become an excellent tool for studying the transcriptional machinery and gene expression regulation on a genome-wide scale. Chip-seq data can be classified as point source (highly localized signals ~100 bp, such as for transcription factors), broad source (signal spanning large domains ~100 kb, such as for some histone modifications), or mixed source (has elements of both, such as RNA polymerase II) *(122)*. Increasing evidence suggests that alterations in both genomic distribution and global level of histone modifications play important roles during tumorigenesis *(123)*. In addition, ChIP-seq has disclosed that besides DNA methylation and histone modifications, nucleosome repositioning and chromosomal looping are important epigenetic regulation mechanisms *(123)*.

7 INTEGRATING OMICS DATA

Each of the above-mentioned techniques can provide valuable data for understanding biological systems. However, such value can be greatly enhanced when multiple types of data are generated on the same system, a property inherent in many sequencing approaches. The recent availability of genome-wide expression data coupled with genotype data has already shown the value of integrating genetic and transcriptional approaches *(124)*. Also, using NGS,

RNA expression levels can be treated as quantitative traits and combined with genotyping or sequencing data to identify quantitative trait loci *(125)*.

Ideally, in the next 5 years, it will be possible to sequence a human genome for less than $1000, the cost that everyone considers will be covered by medical insurance companies, so further significant drops below that price cannot be expected. But even if this price means that NGS will be used broadly, high-throughput sequencing is not without issues. Even if sequencing costs are starting to be reasonably low, there are additional limitations that restrict the benefits derived from cheaper data generation. The huge quantity of sequences has shifted the bottleneck of molecular biology from the generation of data to its analysis. Nevertheless, the challenges for the analysis are not the main issue. After analyzing the NGS data, there is a need for combining the massive datasets and then applying those results to various types of biological information. For a correct interpretation of the variants found in a genomic/exomic survey, we need to develop up-to-date, continually revised databases to ascribe accurately every rare or novel variant as disease-causing and avoid overcalling rare variants as causative of disease. Clinicians will need additional tools to image, storage, and assist them in a better understanding of genetic studies. For example, algorithms for combining gene-by-gene interactions for common variants such as the study that combined sugar-sweetened beverage consumption and a genetic-predisposition to obesity, calculated on the basis of the 32 BMI loci *(126)*, are still far from clinical application, and also require the addition of environmental exposures in making disease prediction. It is important to know that even taking into account epigenetic mechanisms, the dynamics of gene transcription, and translation and of posttranslational changes, environmental factors can determine phenotypic manifestations as well.

In this particular scenario, sequencing-based tests will be routinely used with lower costs than many other high-technology tests in what has been called personalized medicine. The education of physicians and other clinical staff in genomic medicine is therefore an essential need to understand and correctly use NGS. Nevertheless, it is necessary to consider that except when dealing with highly penetrant heritable disorders, genomic information will not be the only determinant of disease. Finally, it is important to consider that the appropriate use of these powerful techniques has to involve careful consideration of the ethical and legal issues in the management of the information obtained *(127,128)*.

REFERENCES

1. Ong, F. S.; Lin, J. C.; Das, K.; Grosu, D. S.; Fan, J. B. *Genomics* **2013**, *102*, 137–139.
2. Ku, C. S.; Cooper, D. N.; Roukos, D. H. *World J. Gastroenterol.* **2013**, *19*, 2011–2018.
3. Ansorge, W. J. *Nat. Biotechnol.* **2009**, *25*, 195–203.
4. Ocana, A.; Pandiella, A. *Mol. Cancer* **2010**, *9*, 202.
5. Ware, J. S.; Roberts, A. M.; Cook, S. A. *Heart* **2012**, *98*, 276–281.

6. Altelaar, A. F.; Munoz, J.; Heck, A. J. *Nat. Rev. Genet.* **2013**, *14*, 35–48.
7. Duan, J.; Zhang, J. G.; Deng, H. W.; Wang, Y. P. *PLoS One* **2013**, *8*, e59128.
8. Hayes, J. L.; Tzika, A.; Thygesen, H.; Berri, S.; Wood, H. M.; Hewitt, S.; Pendlebury, M.; Coates, A.; Willoughby, L.; Watson, C. M.; Rabbitts, P.; Roberts, P.; Taylor, G. R. *Genomics* **2013**, *102*, 174–181.
9. Kadakkuzha, B. M.; Puthanveettil, S. V. *Mol. BioSyst.* **2013**, *9*, 1807–1821.
10. Stevens, H. *Bioassays* **2012**, *34*, 103–105.
11. Xiong, M.; Zhao, Z.; Arnold, J.; Yu, F. *J. Biomed. Biotechnol.* **2010**, *370710*, 2010.
12. ElSharawy, A.; Forster, M.; Schracke, N.; Keller, A.; Thomsen, I.; Petersen, B. S.; Stade, B.; Stahler, P.; Schreiber, S.; Rosenstiel, P.; Franke, A. *BMC Genomics* **2012**, *13*, 417.
13. Flannick, J.; Korn, J. M.; Fontanillas, P.; Grant, G. B.; Banks, E.; DePristo, M. A.; Altshuler, D. *PLoS Comput. Biol.* **2012**, *8*, e1002604.
14. Idris, S. F.; Ahmad, S. S.; Scott, M. A.; Vassiliou, G. S.; Hadfield, J. *Expert Rev. Mol. Diagn.* **2013**, *13*, 167–181.
15. Kohlmann, A.; Grossmann, V.; Haferlach, T. *Semin. Oncol.* **2012**, *39*, 26–36.
16. Kumar, S.; Banks, T. W.; Cloutier, S. *Int. J. Plant Genomics* **2012**, *831460*, 2012.
17. Mardis, E. R. *Annu. Rev. Genomics Hum. Genet.* **2008**, *9*, 387–402.
18. Zhu, Z.; Jenkins, G.; Zhang, W.; Zhang, M.; Guan, Z.; Yang, C. J. *Anal. Bioanal. Chem.* **2012**, *403*, 2127–2143.
19. Shao, K.; Ding, W.; Wang, F.; Li, H.; Ma, D.; Wang, H. *PLoS One* **2011**, *6*, e24910.
20. Metzker, M. L. *Nat. Rev. Genet.* **2010**, *11*, 31–46.
21. Castellana, S.; Romani, M.; Valente, E. M.; Mazza, T. *Brief. Bioinform.* **2013**, *14*, 684–695.
22. Adessi, C.; Matton, G.; Ayala, G.; Turcatti, G.; Mermod, J. J.; Mayer, P.; Kawashima, E. *Nucleic Acids Res.* **2000**, *28*, E87.
23. Hoffmann, J.; Trotter, M.; von Stetten, F.; Zengerle, R.; Roth, G. *Lab Chip* **2012**, *12*, 3049–3054.
24. Pareek, C. S.; Smoczynski, R.; Tretyn, A. *J. Appl. Genet.* **2011**, *52*, 413–435.
25. Gupta, P. K. *Trends Biotechnol.* **2008**, *26*, 602–611.
26. Carneiro, M. O.; Russ, C.; Ross, M. G.; Gabriel, S. B.; Nusbaum, C.; DePristo, M. A. *BMC Genomics* **2012**, *13*, 375.
27. Bowers, J.; Mitchell, J.; Beer, E.; Buzby, P. R.; Causey, M.; Efcavitch, J. W.; Jarosz, M.; Krzymanska-Olejnik, E.; Kung, L.; Lipson, D.; Lowman, G. M.; Marappan, S.; McInerney, P.; Platt, A.; Roy, A.; Siddiqi, S. M.; Steinmann, K.; Thompson, J. F. *Nat. Methods* **2009**, *6*, 593–595.
28. Ozsolak, F.; Milos, P. M. *Wiley Interdiscip. Rev. RNA* **2011**, *2*, 565–570.
29. Hyman, E. D. *Anal. Biochem.* **1988**, *174*, 423–436.
30. Ronaghi, M.; Nygren, M.; Lundeberg, J.; Nyren, P. *Anal. Biochem.* **1999**, *267*, 65–71.
31. Merriman, B.; Rothberg, J. M. *Electrophoresis* **2012**, *33*, 3397–3417.
32. Gong, M.; Foo, S. H.; Lin, L.; Liu, E. T.; Gharizadeh, B.; Goel, S. *Biochem. Biophys. Res. Commun.* **2010**, *401*, 117–123.
33. Bentley, D. R.; Balasubramanian, S.; Swerdlow, H. P.; Smith, G. P.; Milton, J.; Brown, C.; Hall, K. P.; Evers, D. J.; Barnes, C. L.; Bignell, H. R.; Boutell, J. M.; Bryant, J.; Carter, R. J.; Keira, R. Cheetham; Cox, A. J.; Ellis, D. J.; Flatbush, M. R.; Gormley, N. A.; Humphray, S. L.; Irving, L.; Karbelashvili, M. S.; Kirk, S. M.; Li, H.; Liu, X.; Maisinger, K. S.; Murray, L. J.; Obradovic, B.; Ost, T.; Parkinson, M. L.; Pratt, M. R.; Rasolonjatovo, I. M.; Reed, M. T.; Rigatti, R.; Rodighiero, C.; Ross, M. T.; Sabot, A.; Sankar, S. V.; Scally, A.; Schroth, G. P.; Smith, M. E.; Smith, V. P.; Spiridou, A.; Torrance, P. E.; Tzonev, S. S.; Vermaas, E. H.; Walter, K.; Wu, X.; Zhang, L.;

Alam, M. D.; Anastasi, C.; Aniebo, I. C.; Bailey, D. M.; Bancarz, I. R.; Banerjee, S.; Barbour, S. G.; Baybayan, P. A.; Benoit, V. A.; Benson, K. F.; Bevis, C.; Black, P. J.; Boodhun, A.; Brennan, J. S.; Bridgham, J. A.; Brown, R. C.; Brown, A. A.; Buermann, D. H.; Bundu, A. A.; Burrows, J. C.; Carter, N. P.; Castillo, N.; Catenazzi, C. E.; Chang, S.; Cooley, N. R.; Crake, N. R.; Dada, O. O.; Diakoumakos, K. D.; Dominguez-Fernandez, B.; Earnshaw, D. J.; Egbujor, U.; Elmore, D. W.; Etchin, S. S.; Ewan, M. R.; Fedurco, M.; Fraser, L. J.; Fuentes Fajardo, K. V.; Scott, W.; Furey, W.; George, D.; Gietzen, K. J.; Goddard, C. P.; Golda, G. S.; Granieri, P. A.; Green, D. E.; Gustafson, D. L.; Hansen, N. F.; Harnish, K.; Haudenschild, C. D.; Heyer, N. I.; Hims, M. M.; Ho, J. T.; Horgan, A. M.; Hoschler, K.; Hurwitz, S.; Ivanov, D. V.; Johnson, M. Q.; James, T.; Huw Jones, T. A.; Kang, G.; Kerelska, T. H.; Kersey, A. D.; Khrebtukova, I.; Kindwall, A. P.; Kingsbury, Z.; Kokko-Gonzales, P. I.; Kumar, A.; Laurent, M. A.; Lawley, C. T.; Lee, S. E.; Lee, X.; Liao, A. K.; Loch, J. A.; Lok, M.; Luo, S.; Mammen, R. M.; Martin, J. W.; McCauley, P. G.; McNitt, P.; Mehta, P.; Moon, K. W.; Mullens, J. W.; Newington, T.; Ning, Z.; Ng, B.; Novo, S. M.; O'Neill, M. J.; Osborne, M.; Osnowski, A.; Ostadan, O.; Paraschos, L. L.; Pickering, L.; Pike, A. C.; Pike, A. C.; Pinkard, D.; Pliskin, D.; Podhasky, J.; Quijano, V. J.; Raczy, C.; Rae, V.; Rawlings, S. R.; Chiva, A.; Rodriguez, A.; Roe, P. M.; Rogers, J.; Rogert Bacigalupo, - M. C.; Romanov, N.; Romieu, A.; Roth, R. K.; Rourke, N. J.; Ruediger, S. T.; Rusman, E.; Sanches-Kuiper, R. M.; Schenker, M. R.; Seoane, J. M.; Shaw, R. J.; Shiver, M. K.; Short, S. W.; Sizto, N. L.; Sluis, J. P.; Smith, M. A.; Ernest Sohna Sohna, J.; Spence, E. J.; Stevens, K.; Sutton, N.; Szajkowski, L.; Tregidgo, C. L.; Turcatti, G.; Vandevondele, S.; Verhovsky, Y.; Virk, S. M.; Wakelin, S.; Walcott, G. C.; Wang, J.; Worsley, G. J.; Yan, J.; Yau, L.; Zuerlein, M.; Rogers, J.; Mullikin, J. C.; Hurles, M. E.; McCooke, N. J.; West, J. S.; Oaks, F. L.; Lundberg, P. L.; Klenerman, D.; Durbin, R.; Smith, A. J. *Nature* **2008**, *456*, 53–59.

34. Hughes, G. M.; Gang, L.; Murphy, W. J.; Higgins, D. G.; Teeling, E. C. *Mol. Ecol. Resour.* **2013**, *13*, 510–521.
35. Kozarewa, I.; Turner, D. J. *Methods Mol. Biol.* **2011**, *733*, 257–266.
36. van den Oever, J. M.; Balkassmi, S.; Verweij, E. J.; van Iterson, I. M.; van Scheltema, P. N.; Oepkes, D.; van Lith, J. M.; Hoffer, M. J.; den Dunnen, J. T.; Bakker, E.; Boon, E. M. *Clin. Chem.* **2012**, *58*, 699–706.
37. Gulig, P. A.; de Crecy-Lagard, V.; Wright, A. C.; Walts, B.; Telonis-Scott, M.; McIntyre, L. *BMC Genomics* **2010**, *11*, 512.
38. Shendure, J.; Porreca, G. J.; Reppas, N. B.; Lin, X.; McCutcheon, J. P.; Rosenbaum, A. M.; Wang, M. D.; Zhang, K.; Mitra, R. D.; Church, G. M. *Science* **2005**, *309*, 1728–1732.
39. Korlach, J.; Bibillo, A.; Wegener, J.; Peluso, P.; Pham, T. T.; Park, I.; Clark, S.; Otto, G.; Turner, S. W. *Nucleosides Nucleotides Nucleic Acids* **2008**, *27*, 1072–1083.
40. Lundquist, P. M.; Zhong, C. F.; Zhao, P.; Tomaney, A. B.; Peluso, P. S.; Dixon, J.; Bettman, B.; Lacroix, Y.; Kwo, D. P.; McCullough, E.; Maxham, M.; Hester, K.; McNitt, P.; Grey, D. M.; Henriquez, C.; Foquet, M.; Turner, S. W.; Zaccarin, D. *Opt. Lett.* **2008**, *33*, 1026–1028.
41. Yeh, H. C.; Chao, S. Y.; Ho, Y. P.; Wang, T. H. *Curr. Pharm. Biotechnol.* **2005**, *6*, 453–461.
42. Maitra, R. D.; Kim, J.; Dunbar, W. B. *Electrophoresis* **2012**, *33*, 3418–3428.
43. Ku, C. S.; Roukos, D. H. *Expert Rev. Med. Devices* **2013**, *10*, 1–6.
44. Schadt, E. E.; Turner, S.; Kasarskis, A. *Hum. Mol. Genet.* **2010**, *19*, R227–R240.
45. Luan, B.; Peng, H.; Polonsky, S.; Rossnagel, S.; Stolovitzky, G.; Martyna, G. *Phys. Rev. Lett.* **2010**, *104*, 238103.
46. Bateman, A.; Quackenbush, J. *Bioinformatics* **2009**, *25*, 429.

47. Tintle, N.; Aschard, H.; Hu, I.; Nock, N.; Wang, H.; Pugh, E. *Genet. Epidemiol.* **2011**, *35*(Suppl. 1), S56–S60.
48. Kumar, N.; Mukhopadhyay, A. K.; Patra, R.; De, R.; Baddam, R.; Shaik, S.; Alam, J.; Tiruvayipati, S.; Ahmed, N. *J. Bacteriol.* **2012**, *194*, 5963–5964.
49. Iorizzo, M.; Senalik, D.; Szklarczyk, M.; Grzebelus, D.; Spooner, D.; Simon, P. *BMC Plant Biol.* **2012**, *12*, 61. http://dx.doi.org/10.1186/1471-2229-12-61.
50. Narzisi, G.; Mishra, B. *PLoS One* **2011**, *6*, e19175.
51. Drogemoller, B. I.; Wright, G. E.; Niehaus, D. J.; Emsley, R. A.; Warnich, L. *Pharmacogenomics* **2011**, *12*, 1717–1728.
52. Huang, X.; Feng, Q.; Qian, Q.; Zhao, Q.; Wang, L.; Wang, A.; Guan, J.; Fan, D.; Weng, Q.; Huang, T.; Dong, G.; Sang, T.; Han, B. *Genome Res.* **2009**, *19*, 1068–1076.
53. Li, R.; Li, Y.; Fang, X.; Yang, H.; Wang, J.; Kristiansen, K.; Wang, J. *Genome Res.* **2009**, *19*, 1124–1132.
54. Johansson, H.; Isaksson, M.; Sorqvist, E. F.; Roos, F.; Stenberg, J.; Sjoblom, T.; Botling, J.; Micke, P.; Edlund, K.; Fredriksson, S.; Kultima, H. G.; Ericsson, O.; Nilsson, M. *Nucleic Acids Res.* **2011**, *39*, e8.
55. Keller, A.; Harz, C.; Matzas, M.; Meder, B.; Katus, H. A.; Ludwig, N.; Fischer, U.; Meese, E. *PLoS One* **2011**, *6*, e18158.
56. Harismendy, O.; Ng, P. C.; Strausberg, R. L.; Wang, X.; Stockwell, T. B.; Beeson, K. Y.; Schork, N. J.; Murray, S. S.; Topol, E. J.; Levy, S.; Frazer, K. A. *Genome Biol.* **2009**, *10*, R32–R.
57. Yilmaz, S.; Singh, A. K. *Curr. Opin. Biotechnol.* **2012**, *23*, 437–443.
58. Werner, T. *Brief. Bioinform.* **2010**, *11*, 499–511.
59. Forrest, A. R.; Carninci, P. *RNA Biol.* **2009**, *6*, 107–112.
60. Takechi, M.; Takeuchi, M.; Ota, K. G.; Nishimura, O.; Mochii, M.; Itomi, K.; Adachi, N.; Takahashi, M.; Fujimoto, S.; Tarui, H.; Okabe, M.; Aizawa, S.; Kuratani, S. *J. Exp. Zool. B Mol. Dev. Evol.* **2011**, *316*, 526–546.
61. Adomas, A. B.; Lopez-Giraldez, F.; Clark, T. A.; Wang, Z.; Townsend, J. P. *BMC Genomics* **2010**, *11*, 477. http://dx.doi.org/10.1186/1471-2164-11-477.
62. Cloonan, N.; Waddell, N.; Grimmond, S. M. *IDrugs* **2010**, *13*, 778–781.
63. Galasso, M.; Sana, M. E.; Volinia, S. *Genome Med.* **2010**, *2*, 12.
64. Fittipaldi, R.; Caretti, G. *Comp. Funct. Genomics* **2012**, *979168*, 2012.
65. Salbaum, J. M.; Kappen, C. *Birth Defects Res. A Clin. Mol. Teratol.* **2011**, *91*, 770–780.
66. Neff, T.; Armstrong, S. A. *Leukemia* **2009**, *23*, 1243–1251.
67. Liang, S.; Lu, Y.; Jelinek, J.; Estecio, M.; Li, H.; Issa, J. P. *Conf. Proc. IEEE Eng. Med. Biol. Soc.* **2009**, *2009*, 6730.
68. Simon, S. A.; Meyers, B. C. *Curr. Opin. Plant Biol.* **2011**, *14*, 148–155.
69. Fouse, S. D.; Nagarajan, R. O.; Costello, J. F. *Epigenomics* **2010**, *2*, 105–117.
70. Nagarajan, R. P.; Fouse, S. D.; Bell, R. J.; Costello, J. F. *Adv. Exp. Med. Biol.* **2013**, *754*, 313–338.
71. Emes, R. D.; Farrell, W. E. *J. Mol. Endocrinol.* **2012**, *49*, R19–R27.
72. Kelly, T. K.; Liu, Y.; Lay, F. D.; Liang, G.; Berman, B. P.; Jones, P. A. *Genome Res.* **2012**, *22*, 2497–2506.
73. Costa, M. C.; Weese, J. S. *Anim. Health Res. Rev.* **2012**, *13*, 121–128.
74. Fukatsu, T. *Mol. Ecol.* **2012**, *21*, 5908–5910.
75. Fettweis, J. M.; Serrano, M. G.; Girerd, P. H.; Jefferson, K. K.; Buck, G. A. *Chem. Biodivers.* **2012**, *9*, 965–976.
76. Pflughoeft, K. J.; Versalovic, J. *Annu. Rev. Pathol.* **2012**, *7*, 99–122.

77. Barzon, L.; Lavezzo, E.; Costanzi, G.; Franchin, E.; Toppo, S.; Palu, G. *J. Clin. Virol.* **2013**, *10*.
78. Li, R.; Fan, W.; Tian, G.; Zhu, H.; He, L.; Cai, J.; Huang, Q.; Cai, Q.; Li, Q.; Bai, Y.; Zhang, Z.; Zhang, Y.; Wang, W.; Li, J.; Wei, F.; Li, H.; Jian, M.; Li, J.; Zhang, Z.; Nielsen, R.; Li, D.; Gu, W.; Yang, Z.; Xuan, Z.; Ryder, O. A.; Leung, F. C.; Zhou, Y.; Cao, J.; Sun, X.; Fu, Y.; Fang, X.; Guo, X.; Wang, B.; Hou, R.; Shen, F.; Mu, B.; Ni, P.; Lin, R.; Qian, W.; Wang, G.; Yu, C.; Nie, W.; Wang, J.; Wu, Z.; Liang, H.; Min, J.; Wu, Q.; Cheng, S.; Ruan, J.; Wang, M.; Shi, Z.; Wen, M.; Liu, B.; Ren, X.; Zheng, H.; Dong, D.; Cook, K.; Shan, G.; Zhang, H.; Kosiol, C.; Xie, X.; Lu, Z.; Zheng, H.; Li, Y.; Steiner, C. C.; Lam, T. T.; Lin, S.; Zhang, Q.; Li, G.; Tian, J.; Gong, T.; Liu, H.; Zhang, D.; Fang, L.; Ye, C.; Zhang, J.; Hu, W.; Xu, A.; Ren, Y.; Zhang, G.; Bruford, M.; Li, Q.; Ma, L.; Guo, Y.; An, N.; Hu, Y.; Zheng, Y.; Shi, Y.; Li, Z.; Liu, Q.; Chen, Y.; Zhao, J.; Qu, N.; Zhao, S.; Tian, F.; Wang, X.; Wang, H.; Xu, L.; Liu, X.; Vinar, T.; Wang, Y.; Lam, T. W.; Yiu, S. M.; Liu, S.; Zhang, H.; Li, D.; Huang, Y.; Wang, X.; Yang, G.; Jiang, Z.; Wang, J.; Qin, N.; Li, L.; Li, J.; Bolund, L.; Kristiansen, K.; Wong, G. K.; Olson, M.; Zhang, X.; Li, S.; Yang, H.; Wang, J.; Wang, J. *Nature* **2010**, *463*, 311–317.
79. Asan, C. Geng; Chen, Y.; Wu, K.; Cai, Q.; Wang, Y.; Lang, Y.; Cao, H.; Yang, H.; Wang, J.; Zhang, X. *PLoS One* **2012**, *7*, e46211.
80. Desai, A.; Marwah, V. S.; Yadav, A.; Jha, V.; Dhaygude, K.; Bangar, U.; Kulkarni, V.; Jere, A. *PLoS One* **2013**, *8*, e60204.
81. Zhang, W.; Chen, J.; Yang, Y.; Tang, Y.; Shang, J.; Shen, B. *PLoS One* **2011**, *6*, e17915.
82. Underwood, A. P.; Dallman, T.; Thomson, N. R.; Williams, M.; Harker, K.; Perry, N.; Adak, B.; Willshaw, G.; Cheasty, T.; Green, J.; Dougan, G.; Parkhill, J.; Wain, J. *J. Clin. Microbiol.* **2013**, *51*, 232–237.
83. Desai, A. N.; Jere, A. *Clin. Genet.* **2012**, *81*, 503–510.
84. Jimenez-Escrig, A.; Gobernado, I.; Sanchez-Herranz, A. *Rev. Neurol.* **2012**, *54*, 692–698.
85. Kuczynski, J.; Stombaugh, J.; Walters, W. A.; Gonzalez, A.; Caporaso, J. G.; Knight, R. *Curr. Protoc. Microbiol.* **2012**, *27*, 1E.5.1–1E.5.20.
86. Mamanova, L.; Coffey, A. J.; Scott, C. E.; Kozarewa, I.; Turner, E. H.; Kumar, A.; Howard, E.; Shendure, J.; Turner, D. J. *Nat. Methods* **2010**, *7*, 111–118.
87. Xu, F.; Wang, Q.; Zhang, F.; Zhu, Y.; Gu, Q.; Wu, L.; Yang, I.; Yang, X. *Cardiovasc. Diagn. Ther.* **2012**, *2*, 138–146.
88. Bick, D.; Dimmock, D. *Curr. Opin. Pediatr.* **2011**, *23*, 594–600.
89. Sikkema-Raddatz, B.; Johansson, L. F.; de Boer, E. N.; Almomani, R.; Boven, L. G.; van den Berg, M. P.; van Spaendonck-Zwarts, K. Y.; van Tintelen, J. P.; Sijmons, R. H.; Jongbloed, J. D.; Sinke, R. J. *Hum. Mutat.* **2013**, *10*.
90. Klee, W. W.; Hoppman-Chaney, N. L.; Ferber, M. J. *Expert Rev. Mol. Diagn.* **2011**, *11*, 703–709.
91. Inaki, K.; Liu, E. T. *Trends Genet.* **2012**, *28*, 550–559.
92. Welch, J. S.; Westervelt, P.; Ding, L.; Larson, D. E.; Klco, J. M.; Kulkarni, S.; Wallis, J.; Chen, K.; Payton, J. E.; Fulton, R. S.; Veizer, J.; Schmidt, H.; Vickery, T. L.; Heath, S.; Watson, M. A.; Tomasson, M. H.; Link, D. C.; Graubert, T. A.; Dipersio, J. F.; Mardis, E. R.; Ley, T. J.; Wilson, R. K. *JAMA* **2011**, *305*, 1577–1584.
93. Thomas, D. M. *J. Pathol.* **2011**, *224*, 306–308.
94. Wain, J.; Keddy, K. H.; Hendriksen, R. S.; Rubino, S. *J. Infect. Dev. Ctries.* **2013**, *7*, 1–5.
95. Dutton, R. J.; Turnbaugh, P. J. *Curr. Opin. Clin. Nutr. Metab. Care* **2012**, *15*, 448–454.
96. Gosalbes, M. J.; Abellan, J. J.; Durban, A.; Perez-Cobas, A. E.; Latorre, A.; Moya, A. *Clin. Microbiol. Infect.* **2012**, *18*(Suppl. 4), 47–49.

97. Wain, J.; Mavrogiorgou, E. *Expert Rev. Mol. Diagn.* **2013**, *13*, 225–227.
98. Dave, M.; Higgins, P. D.; Middha, S.; Rioux, K. P. *Transl. Res.* **2012**, *160*, 246–257.
99. Archer, J.; Weber, J.; Henry, K.; Winner, D.; Gibson, R.; Lee, L.; Paxinos, E.; Arts, E. J.; Robertson, D. L.; Mimms, L.; Quinones-Mateu, M. E. *PLoS One* **2012**, *7*, e49602.
100. Mellmann, A.; Harmsen, D.; Cummings, C. A.; Zentz, E. B.; Leopold, S. R.; Rico, A.; Prior, K.; Szczepanowski, R.; Ji, Y.; Zhang, W.; McLaughlin, S. F.; Henkhaus, J. K.; Leopold, B.; Bielaszewska, M.; Prager, R.; Brzoska, P. M.; Moore, R. L.; Guenther, S.; Rothberg, J. M.; Karch, H. *PLoS One* **2011**, *6*, e22751.
101. Daly, G. M.; Bexfield, N.; Heaney, J.; Stubbs, S.; Mayer, A. P.; Palser, A.; Kellam, P.; Drou, N.; Caccamo, M.; Tiley, L.; Alexander, G. J.; Bernal, W.; Heeney, J. L. *PLoS One* **2011**, *6*, e28879.
102. Conway, C.; Chalkley, R.; High, A.; Maclennan, K.; Berri, S.; Chengot, P.; Alsop, M.; Egan, P.; Morgan, J.; Taylor, G. R.; Chester, J.; Sen, M.; Rabbitts, P.; Wood, H. M. *J. Mol. Diagn.* **2012**, *14*, 104–111.
103. Liu, P.; Fang, X.; Feng, Z.; Guo, Y. M.; Peng, R. J.; Liu, T.; Huang, Z.; Feng, Y.; Sun, X.; Xiong, Z.; Guo, X.; Pang, S. S.; Wang, B.; Lv, X.; Feng, F. T.; Li, D. J.; Chen, L. Z.; Feng, Q. S.; Huang, W. L.; Zeng, M. S.; Bei, J. X.; Zhang, Y.; Zeng, Y. X. *J. Virol.* **2011**, *85*, 11291–11299.
104. Van Verk, M. C.; Hickman, R.; Pieterse, C. M.; Van Wees, S. C. *Trends Plant Sci.* **2013**, *18*, 175–179.
105. Wang, Z.; Gerstein, M.; Snyder, M. *Nat. Rev. Genet.* **2009**, *10*, 57–63.
106. Yoshida, K.; Sanada, M.; Ogawa, S. *Jpn. J. Clin. Oncol.* **2013**, *43*, 110–115.
107. Hitzemann, R.; Bottomly, D.; Darakjian, P.; Walter, N.; Iancu, O.; Searles, R.; Wilmot, B.; McWeeney, S. *Genes Brain Behav.* **2013**, *12*, 1–12.
108. Mercer, T. R.; Gerhardt, D. J.; Dinger, M. E.; Crawford, J.; Trapnell, C.; Jeddeloh, J. A.; Mattick, J. S.; Rinn, J. L. *Nat. Biotechnol.* **2011**, *30*, 99–104.
109. Lawless, N.; Foroushani, A. B.; McCabe, M. S.; O'Farrelly, C.; Lynn, D. J. *PLoS One* **2013**, *8*, e57543.
110. Chaudhry, M. A.; Omaruddin, R. A.; Brumbaugh, C. D.; Tariq, M. A.; Pourmand, N. *J. Radiat. Res.* **2013**, *55*, 822–880.
111. Gutschner, T.; Diederichs, S. *RNA Biol.* **2012**, *9*, 703–719.
112. Kandpal, R. P.; Rajasimha, H. K.; Brooks, M. J.; Nellissery, J.; Wan, J.; Qian, J.; Kern, T.; Swaroop, A. *Mol. Vis.* **2012**, *18*, 1123–1146.
113. Lista, S.; Garaci, F. G.; Toschi, N.; Hampel, H. *Curr. Pharm. Des.* **2013**, *19*, 6393–6415.
114. Rodriguez, B. A.; Frankhouser, D.; Murphy, M.; Trimarchi, M.; Tam, H. H.; Curfman, J.; Huang, R.; Chan, M.; Lai, H. C.; Parikh, D.; Ball, B.; Schwind, S.; Blum, W.; Marcucci, G.; Yan, P.; Bundschuh, R. *BMC Genomics* **2012**, *13*(Suppl. 6), S14.
115. Baranzini, S. E.; Nickles, D. *Curr. Opin. Neurol.* **2012**, *25*, 239–245.
116. Dudley, K. J.; Revill, K.; Clayton, R. N.; Farrell, W. E. *J. Mol. Endocrinol.* **2009**, *42*, 461–468.
117. Jin, B.; Robertson, K. D. *Adv. Exp. Med. Biol.* **2013**, *754*, 3–29.
118. Uzoma, I.; Zhu, H. *Genomics Proteomics Bioinformatics* **2013**, *11*, 18–28.
119. Hwang, Y. C.; Zheng, Q.; Gregory, B. D.; Wang, L. S. *Nucleic Acids Res.* **2013**, *41*, 4835–4846.
120. Hao, H. *Adv. Exp. Med. Biol.* **2012**, *723*, 753–759.
121. Kidder, B. L.; Hu, G.; Zhao, K. *Nat. Immunol.* **2011**, *12*, 918–922.
122. Furey, T. S. *Nat. Rev. Genet.* **2012**, *13*, 840–852.
123. Chen, Z.; Wang, L.; Wang, Q.; Li, W. *Epigenomics* **2010**, *2*, 551–560.

124. Pei, Y.; Zhang, T.; Renault, V.; Zhang, X. *Acta Biochim. Biophys. Sin.* **2009**, *41*, 1–15.
125. Weller, J. I.; Ron, M. J. *Dairy Sci.* **2011**, *94*, 1082–1090.
126. Qi, Q.; Chu, A. Y.; Kang, J. H.; Jensen, M. K.; Curhan, G. C.; Pasquale, L. R.; Ridker, P. M.; Hunter, D. J.; Willett, W. C.; Rimm, E. B.; Chasman, D. I.; Hu, F. B.; Qi, L. *Engl. J. Med.* **2012**, *367*, 1387–1396.
127. Curren, L.; Boddington, P.; Gowans, H.; Hawkins, N.; Kanellopoulou, N.; Kaye, J.; Melham, K. *Eur. J. Health Law* **2010**, *17*, 329–344.
128. Kaye, J.; Boddington, P.; Vries, J.; Hawkins, N.; Melham, K. *Eur. J. Hum. Genet.* **2010**, *18*, 398–403.

Chapter 4

Omics Tools for the Genome-Wide Analysis of Methylation and Histone Modifications

Josep C. Jiménez-Chillarón, Rubén Díaz and Marta Ramón-Krauel
Hospital Sant Joan de Deu, Endocrinology, Fundacio per la Recerca Sant Joan de Deu, Barcelona, Spain

Chapter Outline

1 OMICS MEETS EPIGENETICS

1.1 Omics

The term "ome" derives from the Sanskrit word OM, which means "completeness, entirety, or totality" *(1)*. Nowadays, *-omics* is an English suffix that, incorporated to the end of already existing areas in life sciences and technology, denotes studies undertaken on a large scale. Historically, the first areas to become part of the new Omics family were genomics, transcriptomics, and proteomics *(2,3)*. They define large-scale gene expression profiles at the level of RNA and protein, respectively. The advent of such new areas came with the appearance of (1) cutting-edge technologies that allowed the acquisition of high-throughput data and (2) development of the bioinformatics tools necessary to handle and analyze the enormous amount of information generated through these approaches *(4,5)*.

The number of Omics has been growing ever since and, as an analogy from the first genomic studies, we can now find such terms as metabolomics,

Comprehensive Analytical Chemistry, Vol. 63. http://dx.doi.org/10.1016/B978-0-444-62651-6.00004-0

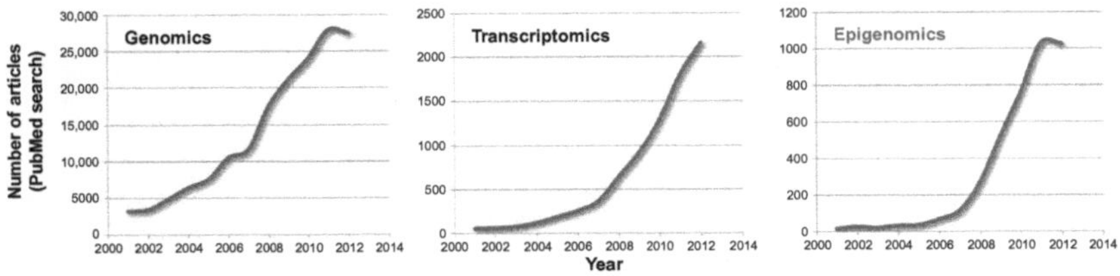

FIGURE 1 PubMed search examples for the keywords genomics, transcriptomics, and epigenomics. Graphs detail the number of publications per year during the period 2001–2012. Search was performed in May 2013. (For color version of this figure, the reader is referred to the online version of this chapter.)

lipomics and lipidomics, nutrigenomics, fluxomics, ORFeomics, microbiomics, or epigenomics, among many others. Several websites maintain an updated list of Omics (www.genomicglossaries.com/content/omes.asp). The appearance and inclusion of new Omics largely depends on the development of novel platforms and technologies that allow the collection of large datasets for a specific biological measurement.

Although not everybody agrees with the general utilization of the term Omics, it has become widely used by the scientific community *(6)*. Accordingly, there is now a growing body of literature referring to these different Omics (Figure 1). This explosion of literature culminated with the launch of a journal devoted to the combination of Omics and systems biology (*Omics Journal*; www.omics.org) that is still running healthy in strong competition with the growing number of scientific journals *(1)*.

1.2 Epigenetics

Epigenetics refers to the inheritance of information from a single cell to another, independent of the DNA sequence. Epigenetic mechanisms include DNA methylation, histone modifications, and, more recently, a variety of noncoding RNAs (ncRNAs). These mechanisms have been extensively reviewed elsewhere *(7)*. Here, we briefly summarize them and emphasize those features relevant in the field of epigenomics.

1.2.1 DNA Methylation

DNA methylation is a covalent modification that consists of the addition of a methyl group at carbon 5 in cytosine residues (5mC) of the DNA template (Figure 2A). The incorporation of methyl groups occurs at dinucleotides CG (CpG). Interestingly, there are many regions in the genome that accumulate a disproportional abundance of CpGs. These areas are known as CpG islands and may play a regulatory role depending on the number and position of methylated–demethylated cytosines within a given island (Figure 2B). Typically, in mammals, CpG islands exist around promoter regions of the genes

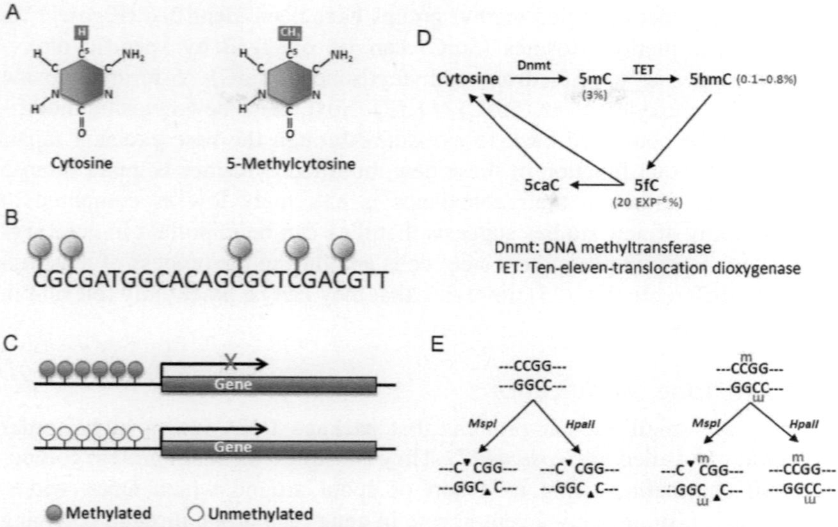

FIGURE 2 DNA methylation. (A) Molecular structure of unmethylated and methylated cytosine at carbon 5. (B) Example of methylation at dinucleotides CpG. (C) CpG islands in promoter regions with various degrees of methylation. Hypermethylated CpG island is associated with repression of gene expression, whereas the hypomethylated CpG island is associated with gene expression. (D) Diagram describing the conversion of methylcytosine to other species, including 5-hydoxymethylcytosine (5hmC), 5-formylcytosine (5fC), and 5-carboxylcytosine (5caC). (E) Isoschizomers MspI/HpaII. They recognize and cut the same target if it is unmethylated. If the cytosine is methylated, only MspI can cut the DNA. (For color version of this figure, the reader is referred to the online version of this chapter.)

and remain mostly hypomethylated. It has been shown that changes in methylation of these areas influence gene expression *(8,9)* (Figure 2C).

There are now many examples showing a causal relationship between DNA methylation and changes in gene expression. In general, the hypermethylation of DNA results in reduced binding of transcription factors and a concomitant reduction in gene expression (Figure 2C). Conversely, hypomethylation leads to readily accessible DNA to transcription factors that can stimulate gene expression *(9)*. The best-characterized examples of this type of regulation come from the field of cancer: It is well established now that in many cancers there exist hypomethylation of oncogenes and hypermethylation of tumor suppressor genes *(10)*. Nevertheless, new data show many examples with a positive correlation between DNA methylation and gene expression: more methylation is associated with higher expression and less methylation with reduced gene expression. Thus, these data suggest that a more complex regulatory mechanism exists in the way DNA methylation influences gene expression. In this regard, epigenomic approaches are definitively helping to sort out such biological complexity (see later).

Recently, other modified methyl groups have been identified (Figure 2D). For example, methylcytosines (5mC) can be oxidized by specific dioxygenases stepwise to 5-hydroxymethylcytosine (5hmC), 5-formylcytosine (5fC), and carboxylcytosine (5caC) *(11,12)*. Next, these new cytosine modifications can be converted back to cytosines through the base excision repair process. The exact function of these new modified cytosines is under intense investigation. Although their abundance is extremely low as compared to 5mC, an array of new studies suggests that they can be important in gene regulation, primarily in pluripotent stem cells and during the process of development and differentiation *(13)*, implying that they can be potentially relevant in human diseases.

1.2.2 Histone Modifications

Histones are small nuclear proteins that package DNA into a higher-order structural unit called *nucleosome (7)*. They constitute the main protein components of chromatin, acting as a sort of spool around which DNA winds. Importantly, histones play a central role in gene regulation, through changing the degree of packaging or condensation of DNA. Histones may change chromatin states through a series of post-translational modifications, including methylation (mono-, di-, and tri-methylation), acetylation, SUMOylation, biotinylation, phosphorylation, ubiquitination, and ADP-rybosilation *(14)*. The question is why there are so many modifications. Two hypotheses have been developed in order to explain the function of so many post-translational histone modifications in gene regulation: the histone code hypothesis and the signaling pathway model.

1.2.2.1 Histone Code Hypothesis

The histone code hypothesis proposes that specific histone modifications (or combination of modifications) may confer unique biological functions to regions of the genome where they associate *(15,16)*. Thus, in the simplest model, a single histone mark is associated with either expression or repression (Table 1). But in reality, histones may integrate several modifications in one single protein. Thus, the signal will be repressive or activate depending on its combinatorial effect. Importantly, the combination of modifications results in a complex, *locus*-specific regulation of gene transcription. As a final turn of the screw, an additional layer of complexity must exist since some examples now show that one single mark may be associated with both activation and repression depending on the genomic context *(17)*.

1.2.2.2 The Signaling Pathway Model

The previous uncertainty gave rise to the signaling pathway model. It acts as a more general model to explain the role of so many histone modifications on gene expression *(18)*. It states that histone modifications can serve as

TABLE 1 Representative Histone Code

Histone	Type of Modification	Site	Function
H1	Methylation	K26me	Binding of HP1
	Phosphorylation	T10ph	Activation expression H1B
		S17ph	Activation expression H1B
		T137ph	Activation expression H1B
		S172ph	Activation expression H1B
	Ribosylation	E2ar1	Neurotrophic activity
		K213ar1	Neurotrophic activity
H2A	Acetylation	K5ac	Transcriptional activation
		K7ac	Transcriptional activation
	Biotinilation	K9bio	Cell proliferation and gene silencing
		K126bio	Cell proliferation and gene silencing
	Phosphorylation	S121ph	Telomere silencing
		T125ph	Telomere silencing
	Ubiquitylation	K119ub1	Polycomb silencing
	Sumoylation	K126su	Transcriptional repression
H2B	Acetylation	K5ac	Transcriptional activation
	Phosphorylation	S33ph	Transcriptional activation
	Sumoylation	K16su	Transcriptional repression
		K17su	Transcriptional repression
	Ubiquitylation	K120ub	Transcriptional activation
		K123ub	Telomeric silencing
H3	Acetylation	K9ac	Transcriptional activation
		K14ac	Transcriptional activation
		K23ac	Transcriptional activation
	Methylation	K4me	Transcriptional activation
		K9me	Transcriptional repression
		K27me	Polycomb repression
		K79me	Telomeric silencing
	Phosphorylation	S10ph	Transcriptional activation of early genes
		S28ph	Mitotic chromosome condensation

Continued

TABLE 1 Representative Histone Code—Cont'd

Histone	Type of Modification	Site	Function
H4	Acetylation	K8ac	Transcriptional activation
	Methylation	R3me	Transcriptional activation
		K20me	Transcriptional silencing
		K59me	Silent chromatin formation

Summary of effect of different histone modifications on gene expression.

platforms for other proteins that can then function on chromatin *(19)*. Thus, in this model there is no strict histone code for a specific process. Instead, one set of histone modifications can give rise to expression or repression depending on additional effectors bound to the chromatin complex.

1.2.3 Noncoding RNAs

ncRNAs are RNA molecules that are not translated into proteins *(20)*. Although a few species of ncRNAs are long known from before (ribosomal RNAs, rRNA; transfer RNA, tRNA; small nucleolar RNA, snoRNA), a new plethora of ncRNAs are becoming increasingly relevant due to their potential role in regulating gene expression. They were at the frontline of life sciences when it was clear that mRNAs account for only 1–1.5% of the genome *(21)*. Noteworthy, the ENCODE project uncovered that up to 60% of the genome is actively transcribed into different types of ncRNAs *(22)*. Importantly, they are transcribed in a developmental and tissue-specific manner, thus suggesting that they truly can play a biological function *(23,24)*. Thus, there is now a general interest in understanding the role of these RNAs in normal biology and, therefore their implications on human health and disease *(25)*.

Although ncRNAS are not strictly part of the epigenetic machinery, there is evidence that some of them may actually work as epigenetic regulators since they can interfere with the epigenetic machinery (DNA methylation and histone modifications), thus influencing gene expression (see Section 2.2) *(26,27)*.

1.3 Epigenomics: When *-Omics* Met *Epigenetics*

From the previous sections it is now possible to define epigenomics. It refers to the study of the complete set of epigenetic modifications (DNA methylation, histone modifications, and/or specific ncRNAs) that occur across the entire genome of a cell/tissue at a specific time *(28)*.

As we will further develop in the next Section 2, the advent of epigenomics is determined by the introduction of technologies such as *microarrays* or *next-generation sequencing* tools. They have allowed the identification of epigenetic marks genome-wide in one single sample in a cost-effective manner. Among epigenetic marks, DNA methylation has been studied more extensively because it is easier to work with than proteins or RNA. In addition, there is only one modification (cytosine methylation) versus a myriad of protein modifications or RNA species. Yet, with the introduction of microarrays and next-generation technologies, the number of epigenomic studies involving histone-maps and the transcriptome of ncRNAs has grown dramatically (see later).

2 METHODS IN EPIGENOMICS

As stated in the previous Section 1, the number of methods to analyze the epigenome has grown dramatically over the past few years. The expectations are that the field will continue in doing so. Indeed, several new approaches are under development and will possibly reach the market in the near future. These new approaches share something in common: They aim to be faster, highly reproducible, and cheaper than currently available technologies. Here, for practical purposes, we illustrate the most widely used tools and summarize their pros and cons.

2.1 DNA Methylation

The first genome-wide approach to study epigenetic modifications was DNA methylation profiling *(29)*. There are several historical reasons for this. First, DNA (and DNA methyl marks) is very stable. Therefore, it is relatively easy to isolate genomic DNA and store it until the assay is performed without losing quality of the material. This is not obviously the case with RNA or proteins, which are much more labile and susceptible to degradation (RNA and protein) and may lose appropriate conformation or post-translational modifications (proteins). Second, it is known from before that DNA methylation may influence gene expression in normal and pathological conditions *(25)*. In eukaryotes, the way methyl marks are distributed along the genome confers whether a region is transcriptionally active or repressed. Thus, identification of DNA methylation profiles has important implications in understanding regulation of gene expression during specific developmental stages and disease states, such as cancer, diabetes, asthma, or mental disorders *(25,30)*.

The detection of DNA methylation genome-wide requires methylation-dependent pretreatment *(31,32)*. This is because the methylation occurs at the major groove of DNA rather than at the hydrogen carbons. Therefore, hybridization techniques are unable to distinguish between 5mC and unmethylated cytosines. In addition, amplification methods (PCR) do not

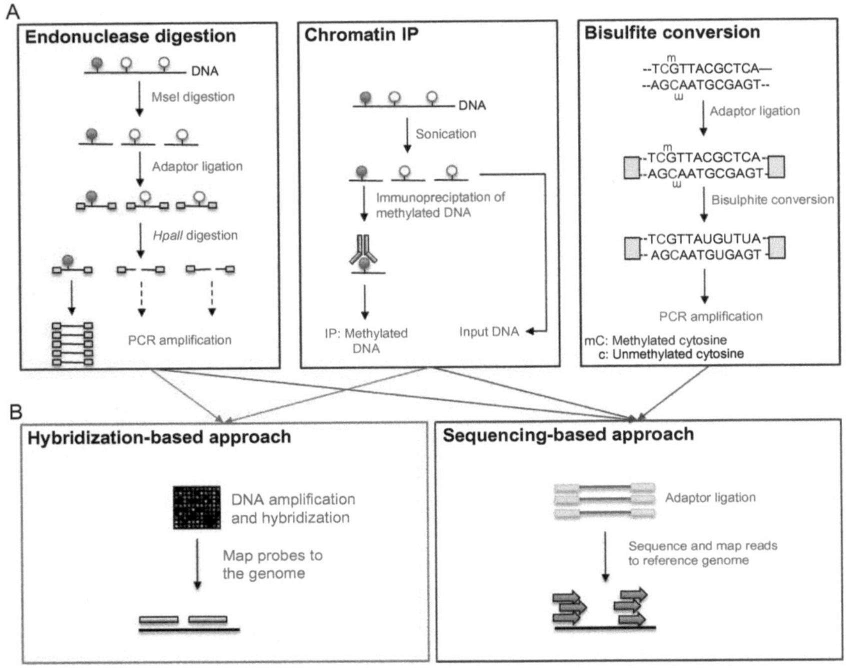

FIGURE 3 Scheme detailing the combination of three different DNA pre-treatments (A) and subsequent high-throughput DNA determination (B). (For color version of this figure, the reader is referred to the online version of this chapter.)

include DNA methyl transferases (Dnmt) and consequently, the methyl groups are erased and diluted during the amplification process. Thus, all existing methods aimed to profile the methylome require the previous treatment of DNA to unmask the existing 5mC. The main approaches are (1) digestion with methylation-sensitive endonucleases, (2) affinity enrichment, and (3) bisulfite conversion (Figure 3A). Once the DNA is treated, whole-genome methylome is obtained through two major approaches: (a) hybridization onto specific microarrays and (b) direct sequencing (DNA-Seq). Obviously, we can get different combinations based on the way the DNA is treated and subsequently analyzed *(29)* (Figure 3B).

2.1.1 Pretreatment of DNA

2.1.1.1 Digestion with Methylation-Sensitive Endonucleases

Restriction enzymes are able to digest DNA in a sequence-specific manner *(33)*. Endonucleases were originally discovered in unicellular organisms that use them as a defense mechanism against exogenous DNA. The endogenous

DNA remains protected against nucleases by methylating the DNA. Endonucleases can be useful for identifying patterns of DNA methylation thanks to the existence of isoschizomers, which are not methylation-sensitive. *Isoschizomers* are pairs of different endonucleases that recognize the same sequences *(33)*. One example of isoschizomers is the tandem formed by *Hpa*II/*Msp*I. Both enzymes recognize the same sequence 5′-CCGG-3′ and cleavage DNA the same way when it is unmethylated (Figure 2E). However, when the second C of the sequence is methylated, *Msp*I can recognize it whereas *Hpa*II cannot.

Now, it is feasible to dissect whether a particular *locus* in the genome containing a target for these endonucleases is differentially methylated by comparing the restriction maps from DNA treated with the restriction-sensitive enzyme with the restriction-insensitive enzyme and mock-treated DNA (Figure 3A). The next step is to couple this approach to an array-based (hybridization) or a sequencing-based analysis to perform a genome-wide study (Figure 3B). We review these options in the next Section 2.1.2. Finally, the endonuclease treatments can be more complex than just the simple utilization of a single enzyme. In fact, combinations of several enzymes have been successfully used for broad DNA methylation profiling *(34)*.

2.1.1.2 Affinity-Enrichment-Based Techniques

Another way to enrich the sample in methylated DNA is by chromatin immunoprecipitation (ChIP) (Figure 3A) *(35)*. ChIP is a technique developed to investigate the interaction between proteins and DNA. It was originally used to study the interaction between transcription factors and histones on specific promoter regions *(36)*. In brief, the method works as follows. First, the DNA is incubated with an antibody that specifically recognizes 5mC regions. The complex antibody-methyl DNA is stabilized with paraformaldehyde. Next, the DNA–protein complex is immunoprecipitated by using the appropriate secondary antibody. Finally, the associated DNA fragments are purified and their sequence determined. Thus, in this context, ChIP assays result in an enrichment of sequences that are methylated in a given sample. Alternatively, to antibodies against 5mC, utilization of methyl-binding proteins (such as methyl-binding protein 2, MECP2 or methyl-CpG-binding-domain protein 2, MBD2) has also been a successful way to pull down methylated DNA *(37–40)*.

Finally, as it happened with the endonuclease-based assays, we can subsequently analyze the enriched DNA through an array-based or a sequencing-based technique to profile the methylome (see the following section) (Figure 3B).

2.1.1.3 Bisulfite Conversion

Bisulfite conversion was discovered in the 1990s and since then it has become the gold standard in the determination of DNA methylation at specific *loci (41)*.

The treatment consists of a chemical reaction where sodium bisulfite involves the deamination of unmethylated cytosines to uracil (Figure 3A). In contrast, the modified cytosines (5mC, 5hmC) remain unmodified. After bisulfite conversion, the uracils are amplified as thymines in a PCR, whereas 5mC or 5hmC are amplified as cytosines. Next, the comparison of sequence information between the reference genome and bisulfite-treated DNA can provide information about patterns of methylation. The major drawback to bisulfite conversion is that it does not distinguish between 5mC and 5hmC DNA (Figure 2D), and some biases may occur in the efficiency of bisulfite conversion.

Bisulfite conversion followed by PCR amplification and sequencing has become the gold standard in determining patterns of DNA methylation. This method excels over all the others because it allows information about methylation at the base-pair resolution. Nevertheless, until recently, its utilization genome-wide has been slow because bisulfite-based methods are not easily adapted to the array-based hybridization approaches. Yet, bisulfite-converted DNA has been rapidly adapted to cheaper everyday deep-sequencing techniques (Figure 3B).

2.1.2 *Genome-Wide Analytical Step*

Once genomic DNA is treated with any of the previous methylation-dependent steps, we can readily apply it to any of the available techniques that allow the scaling-up of the analysis at the entire genome level. As we have been delineating previously, here we focus on two of them: microarrays and next-generation sequencing techniques. Other approaches, based on two-dimensional electrophoresis, have been successfully used in the past (Southern blots, RLGS, MSP, COBRA, etc.), but they have fallen in underuse due precisely to the advent of these two approaches. In fact, it is considered that the genome-scale epigenomic studies actually start with both microarrays and next-generation sequencing techniques.

2.1.2.1 Array-Based Hybridization Approaches (DNA Chip)

DNA microarrays (also named DNA chips) are a collection of DNA spots attached to a solid two-dimensional surface *(42–44)*. Each spot contains a very small amount of a known DNA sequence. The length of these sequences, which are called probes, varies depending on the manufacturer and the way the array is produced, spanning from 20 to $\approx$200 nt. Thus, the probes can be a short fragment of a gene, or any other DNA sequence from the genome. Then, the hybridization between the sample (named target) and the probe occurs under defined-stringency conditions. The sample to be hybridized in the array is the DNA previously treated with any of the methylation-dependent enrichment methods (either enzymatic, affinity enrichment, or bisulfite conversion). The general idea is that the target will hybridize to its complementary probe in a particular spot in the array. To determine the relative abundance of targets bound to its probe, the sample has to be previously labeled with a fluorescent or a chemiluminescent dye. Thus, the intensity of fluorescence or chemiluminescence will be a quantitative determination of the amount of target present in the sample *(45)*. Yet, some specific particularities relate to the coupling of each methylation-enrichment method to the microarray hybridization step.

2.1.2.1.1 Endonuclease Digestion The first techniques that have been developed coupled enzymatic methods with arrays. One of them is called *methylated CpG island amplification* (MCA) *(46)*. In this method, two independent DNA samples (i.e., control and case) are first digested with a methylation-sensitive nuclease, such as *Sma*I. The DNA is purified and deployed of small fragments. Therefore, the sample is now enriched in long-methylated CpG islands. Next, both samples are digested with a methylation-insensitive enzyme (*Xma*I) that renders CCGG overhangs in methylated sites. At this point specific adaptors are ligated to the fragments and amplified by PCR, thus producing enrichment in methylated fragments. Afterward, each sample is labeled with a specific fluorochrome (typically, Cy5 and Cy3). Subsequently, the samples, actually representing the methylated fraction of the genome, are cohybridized in a microarray platform, scanned, and analyzed. This method has proven a valid tool to unravel novel tumor markers in colon, prostate, breast and pancreatic cancers, or lymphocytic leukemia *(47–49)*.

An important issue with MCA is that two samples are required for the analysis. This is sometimes a problem, especially in cancer, where proper control samples are not always available. Thus, a popular alternative approach is *differential methylation hybridization (50)*. In this approach, DNA from one single sample is split into two independent pools of DNA. One of them is treated with a combination of sensitive–insensitive nucleases, in a similar way than MCA. The other one is processed in the absence of nucleases (mock treated). This produces a parallel of DNA that is subsequently amplified by PCR and labeled with fluorescent dyes. In this case, the difference in signal intensity can be used as an assessment of the methylation at different DNA *loci*. This is a relative method that allows determining differences in methylation across different samples. As in MCA, it has been used with success in identifying novel markers in different types of cancer *(51)*.

Finally, over the past years several adaptations of these methods have been published. For example, some investigators have used a methylation-requiring endonuclease, *Mcr*BC *(52,53)*. This nuclease cuts the DNA when it is methylated. *Mcr*BC cleaves all methylated CpG islands in the genome. The utilization of this enzyme provides greater sensitivity in regions that are heavily methylated. A further variation on the utilization of this enzymatic approach is called CHARM (comprehensive high-throughput arrays for relative methylation) *(53)*. The investigators combined novel tiling arrays with an improved algorithm to achieve high sensitivity and specificity in a reliable and relatively inexpensive way.

2.1.2.1.2 Affinity Enrichment So far, affinity-enrichment methods have been mostly coupled to microarrays. Nevertheless, as we discuss in the next sections, the continuous price drop of deep sequencing techniques is now shifting this tendency. Methodologically, sample DNA is split into two pools: input DNA and enriched DNA. As previously described, the immunoprecipitation is performed with an anti-5-methylcytosine antibody. Thus, this approach is known as methyl-DIP or MeDIP. Alternatively to this antibody, methylated DNA can be precipitated with other antibodies that recognize methylated cytosines, such as the human methyl-CpG-binding protein (MECP2) *(10)*. As in the previous described techniques, both pools are labeled with different fluorescent dyes and cohybridized onto a specific microarray. The great asset of this approach is that it allows the rapid and efficient determination of DNA methylation at multiple *loci*. Hence, novel epigenetic markers have been identified in human cancer cells *(54,55)*. Finally, it is possible to obtain the methylome of an entire genome by combining ChIp with tiling arrays *(56)*. The problem is that it gives information about average methylation for the specific regions present in the array, but it does not provide information at the single CpG nucleotide level.

2.1.2.1.3 Bisulfite Conversion Bisulfite genomic sequencing is superior to the other techniques because it produces DNA methylation maps at the single base-pair resolution. Nevertheless, bisulfite-converted samples are not easily adapted to array-based techniques. The reason is that during conversion non-methylated cytosines are converted to uracils. Thus, the microarrays should contain specific sequences complementary for the hybridization to bisulfite-treated genomes or, alternatively, allow for multiple mismatches during the hybridization. Therefore, bisulfite-treated human samples have not been used widely for the analysis of DNA methylation via microarray techniques until recently, with the GoldenGate and the Infinium assays for methylation, both from Illumina *(57–59)*. Briefly, both approaches take advantage of the Illumina bead-based array platform. Each bead contains a pair of probes, which are specific for a particular *locus*. In addition, one of the probes recognizes the target when it is methylated, whereas the other one recognizes it when it is unmethylated. Furthermore, for each CpG site, two sets of primers are designed: The first one is designed to discriminate whether the allele is methylated or unmethylated and a second one recognizes the *locus* through specific binding to a neighbor region to the CpG site. Both primers are then ligated, amplified, and subsequently labeled with fluorescent dyes. Finally, the primers are hybridized with the beads and the degree of intensity results in a good estimate of the degree of methylation of that particular CpG site.

The GoldenGate platform currently offers a panel of CpGs, which are specific for human cancer. It covers up to 1500 CpG sites from 807 genes (http://www.illumina.com). In contrast, the Infinium platform incorporates whole-genome amplification after the bisulfite conversion. The most recent version of the Infinium is the HumanMethylation450 BeadChip, which is able to interrogate more than 485,000 CpG sites at single-nucleotide resolution. Both techniques have been successfully used to identify global patterns of methylation in human samples *(30)*.

There are some drawbacks for the array-based techniques *(60)*: The first one is that in an array it is possible to spot a huge amount of sequences, but certainly it is really far to include the whole genome. Specifically, current commercially available microarrays include promoter microarrays, CpG island microarrays, and, more recently, tiling arrays. Although the degree of coverage of the most recent tiling arrays has grown enormously, the entire human genome is not yet represented in any microarray. Therefore, the analysis of the methylome will be necessarily restricted to the probes spotted on the array. Additionally, the spotting of probes relies on previously known sequences (or the utilization of cloned genomic regions). Thus, those regions that are not well anotated are excluded from the arrays. Finally, there can be biases in the hybridization step, and a potential number of false-positives can result in binding of nonhomologous target–probe hybridizations. Although this possibility has been greatly reduced in novel arrays, it is still an issue in array-based studies.

2.1.2.2 Sequencing-Based Approaches (DNA-Seq)

Automated Sanger sequencing has been the almost exclusive technique to perform genome analysis over the past two decades, and it is considered the "first generation" in automated sequencing technologies *(61)*. The main limitation is that large-scale sequencing of multiple genomes is extremely expensive and requires an enormous amount of time to be completed. Over the past 10 years we have witnessed an explosion of new technologies that allow the sequencing of a large number of genomes in a much faster and cheaper fashion than the Sanger sequencing *(62)*. They are collectively known as *next-generation sequencing*. They have been extensively reviewed elsewhere *(63)*. We comment here on just a few basic principles that apply for large-scale epigenomic studies.

There now exist a few platforms and technologies that differ in many aspects. But they share in common that they perform massive parallel sequencing, thus producing thousands or millions of sequences concurrently. The technical issues concerning the many aspects of next-generation sequencing—including template preparation, sequencing and imaging, and genome alignment and assembly—have been reviewed elsewhere and it is not the main focus of this chapter *(63)*. The current most popular commercial platforms are 454 from Roche, Solexa/Illumina, SOLiD/Life Technologies, and Ion Torrent/Life Technologies. They have some advantages and disadvantages and the selection of one platform over another is primarily based on budgets and the biological question to be addressed (Table 2).

2.1.2.2.1 Endonuclease Reaction Restriction enzyme-enriched samples are all well suited for direct sequencing into the newly available sequencing platforms. Two techniques have been developed to adapt nuclease-treated samples to next-generation sequencing approaches: Methyl-Seq and methylation-sensitive cut counting (MSCC) *(29)*. Methyl-Seq is based on the sequencing of libraries constructed from DNA treated with *Hpa*II or *Msp*I, which are compared to libraries constructed from DNA that has been previously randomly sheared *(64)*. In contrast, MSCC is based on the sequential digestion of genomic DNA with *Hpa*II followed by a second one with *Mme*I *(65)*. First, genomic DNA is treated with *Hpa*II. Next, an adaptor containing an *Mme*I site is ligated. The DNA is then digested with this second nuclease that cuts within the fragment. The key idea is that the amount of digestion is proportional to the number of reads for a given site; the number of reads correlates with the level of methylation in that sequence.

2.1.2.2.2 Affinity Enrichment ChIP-based methods have been mostly combined with microarrays. These affinity-based techniques are now rapidly shifting to analysis by the next-generation techniques *(56,66)*. As it has been described earlier, DNA immunoprecipitated with the specific methylcytosine antibody is de-cross-linked and defragmented. After ligation with specific probes, the sample is applied for on-site deep sequencing.

2.1.2.2.3 Bisulfite Conversion Bisulfite-converted DNA is especially well suited for sequencing-based techniques, including next-generation techniques. In fact, thanks to whole-genome shotgun bisulfite sequencing, coupled to Illumina platforms, it is now possible to obtain single-base-pair resolution DNA methylation profiles for the entire genome *(67)*. The issue with this approach is that around 10% of all CpG dinucleotides in the mammalian libraries do not align with bisulfite-converted reads. So, an important amount of information cannot be properly mapped.

In sum, compared to array-based techniques, sequencing-based approaches are more flexible and powerful because they allow allele-specific, or even single-nucleotide base-pair, resolution *(29)*. It does not require the design of specific probes, as it happens with arrays. It can cover more genomic regions. Input DNA is lower than that used for array-based techniques. It avoids hybridization biases. Downsides are that whole-genome sequencing is still biased by available sequence library. Nowadays, the major downside for the next-generation techniques is that massive accumulation of information requires the parallel development of bioinformatics tools and infrastructure for data storing, data transfer, or quality control *(5,63)*. In fact, it might be

TABLE 2 Comparison of Most Popular Platforms for Next-Generation Sequencing

	454/Roche	Solexa/Illumina	SOLiD/Life Technologies	Ion Torrent/Life Technologies
Total output	0.5 billion nts	5 billion nts	15 billion nts	1 billion nts
Read length	400–700 nts	35 nts	50 nts	200 nts
Number of reads	>1 million	150 million	300 million	5 million
Time to prepare library	2–3 days	2–3 days	2 weeks	2–3 days
Time per sequencing run	24 h	2 days	7 days	2 h
Cost per million bases	$10	$0.05–0.15	$0.13	$1
Pros	Length of sequenced molecules is about 400–700 nts. Thus, sequence assembly is much easier and platform is especially useful for studies on transcript splice variants	A single run produces 150 million sequencing reads of short length, making it an ideal platform for small RNA sequencing	A single run can produce more than 300 million reads. The di-base color system discerns sequencing errors due to SNPs and RNA editing. Thus, this platform is ideal for genome resequencing where identification of SNPs is important	It is extremely fast compared to the other platforms. Equipment is not very expensive
Cons	Number of molecules sequenced in a single run is relatively low. Reagents for a sequencing run are approximately twice as costly as those consumables used in the other two platforms	Rate of sequencing errors increase toward the end of the deep sequencing reads	Preparation of samples is time-consuming. The di-base color encoding system requires dedicated computational tools for the analysis	Homopolymer errors

possible that in the future the cost of data handling and analysis might be even greater than the sequencing itself. But that is next to next-generation sequencing.

This field is evolving at a high speed, and novel technologies are under development. The general perception is that we will have even cheaper and faster methods in the near future. Thus, the important point is to decide which combination of DNA treatment followed by sequencing method is the one that fits best in the experimental design or the questions to be addressed.

2.2 Histone Modifications

As we have previously outlined, histones are small nuclear proteins that contribute to pack the DNA within nucleosomes *(7,33)*. Histones are subject to a number of covalent post-translational modifications that collectively influence chromatin remodeling and gene expression. The question now is how to build genome-wide histone-maps. The answer came recently thanks to the combination of ChIP followed by either microarrays (ChIP-chip) or sequencing methods (ChIP-Seq) *(31)*.

2.2.1 Chromatin Immunoprecipitation

ChIP allows the isolation of chromatin fragments by using specific antibodies that recognize a particular feature of the chromatin *(68)* (Figure 4). In this context, antibodies are suited to recognize a particular histone modification in a specific amino acid (Table 1). Briefly, the basic principle is as follows: intact cells are subjected to cross-linking (DNA-chromatin) with paraformaldehyde or ultraviolet light. After the cross-linking, the nuclear extracts are purified and sonicated to yield fragments around 1 kb in size, which are manageable for hybridization and sequencing techniques. Currently, incubation with paraformaldehyde is the preferred strategy for cross-linking because it is fast, simple, very efficient, and, importantly, fully reversible upon proteinase K digestion and heat treatment.

So far, all methods developed to generate histone-modification maps rely on ChIP. Thus, some cautions need to be made when interpreting ChIP experiments *(31,35)*. First, ChIP experiments rely on the quality of the antibodies that are able to select the target regions versus random genomic DNA in a very specific manner. Thus, different qualities may yield slightly different data. To overcome this problem it is possible to use antibodies from different sources or isotypes to validate the data. Also, the ChIP process results in an enrichment of a specific genomic target through antibody selection. Yet, other genomic DNA is also present in the pulled-down material. Thus, highly specific antibodies will also result in lower unspecific precipitation and lower noise/background. Second, chromatin fragmentation can additionally give rise to potentially different results. For example, chromatin fragmentation can be done by sonication or by digestion with specific nucleases. These methods

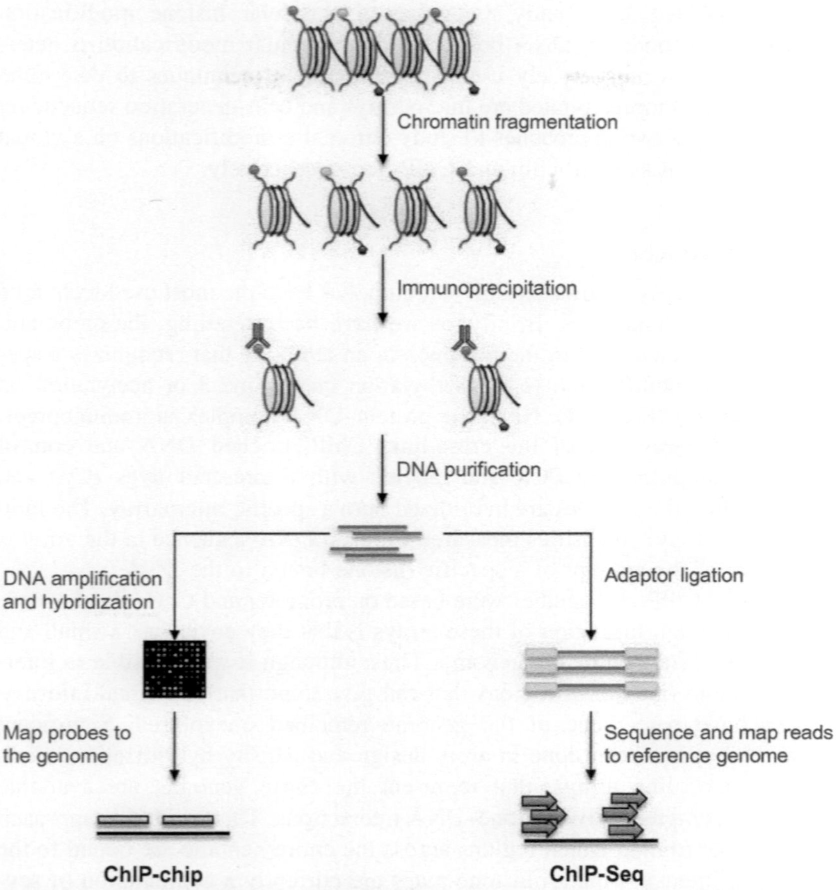

FIGURE 4 Scheme summarizing the fundamental steps to generate histone-maps by ChIP-chip and ChIP-Seq. (For color version of this figure, the reader is referred to the online version of this chapter.)

can render fragments of different sizes that ultimately lead to different immunoprecipitation profiles. Thus, a small-scale trial with different conditions is extremely useful to determine the conditions for a specific sample. Third, the ChIPs require a fairly large amount of biological material. Hence, the result obtained from ChIP experiments are an average of what is happening in thousands or even millions of cells.

2.2.2 Genome-Wide Approaches to Study Chromatin Modifications

As it happened with global-scale approaches to determine DNA methylation, histone-maps require two steps. First, the sample is immunoprecipitated with an antibody that specifically recognizes a particular histone modification (Figure 4). Second, the DNA bound to this particular modification is determined. The two most widely used high-throughput techniques to determine the DNA immunoprecipitated are microarrays and next-generation sequencing tools. Thus, the two approaches to study chromatin modifications on a global scale are known as ChIP-chip and ChIP-Seq, respectively.

2.2.2.1 ChIP-chip

This approach, also known as ChIP-on-chip, has been the most used technique to establish histone-maps. Briefly, as we have been detailing, the chromatin fragments are incubated in the presence of an antibody that recognizes a specific histone modification (e.g., methylation on histone 3 or acetylation on histone 4, etc.) (Figure 4). Next, the protein–DNA complex is immunoprecipitated. After reversal of the cross-link, ChIP-enriched DNA and control DNA are amplified by PCR and labeled with fluorescent dyes (Cy3 and Cy5). Finally, the samples are hybridized onto a specific microarray. The ratio of the Cy5 to Cy3 intensities measured for each DNA sequence in the array is a measure of the amount of a specific histone bound to the DNA.

Initially, ChIP-chip studies were based on promoter and CpG island microarrays. The main limitation of these arrays is that they cover just a small and very specific fraction of the genome. Thus, although it was possible to interrogate a large fraction of regions that can give important biological information, a huge percentage of the genome remained unexplored. Significant advancement has been done in array design and density hybridization. Now, high-density tiling arrays, that represent the entire genome, are available and allow comprehensive histone–DNA interactions. Thus, with this approach it is possible to map which regions across the entire genome are bound to the histone of interest. Finally, histone-maps are currently a combination of several ChIP-chip for the same sample. It is becoming routine to parallel ChIP of several modified histones. Then, the DNA from each ChIP is hybridized onto the same type of microarray. We will obtain a map for each histone included in our panel. This way, we can determine whether different types of modifications tend occur together in a given region or they are incompatible. This type of approach is actually giving support for the histone code hypothesis.

Over the past decade, ChIP-chip has been tremendously useful in obtaining high-resolution histone-maps and identifying many novel-binding sites. Yet this technique has several limitations. First, the major limitation of this approach is that it relies on the sequences spotted onto the array *(35)*. Therefore, the probes included in the arrays are based on known sequences, including partial genomic content, CpG islands, or promoter regions from well-known genes. Hence, information from unknown regions is not included in most microarray designs. Although whole-genome tiling arrays can overcome this limitation, they are extremely expensive and are available for only a few species (human, mouse). Thus, they have not been widely used in this type of approach. Second, the material from the immunoprecipitation is typically subjected to PCR amplification. This process can lead to nonspecific amplification, which will result in higher noise in the subsequent hybridization step. Third, microarrays have a high background and saturation signal. The consequence is that signals in the very low or the very high range are typically not reliable. Fourth, several array designs and genome assemblies are currently available. This fact makes the direct comparison of data from different laboratories difficult and tedious.

2.2.2.2 ChIP-Seq

The ChIP-Seq is based on next-generation sequencing and is now rapidly replacing the previous ChIP-chip assays *(31)*. Briefly, the first immunoprecipitation steps are performed in a similar way than those for the ChIP-chip (Figure 4). But now, after the immunoprecipitation, the DNA is ligated to specific adaptors, depending on the platform by which it will be sequenced, and amplified to generate about 200 ng DNA. The sample is next hybridized to a solid surface containing complementary oligos for the adaptors. Then, the sequence-by-synthesis is performed, which will result in the massive parallel sequencing of about 30–60 million short (20–60 bp) reads. In this technique, the number of reads mapped to a specific region in the genome is directly proportional to the degree of antibody enrichment. Additionally, since the ChIP consists of a significant amount of enriched DNA and nonenriched background DNA, it is possible to sequence control DNA (input). Control DNA will allow the determination of the noise that, when subtracted from the sample, will allow identification of the true positive antibody-enriched regions.

ChIP-Seq has a number of advantages when compared to ChIP-chip. First, ChIP-Seq does not rely on known sequences (probes) in the array. Thus, it is less biased and coverage of the genome is much higher than in the previous approach. Second, ChIP-Seq does include hybridization steps. Therefore, no additional biases are added. Third, it is very flexible and both low and high signals (reads) are reliable. Fourth, the comparison of data across experiments/laboratories is direct and simple. Finally, the cost, which was an issue in the beginning, is much lower compared to the tiling arrays. Therefore, ChIP-Seq is becoming widely adopted by the scientific community to finely map histone binding to the genome.

2.3 Noncoding RNAs

They are not, *sensu stricto*, part of the epigenetic machinery, but contribute to its establishment and, thus, to the regulation of gene expression *(69)*.

2.3.1 *Types of ncRNAs and Their Biological Function*

Many reviews now describe the different types of ncRNAs in depth *(70)*. Thus, here we briefly describe them (Table 3), as well as the high-throughput approaches that are now widely used to analyze their expression.

New members of the ncRNA family include (1) a number of short RNAs, such as microRNAs (miRNA), Piwi-interacting RNA (piRNA), small interfering RNA (siRNA), and enhancer RNA (eRNA), and (2) long noncoding RNAs (lncRNA) (Table 3) *(23,25)*. It is expected that new species of ncRNAs will be uncovered in the near future, thanks in part to the appearance of next-generation sequencing methodologies. Their main characteristics are briefly summarized below.

2.3.1.1 Short ncRNAs

Short ncRNAs include some of the best-characterized ncRNAs: microRNAs and piRNAs.

TABLE 3 Regulatory ncRNAs in Eukaryotes

Type	Name	Length (nt)	Functions
Short ncRNAs			
miRNA	MicroRNA	19–24	Perfect pairing; mRNA degradation Imperfect pairing; translational repression
siRNA	Small interfering RNA	20–24	mRNA cleavage and degradation through perfect sequence matching Heterochromatin formation
piRNA	Piwi-interacting RNA	24–31	Transposon silencing in germ cells DNA methylation
Medium-size ncRNAs			
snoRNA	Small nucleolar RNA	60–300	Modification of rRNA
PAR	Promoter-associated RNA	16–200	Mostly unknown, although possibly implicated in transcriptional activation
Long ncRNAs			
lincRNA	Long intergenic ncRNA	>200	Chromatin remodeling Scaffold DNA-chromatin in enhancer regions
Other		>200	Long RNAs implicated in maintenance of imprinting (*Xist,* Air)

2.3.1.1.1 miRNAs miRNAs are highly conserved, single-stranded RNAs of 20–24 nucleotides *(71,72)*. They are involved in the regulation of almost all biological processes of the cell, including proliferation, differentiation, apoptosis, and development *(72)*. Thus, they have been involved in many different diseases, such as cancer, cardiac hypertrophy, Crohn's disease, and Alzheimer's disease, among many others *(25)*. Computational analysis of the genome proposes that there exist >1500 miRNAs in mammalian cells. Despite these numbers, it has been postulated that they regulate expression of up to 60% of the genome at the post-transcriptional level *(73,74)*. This is because the pairing of a miRNA with its cognate mRNA is imperfect. Thus, a single miRNA can recognize multiple mRNAs. At the same time, a single mRNA can be recognized and regulated by multiple miRNAs. This is the reason why miRNAs can regulate such an enormous amount of targets. General consensus is that when the pairing of miRNA/mRNA is 100% homologous, the mRNA is degraded through similar machinery than the siRNAs. But if the pairing is imperfect, miRNA can stop the expression of a gene through inhibiting translation.

Importantly, as we detail later in Section 2.3.2, miRNAs can also regulate gene expression through translational activation, which involves epigenetic mechanisms *(27,75)*.

2.3.1.1.2 Piwi-Interacting RNAs piRNAs are small ncRNAs (24–31 bp) named for their interaction with the Piwi proteins of the Argonaute family *(76,77)*. They were first discovered in germ cells from *Drosophila* as siRNAs, which are complementary to transposons and repetitive elements. Not surprisingly, their primary role in *Drosophila* was to suppress transposon activity during germline development *(78,79)*. In mammals, piRNAs have been mostly uncovered in germ cells where they also recognize and target the cleavage of transposons *(80,81)*. Importantly, transposon control during spermatogenesis occurs in combination with *de novo* methylation *(27,82,83)*.

2.3.1.2 Long Noncoding RNAs

lncRNAs are a heterogeneous group of noncoding RNAs, which are arbitrarily considered more than 200 nucleotides long *(25,84)*. They are involved in many biological processes and comprise the largest fraction of the noncoding transcriptome *(63)*. Most of them are localized in the nucleus, have very low expression, show poor sequence conservation, and can be both poly A+ and poly A− *(27)*. Given this heterogeneity, lncRNAs have been subsequently classified into several additional categories according to their proximity to coding genes *(84)*: sense, antisense, bidirectional, intronic, and intergenic. For the purposes of this review we focus on just three functional classes that have been involved in epigenetic regulation of gene expression: (a) lncRNAs involved in imprinting, (b) lncRNAs involved in X-chromosome inactivation, and (c) long intergenic ncRNAs (lincRNAs).

2.3.1.2.1 Noncoding RNAs and Genomic Imprinting Genomic imprinting refers to the parental-specific expression in diploid cells *(85)*. Diploid cells contain two copies for all genes, one from each parent. Almost all of them are equally expressed from each parental chromosome at a 50% ratio. The exceptions are imprinted genes, where only one copy is expressed while the other one is silenced. Imprinting is restricted to a few hundred genes, most of them grouped in clusters. The imprinting is controlled by an imprinting control region, which regulates the entire domain through differential methylation. Most genes in imprinting clusters are protein-coding genes. But, strikingly, at least one gene in the cluster encodes an ncRNA *(86)*. These ncRNAs show reciprocal parental expression compared to the imprinted protein-coding genes and there is evidence that they contribute to allele-specific silencing through antisense pairing with mRNA, although the exact mechanisms are not completely elucidated.

2.3.1.2.2 ncRNAs and X-Chromosome Inactivation X-chromosome inactivation in mammals refers to the mechanism by which the expression of X-linked genes in one female X chromosome is silenced *(87,85,88)*. This is accomplished primarily by the action of ncRNAs, *Xist and Tsix* being the most prominent and best characterized (*89*). Silencing of X-linked genes is initiated by the expression of the *Xist* gene. This gene gives rise to the lncRNA *Xist* that initiates gene silencing all along the chromosome *in cis*. Following *Xist* coating, the inactive X-chromosome undergoes a series of changes that contribute to its full inactivation, including (1) reduced acetylation at specific lysines, (2) increased lysine methylation, (3) replacement of the usual variant H2A by the macro-H2A, and (4) massive DNA methylation in selected regions. Collectively, these modifications, initiated by specific ncRNAs, maintain permanent X inactivation in females.

2.3.1.2.3 lincRNAs Another recently characterized class of ncRNAs are long intergenic RNAs (lincRNAs) *(90)*. They are RNAs transcribed from intergenic regions and are highly conserved across species. Their role in biological processes is still under investigation. The first example, lincRNA-21, was shown to be a target of p-53 and was therefore involved in the p-53-mediated response to DNA damage *(91)*. lincRNAs can regulate gene expression of neighbor genes and distant genomic sequences. From the last observation it is inferred that lncRNAs are actually marking enhancer regions and contribute to the recruitment machinery necessary to modify the chromatin and bring together distant regions in the chromosome *(92)*. Thus, they act as scaffold molecules that recruit modified histones to allow the opening of chromatin for transcription.

2.3.2 ncRNAs in Epigenetic Regulation

2.3.2.1 ncRNAs and DNA Methylation

Antisense RNAs (ASRNAs) were among the first class of ncRNA that was shown to influence DNA methylation. They are shown to be involved in the mechanism of DNA methylation for specific transcripts *(93)*. MiRNAs can also influence DNA methylation, primarily through targeting and therefore altering transcription of DNA methyl transferase-3A and -3B *(94)*. In another example, miRNAs can also control *de novo* DNA methylation in embryonic stem cells *(95)*.

Finally, it is under study whether piRNAs can guide DNA methylation in germ cells from mice. Although it deserves further experimentation, it is indicated that Piwi–piRNA complexes play essential roles in establishing *de novo* methylation of transposons, at least in murine germ cells *(78,96)*.

2.3.2.2 Noncoding RNAs and Chromatin Remodeling

Chromatin modification is an epigenetic phenomenon in which chromatin structure is altered through covalent modifications of DNA and histones *(27)*. For example, heterochromatin is found in regions where gene expression is generally repressed. In this regard, it has been shown that RNAi (RNA interference) is essential for maintaining such silencing at heterochromatic *loci* *(97)*. Another example is an ASRNA that induces specific gene silencing of p15 through heterochromatin formation without modifying DNA methylation *(98)*.

2.3.2.3 ncRNAs and Paramutation

Paramutation is defined as the interaction between two alleles in a particular *locus* in which one induces a heritable change on the other one *(99,100)*. Importantly, this change occurs in the absence of any alteration in DNA sequence, indicating that paramutations are mediated by epigenetic modifications. Paramutations were originally described in maize and several of them have already been characterized. Recently, paramutations have been described in mammals (mice) *(101)*. The authors provide evidence that paramutations in mice involve RNAs, since they were not able to detect changes in DNA methylation or histone modifications. Yet, it is unclear whether paramutations in mice are mediated by post-transcriptional or post-translational modifications *(100)*.

2.3.3 Omics and ncRNAs

The transcriptome is defined as the whole set of transcripts present in a cell in a given developmental stage and under a specific physiological condition. As we have been describing, the transcriptome is composed by mRNAs, rRNAs, tRNAs, and a set of ncRNAs *(22,102)*. Over the past decade, the transcriptome has been analyzed by hybridization-based approaches (microarrays). Typically, the RNA is previously converted into cDNA, fluorescently labeled, and hybridized onto a specific microarray (such as Affymetrix GeneChips). They have been widely used over the past 15 years and have given quite precise information about the way the transcriptome is regulated in many specific conditions. Yet, as we have been outlining throughout this chapter, hybridization-based techniques have several limitations *(103)*. They rely on known information of the genome sequence. Microarrays have a low dynamic range of detection due to high background and saturation signal. Thus, both very low and very high expressions are difficult to be determined by this approach.

In contrast to microarray-based approaches, sequence-based methods provide a far more precise determination of the transcriptome *(104)*. This new approach is based on next-generation sequencing techniques and is termed RNA-Seq (Figure 5). Methodologically, the first step consists of conversion of RNA into a library of fragmented cDNA. This can be accomplished by either direct fragmentation of RNA or the newly synthesized cDNA. Next, specific adaptors are attached to the cDNA fragments that are amplified by

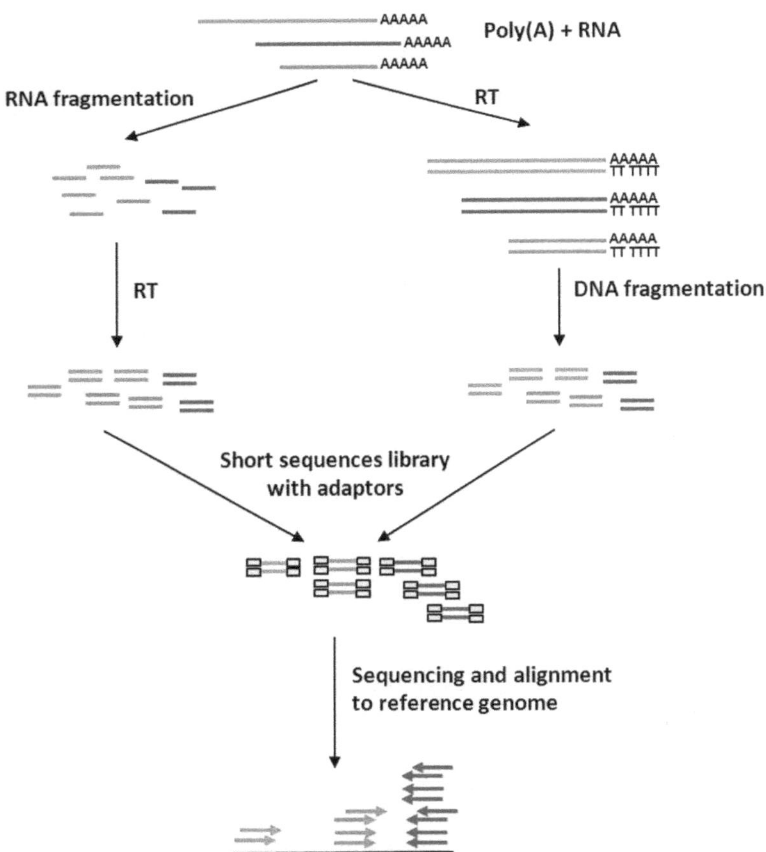

FIGURE 5 Scheme detailing the basics to generate whole transcriptome by RNA-Seq. (For color version of this figure, the reader is referred to the online version of this chapter.)

PCR. The type of adaptor and the amplification depend on the platform to be used for sequencing. Finally, the sample can be sequenced through either of the commercially available sequencing platforms (Illumina/Sollexa, Life-Technologies SOLiD, Roche 454, Helicos Bioscience). At the end, a massive amount of reads can be either aligned to a reference genome or assembled without the reference. In both cases it will be possible to quantitatively assess the level of expression for each transcript. In addition, in the experiments where the reads are aligned against a genome of reference, it will be possible to determine the different splice variants (isoforms) for a given gene and distinguish allelic expression.

RNA-Seq offers a series of advantages that makes it the method of choice for measuring ncRNAs *(104)*. (1) RNA-Seq has very low, if any, background. (2) Likewise, RNA-Seq has no upper limit of detection (that correlates with the number of sequences present in a sample). (3) The amount of RNA needed is very low. (4) Most importantly, it does not rely on previous knowledge of genome sequence. This characteristic is the one that has made possible the massive identification of novel ncRNAs, as well as the identification of many different splicing variants in the sample *(22,102)*. (5) Small RNAs (miRNAs, piRNAs, siRNAs, etc.) can be sequenced directly without previous fragmentation, which avoids a previous step that can introduce some bias in the sample.

The main issues for RNA-Seq are related to (1) the fragmentation process, (2) bioinformatics limitations, and (3) the balance between coverage and cost.

2.3.3.1 Fragmentation

Large RNA molecules have to be fragmented into smaller size sequences (about 200 nt long). Fragmentation can be performed at the RNA level, through RNA hydrolysis or nebulization, or at the cDNA level (DNAse I treatment or sonication). The problem is that fragmentation can cause an important bias in the sample.

2.3.3.2 Bioinformatics

RNA-Seq generates an enormous amount of data that requires specific methods for storing and retrieving, as well as analysis. This is especially challenging when the transcriptome is captured from a complex organism where both sense and antisense transcription occur in the same *locus* and different patterns of splicing occur as well *(105)*.

2.3.3.3 Coverage and Cost

Finally, an important matter to take into account is the coverage of the sample. That is, it is important to design the percentage of transcripts that we want to survey, including very small copy number. The greater the coverage, the greater the sequencing depth (and the greater the cost). In general, complex genomes, such as human, require more depth than simple genomes, such as yeast. Also, it is important to note that depth and coverage do not maintain a linear relationship. For example, jumping from 80% to 90% coverage (10% increase in coverage), which requires doubling the sequencing depth. Thus, it is really important to bear in mind the biological question in order to have a good design in addressing deepness–coverage–cost.

3 CONCLUDING REMARKS

The advent of novel technologies has allowed the determination of epigenetic markers at the entire genome level. This has revolutionized the understanding of how these epigenetic phenomena (DNA methylation, histone modifications, ncRNAs) contribute to regulate gene expression. In addition to understanding such higher order of complexity in regulating expression, there are three other issues to emphasize. They actually apply not only to epigenomics, but also to all other Omics fields.

First, the main challenge that Omics has to face in the near future is the handling, storing, and retrieval of the massive amount of information generated. Such information is currently testing the available bionformatics tools. Thus, new ways for analysis and storing are under implementation to allow individual labs to access meaningful information from the large datasets obtained through microarray and next-generation sequencing techniques.

Second, Omics approaches have changed the way we are currently conducting science. We have shifted from classic *hypothesis-driven* research approaches to *hypothesis-generating* research. Nowadays, many experimental designs focus on the analysis of whole transcriptome, epigenome, or metabolome, and so on, in a particular biological condition or disease estate. The main advantage of this approach is that the search for biological mechanisms is not biased by previous knowledge. In this sense, Omics has constituted a way to open the frontiers of novel mechanisms to better understand fundamental biological processes.

Third, and linked to the previous issue, Omics has opened the door to personalized medicine. We can, for the first time, think about realistically obtaining the whole genome, epigenome, and so on, from a single individual in a cost-effective manner. This knowledge will allow us to determine disease risk for particular diseases or whether a specific treatment can be effective for a certain illness. It opens the real possibility to decide prevention programs and treatments based on the characteristics of a single individual (i.e., a given genome or epigenome).

REFERENCES

1. Ozdemir, V. *OMICS* **2013**, *17*, 1–4.
2. Wilkins, M. R.; Pasquali, C.; Appel, R. D.; Ou, K.; Golaz, O.; Sanchez, J. C. *Biotechnology (N.Y.)* **1996**, *14*, 61–65.
3. Yadav, S. P. *J. Biomol. Tech.* **2007**, *18*, 277.

4. Choi, H.; Pavelka, N. *Front. Genet.* **2011**, *2*, 105.
5. Berger, B.; Peng, J.; Singh, M. *Nat. Rev. Genet.* **2013**, *14*, 333–346.
6. Pereira, D. M. *Curr. Pharm. Biotechnol.* **2012**, *13*, 623.
7. Allis, C. D.; Jenuwein, T.; Reinberg, D.; Caparros, M.-L., Eds.; *Epigenetics*; 2007, Cold Spring Harbor Laboratory Press: Cold Spring Harbor, New York.
8. Jaenisch, R.; Bird, A. *Nat. Genet.* **2003**, *33*(Suppl.), 245–254.
9. Deaton, A. M.; Bird, A. *Genes Dev.* **2011**, *25*, 1010–1022.
10. Esteller, M. *Nat. Rev. Genet.* **2007**, *8*, 286–298.
11. Tahiliani, M.; Koh, K. P.; Shen, Y.; Pastor, W. A.; Bandukwala, H.; Brudno, Y.; et al. *Science* **2009**, *324*, 930–935.
12. Kriaucionis, S.; Heintz, N. *Science* **2009**, *324*, 929–930.
13. Branco, M. R.; Ficz, G.; Reik, W. *Nat. Rev. Genet.* **2011**, *13*, 7–13.
14. Zentner, G. E.; Henikoff, S. *Nat. Struct. Mol. Biol.* **2013**, *20*, 259–266.
15. Strahl, B. D.; Allis, C. D. *Nature* **2000**, *403*, 41–45.
16. Jenuwein, T.; Allis, C. D. *Science* **2001**, *293*, 1074–1080.
17. Kouzarides, T. *Cell* **2007**, *128*, 693–705.
18. Schreiber, S. L.; Bernstein, B. E. *Cell* **2002**, *111*, 771–778.
19. Li, B.; Carey, M.; Workman, J. L. *Cell* **2007**, *128*, 707–719.
20. Huttenhofer, A.; Schattner, P.; Polacek, N. *Trends Genet.* **2005**, *21*, 289–297.
21. Washietl, S.; Pedersen, J. S.; Korbel, J. O.; Stocsits, C.; Gruber, A. R.; Hackermuller, J.; et al. *Genome Res.* **2007**, *17*, 852–864.
22. Djebali, S.; Davis, C. A.; Merkel, A.; Dobin, A.; Lassmann, T.; Mortazavi, A.; et al. *Nature* **2012**, *489*, 101–108.
23. Carninci, P.; Kasukawa, T.; Katayama, S.; Gough, J.; Frith, M. C.; Maeda, N.; et al. *Science* **2005**, *309*, 1559–1563.
24. Kapranov, P.; Cheng, J.; Dike, S.; Nix, D. A.; Duttagupta, R.; Willingham, A. T.; et al. *Science* **2007**, *316*, 1484–1488.
25. Esteller, M. *Nat. Rev. Genet.* **2011**, *12*, 861–874.
26. Zhou, H.; Hu, H.; Lai, M. *Biol. Cell* **2010**, *102*, 645–655.
27. Kaikkonen, M. U.; Lam, M. T.; Glass, C. K. *Cardiovasc. Res.* **2011**, *90*, 430–440.
28. Shakya, K.; O'Connell, M. J.; Ruskin, H. J. *Epigenetics* **2012**, *7*, 982–986.
29. Laird, P. W. *Nat. Rev. Genet.* **2010**, *11*, 191–203.
30. Heyn, H.; Esteller, M. *Nat. Rev. Genet.* **2012**, *13*, 679–692.
31. Schones, D. E.; Zhao, K. *Nat. Rev. Genet.* **2008**, *9*, 179–191.
32. Pomraning, K. R.; Smith, K. M.; Freitag, M. *Methods* **2009**, *47*, 142–150.
33. Cox, M.; Nelson, D. R.; Lehninger, A. L., Eds.; *Lehninger Principles of Biochemistry*; 2005, V.H. Freeman and Company: New York.
34. Ordway, J. M.; Bedell, J. A.; Citek, R. W.; Nunberg, A.; Garrido, A.; Kendall, R.; et al. *Carcinogenesis* **2006**, *27*, 2409–2423.
35. Wong, E.; Wei, C. L. *Genome Med.* **2009**, *1*, 89.
36. Collas, P. *Mol. Biotechnol.* **2010**, *45*, 87–100.
37. Cross, S. H.; Charlton, J. A.; Nan, X.; Bird, A. P. *Nat. Genet.* **1994**, *6*, 236–244.
38. Mukhopadhyay, R.; Yu, W.; Whitehead, J.; Xu, J.; Lezcano, M.; Pack, S.; et al. *Genome Res.* **2004**, *14*, 1594–1602.
39. Jorgensen, H. F.; Adie, K.; Chaubert, P.; Bird, A. P. *Nucleic Acids Res.* **2006**, *34*, e96.
40. Rauch, T.; Pfeifer, G. P. *Lab. Invest.* **2005**, *85*, 1172–1180.
41. Frommer, M.; McDonald, L. E.; Millar, D. S.; Collis, C. M.; Watt, F.; Grigg, G. W.; et al. *Proc. Natl. Acad. Sci. U.S.A.* **1992**, *89*, 1827–1831.

42. Schena, M.; Shalon, D.; Davis, R. W.; Brown, P. O. *Science* **1995**, *270*, 467–470.
43. Abdullah-Sayani, A.; Bueno-de-Mesquita, J. M.; van de Vijver, M. J. *Nat. Clin. Pract. Oncol.* **2006**, *3*, 501–516.
44. Sobek, J.; Bartscherer, K.; Jacob, A.; Hoheisel, J. D.; Angenendt, P. *Comb. Chem. High Throughput Screen.* **2006**, *9*, 365–380.
45. Kolbert, C. P.; Taylor, W. R.; Krajnik, K. L.; O'Kane, D. J. *Methods Mol. Med.* **2003**, *85*, 239–255.
46. Estecio, M. R.; Yan, P. S.; Ibrahim, A. E.; Tellez, C. S.; Shen, L.; Huang, T. H.; et al. *Genome Res.* **2007**, *17*, 1529–1536.
47. Chung, W.; Kwabi-Addo, B.; Ittmann, M.; Jelinek, J.; Shen, L.; Yu, Y.; et al. *PLoS One* **2008**, *3*, e2079.
48. Kuang, S. Q.; Tong, W. G.; Yang, H.; Lin, W.; Lee, M. K.; Fang, Z. H.; et al. *Leukemia* **2008**, *22*, 1529–1538.
49. Omura, N.; Li, C. P.; Li, A.; Hong, S. M.; Walter, K.; Jimeno, A.; et al. *Cancer Biol. Ther.* **2008**, *7*, 1146–1156.
50. Yan, P. S.; Potter, D.; Deatherage, D. E.; Huang, T. H.; Lin, S. *Methods Mol. Biol.* **2009**, *507*, 89–106.
51. Yan, P. S.; Chen, C. M.; Shi, H.; Rahmatpanah, F.; Wei, S. H.; Caldwell, C. W.; et al. *Cancer Res.* **2001**, *61*, 8375–8380.
52. Ordway, J. M.; Budiman, M. A.; Korshunova, Y.; Maloney, R. K.; Bedell, J. A.; Citek, R. W.; et al. *PLoS One* **2007**, *2*, e1314.
53. Irizarry, R. A.; Ladd-Acosta, C.; Carvalho, B.; Wu, H.; Brandenburg, S. A.; Jeddeloh, J. A.; et al. *Genome Res.* **2008**, *18*, 780–790.
54. Weber, M.; Davies, J. J.; Wittig, D.; Oakeley, E. J.; Haase, M.; Lam, W. L.; et al. *Nat. Genet.* **2005**, *37*, 853–862.
55. Keshet, I.; Schlesinger, Y.; Farkash, S.; Rand, E.; Hecht, M.; Segal, E.; et al. *Nat. Genet.* **2006**, *38*, 149–153.
56. Zhang, X.; Yazaki, J.; Sundaresan, A.; Cokus, S.; Chan, S. W.; Chen, H.; et al. *Cell* **2006**, *126*, 1189–1201.
57. Bibikova, M.; Lin, Z.; Zhou, L.; Chudin, E.; Garcia, E. W.; Wu, B.; et al. *Genome Res.* **2006**, *16*, 383–393.
58. Sandoval, J.; Heyn, H.; Moran, S.; Serra-Musach, J.; Pujana, M. A.; Bibikova, M.; et al. *Epigenetics* **2011**, *6*, 692–702.
59. Bibikova, M.; Barnes, B.; Tsan, C.; Ho, V.; Klotzle, B.; Le, J. M.; et al. *Genomics* **2011**, *98*, 288–295.
60. Chittur, S. V. *Comb. Chem. High Throughput Screen.* **2004**, *7*, 531–537.
61. Hutchison, C. A., 3rd. *Nucleic Acids Res.* **2007**, *35*, 6227–6237.
62. Schuster, S. C. *Nat. Methods* **2005**, *5*, 16–18.
63. Metzker, M. L. *Nat. Rev. Genet.* **2010**, *11*, 31–46.
64. Brunner, A. L.; Johnson, D. S.; Kim, S. W.; Valouev, A.; Reddy, T. E.; Neff, N. F.; et al. *Genome Res.* **2009**, *19*, 1044–1056.
65. Ball, M. P.; Li, J. B.; Gao, Y.; Lee, J. H.; LeProust, E. M.; Park, I. H.; et al. *Nat. Biotechnol.* **2009**, *27*, 361–368.
66. Down, T. A.; Rakyan, V. K.; Turner, D. J.; Flicek, P.; Li, H.; Kulesha, E.; et al. *Nat. Biotechnol.* **2008**, *26*, 779–785.
67. Lister, R.; Pelizzola, M.; Dowen, R. H.; Hawkins, R. D.; Hon, G.; Tonti-Filippini, J.; et al. *Nature* **2009**, *462*, 315–322.
68. Furey, T. S. *Nat. Rev. Genet.* **2012**, *13*, 840–852.

69. Lee, J. T. *Science* **2012**, *338*, 1435–1439.
70. Barciszewski, J.; Erdmann, V. A., Eds.; *Noncoding RNAs: Molecular Biology and Molecular Medicine*; 2003, Landes-Springer: Georgetown, Washington, DC.
71. He, L.; Hannon, G. J. *Nat. Rev. Genet.* **2004**, *5*, 522–531.
72. Bartel, D. P. *Cell* **2009**, *136*, 215–233.
73. Farh, K. K.; Grimson, A.; Jan, C.; Lewis, B. P.; Johnston, W. K.; Lim, L. P.; et al. *Science* **2005**, *310*, 1817–1821.
74. Krol, J.; Loedige, I.; Filipowicz, W. *Nat. Rev. Genet.* **2010**, *11*, 597–610.
75. Kim, D. H.; Saetrom, P.; Snove, O., Jr.; Rossi, J. J. *Proc. Natl. Acad. Sci. U.S.A.* **2008**, *105*, 16230–16235.
76. Aravin, A. A.; Lagos-Quintana, M.; Yalcin, A.; Zavolan, M.; Marks, D.; Snyder, B.; et al. *Dev. Cell* **2003**, *5*, 337–350.
77. Aravin, A. A.; Sachidanandam, R.; Girard, A.; Fejes-Toth, K.; Hannon, G. J. *Science* **2007**, *316*, 744–747.
78. Aravin, A. A.; Hannon, G. J.; Brennecke, J. *Science* **2007**, *318*, 761–764.
79. Brennecke, J.; Aravin, A. A.; Stark, A.; Dus, M.; Kellis, M.; Sachidanandam, R.; et al. *Cell* **2007**, *128*, 1089–1103.
80. Reuter, M.; Chuma, S.; Tanaka, T.; Franz, T.; Stark, A.; Pillai, R. S. *Nat. Struct. Mol. Biol.* **2009**, *16*, 639–646.
81. Chuma, S.; Nakano, T. *Philos. Trans. R. Soc. Lond. B Biol. Sci.* **2013**, *368*, 20110338.
82. Aravin, A. A.; Sachidanandam, R.; Bourc'his, D.; Schaefer, C.; Pezic, D.; Toth, K. F.; et al. *Mol. Cell* **2008**, *31*, 785–799.
83. Watanabe, T.; Tomizawa, S.; Mitsuya, K.; Totoki, Y.; Yamamoto, Y.; Kuramochi-Miyagawa, S.; et al. *Science* **2011**, *332*, 848–852.
84. Ponting, C. P.; Oliver, P. L.; Reik, W. *Cell* **2009**, *136*, 629–641.
85. Barlow, D. P.; Bartolomei, M. S. In: *Epigenetics;* Allis, C. D.; Jenuwein, T.; Reinberg, D., Eds.; 2007, The Cold Spring Harbor Laboratory Press: Cold Spring Harbor, New York.
86. Wan, L. B.; Bartolomei, M. S. *Adv. Genet.* **2008**, *61*, 207–223.
87. Boumil, R. M.; Lee, J. T. *Hum. Mol. Genet.* **2001**, *10*, 2225–2232.
88. Nesterova, T. B.; Brockdorff, N.; Turner, B. M. In: *Epigenetics;* Allis, C. D.; Jenuwein, T.; Reinberg, D., Eds.; 2007, The Cold Spring Harbor Laboratory Press: Cold Spring Harbor, New York.
89. Erwin, J. A.; Lee, J. T. *Curr. Opin. Cell Biol.* **2008**, *20*, 349–355.
90. Guttman, M.; Amit, I.; Garber, M.; French, C.; Lin, M. F.; Feldser, D.; et al. *Nature* **2009**, *458*, 223–227.
91. Huarte, M.; Guttman, M.; Feldser, D.; Garber, M.; Koziol, M. J.; Kenzelmann-Broz, D.; et al. *Cell* **2010**, *142*, 409–419.
92. Orom, U. A.; Derrien, T.; Guigo, R.; Shiekhattar, R. *Cold Spring Harb. Symp. Quant. Biol.* **2010**, *75*, 325–331.
93. Mattick, J. S.; Makunin, I. V. *Hum. Mol. Genet.* **2005**, *14*(Spec. No. 1), R121–R132.
94. Fabbri, M.; Garzon, R.; Cimmino, A.; Liu, Z.; Zanesi, N.; Callegari, E.; et al. *Proc. Natl. Acad. Sci. U.S.A.* **2007**, *104*, 15805–15810.
95. Sinkkonen, L.; Hugenschmidt, T.; Berninger, P.; Gaidatzis, D.; Mohn, F.; Artus-Revel, C. G.; et al. *Nat. Struct. Mol. Biol.* **2008**, *15*, 259–267.
96. Kuramochi-Miyagawa, S.; Watanabe, T.; Gotoh, K.; Totoki, Y.; Toyoda, A.; Ikawa, M.; et al. *Genes Dev.* **2008**, *22*, 908–917.
97. Fukagawa, T.; Nogami, M.; Yoshikawa, M.; Ikeno, M.; Okazaki, T.; Takami, Y.; et al. *Nat. Cell Biol.* **2004**, *6*, 784–791.

98. Yu, W.; Gius, D.; Onyango, P.; Muldoon-Jacobs, K.; Karp, J.; Feinberg, A. P.; et al. *Nature* **2008**, *451*, 202–206.
99. Chandler, V. L. *Cell* **2007**, *128*, 641–645.
100. Cuzin, F.; Grandjean, V.; Rassoulzadegan, M. *Curr. Opin. Genet. Dev.* **2008**, *18*, 193–196.
101. Rassoulzadegan, M.; Grandjean, V.; Gounon, P.; Vincent, S.; Gillot, I.; Cuzin, F. *Nature* **2006**, *441*, 469–474.
102. ENCODE Project Consortium; Birney, E.; Stamatoyannopoulos, J. A.; Dutta, A.; Guigo, R.; Gingeras, T. R.; et al. *Nature* **2007**, *447*, 799–816.
103. Royce, T. E.; Rozowsky, J. S.; Gerstein, M. B. *Nucleic Acids Res.* **2007**, *35*, e99.
104. Wang, Z.; Gerstein, M.; Snyder, M. *Nat. Rev. Genet.* **2009**, *10*, 57–63.
105. Morin, R.; Bainbridge, M.; Fejes, A.; Hirst, M.; Krzywinski, M.; Pugh, T.; et al. *Biotechniques* **2008**, *45*, 81–94.

Chapter 5

An Overview of Quantitative Proteomic Approaches

Adam J. McShane, Vahid Farrokhi, Reza Nemati, Song Li and Xudong Yao

Department of Chemistry, University of Connecticut, Storrs, Connecticut, USA

Chapter Outline

1 INTRODUCTION

Proteomics is the study of the protein complement from a particular biological system such as tissue, a cell type, or an organism. The study includes protein identity, quantity, localization, modification, interaction, and function. The focus of this chapter is on current quantitative proteomic approaches including immuno-based detection and mass spectrometry-based analysis. With the impact of quantitative proteomics in biological sciences, the goal of the chapter is to take the novice in proteomic quantitation through the fundamentals of the field to current, cutting-edge methods, as well as to provide a collection of current literature on the topic for proteomics practitioners. The bulk of the material focuses on mass spectrometry-based detection methods with

Comprehensive Analytical Chemistry, Vol. 63. http://dx.doi.org/10.1016/B978-0-444-62651-6.00005-2

sections on derivatization-free and derivatization-based quantitation, activity probes, mass spectrometers, sample enrichment techniques, software, and throughput considerations.

2 IMMUNO-BASED DETECTION METHODS

2.1 Gel-Based Approaches

In the 1970s, one-dimensional gel electrophoresis was introduced as a separation method for the determination of protein molecular weight and for assessing protein purity *(1)*. The most popular electrophoretic techniques for protein separation are isoelectric focusing (IEF) using immobilized pH gradients and sodium dodecyl sulfate-polyacrylamide gel electrophoresis (SDS-PAGE). Proteins are separated based on their isoelectric point (p*I*) in IEF, whereas SDS-PAGE separates based on size *(2)*. In 1975, two-dimensional gel electrophoresis (2-DE) was developed to separate proteins using IEF as the first dimension and SDS-PAGE as the second dimension. Modern 2-DE methods still have similar designs *(2,3)*.

Different methods have been used to visualize proteins after SDS-PAGE separation. Silver staining due to its excellent detectability is among the most routinely used technique; however, it is not quantitative because it has varying interactions with different proteins and a limited dynamic range. Staining gels using Colloidal Coomassie blue leads to a better linear dynamic range but a relatively compromised limit of detection. Fluorescent labeling has gained popularity due to its sensitivity and wide linear dynamic range *(4)*.

2.1.1 Fluorescent Two-Dimensional Difference Gel Electrophoresis

Despite its high-resolving power, 2-DE has inherent limitations. It suffers from poor sensitivity, low solubilization of membrane proteins, low throughput, and limited gel-to-gel reproducibility *(4–6)*. Other reported drawbacks are labor-intensive image analysis and poor representation of extremely acidic or basic proteins as well as small proteins *(7,8)*. The inherent variability of 2-DE has been addressed by the emergence of fluorescent two-dimensional difference gel electrophoresis (2-D DIGE) *(4–6)*. In 2-D DIGE, amine groups of proteins are covalently labeled with two spectrally resolvable *n*-hydroxysuccinimide (NHS) fluorescent dyes, cyanine dyes (Cy3 and Cy5), prior to protein separation. To ensure that the same proteins in both samples for comparison have the same relative 2-DE mobility, the dyes have the same mass and charge *(4)*. Fluorescent labeling also renders 2-D DIGE much more quantitative than colorimetric methods. It has a linear dynamic range of four to five orders of magnitude, compared to one to three orders with staining methods *(9)*.

2.1.2 *Western Blots (Immunoblotting)*

Western blots are a fundamental tool for the measurement of proteins in biomedical labs. The Western blot utilizes antibodies that are specific to the protein of interest for detection. From 1979 to 1981, three papers were published with the first use of this method *(10–12)*. Generally, gel electrophoresis is performed first to separate proteins. The proteins are then transferred to a membrane (polyvinylidene fluoride and nitrocellulose membranes are commonly used). After the transfer, primary and secondary antibodies are added consecutively. Finally, the signal is produced by adding detection reagents (e.g., chromogenic and colorimetric) to the secondary antibodies. A general procedure is displayed in Figure 1. Recent technological updates to Western blotting include quantum dot conjugation *(13–15)*, microfluidic engineering *(16–20)*, and imagers (e.g., ChemiDoc™ and G:Box).

2.2 Non-Gel-Based Approaches

2.2.1 *Enzyme-Linked Immunosorbent Assays and Aptamer-Based Assays*

Enzyme-linked immunosorbent assay (ELISA) uses antibodies and surface binding to accomplish detection. This can be done by indirect antigen binding to the surface or direct antigen binding to antibodies on the surface *(21)*. Like Western blotting, the antigen is recognized with an antibody that has an enzyme covalently linked (e.g., horseradish peroxidase). The predecessor to ELISA was described in 1960 *(22)*. Recent advancements push the detection limit of ELISA to subfemtomolar concentrations *(23,24)*.

The popularity of aptamer-based assays has risen over the last several years in proteomics *(25)*. Typical aptamers are single-stranded nucleic acids that can recognize target molecules. The general basis is similar to immuno-based methods. These aptamers are found by screening large libraries of nucleic acid sequences. This technology has been applied to several proteome-level detections *(26–29)*.

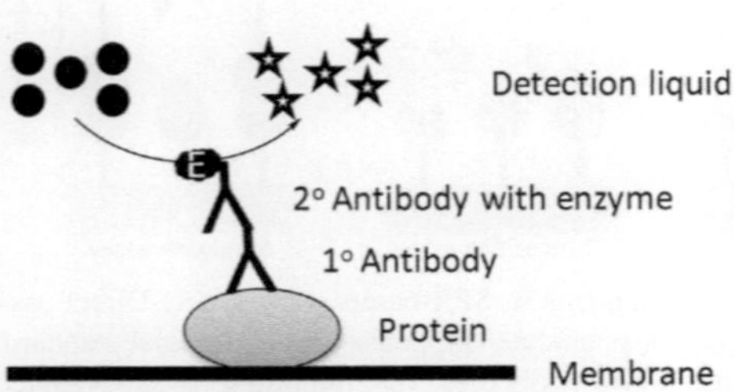

FIGURE 1 Generic Western blotting steps.

2.2.2 *Surface Plasmon Resonance Spectroscopy*

Surface plasmon resonance (SPR) spectroscopy is a phenomenon that occurs on an electron-rich metal surface (active surface) upon the impact of an incident light with a certain frequency. Techniques based on SPR detect any change of the refractive index on the active surface due to the adsorption of a target analyte *(30)*. For protein detection, antibodies are extensively used to modify the SPR-active surface. As detailed in Figure 2, the four major mechanisms of the antibody-based SPR reported are direct *(31)*, competitive *(32)*, sandwich *(33)*, and inhibition assays *(34)*. Using antibody arrays leads to an extension of the technique that is referred to as *surface plasmon resonance imaging*. This technique is a high-throughput and multiplexible version of SPR. Biological SPR is categorized as a label-free quantitation method. Thus, there is no need for elaborate sample preparation, which allows for fast, real-time, and reusable sensing *(35)*. This technique suffers from two common antibody-based detection problems: the high cost of immunograde antibodies and nonspecific binding of proteins. A vast variety of modifications and alterations have been made to the active surface of SPR to enhance the selectivity and sensitivity of the SPR sensor. The use of nanoparticles together with an immunodetection sandwich compartment is one popular approach *(36–38)*. This approach has been applied for amplified detection of total prostate-specific antigen *(39)*. Aptamers are an alternative to protein-based antibodies *(40)*. Nanostructures (i.e., nanowires, nanoholes, and nanolayers) on the active surface have been used to increase antigen binding *(41,42)*. Using these versatile signal enhancement techniques, achieving an attomolar limit of detection for proteins *(43)* is possible.

2.2.3 *Lab-on-a-Chip*

The incorporation of microfluidics into small devices that are capable of analysis or experimentation is commonly referred to as "lab-on-a-chip" *(44)*.

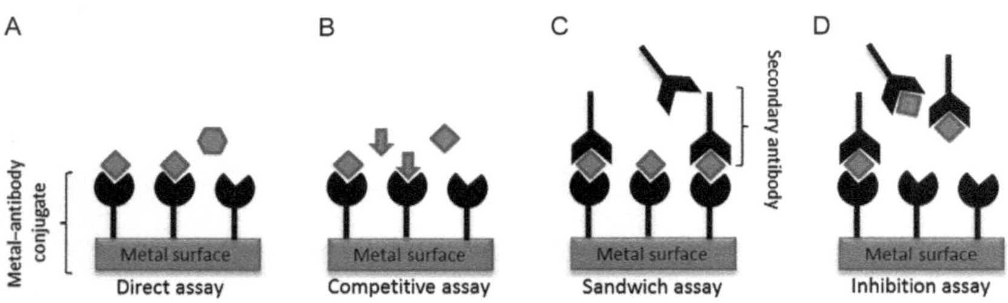

FIGURE 2 Detection mechanisms in SPR-based sensors. (A) Direct assay: direct detection of analyte. (B) Competitive assay: analyte competes with an internal standard. (C) Sandwich assay: analyte is trapped between two antibodies. (D) Inhibition assay: analyte is pretreated with an antibody.

These small devices have been used in a variety of ways. Liquid chromatography (LC) on-chip approaches are well represented in the literature *(45,46)*. These chips have also been used to study the membrane proteome *(47)* and protein–protein interaction via fluorescence resonance energy transfer *(48)*. Further examples of technology miniaturization can be found in journals dedicated to the on-chip methodology (i.e., *Lab-on-a-Chip*).

3 MASS SPECTROMETRY-BASED DETECTION METHODS

3.1 Introduction

Among various technologies used in quantitative proteomics, mass spectrometry (MS) steps out as the major player in the field. Fast, reproducible, and capable of analyzing highly complex samples are some advantages of MS-based quantitation methods *(49)*. Two major approaches in MS-based proteomic quantitation are derivatization-based and derivatization-free. Selecting a particular approach for proteomic analysis depends on many factors *(50)*. In a derivatization-free approach, there is no concern of the completion and efficiency of chemical reactions or subsequent sample cleanup. Thus, this approach delivers a faster and more straightforward solution for a large-scale primary study. Derivatization-based approaches, however, deliver more flexibility and improved sample throughput. Figure 3 categorizes different derivatization-free and derivatization-based quantitation methods. The two schools of proteomic research use either direct MS analysis of proteins (top-down proteomics) or analysis of peptides that resulted from proteolytic digestion of proteome samples (bottom-up proteomics or shotgun proteomics) *(51)*. However, the majority of quantitative analysis in proteomics uses the bottom-up principle, in which the peptide quantity is representative of the protein amount *(52)*. This chapter focuses only on bottom-up techniques for proteomic quantitation.

3.2 Mass Spectrometers

Mass spectrometers are composed of three main components: the ionization source, the mass analyzer, and the detector. With the development of different mass analyzers, the choice of mass spectrometer for a specific analysis becomes critical *(53)*, for example, if the analysis is to quantitatively profile all proteins in a sample or quantify a preselected set of protein targets. One or a combination of mass analyzers, such as time of flight (TOF), quadrupole (Q), ion trap (IT), ion cyclotron resonance, and Orbitrap^{TM}, can be implemented in MS or tandem MS (MS/MS).

For nontargeted quantitative proteomics, in which the protein profiling and relative quantitation are performed simultaneously, mass spectrometers with high accuracy and resolution such as Q-TOF, TOF-TOF, or Orbitrap^{TM} are

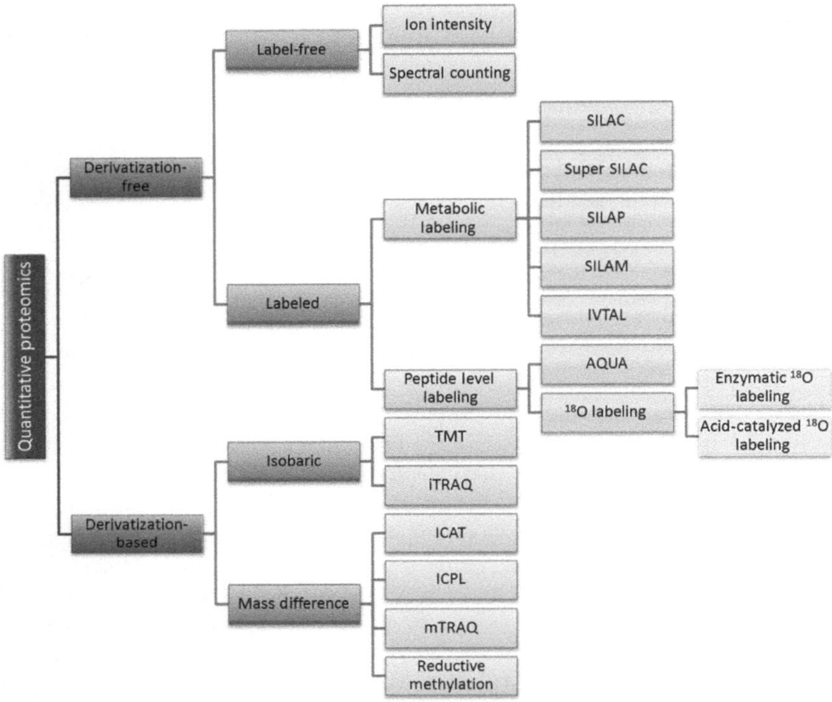

FIGURE 3 Categorization of MS-based proteomics quantitation approaches.

preferred. Mass-resolving power and accuracy, which are continuously being improved in new commercial instruments, are the important factors necessary for software-assisted protein identification with high statistical confidence *(54)*. TOF analyzers are the most traditional analyzer for delivering high mass accuracy and resolution, together with great speed of analysis *(55)*. An Orbitrap^{TM} analyzer works based on complex electrostatic fields applied between inner and outer electrodes to trap ions. Ions are resolved around the inner electrode depending on the applied electrostatic frequency and the mass-to-charge ratio (m/z) of ions *(56)*. Hybrid versions of the Orbitrap (e.g., linear ion trap-Orbitrap [LTQ-Orbitrap^{TM}]) have been utilized for protein identification and quantitation *(57)*. Based on the fundamentals of the IT, ion movements are controlled by alternating current (AC) and direct current (DC) electric fields that are dynamically trapping, stabilizing, or filtering out the ions depending on the mode of the analysis *(58)*. Q-TOF instruments are another type of tandem mass spectrometer for global protein profiling and quantitation *(59)*. The quadrupole is usually used for its ion filtering capability as the first mass analyzer in hybrid instruments. Quadrupole analyzers use AC and DC electric fields to stabilize or destabilize ions with different m/z ratios, thus often referred to as mass filters; these analyzers are typically operated at modes with low mass resolution and accuracy *(60)*.

Mass filtering capability of quadrupole analyzers makes triple quadruple (QqQ) mass spectrometers suitable for targeted proteomic quantitation. These instruments are used in the mode of multiple reaction monitoring (MRM) analysis *(61)*. Alternative approaches for targeted proteomic quantitation have been reported based on newly developed hybrid instruments. A quadrupole-Orbitrap platform (Q Exactive™) is able to deliver pseudo-MRM analysis. It can also perform the method of parallel reaction monitoring in which a full scan of MS/MS data is achievable from all the precursors and fragments in high resolution and the needed information is extracted after data acquisition *(62)*. A recent exploration of the Q-TOF instrument performs parallel selection and fragmentation of a range of precursor ions *(63–65)*.

3.3 Derivatization-Free Techniques

3.3.1 Label-Free Quantitation

Label-free quantitation is the simplest, fastest, and cleanest among all the MS-based quantitation methods. It has been frequently used in shotgun proteomics in which the protein profiling/quantitation is performed on a large scale and in a nontargeted fashion *(66)*. The sample goes through the least manipulation before MS analysis. The label-free quantitation can be done based on (1) ion intensity or peak area measurement in the chromatogram for an intact peptide and (2) spectral counting of the fragment ions from all the peptides (originated from a single protein) in an MS/MS-based analysis. Spectral counting is implemented in a data-dependent acquisition (DDA) mode. In a DDA tandem mode, selected precursor ions are subjected to fragmentation when their abundances are above a user-set, predefined level. DDA mode-based spectral counting shows good correlation with the protein abundance *(67)*. However, this method is not recommended for low-abundance proteins due to poor accuracy. Software to assist in label-free quantitation includes an ion intensity chromatogram measurement called *MS1 filtering*. This increases peak selection accuracy by incorporating the peptide retention time *(68)*. Another software program takes advantage of simultaneous spectral counting and a total ion chromatogram to enhance analysis quality *(69)*.

3.3.2 Label-Assisted Quantitation

With the label-free methods in the repertoire of quantitative proteomics practitioners, there is still a need for label-assisted methods to perform quantitation with improved accuracy and precision. Label-assisted methods incorporate stable isotope labels into internal standards (IS) for quantitation; this MS-based quantitation method is termed *stable isotope dilution*. The incorporation of stable isotope labels can be achieved either with or without chemical derivatization of proteome samples. Use of metabolic labeling, absolute quantitation (AQUA) peptides, and ^{18}O labeling are three common derivatization-free methods.

3.3.2.1 Metabolic Labeling

Metabolic labeling is capable of labeling the whole proteome in cells or an entire organism. Metabolic labeling of cells is accomplished by the addition of isotopically labeled amino acids into the cell culture media, which lacks the same, but nonisotopically labeled amino acids. Each stable isotope label is commonly 1 or 2 Da higher in mass, such as ^{15}N and ^{13}C. *Stable isotope labeling* with amino acids in cell culture (SILAC) is the most popular method of metabolic labeling *(70)*. The comprehensiveness and relative simplicity of SILAC makes it commonly used for the relative quantitation of proteome samples *(71–73)*. Labeling proteins is advantageous because it allows for more accurate quantitation due to the early addition of the IS in sample preparation. This mitigates the sample loss variations from the downstream manipulation, including protein fractionation and separation, digestion, and peptide separation. A drawback is that it is not applicable to tissue samples or biological fluids, such as plasma and urine. However, an altered approach that uses stable isotope-labeled proteome (SILAP) as the IS applies the same metabolic labeling concept to the secretome of a cell line(s) *(72)*. It was used to make a secretome IS from CAPAN-2 human pancreatic cancer-derived cells for relative quantitation of sera proteins from early-stage pancreatic cancer patients. SILAP has also been utilized to investigate different biomarkers for prostate cancer *(74)*. Stable isotope labeling of mammals (SILAM) is an adapted approach to make the metabolic labeling compatible with tissue analysis *(75)*. This was accomplished in rats with an ^{15}N-enriched spirulina diet. Providing an IS for tissue allows for quantitative pathway analysis *(76,77)*. Another developed solution for the tissue proteome analysis is super-SILAC *(71)*. In this approach, labeled proteins are provided from multiple cell lines and used as a combined IS. The cell lines are pooled together to better represent a tissue proteome. Another SILAC derivative is the method of *in vivo* termini amino acid labeling *(78)*. This method couples two different endoproteases (Arg-C and Lys-N) and two different cell culture media (heavy arginine and heavy lysine) to produce isobaric peptides that have varied MS fragment ions *(78)*.

3.3.2.2 Absolute Quantitation Peptides

AQUA allows for direct comparison of proteomic data obtained on different instruments and in different laboratories. The need to measure the absolute abundance of proteins is achievable using a targeted quantitation approach. A synthetic, isotopically labeled peptide is used as the IS. This peptide is the isotopic counterpart of a signature peptide, which represents a 1:1 ratio with the corresponding protein. There are a number of cautions that need to be considered to choose an endogenous peptide as a signature of the target protein. The signature peptide quantitatively represents the target; therefore, it should be a stoichiometric equivalence of the protein (i.e., lacking co/post-translational modifications [PTMs]). Digestion efficiency, peptide uniqueness, and a number of other peptide selection criteria should be considered in choosing the signature peptide *(79)*. This type of quantitation is often performed via tandem MS specifically on an MRM platform *(80)*. Unlike metabolic labeling, AQUA is applicable to quantitation of proteomes of almost every biological source such as tissues, plasma, or cells. AQUA using synthetic heavy peptides is among the most popular methods for quantitative analysis in system biology and therapeutics and biomarker development *(81–84)*.

3.3.2.3 ^{18}O Labeling

Isotope labeling of proteome samples can be achieved via enzymatic ^{18}O replacement of ^{16}O on carboxylic groups at C-termini of peptides from proteome digests *(85,86)*. The source of ^{18}O is $H_2^{18}O$ comprised buffer. Enzymatic ^{18}O labeling globally incorporates two ^{18}O atoms at the C-terminus carboxyl group of almost all of the peptides of a proteome digest. Use of enzymes in immobilized formats allows for efficient postlabeling removal and thus eliminating back-exchange of ^{18}O labels *(87)*. In-gel ^{18}O labeling has been found to facilitate protein identification and quantification *(88)*. ^{18}O labeling initially developed for relative quantitation has also been adapted for AQUA of proteins *(89,90)*. Studies of protein PTMs are another application area. ^{18}O labeling has been utilized for quantitation of phosphorylation *(91)* and deamidation *(92,93)*. Acid catalysis can also exchange peptidyl carboxylate oxygen atoms, but in a nonselective manner *(94,95)*. Due to drawbacks like residue and peptide sequence dependence of the labeling and chemical changes of peptides under acidic conditions (e.g., deamidation), the acid labeling should be limited to special applications *(90)*.

3.4 Derivatization-Based Techniques

3.4.1 Isobaric Tagging

Isobaric mass tagging reagents are the most widely used for derivatizing peptides in quantitative proteomics *(96)*. While the structure varies, most isobaric mass tagging reagents are comprised of three elements: (1) a mass reporter moiety with a unique number of heavy isotope (^{13}C or ^{15}N) substitutions, (2) a mass balancer moiety with complementary labels of heavy isotopes to balance the total mass of the reporter and balancer moieties, and (3) a reactive moiety to tag peptidyl amines. Two sets of amine-reactive isobaric mass tagging reagents, with a common reactive moiety of NHS ester, are

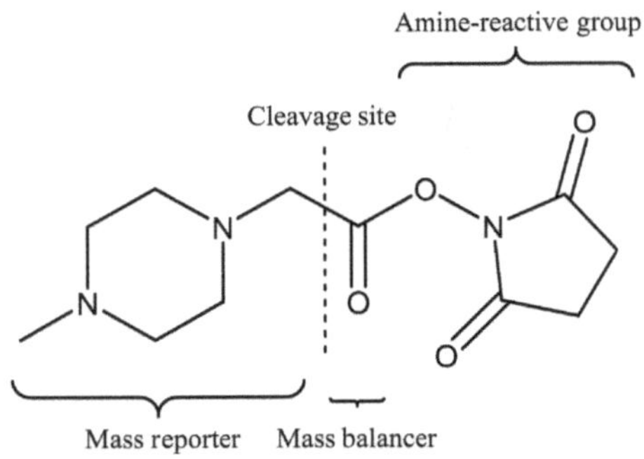

FIGURE 4 The structure of iTRAQ® reagents.

commercially available: tandem mass tags (TMT™) *(97)* and isobaric tags for relative and absolute quantitation (iTRAQ®) *(98)*. They are designed to label primary amines (N-termini and lysine's side chain amines) in peptides. One example of isobaric tags is shown in Figure 4.

A set of isobaric tags has identical intact masses and chemical properties that allow for coelution of differentially derivatized peptides during chromatography. Ions for the derivatized peptides have a common m/z and they can be co-selected by the first mass analyzer of tandem MS. The isobaric tags on derivatized peptide ions are then cleaved at a specific site via gas-phase dissociation. A set of reporter ions with unique numbers of heavy isotope labels are then detected. The relative signal intensities of the reporter ions are used to quantify the differentially labeled peptides and thus proteins. Up to eight samples can be analyzed in a single experiment using isobaric derivatization approaches. Recently, thiol-reactive TMT™ reagents coupled with immuno-purification techniques were reported *(99)*. Reagents that are structurally similar to iTRAQ®/TMT™, but use less-expensive deuterium labels, have the potential for further increase in sample throughput *(100)*.

3.4.2 Mass-Difference Reagents

Isotope-coded affinity tag (ICAT®), first invented in 1999, is one of the earliest chemical reagents introduced for quantitative proteomics at the protein level *(101)*. The original ICAT® reagents include a thiol-specific reactive group, an isotopically modified linker, and a biotin tag for the enrichment of the tagged peptides. After derivatization, both heavy- and light-labeled proteins are pooled and enzymatically digested into peptides. The tagged, cysteine-containing peptides are then enriched with avidin affinity chromatography and quantified via MS. Cleavable versions of this reagent were later developed to improve the recovery of the tagged peptides *(102)*. The heavy version of the so-called cICAT reagents contains nine ^{13}C instead of eight

^{2}H and an acid-cleavable biotin moiety. Insertion of an acid-labile linker greatly decreases the elution of nonspecifically bound peptides from avidin affinity materials. Isotope-coded protein labels differentially label all of the free amino groups on proteins for relative quantitation *(103)*. This method does not require an affinity tag.

The so-called mTRAQ® reagents are developed for introducing differential isotope labels in a mass different manner and are used in MRM-MS quantitation *(104)*. These reagents have the same chemical structure as those for iTRAQ® reagents, but carry redesigned stable isotope labeling states. They can also be used for peptide quantitation based on the signal intensity of precursor ions. Reductive methylation of peptidyl amines is another commonly applied technique to introduce differential isotope labels to peptides. Derivatized peptides preferentially produce a_1 ions serving as quantitation reporters *(105)*. These reporter ions are peptide specific, thus lessening interference from coeluting peptides with the same or similar mass. Reductive dimethylation has been applied to analyze low-abundance phosphopeptides *(106)*.

3.5 Activity-Based Probes

Activity-based protein profiling (ABPP) technology utilizes active site-directed chemical probes, combined with a range of analytical methods, for functional proteome analysis. These small molecules are commonly referred to as activity-based probes (ABPs). A typical structure of ABPs is comprised of three elements: (1) a reactive group for forming covalent adducts with functional groups at the active site of enzymes, (2) a recognizing moiety for noncovalent interactions with proteins, and (3) a reporting tag for detection and/or separation. Changes in recognizing structural moiety can vary the reactivity and/or selectivity of the probes *(107)*. Small "clickable" handles such as an alkyne can also be incorporated as dormant reporting groups for coupling with quantitation or separation tags after the *in vivo* reaction of the probes with living cells *(108)*.

One major advantage of ABPP *(109,110)* over the previous techniques *(111,112)* is the ability to directly monitor active sites on a multitude of enzyme classes, and the analysis depends on the protein activity not on concentration or abundance. Furthermore, ABPP can be combined with tandem MS analysis for enriched enzyme identification. One approach, termed *activity-based protein profiling by multidimensional protein identification technology* (ABPP-MudPIT), was used to profile inhibitor selectivity against hundreds of targets simultaneously via *in vitro* or *in vivo* reactions *(113)*. The integrated design of stable isotopic ABPs allows for easy implementation of quantitative MS measurements in biological samples *(114,115)*. Another approach, termed *isotopic tandem orthogonal proteolysis–activity-based protein profiling*, was used to quantitatively profile the intrinsic reactivity of cysteine residues across multiple proteomes by using a heavy isotope-labeled valine and a protease-cleavable linker *(116)*.

3.6 Sample Preparation

Generically, sample preparation in bottom-up proteomics involves the solubilization of proteins from the biological source, protein fractionation, subsequent enzymatic digestion of the proteins, and separation of the resulting peptides. Unique enrichment techniques can be applied in many of these preparatory steps (Figure 5).

3.6.1 Pre-Lysis Techniques

3.6.1.1 Recombinant Fusion Proteins

The use of recombinant fusion proteins for protein expression determination and purification is a well-established technique *(117)*. It uses recombinant DNA technology to fuse a "tag" to a protein substrate. Commonly used tags are green fluorescent protein (GFP), polyhistidine (His-tag), glutathione *s*-transferase, and FLAG®. A comprehensive list can be found in the literature *(118)*. Using these proteins for MS-based quantitation of their tagged substrates is possible. GFP is a small protein that can produce green fluorescence when illuminated under certain wavelengths. This qualitative measurement

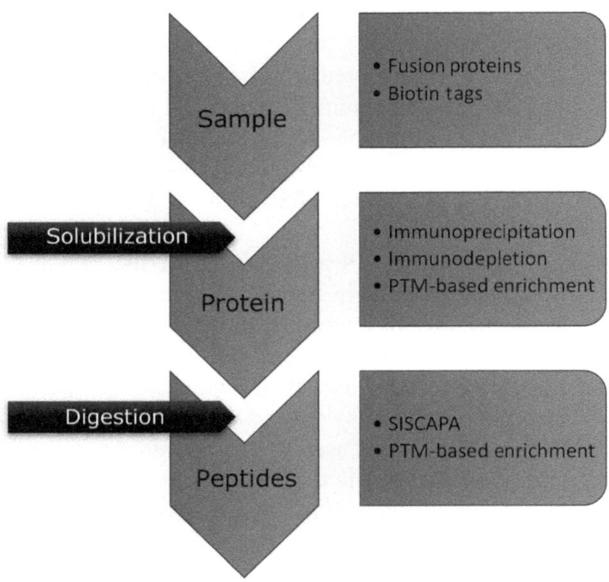

FIGURE 5 Enrichment techniques in proteome analysis.

has been transcended into a quantitative measurement in MS by introducing SILAC *(119)* or an amine-specific isotope tag termed *iTAG (120)*. The inherit fluorescence of GFP was also quantitatively measured and protein-binding partners were determined with shotgun proteomics *(121)*. All of these measurements exploit immunoprecipitation (IP) of the GFP fusion protein. There are commercial products (e.g., GFP-Trap®) that recognize GFP and pull-down protein complexes. His-tag can also be used similarly. The original use of recombinant polyhistidine protein with immobilized metal ion affinity chromatography was described in 1988 *(122)*. This was applied to purify proteins by adding a small tag. It can also be used for the same purpose, but with subsequent MS analysis. There are, however, few quantitative examples of this *(123)*.

3.6.1.2 Biotin Tagging

Chemical derivatization of proteins can be performed at the cellular level. There are many forms of chemical "tagging," but biotin tagging is commonly used. Biotin has a very low dissociation rate ($K_D \approx 10^{-15}$ M) from avidin *(124)*. This kinetic property is the basis for enrichment of biotin-tagged proteins with avidin on a solid support (e.g., agarose beads). The nonbound proteins can then be washed away, and the bound proteins eluted. By functionalizing the biotin moiety with reactive groups, biotin tagging can be performed on specific sites of proteins. Two examples that use functionalized biotin in quantitative proteomics are experiments measuring lysine exposure on a peptide *(125)* and cell surface protein expression *(126)*.

3.6.2 Protein-Level Enrichment

3.6.2.1 Antibody-Based Enrichment

Co-immunoprecipitation is a common technique to enrich protein interaction complexes. This method uses an antibody to bind a known protein, which in turn has known or unknown binding partners. Shotgun proteomics is then performed to determine the enriched proteins. Quantitative MS is used to validate bound proteins and measure binding stoichiometry for proteins in the complexes prepared via IP *(127,128)*. Quantitation approaches used include SILAM *(129)*, label-free *(130)*, and AQUA peptides with a normalization step *(131)*.

3.6.2.2 Immunodepletion

Proteome samples often have very large ranges in protein concentration. Dominant, high-abundance proteins can compromise the analysis of minor protein components. For instance, plasma is composed of a complex mixture of proteins and albumin and immunoglobulin G compose over 60% of the total protein amount in plasma *(132)*. These abundant proteins can make the analysis of medium- to low-abundance proteins very difficult *(133,134)*.

Depletion of these and several other abundant proteins allow for in-depth analysis of low-abundance ones. There are many commercial products for protein immunodepletion. Applications of the products are reported for a Seppro® column in a label-free method *(135)*, Seppro® column in a SILAP method *(136)*, a Hu-6 Multiple Affinity Removal System (MARS) kit in an iTRAQ® method *(137)*, and a Hu-6 MARS kit in another iTRAQ®-based method *(138)*. Caution should be observed if a low-abundance protein has affinity to a high-abundance protein because it can be co-depleted.

3.6.2.3 Techniques Based on PTMs

PTMs of proteins enable many important functions in organisms *(139)*. Proteins with PTMs can be enriched in a variety of ways: antibody-based, chemical derivatization, and ionic interaction-based *(89)*. Antibodies that are highly specific to a particular PTM site help functional assays greatly. Antibodies that preferably bind a class of PTM (e.g., proteins carrying phosphotyrosine residues) make possible the in-depth quantitative proteomic analysis of these proteins. The substoichiometric nature of PTMs yields a low concentration of modified proteins. Enrichment of these proteins allows for improved analysis.

3.6.3 Peptide-Level Enrichment

3.6.3.1 Stable Isotope Standards and Capture by Anti-Peptide Antibodies

Immunoenrichment can be applied at the peptide level, particularly in the stable isotope standards and capture by anti-peptide antibodies (SISCAPA) method *(140)*. Target peptides together with their stable isotope-labeled counterparts (used as quantitation references) are enriched by antibodies that are developed against antigens with almost the same sequences. The antibodies are often covalently attached to magnetic beads for sample cleanup. This technique boasts an improved limit of quantitation and increased inter-laboratory reproducibility *(141)*.

3.6.3.2 Techniques Based on PTMs

Many of the techniques used to enrich PTMs at the protein level are applicable at the peptide level. A popular method for enrichment of phosphopeptides is to use an immobilized metal affinity column. The molecular basis for the enrichment is the phosphate affinity to transition metal ions, such as copper, nickel, cobalt, iron, aluminum, gallium, and zinc *(142)*. A comprehensive source for phosphopeptide enrichment strategies can be found in a recent review *(143)*. Lectin-based chromatography is used to enrich for glycopeptides *(144)*. Lectins are proteins that have an affinity for carbohydrates. With multiple available enrichment options, choosing the right one for the experiment is important. For determining which proteins have a certain PTM, protein-level enrichment can be performed. However, if a type of PTM is of interest, then peptide-level enrichment will yield less sample complexity. The use of multiple antibodies to enrich for multiple PTMs simultaneously has led to the development of PTMScan® technology *(145)*. A monoclonal antibody specific to the diglycine tag on lysine residues has also been used to enrich ubiquitinated tryptic peptides *(146)*.

3.7 Liquid Chromatography

In shotgun (bottom-up) proteomics workflows, proteins are digested by proteases and the resulting peptides are separated by LC before MS analysis *(147)*. With the rapid advancement in LC and MS instrumentation, LC-based proteomics overwhelms the 2-DE-based approaches. The shotgun proteomics pipeline can provide improved throughput, sensitive protein profiling, and accurate protein quantitation. When working with complex samples, such as human serum, the bottom-up approach produces thousands of peptides with very different abundances and some of the peptides have similar *m*/*z* ratios. This requires high-resolution separation of peptides prior to MS analysis. LC reduces ion suppression of coeluting peptides during ionization and increases the likelihood of detecting low-abundance peptides *(147)*. Different LC methods have been developed for proteomic analyses such as reversed-phase (RP), ion exchange (IEX), and hydrophilic interaction chromatography (HILIC). RP separations are the dominant choice due to the very high-resolving power and electrospray ionization (ESI) MS-friendly solvents.

3.7.1 Common Separation Modes

3.7.1.1 RP Chromatography

Introduced in 1976, RP chromatography is the most-used approach for peptide separation *(147,148)*. Peptides are separated in an RP column upon partitioning between the hydrophobic stationary phase and the mobile phase. Peptides are loaded onto the column using low organic solvent, allowing for initial desalting and concentration of the sample. Separation and elution of the peptides occur by increasing the percentage of organic solvent in the mobile phase; commonly used solvents are acetonitrile and methanol due to their appropriate solvent strength, volatility, viscosity, and MS compatibility. Acidic modifiers help improve resolution of RP chromatography of peptides by acting as ion pairing agents for positively charged groups of peptides. RP stationary phases are mostly silica-based C18 resin and acid additives help improve peak resolution by keeping residual silanol groups protonated. From a separation perspective, trifluoroacetic acid has a higher hydrophobicity and enables narrower peaks than other acids like acetic acid and formic acid (FA). FA, however, is the common choice for online use with ESI-MS to minimize ion suppression.

3.7.1.2 IEX Chromatography

Peptide separation in IEX chromatography is based on electrostatic interactions between charged groups on the peptides and the charged stationary phase. The mobile phase can be a buffer whose salt strength changes during separation. Separation can also be achieved by changing the pH value of the mobile phase to neutralize the charge of the analytes and/or stationary phase. IEX chromatography is classified into cation exchange (CX) and anion exchange chromatography *(149)*. CX is divided into two categories: strong cation exchange (SCX) and weak cation exchange (WCX). SCX chromatography utilizes strong acids such as sulfonic acid derivatives and is the second most-used chromatographic method in LC-based proteomic analysis. WCX, on the other hand, is restricted to a smaller pH range and its application is limited *(147,150,151)*.

3.7.1.3 Hydrophilic Interactions Chromatography

HILIC uses a polar stationary phase to separate proteins and peptides. Samples are loaded onto a column using a low aqueous mobile phase and are eluted by increasing the water content of the mobile phase *(152)*. HILIC is very suitable for separation of polar species such as sialic acid–containing glycopeptides, which have limited retention on most RP materials *(153)*. A variant of HILIC is electrostatic repulsion–hydrophobic interactions chromatography and it utilizes IEX stationary phase and a high organic mobile phase *(152,154)*.

3.7.2 Multidimensional LC Strategies

When working with complex peptide mixtures like a proteome digest, multiple separations help with analysis of minor components in the mixture. Coupling two different modes of LC results in the improved separation of peptides and thus a reduced number of coeluting peptides for MS analysis. To utilize the maximum separation space afforded by multidimensional LC, orthogonality of different chromatography modes that separate peptides based on largely different properties is important *(155)*. For example, SCX-RP is more orthogonal than an RP–RP combination. Multidimensional LC can be achieved by coupling two or more one-dimensional techniques either online or offline *(147)*. LC separation of peptides that utilizes different protonation states of peptides at low and high pH in an RP–RP combination is also attractive. This combination avoids the use of MS-unfriendly salts used in SCX-RP separations *(156,157)*. Figure 6 illustrates the first MudPIT system for the direct analysis of large protein complexes *(158,159)*.

3.7.3 LC Using New Separation Media

Ultra performance liquid chromatography (UPLC) is an important advancement in chromatography due to the significantly improved speed, resolution, and sensitivity *(160)*. UPLC typically uses separation media with a particle

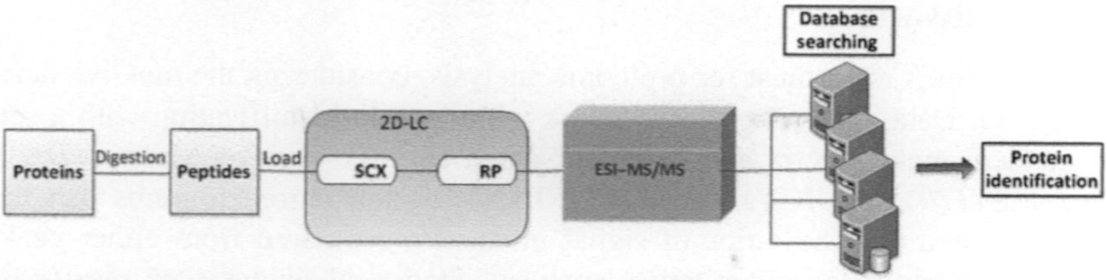

FIGURE 6 Online two-dimensional chromatography systems coupled with MS/MS.

size less than 2 μm *(161,162)*. UPLC systems operate under very high pressure to accommodate the use of small particles. This technology allows for faster flow rates while maintaining the separation efficiency and thus increasing the sample throughput of analysis *(161,163)*.

As an alternative to wholly porous sub-2-μm particles, 1.7-μm fused-core particles surrounded by a 0.5-μm porous silica layer with 90 Å pores, have emerged. In a comparative study, substantially lower back pressure was reported when the fused-core particles were used. This allowed for columns to be coupled in series, which increased the peak efficiency up to 92,750 plates *(164)*. Wider-pore fused-core particles have an average pore diameter of 160 Å. The wider-pore particles are particularly useful for increased sample loading and the rapid separation of peptides using volatile mobile phases *(165)*.

3.7.4 Nanoscale LC

Nanoscale liquid chromatography (nano-LC) using RP separation materials is the default choice for LC-MS of peptides from proteome digests. Nano-LC has excellent sensitivity using 50–100 μm diameter columns and 200–300 nL/min flow rates *(166)*. An all-inclusive definition for nano-LC is an LC that separates nanoliter amounts of sample, with nanoliter amounts of mobile phase, and yields detection limits in the range of nanograms per milliliter *(167)*.

3.7.5 GeLC Method

In this method, proteins in cell lysates or complex mixtures are typically first separated by SDS-PAGE. Followed by in-gel enzymatic digestion of the separated proteins, the resulting peptides are extracted for one-dimensional or multidimensional LC separation before MS analysis. This approach takes advantage of high-resolution SDS-PAGE for proteins under denaturing condition, thus reducing the sample complexity at the protein level. LC separations further reduce sample complexity at the peptide level. Also, with staining gel bands, it is possible to separate gel pieces containing high-abundance proteins *(158)*. This technique can be applied to two-dimensional gels as well *(168)*.

3.8 Software

Software tools are a must for proteome analysis, considering the massive data produced. Data analysis typically starts with peptide identification with a set of tandem mass spectra and database search engines, such as Mascot *(169)*, Sequest *(170)*, and X!Tandem *(171)*. Extracted ion chromatograms can be reconstructed and the ratios of signal intensities estimated from either peak heights or areas for quantitative analysis. Statistical analysis of results is essential, including assignment of significant scores, normalization of peptide abundances, and statistical evaluation of quantitation results, as well as function and location of proteins. Significant development of MS-based proteomic technologies in the past decade encouraged the concurrent advance in data analysis tools *(172–174)*. Further reading on bioinformatics and other computing concerns for proteomics can be found elsewhere in this book.

3.9 Analyte Multiplexing and Sample Throughput

Contemporary proteomics demands high-throughput and multiplexed analytical strategies and platforms. In the early stages of the biomarker discovery pipeline, for instance, multiplexed analysis of peptides from thousands of proteins in a few samples is required. However, further down the pipeline there is a shift toward high-throughput analysis because the number of samples increases dramatically and the number of peptides or analytes decreases *(175)*. MRM-MS methods currently dominate MS-based high-throughput quantitative proteomics *(176,177)*. Efforts in developing MRM approaches to perform both multiplexed and high-throughput analysis via either immuno-enrichment *(178)* or ultra-throughput MRM *(179)* are reported. New MS strategies to improve the sample throughput of proteomic analysis are emerging using recently available mass spectrometers.

4 CONCLUSIONS

The field of proteomics has evolved significantly thanks to collective efforts from practitioners with eclectic backgrounds. Technological advances in every step of the proteome analysis pipeline, from sample preparation to MS analysis to protein identification and quantitation to protein functional denotation, have significantly reduced the induction period of initial learning and encouraged the widespread adaption of proteomics. Fruitful applications continuously emerge in the study of biomarkers, disease, medicine, food and plants, and the environment, just to name a few. The inherit sample complexity and diversity still present both technical challenges and opportunities for in-depth analysis of the entire proteome.

ACKNOWLEDGMENTS

Financial supports from the National Cancer Institute/National Institutes of Health and the Cystic Fibrosis Foundation are greatly appreciated.

REFERENCES

1. Laemmli, U. K. *Nature* **1970**, *227*, 680–685.
2. Righetti, P. G.; Candiano, G. *J. Chromatogr. A* **2011**, *1218*, 8727–8737.
3. O'Farrell, P. H. *J. Biol. Chem.* **1975**, *250*, 4007–4021.
4. Monteoliva, L.; Albar, J. P. *Brief. Funct. Genomic. Proteomic.* **2004**, *3*, 220–239.
5. Ünlü, M.; Morgan, M. E.; Minden, J. S. *Electrophoresis* **1997**, *18*, 2071–2077.
6. Minden, J. S.; Dowd, S. R.; Meyer, H. E.; Stühler, K. *Electrophoresis* **2009**, *30*, S156–S161.
7. Tannu, N. S.; Hemby, S. E. *Nat. Protoc.* **2006**, *1*, 1732–1742.
8. Van den Bergh, G.; Clerens, S.; Vandesande, F.; Arckens, L. *Electrophoresis* **2003**, *24*, 1471–1481.
9. Tonge, R.; Shaw, J.; Middleton, B.; Rowlinson, R.; Rayner, S.; Young, J.; Pognan, F.; Hawkins, E.; Currie, I.; Davison, M.; Young, J.; Pognan, F.; Hawkins, E.; Currie, I.; Davison, M. *Proteomics* **2001**, *1*, 377–396.
10. Burnette, W. *Anal. Biochem.* **1981**, *112*, 195–203.
11. Renart, J.; Reiser, J.; Stark, G. R. *Proc. Natl. Acad. Sci. U. S. A.* **1979**, *76*, 3116–3120.
12. Towbin, H.; Staehelin, T.; Gordon, J. *Proc. Natl. Acad. Sci. U. S. A.* **1979**, *76*, 4350–4354.
13. Kale, S.; Kale, A.; Gholap, H.; Rana, A.; Desai, R.; Banpurkar, A.; Ogale, S.; Shastry, P. *J. Nanopart. Res.* **2012**, *14*, 732/1–732/15.
14. Chen, W.; Xu, D.; Liu, L.; Peng, C.; Zhu, Y.; Ma, W.; Bian, A.; Li, Z.; Yuanyuan; Jin, Z.; Zhu, S.; Xu, C.; Wang, L. *Anal. Chem.* **2009**, *81*, 9194–9198.
15. Ornberg, R. L.; Harper, T. F.; Liu, H. *Nat. Methods* **2005**, *2*, 79–81.
16. Pan, W.; Chen, W.; Jiang, X. *Anal. Chem.* **2010**, *82*, 3974–3976.
17. Hou, C.; Herr, A. E. *Analyst* **2013**, *138*, 158–163.
18. Hughes, A. J.; Herr, A. E. *Proc. Natl. Acad. Sci. U. S. A.* **2012**, *109*, 21450–21455.
19. Kim, D.; Karns, K.; Tia, S. Q.; He, M.; Herr, A. E. *Anal. Chem.* **2012**, *84*, 2533–2540.
20. Tia, S. Q.; He, M.; Kim, D.; Herr, A. E. *Anal. Chem.* **2011**, *83*, 3581–3588.
21. Voller, A.; Bartlett, A.; Bidwell, D. E. *J. Clin. Pathol.* **1978**, *31*, 507–520.
22. Yalow, R. S.; Berson, S. A. *J. Clin. Invest.* **1960**, *39*, 1157–1175.
23. Rissin, D. M.; Kan, C. W.; Campbell, T. G.; Howes, S. C.; Fournier, D. R.; Song, L.; Piech, T.; Patel, P. P.; Chang, L.; Rivnak, A. J.; Ferrell, E. P.; Randall, J. D.; Provuncher, G. K.; Walt, D. R.; Duffy, D. C. *Nat. Biotechnol.* **2010**, *28*, 595–599.
24. Rissin, D. M.; Fournier, D. R.; Piech, T.; Kan, C. W.; Campbell, T. G.; Song, L.; Chang, L.; Rivnak, A. J.; Patel, P. P.; Provuncher, G. K.; Ferrell, E. P.; Howes, S. C.; Pink, B. A.; Minnehan, K. A.; Wilson, D. H.; Duffy, D. C. *Anal. Chem.* **2011**, *83*, 2279–2285.
25. Wilson, R. *J. Proteomics* **2011**, *74*, 1852–1854.
26. Bruno, J. G.; Carrillo, M. P.; Phillips, T.; Edge, A. *J. Biomol. Technol.* **2011**, *22*, 27–36.
27. Gold, L.; Ayers, D.; Bertino, J.; Bock, C.; Bock, A.; Brody, E. N.; Carter, J.; Dalby, A. B.; Eaton, B. E.; Fitzwater, T.; Flather, D.; Forbes, A.; Foreman, T.; Fowler, C.; Gawande, B.; Goss, M.; Gunn, M.; Gupta, S.; Halladay, D.; Heil, J.; Heilig, J.; Hicke, B.; Husar, G.; Janjic, N.; Jarvis, T.; Jennings, S.; Katilius, E.; Keeney, T. R.; Kim, N.; Koch, T. H.; Kraemer, S.; Kroiss, L.; Le, N.; Levine, D.; Lindsey, W.; Lollo, B.; Mayfield, W.; Mehan, M.; Mehler, R.; Nelson, S. K.; Nelson, M.; Nieuwlandt, D.; Nikrad, M.;

Ochsner, U.; Ostroff, R. M.; Otis, M.; Parker, T.; Pietrasiewicz, S.; Resnicow, D. I.; Rohloff, J.; Sanders, G.; Sattin, S.; Schneider, D.; Singer, B.; Stanton, M.; Sterkel, A.; Stewart, A.; Stratford, S.; Vaught, J. D.; Vrkljan, M.; Walker, J. J.; Watrobka, M.; Waugh, S.; Weiss, A.; Wilcox, S. K.; Wolfson, A.; Wolk, S. K.; Zhang, C.; Zichi, D. *PLoS One* **2010**, *5*, e15004.

28. Janssen, K. P.; Knez, K.; Spasic, D.; Schrooten, J.; Lammertyn, J. *Anal. Bioanal. Chem.* **2012**, *404*, 2073–2081.
29. Srinivas, R. L.; Chapin, S. C.; Doyle, P. S. *Anal. Chem.* **2011**, *83*, 9138–9145.
30. Abdulhalim, I.; Zouroba, M.; Lakhtakia, A. *Electromagnetics* **2008**, *28*, 214–242.
31. Retra, K.; Geitmann, M.; Kool, J.; Smit, A. B.; de Esch, I. J.; Danielson, U. H.; Irth, H. *Anal. Biochem.* **2010**, *407*, 58–64.
32. Estevez, M.; Belenguer, J.; Gomez-Montes, S.; Miralles, J.; Escuela, A. M.; Montoya, A.; Lechuga, L. M. *Analyst* **2012**, *137*, 5659–5665.
33. Gnedenko, O. V.; Mezentsev, Y. V.; Molnar, A. A.; Lisitsa, A. V.; Ivanov, A. S.; Archakov, A. I. *Anal. Chim. Acta* **2012**, *759*, 105–109.
34. Alterman, M.; Sjöbom, H.; Säfsten, P.; Markgren, P.; Danielson, U. H.; Hämäläinen, M.; Löfås, S.; Hultén, J.; Classon, B.; Samuelsson, B.; Hallberg, A. *Eur. J. Pharm. Sci.* **2001**, *13*, 203–212.
35. Lausted, C.; Hu, Z.; Hood, L. *Mol. Cell. Proteomics* **2008**, *7*, 2464–2474.
36. Law, W. C.; Yong, K. T.; Baev, A.; Prasad, P. N. *ACS Nano* **2011**, *5*, 4858–4864.
37. Hong, X.; Hall, E. A. *Analyst* **2012**, *137*, 4712–4719.
38. Yang, C.; Brooks, E.; Li, Y.; Denny, P.; Ho, C.; Qi, F.; Shi, W.; Wolinsky, L.; Wu, B.; Wong, D. T. W.; Montemagno, C. D. *Lab Chip* **2005**, *5*, 1017–1023.
39. Uludag, Y.; Tothill, I. E. *Anal. Chem.* **2012**, *84*, 5898–5904.
40. Li, Y.; Lee, H. J.; Corn, R. M. *Anal. Chem.* **2007**, *79*, 1082–1088.
41. Byun, K. M.; Yoon, S. J.; Kim, D.; Kim, S. J. *Opt. Lett.* **2007**, *32*, 1902–1904.
42. Ferreira, J.; Santos, M. J.; Rahman, M. M.; Brolo, A. G.; Gordon, R.; Sinton, D.; Girotto, E. M. *J. Am. Chem. Soc.* **2008**, *131*, 436–437.
43. Sim, H. R.; Wark, A. W.; Lee, H. J. *Analyst* **2010**, *135*, 2528–2532.
44. Becker, H.; Gartner, C. *Sci. Prog.* **2012**, *95*, 175–198.
45. Gama, M. R.; Collins, C. H.; Bottoli, C. B. *J. Chromatogr. Sci.* **2013**, *51*, 694–703.
46. Li, Y.; Zhang, Y.; Qiu, F.; Qiu, Z. *Electrophoresis* **2011**, *32*, 1976–1983.
47. Ohlsson, G.; Tabaei, S. R.; Beech, J.; Kvassman, J.; Johanson, U.; Kjellbom, P.; Tegenfeldt, J. O.; Höök, F. *Lab Chip* **2012**, *12*, 4635–4643.
48. Benz, C.; Retzbach, H.; Nagl, S.; Belder, D. *Lab Chip* **2013**, *13*, 2808–2814.
49. Pailleux, F.; Beaudry, F. *Biomed. Chromatogr.* **2012**, *26*, 881–891.
50. Yao, X. *Anal. Chem.* **2011**, *83*, 4427–4439.
51. McLafferty, F. W.; Breuker, K.; Jin, M.; Han, X.; Infusini, G.; Jiang, H.; Kong, X.; Begley, T. P. *FEBS J.* **2007**, *274*, 6256–6268.
52. Yates, J. R.; Ruse, C. I.; Nakorchevsky, A. *Annu. Rev. Biomed. Eng.* **2009**, *11*, 49–79.
53. Domon, B.; Aebersold, R. *Sci. Signal.* **2006**, *312*, 212.
54. Thompson, J. D.; Schaeffer-Reiss, C.; Ueffing, M., Eds.; *Functional Proteomics*; Totowa: New Jersey, 2008.
55. Standing, K. *Int. J. Mass Spectrom.* **2000**, *200*, 597–610.
56. Yates, J. R.; Cociorva, D.; Liao, L.; Zabrouskov, V. *Anal. Chem.* **2006**, *78*, 493–500.
57. Lu, B.; Motoyama, A.; Ruse, C.; Venable, J.; Yates, J. R. *Anal. Chem.* **2008**, *80*, 2018–2025.
58. March, R. E. *J. Mass Spectrom.* **1997**, *32*, 351–369.
59. Chernushevich, I. V.; Loboda, A. V.; Thomson, B. A. *J. Mass Spectrom.* **2001**, *36*, 849–865.

60. Miller, P. E.; Denton, M. B. *J. Chem. Educ.* **1986**, *63*, 617.
61. Kuzyk, M. A.; Smith, D.; Yang, J.; Cross, T. J.; Jackson, A. M.; Hardie, D. B.; Anderson, N. L.; Borchers, C. H. *Mol. Cell. Proteomics* **2009**, *8*, 1860–1877.
62. Michalski, A.; Damoc, E.; Hauschild, J.; Lange, O.; Wieghaus, A.; Makarov, A.; Nagaraj, N.; Cox, J.; Mann, M.; Horning, S. *Mol. Cell. Proteomics* **2011**, *10*, M111.011015. http://dx.doi.org/10.1074/mcp.M111.011015.
63. Liu, Y.; Hüttenhain, R.; Surinova, S.; Gillet, L. C.; Mouritsen, J.; Brunner, R.; Navarro, P.; Aebersold, R. *Proteomics* **2013**, *8*, 1247–1256.
64. Gillet, L. C.; Navarro, P.; Tate, S.; Röst, H.; Selevsek, N.; Reiter, L.; Bonner, R.; Aebersold, R. *Mol. Cell. Proteomics* **2012**, *11*. http://dx.doi.org/10.1074/mcp.O111.016717.
65. Held, J. M.; Schilling, B.; D'Souza, A. K.; Srinivasan, T.; Behring, J. B.; Sorensen, D. J.; Benz, C. C.; Gibson, B. W. *Int. J. Proteomics* **2013**, *2013*. http://dx.doi.org/10.1155/2013/791985.
66. Zhu, W.; Smith, J. W.; Huang, C. M. *J. Biomed. Biotechnol.* **2010**, *2010*. http://dx.doi.org/10.1155/2010/840518.
67. Liu, H.; Sadygov, R. G.; Yates, J. R. *Anal. Chem.* **2004**, *76*, 4193–4201.
68. Schilling, B.; Rardin, M. J.; MacLean, B. X.; Zawadzka, A. M.; Frewen, B. E.; Cusack, M. P.; Sorensen, D. J.; Bereman, M. S.; Jing, E.; Wu, C. C.; Verdin, E.; Kahn, C. R.; Maccoss, M. J.; Gibson, B. W. *Mol. Cell. Proteomics* **2012**, *11*, 202–214.
69. Freund, D. M.; Prenni, J. E. *J. Proteome Res.* **2013**, *12*, 1996–2004.
70. Oda, Y.; Huang, K.; Cross, F. R.; Cowburn, D.; Chait, B. T. *Proc. Natl. Acad. Sci. U. S. A.* **1999**, *96*, 6591–6596.
71. Geiger, T.; Madden, S. F.; Gallagher, W. M.; Cox, J.; Mann, M. *Cancer Res.* **2012**, *72*, 2428–2439.
72. Yu, K. H.; Barry, C. G.; Austin, D.; Busch, C. M.; Sangar, V.; Rustgi, A. K.; Blair, I. A. *J. Proteome Res.* **2009**, *8*, 1565–1576.
73. Wang, C. I.; Chien, K. Y.; Wang, C. L.; Liu, H. P.; Cheng, C. C.; Chang, Y. S.; Jau-Song, Y.; Chia-Jung, Y. *Mol. Cell. Proteomics* **2012**, *11*, 1105–1122.
74. Zhao, T.; Zeng, X.; Bateman, N. W.; Sun, M.; Teng, P. N.; Bigbee, W. L.; Dhir, R.; Nelson, J. B.; Conrads, T. P.; Hood, B. L. *J. Proteome Res.* **2012**, *11*, 1089–1099.
75. McClatchy, D. B.; Yates, J. R. *Cold Spring Harb. Protoc.* **2008**, *3*. http://dx.doi.org/10.1101/pdb.prot4940.
76. Rauniyar, N.; McClatchy, D. B.; Yates, J. R. *Methods* **2013**, *61*, 260–268.
77. Rayavarapu, S.; Coley, W.; Cakir, E.; Jahnke, V.; Takeda, S.; Aoki, Y.; Grodish-Dressman, H.; Jaiswal, J. K.; Hoffman, E. P.; Brown, K. J.; Hathout, Y.; Nagaraju, K. *Mol. Cell. Proteomics* **2013**, *12*, 1061–1073.
78. Nie, A. Y.; Zhang, L.; Yan, G. Q.; Yao, J.; Zhang, Y.; Lu, H. J.; Yang, P. Y.; He, F. C. *Anal. Chem.* **2011**, *83*, 6026–6033.
79. Kamiie, J.; Ohtsuki, S.; Iwase, R.; Ohmine, K.; Katsukura, Y.; Yanai, K.; Sekine, Y.; Uchida, Y.; Ito, S.; Terasaki, T. *Pharm. Res.* **2008**, *25*, 1469–1483.
80. Schmidt, C.; Urlaub, H. *Methods Mol. Biol.* **2012**, *893*, 249–265.
81. Gerber, S. A.; Rush, J.; Stemman, O.; Kirschner, M. W.; Gygi, S. P. *Proc. Natl. Acad. Sci. U. S. A.* **2003**, *100*, 6940–6945.
82. Percy, A. J.; Chambers, A. G.; Parker, C. E.; Borchers, C. H. *Methods Mol. Biol.* **2013**, *1000*, 167–189.
83. Percy, A. J.; Chambers, A. G.; Yang, J.; Borchers, C. H. *Proteomics* **2013**, *13*, 2202–2215.
84. Guingab-Cagmat, J. D.; Newsom, K.; Vakulenko, A.; Cagmat, E. B.; Kobeissy, F. H.; Zoltewicz, S.; Wang, K. K.; Anagli, J. *Electrophoresis* **2012**, *33*, 3786–3797.

85. Yao, X.; Freas, A.; Ramirez, J.; Demirev, P. A.; Fenselau, C. *Anal. Chem.* **2001**, *73*, 2836–2842.
86. Fenselau, C.; Yao, X. *Methods Mol. Biol.* **2007**, *359*, 135–142.
87. Chen, X.; Cushman, S. W.; Pannell, L. K.; Hess, S. *J. Proteome Res.* **2005**, *4*, 570–577.
88. Broedel, O.; Krause, E.; Stephanowitz, H.; Schuemann, M.; Eravci, M.; Weist, S.; Brunkau, C.; Wittke, J.; Eravci, S.; Baumgartner, A. *J. Proteome Res.* **2009**, *8*, 3771–3777.
89. Zhao, Y.; Jensen, O. N. *Proteomics* **2009**, *9*, 4632–4641.
90. Jiang, H.; Ramos, A. A.; Yao, X. *Anal. Chem.* **2010**, *82*, 336–342.
91. Alghamdi, W. M.; Gaskell, S. J.; Barber, J. *Anal. Chem.* **2012**, *84*, 7384–7392.
92. Liu, H.; Wang, F.; Xu, W.; May, K.; Richardson, D. *Anal. Biochem.* **2013**, *432*, 16–22.
93. Fukuda, M.; Takao, T. *Anal. Chem.* **2012**, *84*, 10388–10394.
94. Niles, R.; Witkowska, H. E.; Allen, S.; Hall, S. C.; Fisher, S. J.; Hardt, M. *Anal. Chem.* **2009**, *81*, 2804–2809.
95. Haaf, E.; Schlosser, A. *Anal. Chem.* **2012**, *84*, 304–311.
96. Dayon, L.; Hainard, A.; Licker, V.; Turck, N.; Kuhn, K.; Hochstrasser, D. F.; Burkhard, P. R.; Sanchez, J. C. *Anal. Chem.* **2008**, *80*, 2921–2931.
97. Thompson, A.; Schäfer, J.; Kuhn, K.; Kienle, S.; Schwarz, J.; Schmidt, G.; Neumann, T.; Hamon, C. *Anal. Chem.* **2003**, *75*, 1895–1904.
98. Ross, P. L.; Huang, Y. N.; Marchese, J. N.; Williamson, B.; Parker, K.; Hattan, S.; Khainovski, N.; Pillai, S.; Dey, S.; Daniels, S.; Purkayastha, S.; Juhasz, P.; Martin, S.; Bartlet-Jones, M.; He, F.; Jacobson, A.; Pappin, D. J. *Mol. Cell. Proteomics* **2004**, *3*, 1154–1169.
99. Murray, C. I.; Uhrigshardt, H.; O'Meally, R. N.; Cole, R. N.; Van Eyk, J. E. *Mol. Cell. Proteomics* **2012**, *11*, M111.013441. http://dx.doi.org/10.1074/mcp.M111.013441.
100. Sohn, C. H.; Lee, J. E.; Sweredoski, M. J.; Graham, R. L.; Smith, G. T.; Hess, S.; Czerwieniec, G.; Loo, J. A.; Deshaies, R. J.; Beauchamp, J. L. *J. Am. Chem. Soc.* **2012**, *134*, 2672–2680.
101. Gygi, S. P.; Rist, B.; Gerber, S. A.; Turecek, F.; Gelb, M. H.; Aebersold, R. *Nat. Biotechnol.* **1999**, *17*, 994–999.
102. Hansen, K. C.; Schmitt-Ulms, G.; Chalkley, R. J.; Hirsch, J.; Baldwin, M. A.; Burlingame, A. *Mol. Cell. Proteomics* **2003**, *2*, 299–314.
103. Schmidt, A.; Kellermann, J.; Lottspeich, F. *Proteomics* **2005**, *5*, 4–15.
104. DeSouza, L. V.; Krakovska, O.; Darfler, M. M.; Krizman, D. B.; Romaschin, A. D.; Colgan, T. J.; Siu, K. W. *Proteomics* **2010**, *10*, 3108–3116.
105. Hsu, J.; Huang, S.; Shiea, J.; Huang, W.; Chen, S. *J. Proteome Res.* **2005**, *4*, 101–108.
106. Boersema, P. J.; Foong, L. Y.; Ding, V. M.; Lemeer, S.; van Breukelen, B.; Philp, R.; Boekhorst, J.; Snel, B.; den Hertog, J.; Choo, A. B.; Heck, A. J. *Mol. Cell. Proteomics* **2010**, *9*, 84–99.
107. Wang, Z.; Gu, C.; Colby, T.; Shindo, T.; Balamurugan, R.; Waldmann, H.; Kaiser, M.; Van der Hoorn, R. L. *Nat. Chem. Biol.* **2008**, *4*, 557–563.
108. Kolb, H. C.; Finn, M.; Sharpless, K. B. *Angew. Chem. Int. Ed. Engl.* **2001**, *40*, 2004–2021.
109. Cutter, J. L.; Cohen, N. T.; Wang, J.; Sloan, A. E.; Cohen, A. R.; Panneerselvam, A.; Schluchter, M.; Blum, G.; Bogyo, M.; Basilion, J. P. *PLoS One* **2012**, *7*. http://dx.doi.org/10.1371/journal.pone.00330606.
110. Adibekian, A.; Martin, B. R.; Chang, J. W.; Hsu, K.; Tsuboi, K.; Bachovchin, D. A.; Speers, A. E.; Brown, S. J.; Spicer, T.; Fernandez-Vega, V.; Ferguson, J.; Hodder, P. S.; Rosen, H.; Cravatt, B. F. *J. Am. Chem. Soc.* **2012**, *134*, 10345–10348.
111. Kam, C. M.; Abuelyaman, A. S.; Li, Z.; Hudig, D.; Powers, J. C. *Bioconjug. Chem.* **1993**, *4*, 560–567.

112. Abuelyaman, A. S.; Hudig, D.; Woodard, S. L.; Powers, J. C. *Bioconjug. Chem.* **1994**, *5*, 400–405.
113. Jessani, N.; Niessen, S.; Wei, B. Q.; Nicolau, M.; Humphrey, M.; Ji, Y.; Han, W.; Noh, D. Y.; Yates, J. R., 3rd.; Jeffrey, S. S.; Cravatt, B. F. *Nat. Methods* **2005**, *2*, 691–697.
114. Hekmat, O.; He, S.; Warren, R. A. J.; Withers, S. G. *J. Proteome Res.* **2008**, *7*, 3282–3292.
115. Altun, M.; Kramer, H. B.; Willems, L. I.; McDermott, J. L.; Leach, C. A.; Goldenberg, S. J.; Suresh Kumar, K. G.; Konietzny, R.; Fischer, R.; Kogan, E.; Mackeen, M. M.; McGouran, J.; Khoronenkova, S. V.; Parsons, J.; Dianov, G. L.; Nicholson, B.; Kessler, B. M. *Chem. Biol.* **2011**, *18*, 1401–1412.
116. Weerapana, E.; Wang, C.; Simon, G. M.; Richter, F.; Khare, S.; Dillon, M. B.; Bachovchin, D. A.; Mowen, K.; Baker, D.; Cravatt, B. F. *Nature* **2010**, *468*, 790–795.
117. Young, C. L.; Britton, Z. T.; Robinson, A. S. *Biotechnol. J.* **2012**, *7*, 620–634.
118. Terpe, K. *Appl. Microbiol. Biotechnol.* **2003**, *60*, 523–533.
119. Trinkle-Mulcahy, L.; Boulon, S.; Lam, Y. W.; Urcia, R.; Boisvert, F.; Vandermoere, F.; Morrice, N. A.; Swift, S.; Rothbauer, U.; Leonhardt, H.; Lamond, A. *J. Cell Biol.* **2008**, *183*, 223–239.
120. Galan, J. A.; Paris, L. L.; Zhang, H.; Adler, J.; Geahlen, R. L.; Tao, W. A. *J. Am. Soc. Mass Spectrom.* **2011**, *22*, 319–328.
121. Neumuller, R. A.; Wirtz-Peitz, F.; Lee, S.; Kwon, Y.; Buckner, M.; Hoskins, R. A.; Venken, K. J.; Bellen, H. J.; Mohr, S. E.; Perrimon, N. *Genetics* **2012**, *190*, 931–940.
122. Hochuli, E.; Bannwarth, W.; Dobeli, H.; Gentz, R.; Stuber, D. *Bio/Technology* **1988**, *6*, 1321–1325.
123. Song, M. N.; Moon, P. G.; Lee, J. E.; Na, M.; Kang, W.; Chae, Y. S.; Park, J. Y.; Park, H.; Baek, M. C. *Arch. Pharm. Sci. Res.* **2012**, *35*, 1839–1847.
124. Green, N. M. *Biochem. J.* **1963**, *89*, 585–591.
125. Ovod, V.; Scott, E. A.; Flake, M. M.; Parker, S. R.; Bateman, R. J.; Elbert, D. L. *J. Biomed. Mater. Res. A* **2012**, *100*, 622–631.
126. Karhemo, P. R.; Ravela, S.; Laakso, M.; Ritamo, I.; Tatti, O.; Makinen, S.; Goodison, S.; Stenman, U. H.; Hölttä, E.; Hautaniemi, S.; Valmu, L.; Lehti, K.; Laakkonen, P. *J. Proteomics* **2012**, *77*, 87–100.
127. Vermeulen, M.; Hubner, N. C.; Mann, M. *Curr. Opin. Biotechnol.* **2008**, *19*, 331–337.
128. Gingras, A. C.; Gstaiger, M.; Raught, B.; Aebersold, R. *Nat. Rev. Mol. Cell Biol.* **2007**, *8*, 645–654.
129. Lu, X. M.; Tompkins, R. G.; Fischman, A. J. *Int. J. Mol. Med.* **2012**, *29*, 461–471.
130. Smaczniak, C.; Li, N.; Boeren, S.; America, T.; van Dongen, W.; Goerdayal, S. S.; de Vries, S.; Angenent, G. C.; Kaufmann, K. *Nat. Protoc.* **2012**, *7*, 2144–2158.
131. Rogstad, S. M.; Sorkina, T.; Sorkin, A.; Wu, C. C. *Anal. Chem.* **2013**, *85*, 4301–4306.
132. Tirumalai, R. S.; Chan, K. C.; Prieto, D. A.; Issaq, H. J.; Conrads, T. P.; Veenstra, T. D. *Mol. Cell. Proteomics* **2003**, *2*, 1096–1103.
133. Faulkner, S.; Elia, G.; Hillard, M.; O'Boyle, P.; Dunn, M.; Morris, D. *Proteomics* **2011**, *11*, 2329–2335.
134. Shi, T.; Su, D.; Liu, T.; Tang, K.; Camp, D. G.; Qian, W. J.; Smith, R. D. *Proteomics* **2012**, *12*, 1074–1092.
135. Hao, P.; Ren, Y.; Xie, Y. J. Chromatogr. B: Anal. Technol. Biomed. *Life Sci.* **2009**, *877*, 1657–1666.
136. Wehr, A. Y.; Hwang, W. T.; Blair, I. A.; Yu, K. H. *J. Proteome Res.* **2012**, *11*, 1749–1758.
137. Ortea, I.; Roschitzki, B.; Ovalles, J. G.; Longo, J. L.; de la Torre, I.; Gonzalez, I.; Gómez-Reino, J. J.; González, A. *J. Proteomics* **2012**, *77*, 372–382.

138. Cao, Z.; Yende, S.; Kellum, J. A.; Robinson, R. A. *Int. J. Proteomics* **2013**, *2013*. http://dx.doi.org/10.1155/2013/654356.
139. Prabakaran, S.; Lippens, G.; Steen, H.; Gunawardena, J. *Wiley Interdiscip. Rev.: Syst. Biol. Med.* **2012**, *4*, 565–583.
140. Anderson, N. L.; Anderson, N. G.; Haines, L. R.; Hardie, D. B.; Olafson, R. W.; Pearson, T. W. *J. Proteome Res.* **2004**, *3*, 235–244.
141. Kuhn, E.; Whiteaker, J. R.; Mani, D. R.; Jackson, A. M.; Zhao, L.; Pope, M. E.; Smith, D.; Rivera, K. D.; Anderson, N. L.; Skates, S. J.; Pearson, T. W.; Paulovich, A. G.; Carr, S. A. *Mol. Cell. Proteomics* **2012**, *11*, M111.013854. http://dx.doi.org/10.1074/mcp.M111.013854.
142. Gaberc-Porekar, V.; Menart, V. *J. Biochem. Biophys. Methods* **2001**, *49*, 335–360.
143. Beltran, L.; Cutillas, P. R. *Amino Acids* **2012**, *43*, 1009–1024.
144. Wuhrer, M.; Catalina, M. I.; Deelder, A. M.; Hokke, C. H. *J. Chromatogr. B: Analyt. Technol. Biomed. Life Sci.* **2007**, *849*, 115–128.
145. Stokes, M. P.; Farnsworth, C. L.; Moritz, A.; Silva, J. C.; Jia, X.; Lee, K. A.; Guo, A.; Polakiewicz, R. D.; Comb, M. J. *Mol. Cell. Proteomics* **2012**, *11*, 187–201.
146. Na, C. H.; Jones, D. R.; Yang, Y.; Wang, X.; Xu, Y.; Peng, J. *J. Proteome Res.* **2012**, *11*, 4722–4732.
147. Di Palma, S.; Hennrich, M. L.; Heck, A. J.; Mohammed, S. *J. Proteomics* **2012**, *75*, 3791–3813.
148. Gruber, K. A.; Stein, S.; Brink, L.; Radhakrishnan, A.; Udenfriend, S. *Proc. Natl. Acad. Sci. U. S. A.* **1976**, *73*, 1314–1318.
149. Donato, P.; Cacciola, F.; Mondello, L.; Dugo, P. *J. Chromatogr. A* **2011**, *1218*, 8777–8790.
150. Motoyama, A.; Yates, J. R. *Anal. Chem.* **2008**, *80*, 7187–7193.
151. Nie, S.; Dai, J.; Ning, Z.; Cao, X.; Sheng, Q.; Zeng, R. *J. Proteome Res.* **2010**, *9*, 4585–4594.
152. Alpert, A. J. *J. Chromatogr. A* **1990**, *499*, 177–196.
153. Palmisano, G.; Lendal, S. E.; Engholm-Keller, K.; Leth-Larsen, R.; Parker, B. L.; Larsen, M. R. *Nat. Protoc.* **2010**, *5*, 1974–1982.
154. Zhang, H.; Guo, T.; Li, X.; Datta, A.; Park, J. E.; Yang, J.; Lim, S. K.; Tam, J.; Sze, S. K. *Mol. Cell. Proteomics* **2010**, *9*, 635–647.
155. Gilar, M.; Olivova, P.; Daly, A. E.; Gebler, J. C. *Anal. Chem.* **2005**, *77*, 6426–6434.
156. Kong, R. P. W.; Siu, S. O.; Lee, S. S. M.; Lo, C.; Chu, I. K. *J. Chromatogr., A* **2011**, *1218*, 3681–3688.
157. Lau, E.; Lam, M. P. Y.; Siu, S. O.; Kong, R. P. W.; Chan, W. L.; Zhou, Z.; Huang, J.; Lo, C.; Chu, I. H. *Mol. BioSyst.* **2011**, *7*, 1399–1408.
158. Stasyk, T.; Huber, L. A. *Proteomics* **2004**, *4*, 3704–3716.
159. Link, A. J.; Eng, J.; Schieltz, D. M.; Carmack, E.; Mize, G. J.; Morris, D. R.; Garvik, B. M.; Yates, J. R., 3rd. *Nat. Biotechnol.* **1999**, *17*, 676–682.
160. Pratima, N. A.; Shraddha, B.; Zibran, S. *Int. J. Res. Pharm. Sci.* **2013**, *3*, 19–40.
161. Swartz, M. E. *J. Liq. Chromatogr. Relat. Technol.* **2005**, *28*, 1253–1263.
162. Sherathia, B.; Prajapati, G.; Singh, R.; Jat, R. *Int. J. Pharm. Sci. Rev. Res.* **2012**, *15*, 54–57.
163. Roumeliotis, T. I.; Halabalaki, M.; Alexi, X.; Ankrett, D.; Giannopoulou, E.; Skaltsounis, A.; Sayan, B. S.; Alexis, M. N.; Townsend, P. A.; Garbis, S. D. *J. Proteome Res.* **2013**, *12*, 1591–1603.
164. Cunliffe, J. M.; Maloney, T. D. *J. Sep. Sci.* **2007**, *30*, 3104–3109.
165. Schuster, S.; Wagner, B.; Boyes, B.; Kirkland, J. *J. Chromatogr. Sci.* **2010**, *48*, 566–571.

166. Toms, S. A.; Weil, R. J., Eds.; *Nanoproteomics: Methods and Protocols*; Humana Press: New York, 2011.
167. Ali, I.; Aboul-Enein, H. Y.; Cazes, J. *J. Liq. Chromatogr. Relat. Technol.* **2010**, *33*, 645–653.
168. Fröhlich, T.; Helmstetter, D.; Zobawa, M.; Crecelius, A. C.; Arzberger, T.; Kretzschmar, H. A.; Arnold, G. J. *Proteomics* **2006**, *6*, 4950–4966.
169. Eng, J. K.; McCormack, A. L.; Yates, J. R. *J. Am. Soc. Mass Spectrom.* **1994**, *5*, 976–989.
170. Cottrell, J.; London, U. *Electrophoresis* **1999**, *20*, 3551–3567.
171. Craig, R.; Beavis, R. C. *Bioinformatics* **2004**, *20*, 1466–1467.
172. Ong, S.; Mann, M. *Nat. Chem. Biol.* **2005**, *1*, 252–262.
173. Cox, J.; Mann, M. *Annu. Rev. Biochem.* **2011**, *80*, 273–299.
174. Escher, C.; Reiter, L.; MacLean, B.; Ossola, R.; Herzog, F.; Chilton, J.; MacCoss, M. J.; Rinner, O. *Proteomics* **2012**, *12*, 1111–1121.
175. Rifai, N.; Gillette, M. A.; Carr, S. A. *Nat. Biotechnol.* **2006**, *24*, 971–983.
176. Cho, C. J.; Drabovich, A. P.; Batruch, I.; Diamandis, E. P. *J. Proteomics* **2011**, *74*, 2052–2059.
177. Gallien, S.; Duriez, E.; Domon, B. *J. Mass Spectrom.* **2011**, *46*, 298–312.
178. Whiteaker, J. R.; Zhao, L.; Anderson, L.; Paulovich, A. G. *Mol. Cell. Proteomics* **2010**, *9*, 184–196.
179. Yao, X.; Bajrami, B.; Shi, Y. *Anal. Chem.* **2010**, *82*, 794–797.

Chapter 6

Emerging Nanotechniques in Proteomics

Noelia Dasilva*, Maria Gonzalez-Gonzalez*, Paula Diez*, Ricardo Jara-Acevedo[†], Lucia Lourido[‡], J.M. Sayagues*, Alberto Orfao* and Manuel Fuentes*

*Centro de Investigación del Cáncer/IBMCC (USAL/CSIC), IBSAL, Departamento de Medicina Unidad de Proteomica & Servicio General de Citometría, University of Salamanca, Salamanca, Spain

[†]ImmunoStep, Edificio Centro de Investigación del Cáncer, Avda. Coimbra s/n, Campus Miguel de Unamuno, Salamanca, Spain

[‡]Instituto de Investigación Biomedica da Coruña (INIBIC), Hospital Universitario A Coruña, A Coruña, Spain

Chapter Outline

1 INTRODUCTION

The human genome opened new scientific fields in biological and clinical scientific areas *(1,2)*; however, some biological information is still unknown that cannot be deciphered by genomics tools, particularly the dynamics of

Comprehensive Analytical Chemistry, Vol. 63. http://dx.doi.org/10.1016/B978-0-444-62651-6.00006-4

biological processes involved in diseases. For this reason, recently, proteomics tools have shown rapid growth, specifically on the development of robust and high-throughput (HT) technologies, in understanding the dynamic proteome changes related to human pathologies, such as cancer, diabetes, and other diseases *(1–4)*.

Currently, the human proteome has been started as a global project with the aim to identify the entire protein counterpart of the human genome in a biological system *(1–4)*. Proteomics research has been focused on several clinical applications for accurate diagnosis, early detection, and/or prognostic or candidate proteins or peptides associated with disease status *(1–4)*.

The identification of biomarkers in complex biological fluids (e.g., urine and blood) requires HT detection methods. Hence, protein microarray seems to be a suitable tool for these studies because it can handle large amounts of data. Also, microarray technologies are currently making huge progress in order to increase sensitivity and reliability. Mainly, protein detection systems have progressively undergone a transition from label-free-based methods to label-free technologies, but the last ones in many cases are not still completely implemented for HT studies *(1,2,5,6)*.

Bearing in mind the interest in the development of highly specific and HT approaches in proteomics, the introduction of nanotechniques offers a wide spectrum of novel techniques and methods to fulfill HT requirements. In this regard, the integration of nanotechniques and proteomics has led to the development of nanoproteomics, which allow sensitive and real-time detection of proteins or peptides presented in a complex biological sample *(1,2,5,6)*. In summary, nanoproteomics could perform highly sensitive analysis with minimal amounts of sample in real-time and multiplex format, thereby finding an increasing number of potential applications in biomedical research.

In this chapter, we briefly describe advantages, challenges, applications, and promises of diverse current nanotechniques with applications in proteomics.

2 OVERVIEW OF PROTEIN MICROARRAYS

Basically protein microarrays are generated by hundreds to thousands of proteins that are printed and immobilized onto chemically activated glass slides or polymer-coated glass slides, offering huge potential for HT and complete characterization of a sample of interest. In general, protein microarrays are characterized by requirement of low amounts of sample, reagent consumption, and HT detection of a high number of target and query molecules. At present, several microarrays formats have been developed and implemented for different applications, such as biomarker detection, protein profiles, protein interactions, and vaccine development *(5–8)*.

Briefly, protein microarrays are basically classified as target microarrays, reverse-phase protein microarrays (RPP), and *in situ* expressed arrays (Figure 1) *(5,6)*.

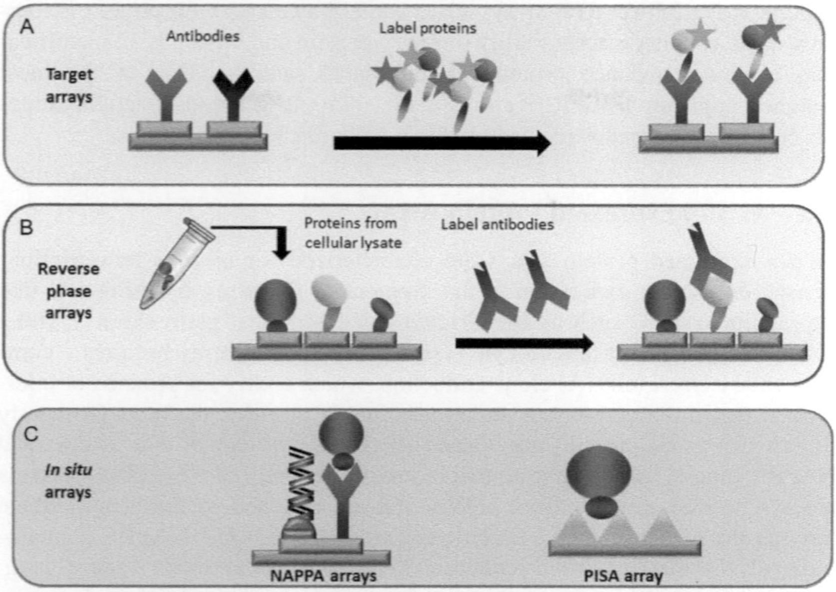

FIGURE 1 Major types of protein microarrays: (A) targeted protein arrays, (B) reverse-phase protein arrays, and (C) *in situ* expressed protein arrays. (For color version of this figure, the reader is referred to the online version of this chapter.)

2.1 Target Microarray

Antibody microarrays are the most common target microarrays. Target microarrays consist of different proteins printed onto planar-solid surfaces *(7,8)*. In general, these arrays have found several applications in clinical studies because they allow identification of specific protein profiles. One of the major drawbacks of antibody microarrays is the need for oriented immobilization of the capture reagent onto the surface in order to avoid steric hindrance with the specific proteins presented in the sample. Therefore, to solve this problem, several strategies have been described that implied that chemical groups promote orientation and avoid cross-linkers that can interact nonspecifically with the query protein. Another important disadvantage is the loss of activity of proteins upon immobilization *(9–11)*.

2.2 Reverse-Phase Protein Array (RPP)

RPPs have been designed to capture proteins from cellular or tissue lysate samples or present in the serum or plasma onto surfaces *(12)*, which are further identified by an affinity reagent such as antibody, single-chain variable fragments, and aptamers, among others. Thus, the signal could be related to specific protein detection and it could be affected by antibody affinity and epitope accessibility. RPP arrays share some drawbacks with target microarrays, such as epitope accessibility of query protein and difficulty in identification of low-abundance proteins in biological samples. One of the most common applications of RPP arrays is the ability to generate functional maps for cell signaling pathways or individual cells or tissues *(5)*.

2.3 *In Situ* Expressed Protein Array

In situ expressed protein arrays are characterized for *in vitro* transcription-translation protein expression at the moment of the assay by using cell-free expression systems such as *Escherichia coli* 30s, wheat germ extracts, HeLa cell lysates, or rabbit reticulocyte lysates. This type of array requires a complementary DNA (cDNA) clone collection, which codify for proteins of interest and tag in the carboxy or amino terminus. Currently, the most frequently *in situ* expressed protein arrays used are PISA (protein *in situ* arrays) and NAPPA (nucleic acids programmable protein arrays) *(13–16)*. PISA proteins are synthesized directly from cDNA and are attached to the array surface through the recognition of a specific tag sequence *(13–15)*; NAPPA coimmobilize cDNA that encodes a protein and a tag with an antibody against a tag. Thus, synthesized proteins are captured onto array surfaces by the interaction between the tag and the specific antibody *(16)*. The most important advantage of these arrays is that proteins are expressed just at the moment of the assay, which eliminates the problem of protein degradation derived from storage and drastic washing steps. These protein arrays have been successfully used for identifying auto-antigens, tumor antigen proteins, and protein–protein interactions, among others *(17–19)*.

3 DETECTION PLATFORMS IN NANOPROTEOMICS

During the last years, several detection methodologies have been significantly developed for their application in nanoproteomics to analyze complex biological samples, providing an increment in the multiplexing capacity, dynamic range, and sensitivity *(1,2,20)*. Basically, the two detection techniques commonly used for protein microarrays analysis are label-based and label-free detection systems. The label-based approaches are mainly based on the use of specific tags for target molecules, such as conventional fluorescent dyes and/or radioisotopes. In addition, other molecules and materials such as quantum dots (QDs), silica nanoparticles, gold nanoparticles (AuNPs), and carbon nanotubes (CNTs), among others, have been introduced in order to improve sensitivity *(1,2)*. Moreover, flow cytometry and magnetic beads-based detection systems are good examples of dye-bead-based techniques already implemented in biomedical research with promising results *(1,2)*. Label-free detection methods measure an inherent property of the query such as mass, dielectric, or optical properties.

In the field of proteomics, especially protein microarrays, there have been numerous advancements in both label-based and label-free technologies, which are discussed in the following sections.

3.1 Label Detection Techniques

For protein microarrays, several label detection techniques are already used because of their simplicity, broad availability of fluorescence dyes and chemical functional groups for conjugation, and scanning instruments.

3.1.1 Quantum Dots

QDs are a semiconductor fluorescence core (e.g., Cd or Se) that is coated with another semiconductor shell (i.e., Cd or zinc sulfide) with a larger spectral band gap and stable light-scattering or -emitting properties. In addition, these nanoparticles can be conjugated by covalent link with biospecific molecules (e.g., antibodies, oligonucleotides, peptides, etc.). Despite the advantages of QDs, they present serious disadvantages such as critical lack of stability and high oxidability. Lately, QDs have been applied in clinical research projects *(21,22)*.

3.1.2 Gold Nanoparticles

AuNPs have been considered suitable for *in vivo* studies, due to well-known characterized biocompatibility. For AuNPs, many different strategies have been described for surface functionalization with drugs, peptides, and so on, which are very useful to perform systematic cytotoxic studies. AuNPs are used in proteomics because of its physicochemical properties (e.g., optical features, electromagnetic or photothermal properties), which are very useful for label-based detection methods *(23,24)*.

3.2 Label-Free Detection Methods

Despite wide use of label-based techniques, the interest of label-free techniques has been increased because they are cleaner, faster, and simpler; moreover, they allow real-time detection *(1,2,20,25)*. Historically, label-free detection techniques appeared mainly to solve the disadvantages and limitations of the label-based techniques, such as chemical modification of query molecules, alteration of molecule properties by labeling procedures, and difficulties for multiplex sensing *(1,2,20,25)*, among others. Frequently, label-free techniques are very useful for real-time interaction studies, avoiding steric hindrance related to tag molecules *(1,2,20,25)*.

At present, a variety of label-free detection strategies are available for biomedical research, such as surface plasmon resonance (SPR), CNTs, cantilevers, surface-enhanced laser desorption ionization (SELDI)-time of flight (TOF)-mass spectrometry (MS), microfluidic purification chips, immunosensors based on channels of mesoporous silica, functionalized nanopipette probes, nanostructured electromechanical immunosensors featuring single-wall nanotubes, and AuNPs *(1,20,25,26)*. It has been recently published that a combination of self-assembled protein arrays with atomic-force microscopy (AFM), nanogravimetry, and anodic porous alumina for protein–protein interaction studies are related to several cell processes, for example, cell proliferation, cell differentiation, and neoplastic transformation *(20,25,26)*.

In this chapter, a few label-free techniques (SPR, MS, and AFM microcantilevers), suitable for protein microarrays, are described further because they are fully implemented and accepted in many biomedical research projects.

3.2.1 Surface Plasmon Resonance

The principle of SPR technology is grounded on the oscillations of free electrons, which are propagated in parallel to a metal/dielectric interface. These oscillations are directly related to variations of index refraction close to the sensor surface. The major application of the SPR technique is the determination of binding parameters (kon, koff) for biomolecular interactions because this technique allows for studying kinetics *(27–29)*. In fact, SPR enables the evaluation of the strength of the binding and interaction specificity with a very high sensitivity *(27–29)*.

A few years ago, surface plasmon resonance imaging (SPRi) has been implemented in biomarker discovery *(30,31)*. Ladds *et al.* employed SPRi for detection of candidate diagnostic biomarkers in cancer in combination with antibody arrays *(32)*. Basically, SPRi is a SPR device coupled with a CCD camera, which obtains a slide image when a polarized light is reflected depending on the interaction on the surface. In fact, it is based on surface plasmons as SPR; however, SPRi can analyze hundreds of samples on a single experiment.

3.2.2 AFM Microcantilevers

Microcantilevers are associated with AFM, which basically are gold-coated surfaces based on a silicon core, which are associated with nanomechanics for biomolecular recognition *(1,2,20,25)*. In AFM technology, a cantilever is in direct contact with the sample surface; then the bending of the cantilever is determined by optical detection of the position of a laser beam *(33)*. In fact, this method is a versatile tool for surface characterization and provides information concerning topological variations at the molecular level.

3.2.3 Mass Spectrometry

MS is the most common label-free technique widely used in proteomics, with many different applications in very different areas from environmental to biomedical research; MS especially has become one of the leading methodologies in the field of biomarker discovery. Before MS was applied to protein quantification and its analysis, it was used for quantification of small molecules in clinical laboratories *(34)*. In fact, recent developments in technology have shown MS as an accurate alternative to ELISA and immunoassays *(35)*.

In general, MS can be divided into three different experimental approaches:

(a) Relative quantification by peak intensity of LC-MS: It is based on the linear correlation between protein concentration and the area under the peaks from LC-MS.
(b) Relative quantification by spectral count: For one specific protein, the quantification is provided from a comparison between the number of MS/MS spectra and databases (LC-MS/MS or LC/LC-MS/MS).
(c) Absolute label-free quantification: Protein abundance index is obtained by this method, which corresponds with the number of identified peptides divided by the number of the theoretical expected tryptic peptides for each protein.

Bearing in mind that MS allows monitoring of multiple peptide transitions, it is also possible to distinguish two major methodological approaches, selected reaction monitoring (SRM) and multiple reaction monitoring (MRM), which are based on the specific identification and quantification of a protein that is represented by one or two prototypic peptides *(36)*, always relevant for identification of a specific protein *(36–38)* (Figure 2). Briefly, MRM can be defined as multiple SRM transitions: In general, SRM is considered a useful technique for detection in complex proteome; however, the number of potential peptides identified through SRM is limited. SRM also overcomes the problems of sensitivity and reproducibility. For identification of specific proteins in a complex background, SRM is the technique of choice because it speeds up analysis time, avoids redundant peptide sampling, and provides high

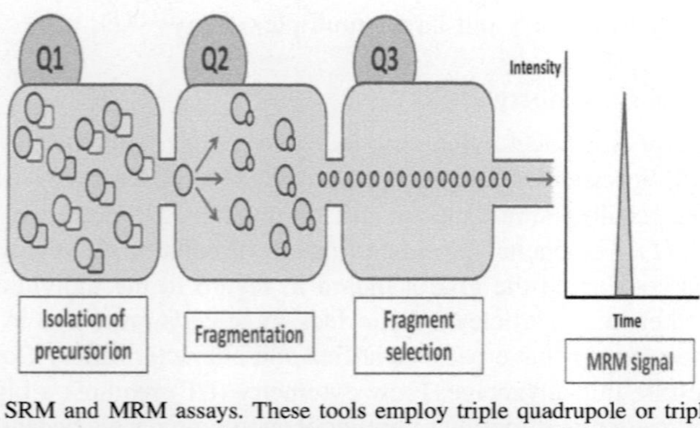

FIGURE 2 SRM and MRM assays. These tools employ triple quadrupole or triple quadrupole ion trap mass spectrometers. In the first quadrupole the precursor ion is isolate; the fragmentation occurs in the second quadrupole and the optimum fragment ions are monitored in the third quadrupole. (For color version of this figure, the reader is referred to the online version of this chapter.)

reproducibility. In summary, SRM provides high-selectivity, sensitivity, and HT quantitation *(37–39)*.

Here a few methodological approaches are described that are based on a combination of MS technology with nanotechnology, such as SELDI-TOF and mass cytometer.

3.2.3.1 SELDI-TOF-MS

This technology makes use of proteins bound on a solid-phase chromatographic surface with specific properties onto which complex biological mixtures are added. The subset of proteins retained by the chip are subsequently ionized and detected by TOF-MS. Then the unique fingerprint is generated for each sample; as a consequence, patterns of masses are produced by SELDI analysis. This technology has been successfully applied for quantitative profiling of tissue proteome and on-chip HT sample processing. Two major disadvantages are that this technique is not suitable for low-molecular-weight proteins (<20 kDa) and its inability to identify proteins after subsequent recovery *(1,2)*.

3.2.3.2 ESI-MS for Multiplex Protein Detection in Array Format by Using Metal–Antibody Conjugates

By using microarray technology, protein quantitative detection on one spot microarray is usually not possible. However, recently analysis of a microarray through laser ablation (LA)-inductively coupled plasma (ICP)-MS has been introduced, increasing the feasibility of protein detection *(40)*.

Tanner and coworkers developed an immunoassay based on multiparametric detection of proteins through metal-tag antibodies coupled to ICP-MS. Since this technique is able to identify low-abundance proteins in complex samples, this group achieved an increase in signal amplification and as a consequence its sensitivity. The new possibilities offered by the improvements allow them to carry out larger multiplex assays *(41)*.

3.2.3.3 Mass Cytometry

Mass cytometry is a novel cytometric technology that couples flow cytometry with ICP-MS. Specific and selective phenotypical identification of cells in complex samples requires simultaneous and multiparametric detection of several biomarkers *(41)*. For phenotypic identification of cells or microparticles, flow cytometry is considered the gold standard in regard to the individual analysis of cells or other microparticles *(41)*; in fact, most cells presented in peripheral blood or bone marrow have been identified and characterized by flow cytometry *(42)*. Despite this advantage, Flow cytometry (CF) requires a high level of expertise for designing of optimal immunophenotyping panels because it is very important to avoid overlapping of fluorescence signals (excitation/emission) and the number of different dyes that can be detected simultaneously. In addition, it is necessary to produce new tools that allow exploitation of CF's potential as a multiparametric detection tool *(41–44)*. Owing to the number of potential parameters to analyze, flow cytometry has reached a limit between 15 and 17 different colors avoiding the overlapping among the emission spectra of these fluorochromes *(44,45)*. Due to the narrow spectra of many fluorescence dyes, complex correction algorithms are needed to deconvolute the overlapped regions and facilitate the analysis and data interpretation. Moreover, the optical and acquisition systems of flow cytometers require accurate compensation by using many different panels of control tubes, which complicate the acquisition protocol. The use of isotope tags avoids the need of compensation *(45)*, increasing the number of possible parameters to determine to 60 *(41)*. With this aim, Bandura *et al.* have developed mass cytometry *(41)*. Mass cytometry is based on the binding of stable isotope tags, instead of fluorescent labels, which allow the identification and quantification of each single label by MS *(41,43,46)*. The widely used CyTOF instrument, based on TOF-MS detection, was the first commercial version of mass cytometry (Figure 3).

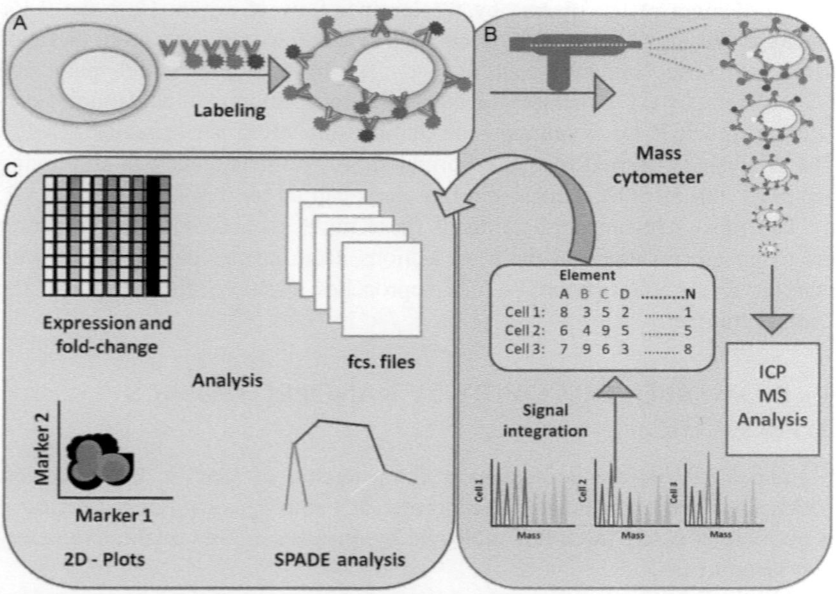

FIGURE 3 Workflow of mass cytometry analysis. (A) The sample is labeled with lanthanides. (B) The sample to examine is introduced as a liquid suspension with a syringe pump into a nebulizer where heat argon gas flow is supplied to dry the droplets, changing the turbulent flow to a laminar one. Then the droplets arrive to the ICP torch through the aerosol splitter to be delivered to the central channel of the plasma by the laminar flow of argon. Here the 0.1% ionization of argon at 7000 K is enough to atomize and ionize the particles. The elemental ions are sampled by a TOF-MS and quantified. (C) The analysis is carried out as a normal flow cytometry analysis employing dot plots, but also heat maps and tree plots can be used. (For color version of this figure, the reader is referred to the online version of this chapter.)

Mass cytometry, however, has important limitations. First of all, the cells employed in the analysis cannot be recovered. Second, the potential huge number of labels complicates the analysis and data interpretation. Finally, the prize of the stable isotope conjugated monoclonal antibodies is quite high, also necessary to test their specificity, affinity, stability, and resistance *(43)*.

Mass cytometry uses stable isotopes as lanthanide series because of the huge number of elements of this family and their similar chemical structure, allowing their incorporation into the same tag *(45,46)*. They also exhibit very low background due to their intrinsic low natural abundance *(45–47)*. These rare earth elements are relatively biocompatible and easily conjugated to biomolecules by broad availability and simple protocols. Finally, they proffer high sensitivity for traces of molecules in biological samples, being attractive markers for their use in the clinical routine *(47)*.

The elements to examine are introduced as a liquid suspension with a syringe pump into a nebulizer where heat argon gas flow is supplied to dry the droplets, changing the turbulent flow to a laminar one. Then the droplets arrive to the ICP torch through the aerosol splitter to be delivered until the central channel of the plasma by the laminar flow of argon. Here the 0.1% ionization of argon at 7000 K is enough to atomize and ionize the particles *(41,42,45)*. The elemental ions are sampled by a TOF-MS and quantified *(42)*. It is important to achieve the complete atomization and ionization of particles in the ICP for obtaining reproducible analytical performance *(41,45)*. The analysis is carried out as a normal flow cytometry analysis employing dot plots; but also, heat maps and tree plots can be used *(42)*.

The most relevant application of these novel technologies in biomarker discovery are reviewed in the next sections, from a brief description of biomarkers to current nanoproteomics approaches already implemented in the clinical routine.

4 BIOMARKER DISCOVERY BY NANOPROTEOMICS APPROACHES

Biomarkers (BM) are defined as a characteristic of specific biomolecules, which could be an indicator of a normal or pathological process because it is possible to easily track this molecule by quantitative and qualitative detection techniques.

Although biomarkers have been used for many centuries (i.e., clinicians were looking for signals from the patient directly or indirectly related to the disease state), it was not until the mapping of the human genome, which provided genomic information, that clinicians and the scientific community realized the potential utility of the information given by biomarkers *(1,2,20,25)*. Since then, medicine has invested great effort in discovery of novel BM, which allow for making an accurate and early diagnosis of diseases and/or predicting their development; indeed, the diagnosis of several pathologies such as cancer are detected in advanced stages of development. Furthermore, the discovery of BM associated with response to treatment could be used as a powerful tool to introduce personalized medicine in the clinical practice, improving safe and efficient therapies and monitoring response to treatments *(1,2,20,25)*.

In general, the identification of BM has been based on comparison of normal versus pathological cellular or tissue lysates, depending on the pathology of interest and the methodology employed in the study; however, lately, the methodology for BM identification in serum and/or plasma has been improved. Serum and/or plasma are simple, noninvasive samples to collect and are easy to store, and quality control protocols are widely available. Despite these advantages, serum and/or plasma are complex biological samples because they contain highly diverse analytes (e.g., proteins, peptides, and metabolites); the range of concentrations varies from mg/mL, the most abundant proteins (e.g., albumin or antibodies), to pg/mL for molecules susceptible to be considered for BM *(48–50)*. This large dynamic range in serum/plasma is dramatically relevant for the choice of detection method to identify or quantify any BM of interest. In addition, it is important to highlight that any alterations of normal physiology or pathology should be reflected with alterations of normal levels of serum biomolecules or even with detections of new molecules characteristic of pathology (BM) that could be presented in low abundance at the early stages of disease.

Recently, nanotechnology has had an important role in BM discovery because of development of HT platforms, which help to characterize thousands of molecules simultaneously using a minimal amount of samples and to increase its sensitivity to detect low-abundance molecules, which can be BM candidates. As well, proteomics allows study of a group of biomolecules (in a complex sample) that are changing all the time, reflecting alterations throughout the progress of a disease, which is not possible by conventional genomics and molecular biology approaches *(1,2,20,25)*. On the other hand, proteomics also contributes in carrying out analysis of the whole proteome, finding differences between the protein expression profile of healthy patients or pathological samples. In consequence, the most promising results in BM discovery will be expected by close collaboration between genomics and nanoproteomics technologies with bioinformatics to manage all the data generated from these two platforms *(1,2,20,25,50)*.

Briefly, BM can be classified as diagnostic biomarkers, which are molecules that help to discriminate between healthy and one or more pathological states; prognostic biomarkers are molecules that follow disease evolution. Predictive biomarkers are defined as the molecules that could provide relevant information to predict response or resistance to therapy; they also could be used to monitor the response, as well as, in some cases, as diagnostic and prognostic biomarkers.

At present there is an urge to validate biomarkers that can be introduced into clinics, even through the efforts in biomarker identification *(50–52)*. Over the last years, many biomedical researchers have focused on identifying useful biomarkers to understand the mechanism and the mode of action, instead of applying these biomarkers to etiological research or directly into clinics *(50–52)*. The National Institutes of Health has designed a strategy that consists of six consecutive steps: 1. Preclinical testing focused on elaboration of hypotheses related to the disease mechanism and outcome. 2. Phase 0 encloses the earlier assays with patient samples based on a preclinical test. 3. Phase 1 determines which biomarker could distinguish among a healthy and a nonhealthy individual in a small group of patients. 4. Phase 2 has to validate the accuracy of the assay in order to determine different intra- and interindividuals. 5. Phase 3 has to support the efficacy in a higher number of patients. 6. Phase 4 involves postapproval controls *(50–52)*. As well, any biomarkers have to fulfill some specific conditions: (i) They must be easy to apply to clinics; (ii) there must be availability of technology, equipment, and well-trained staff; (iii) each novel biomarker should be more informative than the previous one; (iv) there must be rapid detection to avoid any negative consequences to individuals; and (v) the cost, including equipment and assay performance, should be very low.

In the following sections we describe the uses of nanoproteomics technologies for discovery, identification, and characterization of biomarkers useful in diagnosis and personalized medicine (see Table 1).

4.1 AuNPs and QDs

AuNPs and QDs are considered an excellent approach for biomarker discovery due to their intrinsic features, such as low cytotoxicity.

By using a combination of antibody microarrays and streptavidin-QDs, several cytokines (TNF-α, IL-8, IL-6, macrophage inflammatory protein-1β, IL-13, and IL-1β) were detected at 1 pM concentrations in cancer samples *(62)*. In 2006, Ghazani *et al.* increased 30-fold the sensitivity in a few cancer biomarkers, such as CEA and Her-27Neu, in serum and plasma by using QDs instead of conventional fluorochromes *(63)*.

Also, sensitivity to specific detection of prostate-specific antigen (PSA) was improved by employing a scanometric immunoassay with Au-NPs, reporting a detection limit of 3 fmol in 10% serum *(64)*.

4.2 Surface Plasmon Resonance

In 2009, Ladd *et al.* studied diagnostic BMs in cancer by nanoproteomics approaches based on a combination of antibody microarrays and a SPRi detection system. This work also showed the potential of SPRi as a suitable methodology for studying serum profiles in biomarker discovery projects *(32)*.

In 2011, Borroto *et al.* described that the Nck recruitment to the TCR is dramatically important for ZAP70 activation during thymic development.

TABLE 1 Summary of Biomarkers Discovered by Nanoproteomics Approaches

		Nanotechnology	Biomarkers	References
Cancer disorders				
Colorectal cancer	Diagnosis	Reverse-phase protein microarray	GADPH, PK-M2	*(53)*
	Prognosis	LTQ-MS/MS	EB1	*(54)*
		Reverse-phase protein microarray	EGFR, COX2	*(55)*
	Personalized medicine	Protein microarrays	KRAS	*(56)*
Metabolic disorders				
Diabetes mellitus	Diagnosis	Electrostatic layer-by-layer (LBL) nanoassembly of capsules composed of multilayers of polymer films	Glucose	*(2)*
	Prognosis	SELDI-TOF-MS	Ubiquitin, β2 microglobulin, transthyretin, apolipoprotein A1, C1, and C	*(57,58)*
	Personalized medicine	SELDI-TOF-MS	Haptoglobin	*(59)*
Autoimmune disorders				
Rheumatoid arthritis	Diagnosis	TQ-MS by MRM	Peptides of reactive protein (PCR)	*(2)*
		Peptide-coated nanotube-based biosensor	Cyclic citrulline peptide	
	Prognosis	Protein biochip array	MCP-1, EGF, creactive protein, IL6	*(60)*
	Personalized medicine	Reverse-phase peptide separation (nano-LC-MALDI-TOF/TOF)	Fibronectin, kininogen-1, cartilage acidic protein 1, cartilage oligomeric matriz protein	*(61)*

All the protein–protein interactions involved in this study were characterized by SPR *(65)*.

Bearing in mind SPR-featured characteristics, it is becoming increasingly relevant in biomedical research, for example, in antibody screening, implementation of cancer diagnosis, or label-free detection of small molecules on a microarray chip.

4.3 Microcantilevers and AFM

Recently, several bioassays have been developed for the quantitative detection PSA over a wide range of different concentrations using microcantilevers. PSA was also chosen as the model analyte for the development of immunoassays with a two-dimensional microcantilever array. In summary, these experiments demonstrated microcantilevers as a quite specific and selective technique to detect biomarkers in clinical conditions and concentrations *(66)*.

In the same way, cantilever biosensors offer interesting possibilities for label-free detection of specific protein conformations and antigen–antibody binding assays, for example, by using the ligand-binding domain of human estrogen receptor (ERα-LDB) as the model to distinguish between the conformations of estrogen receptors (ERs) in free and estradiol (E2)-bound *(67)*.

4.4 ESI-MS

Microarrays have allowed the detection of multiple proteins in one single assay. As was previously described, protein microarrays are compatible with label-free methods and labeled probe detection methods. Both detection methods have advantages and disadvantages, as was discussed previously. Recently, some groups have introduced the analysis of a microarray through LA-ICP-MS, increasing the feasibility of the detection of proteins *(40)*. In addition, the combination of LA with microarray is able to provide a HT technique that quantifies protein concentration with a low sample consumption and high selectivity. However, LA-ICP-MS needs to be improved to be totally irrefutable in diagnosis *(68)*.

Hu *et al.* have prepared an antibody microarray with monoclonal antibodies against anti-AFP, anti-CEA, and RAH. Antibodies were dissolved in a glycerol and a SDS solution at a concentration of 0.10 mg/mL in a final volume of 0.20 μl for one spot in a diameter of 1.5 mm. The previously blocked arrays were incubated against the proteins—a mixture of α-fetoprotein (AFP), CEA, and human IgG. Then an aliquot of the labeled antibodies Sm^{3+}–anti-AFP, Eu^{3+}–anti-CEA, and Au-GAH was added at a concentration of 0.50 μg/ml. The results showed that each labeled antibody displayed clear signals and good reproducibility. Furthermore, they mixed two or more antibodies in one spot, which could also specifically and selectively detect each individual signal *(40)*.

Seuma *et al.* have analyzed two key proteins involved in breast cancer through tissue microarrays: MUC-1, which is a membrane-associated protein placed in the luminal surface of many columnar epithelia, and HER2, which is an epidermal growth factor receptor member of the EGFR family up-expressed in 25–40% of invasive breast cancer. Tissue microarrays were built by employing samples from normal gastric mucosa; a normal breast and a breast with cancer were embedded in paraffin. They were incubated against a primary antibody—anti-MUC1 or anti-HER2—and against a secondary antibody conjugated with gold. Finally, they were treated with a silver solution. A section of both normal and pathological tissue was tested by LA-ICP-MS for MUC-1, being a technology unable to discriminate against them. However, the distribution maps for silver and gold were in perfect correlation. On the other hand, tissue microarrays of breast cancer were analyzed through LA for HER2, finding differences in signal intensities among the different samples with differential expression of HER2 *(69)*.

Joo *et al.* have developed an analytical method that allows the determination of the holoceruloplasmin concentration in serum through LA-ICP-MS with a microarray chip. Holoceruloplasmin is a copper-containing protein in plasma taking part in fundamental role in some neurodegenerative diseases such as Parkinson's or Alzheimer's disease. The existing methods to quantify this protein are limited due to their low selectivity and reliability in the measures. This group prepared protein microarrays onto polydimethylsiloxane surfaces where IgG were covalently immobilized through a cross-linker. Then serum samples containing holoceruloplasmin were incubated and captured by the antibodies. The bound between antigen–antibody was determined by a quadrupole ICP-MS with a 213 nm LA system, the images obtained through an optical microscope with a CCD camera. This method offers two key requirements of serum analysis: the low sample consumption, which is about 126 pl, and the high selectivity, which is achieved with the use of specific antibodies against holoceruloplasmin unable to detect other proteins found in serum samples. This protein was measured by both LA-ICP-MS and ELISA, finding high coincidences in the values and being concordant with the literature *(68)*.

Zhang *et al.* have designed a protocol to accurately and simultaneously detect both AFP and free β-human chorionic gonadotropin (hCGβ). These proteins are significant biomarkers of cancer and also help in prenatal detection of fetal malformations. Although there are several methods to measure their concentrations, there is a need for a method for the concurrent detection of both AFP and hCGβ. Microtiter plates were covered with both monoclonal anti-AFP and anti-hCGβ (conjugated with Eu^{3+} and Sm^{3+} chelates, respectively) on the well surface, and were incubated against the sample. Finally, the solutions were introduced in the ICP-MS instrument with a peristaltic pump. The ^{153}Eu and ^{152}Sm chelate concentrations are directly proportional to the concentration of AFP and hCGβ. A restricting point of the process is the ratio of labeled lanthanide to its corresponding antibody because of the variations in its sensitivity due to the proportional signal to the number of atoms of the labels. As a consequence, nanoparticles–antibody bioconjugates that solve this problem without being affected by the binding affinity are in development *(47)*.

Quinn *et al.* have accurately determined the concentration of Smad2, Smad4, and phosphorylated Smad2. With this aim, this group has cultured C2C12 cell line, which is known to express both Smad2 and Smad4, and has induced myogenesis. Through ICPMS they noticed that levels of Smad2, Smad4, and Smad2 phosphorylation did not significantly change during the myogenesis *(70)*.

4.5 CyTof Applications

In 2009, Bandura *et al.* presented the mass cytometer, which has allowed higher sensitivity and opened new potential applications in terms of sample characterization *(41)*.

These authors have determined the phenotype of same leukemia samples and characterized two different cell lines through mass cytometry. KG1a (CD34+ primitive leukemia cell line) and Ramos (B-lymphoblastoid cell line) cell lines were cultured and harvested. Frozen samples of mononuclear cells from blood or bone marrow corresponding to acute monoblastic/monocytic leukemia subtypes were also selected for analysis. Both types of cells were marked with a panel of 20 antibodies selected against lymphoid (CD3, CD4, CD5, CD19, CD20, CD28, CD57) and myeloid (CD13, CD14, CD15, CD33), nonlineage (CD34, CD38, CD117, HLA-DR) and differentiation (CD36, CD56, CD64, CD45, CD45RA) markers. Each antibody against every marker was conjugated to a different tag consisting of a stable isotope belonging to the lanthanide family. The whole combination of antibodies was added to each single sample. The cells were fixed and incubated with a DNA staining. Then samples were analyzed through mass cytometry. As a result, KG1a cells, characterized by showing a phenotype similar to that of primitive hematopoietic stem cells, exhibited a high signal of CD34, low HLA-DR, low CD33, and absence of lymphoid markers. Ramos cells, immature B-cells, display a significant signal of lymphoid markers such as CD20, CD19, and high HLA-DRA. Referring to patient samples, three of them were classified as monoblastic and two as monocytic by flow cytometry, this result being concordant with that obtained by mass cytometry *(41)*.

Ornatsky *et al.* worked in the same way, analyzing both sample patients and cell lines. The cell lines—KG1a, HL-60 (promyelocytic leukemia), Jurkat (acute T-cell leukemia), and Ramos—were cultured and stained with a cocktail of 22 antibodies bound to a different metal tag. All the results were consistent with that referred to in the bibliography through other techniques. KG1a exhibited high levels of CD34. Ramos showed very high levels of CD38 and high levels of CD19 and CD20. Jurkat expressed CD2 and CD3. Finally, HL-60 displayed a high level of CD15 and CD33. Moreover, the specificity of each single antibody allows them to continue with the analysis of patient samples through mass cytometry. Bone marrow samples from patients diagnosed with acute myeloid leukemia and patients with B-cell acute lymphoid leukemia were used with the aim of comparing analysis obtained through mass and flow cytometry. Both methods showed similar results. Employing 20 markers, it was determined that each single sample had highly expressed CD45, but displayed differences in CD4, CD13, CD14, CD36, CD64, and HLA-DR expression. However, one of these samples exhibited high levels of CD13, CD14, CD33, and CD64, showing HLA-DR characteristic levels of acute monocytic leukemia. Despite the differences found, 9 out of 13 markers were similarly expressed both in mass and flow cytometry. The differently expressed markers can be due to the compensation, some problems related to the acquisition or data analysis, or even because of the antibody clone differences or antibody alterations caused by the label-conjugation protocol *(45)*.

Bendall *et al.* have carried out an assay with the aim of characterizing healthy bone marrow. With this purpose they designed an immunephenotyping panel with 13 core and 18 cell surface markers and a functional panel that contains the 13 core markers and also 18 intracellular markers. Three additional parameters were added: a DNA intercalator, the cell length, and a cell viability dye. The final panel provided 52 markers able to characterize different cells of the immune system along with their maturation stages and recognize each single cluster of differentiation expressed by the cells. They have hypothesized that the similitude among the cell stages and the continuance on the maturation stages allow the organization of similar cell phenotypes in clusters, which can be shown in two-dimensional plots. This approximation has allowed them to simultaneously measure 34 parameters in each single cell. They have tested this panel stimulating healthy human bone marrow cells, summarizing the data in a heat map. The results concluded that the expression markers change with the cell response in such a way that the signaling capacities are directly related to the cellular lineage *(42)*.

Newell *et al.* analyzed human CD8+ T cells through mass cytometry in order to characterize surface biomarkers, cytokines, and antigen specificity with modified peptide-MHC. They have built a panel of 17 cell surface markers of human CD8+ T cells. They have also added nine additional markers: six intracellular cytokines, CD107, two cytotoxic granule components, and the perforin and granzyme B. The samples were recollected from six healthy individuals and were stimulated to demonstrate their functional abilities. The analysis was carried out through principal component analysis. This group found that common definitions of naive, central memory, effector memory, and terminal effector subsets correspond to phenotypic clusters in their analysis. Moreover, they have found intermediate phenotypes that connect the clusters to form a continuum, but also exceptional cells that lead the classic definition *(44)*.

Leipod *et al.* have employed mass cytometry to characterize bacterial populations whose cell size is very small in comparison with mammalian cells in such a way that conventional flow cytometry may be limited due to the low fluorescence intensity signal. However, QDs have been introduced as fluorophores, being even more sensitive than mass cytometry. This group has substituted the antibodies by lectins, which are cheaper than antibodies and allow the characterization of bacteria based on their surface carbohydrates. The specificity of lectin conjugates for each particular strain was previously proved, being able to discriminate each strain *(46)*.

Ornatsky *et al.* have designed bioassays with metal-encoded beads functionalized in such a way that COOH groups are exposed to covalently bind mouse IgGs. Their detection is carried out through goat anti-mouse antibody conjugated to a lanthanide group *(45)*.

5 CONCLUSIONS

In the area of biomarker discovery, some biological and technological challenges are still completely unsolved, for example, huge differences in dynamic range and low-abundance proteins and the intrinsic variations between individuals. Recently, nanotechniques and proteomics have been combined as a novel methodological approach to overcome most of the challenges in biomarker detection and identification. In fact, this novel strategy, called nanoproteomics, is progressively adopted by clinicians and biomedical researchers because it allows multiplex detection in HT format with high sensitivity and accuracy; moreover, most of the nanoproteomics approaches require a minimal amount of sample and reagent consumption.

All the methodologies described in this chapter may lead to the characterization of new molecular entities and/or disease-associated molecular modifications for improving diagnosis and patient stratification. However, some nanotechniques still face several limitations, to be resolved for widespread application in biomarker discovery, because more efforts are still required to implement the status of these approaches in clinical standardization, clinical trials, and diagnostic routine. As well as the nanotechniques described here, other relevant nanoapproaches could be considered in the future due to their potential in reliability, robustness, sensitivity, and reliability, together with minimal amounts of reagents and biological samples.

In summary, it is possible to anticipate a significant development in the near future that will make nanoproteomics for the biomarker discovery field more robust, sensitive, reliable, and, above all, biocompatible and environmentally friendly.

ACKNOWLEDGMENTS

We gratefully acknowledge financial support from Health Institute Carlos III of Spain (ISCIII-FEDER, FIS PI1102114 and JCyL SA198-A12-2) and Plataforma de Recursos Biomoleculares y BioInformaticos. N. D. and P. D. are supported by a PhD Scholarship of JCyL (Orden EDU 346/2013).

REFERENCES

1. Gonzalez-Gonzalez, M.; Jara-Acevedo, R.; Matarraz, S. *Eur. J. Pharm. Sci.* **2012**, *45*, 499–506.
2. Dasilva, N.; Diez, P.; Matarraz, S.; Gonzales-Gonzales, M.; Paradinas, S.; Orfao, A.; Fuentes, M. *Sensors* **2012**, *12*, 2284–2308.
3. Segura, V.; Medina-Aunon, J. A.; Guruceaga, E.; Gharbi, S. I.; González-Tejedo, C.; Sánchez del Pino, M. M.; Canals, F.; Fuentes, M.; Ignacio Casal, J.; Martínez-Bartolomé, S.; Elortza, F.; Mato, J. M.; Arizmendi, J. M.; Abian, J.; Oliveira, E.; Gil, C.; Vivanco, F.; Blanco, F.; Albar, J. P.; Corrales, F. J. *J. Proteome Res.* **2013**, *12*, 112–122.
4. Nilsson, C. L.; Berven, F.; Selheim, F.; Liu, H.; Moskal, J. R.; Kroes, R. A.; Sulman, E. P.; Conrad, C. A.; Lang, F. F.; Andrén, P. E.; Nilsson, A.; Carlsohn, E.; Lilja, H.; Malm, J.; Fenyö, D.; Subramaniyam, D.; Wang, X.; Gonzales-Gonzales, M.; Dasilva, N.; Diez, P.; Fuentes, M.; Végvári, Á.; Sjödin, K.; Welinder, C.; Laurell, T.; Fehniger, T. E.; Lindberg, H.; Rezeli, M.; Edula, G.; Hober, S.; Marko-Varga, G. *J. Proteome Res.* **2013**, *12*, 135–150.
5. Matarraz, S.; Gonzalez-Gonzalez, M.; Jara, M.; Orgao, A.; Fuentes, M. *Clin. Transl. Oncol.* **2011**, *13*, 156–161.
6. Ramachandran, N.; Srivastava, S.; LaBaer, J. *Proteomics Clin. Appl.* **2008**, *2*, 1444–1459.
7. Yu, X.; Scheneiderhan-Marra, N.; Joos, T. O. *Clin. Chem.* **2010**, *56*, 376–387.
8. Zhu, H.; Snyder, M. *Curr. Opin. Chem. Biol.* **2003**, *7*, 55–63.
9. Rusmini, F.; Zhong, Z.; Feijen, J. *Biomacromolecules* **2007**, *8*, 1775–1789.
10. Stamos, B.; Loredo, L.; Chand, S.; Phan, T. V.; Zhang, Y. B.; Mohapatra, S.; Rajeshwar, K.; Perera, R. *Anal. Biochem.* **2012**, *424*, 114–123.
11. Fu, J.; Reinhold, J.; Woodbury, N. W. *PLoS One* **2011**, *6*, e18692.
12. Mueller, C.; Liotta, L. A.; Espina, V. *Mol. Oncol.* **2010**, *4*, 461–481.
13. He, M.; Stoevesandt, O.; Taussig, M. J. *Curr. Opin. Biotechnol.* **2008**, *19*, 4–9.
14. He, M.; Taussign, M. J. *Methods Mol. Biol.* **2008**, *484*, 207–215.
15. He, M.; Stoevesandt, O. *Methods Mol. Biol.* **2010**, *615*, 345–356.
16. Ramachandran, N.; Raphael, J. V.; Hainsworth, E.; Sibani, S.; Montor, W. R.; Pacek, M.; Wong, J.; Eljanne, M.; Sanda, M. G.; Hu, Y.; Logvinenko, T.; Labaer, J. *Nat. Methods* **2008**, *5*, 535–538.
17. Gibson, D. S.; Qiu, J.; Mendoza, E. A.; Barker, K.; Rooney, M. E.; LaBaer, J. *Arthritis Res. Ther.* **2012**, *14*, R77.
18. Manzano-Román, R.; Díaz-Martín, V.; González-González, M.; Matarraz, S.; Álvarez-Prado, - A. F.; LaBaer, J.; Orfao, A.; Pérez-Sánchez, R.; Fuentes, M. *J. Proteome Res.* **2012**, *11*, 5972–5982.
19. Wallstron, G.; Anderson, K. S.; LaBaer, J. *Cancer Epidemiol. Biomarkers Prev.* **2013**, *22*, 747–755.
20. Lee, J. R.; Magee, D. M.; Gaster, R. S.; LaBaer, J.; Wang, S. X. *Expert Rev. Proteomics* **2013**, *10*, 65–75.
21. Wu, H. F.; Gopal, J.; Abdelhamid, H. N.; Hasan, N. *Proteomics* **2012**, *12*, 2949–2961.
22. Prapainop, K.; Wentworh, P. *Eur. J. Pharm. Biopharm.* **2011**, *77*, 353–359.
23. Kong, H.; Lu, Y.; Wang, H. *Anal. Chem.* **2012**, *84*, 4258–4261.
24. Rothen-Rutishauser, B.; Kuhn, D. A.; Ali, Z.; Gasser, M.; Amin, F.; Parak, W. J.; Vanhecke, D.; Fink, A.; Gehr, P.; Brandenberger, C. *Nanomedicine* **2013**, http://dx.doi.org/10.2217/nnm.13.24, (online).
25. Ray, S.; Chandra, H.; Srivastava, S. *Biosens. Bioelectron.* **2010**, *25*, 2389–2401.

26. Nicolini, C.; Pechova, E. *Nanomedicine (London)* **2010**, *11*, 677–682.
27. Ogiso, M.; Kobayashi, J.; Imai, T.; Matsuoka, K.; Itoh, M.; Imamura, T.; Okada, T.; Miura, H.; Nishiyama, T.; Matsuda, T.; Hirahashi, T.; Fujieda, S.; Minoura, N.; Hatanaka, K. *Biosens. Bioelectron.* **2013**, *41*, 465–470.
28. Fee, C. J. *Methods Mol. Biol.* **2013**, *996*, 287–312.
29. Alvarez, M.; Farina, D.; Escuela, A. M.; Sendra, J. R.; Lechuga, L. M. *Rev. Sci. Instrum.* **2013**, *84*, 015008.
30. Lausted, C.; Zhiyuan, H.; Hood, L. *Mol. Cell. Proteomics* **2008**, *7*, 2464–2474.
31. Wong, C. L.; Ho, H. P.; Suen, Y. K.; Kong, S. K.; Chen, Q. L.; Yuan, W.; Wu, S. Y. *Biosens. Bioelectron.* **2008**, *24*, 606–612.
32. Ladds, J.; Taylor, A. D.; Pilarik, M.; Homola, J.; Jiang, S. *Anal. Bioanal. Chem.* **2009**, *393*, 1157–1163.
33. Brenstein, M.; Holzel, R.; Bier, F. F. *J. Nanobiotechnol.* **2010**, *8*, 10–13.
34. Meng, Z.; Veenstra, T. D. *J. Proteomics* **2011**, *74*, 2650–2659.
35. Krastins, B.; Prakash, A.; Sarracino, D. A.; Nedelkov, D.; Niederkofler, E. E.; Kiernan, U. A.; Nelson, R.; Vogelsang, M. S.; Vadali, G.; Garces, A.; Sutton, J. N.; Peterman, S.; Byram, G.; Darbouret, B.; Pérusse, J. R.; Seidah, N. G.; Coulombe, B.; Gobom, J.; Portelius, E.; Pannee, J.; Blennow, K.; Kulasingam, V.; Couchman, L.; Moniz, C.; Lopez, M. F. *Clin. Biochem.* **2013**, *46*, 399–410.
36. Zhao, Y.; Brasier, A. R. *Methods* **2013**, *61*, 313–322.
37. Maiolica, A.; Junger, M. A.; Ezkurdia, I.; Aebersold, R. *J. Proteomics* **2012**, *75*, 3495–3513.
38. Cho, C. K.; Drabovich, A. P.; Batruch, I. *J. Proteomics* **2011**, *74*, 2052–2059.
39. Giesen, C.; Waentig, L.; Panne, U.; Jakubowski, N. *Spectrochim. Acta B At. Spectrosc.* **2012**, *76*, 27–39.
40. Hu, S. H.; Zhang, S. C.; Hu, Z. C.; Xing, Z.; Zhang, X. R. *Anal. Chem.* **2007**, *79*, 923–929.
41. Bandura, D. R.; Baranov, V. I.; Ornatsky, O. I.; Antonov, A.; Kinach, R.; Lou, X.; Pavlov, S.; Vorobiev, S.; Dick, J. E.; Tanner, S. D. *Anal. Chem.* **2009**, *81*, 6813–6822.
42. Bendall, S. C.; Simonds, E. F.; Qiu, P.; Amir, E. D.; Krutzik, P. O.; Finck, R.; Bruggner, R. V.; Melamed, R.; Trejo, A.; Ornatsky, O. I.; Balderas, R. S.; Plevritis, S. K.; Sachs, K.; Pe'er, D.; Tanner, S. D.; Nolan, G. P. *Science* **2011**, *332*, 687–696.
43. Benoist, C.; Hacohen, N. *Science* **2012**, *332*, 677–678.
44. Newell, E. W.; Sigal, N.; Bendall, S. C.; Nolan, G. P.; Davis, M. M. *Immunity* **2012**, *36*, 142–152.
45. Ornatsky, O.; Bandura, S.; Baranov, V.; Nitz, M.; Winnik, M. A.; Tanner, S. *J. Immunol. Methods* **2010**, *361*, 1–20.
46. Leipold, M. D.; Ornatsky, O.; Baranov, V.; Whitfield, C.; Nitz, M. *Anal. Biochem.* **2011**, *419*, 1–8.
47. Zhang, S.; Zhang, C.; Xing, Z.; Zhang, X. *Clin. Chem.* **2004**, *50*, 1214–1221.
48. Rosenblatt, K. P.; Bryant-Greenwood, P.; Killian, J. K.; Mehta, A.; Geho, D.; Espina, V.; Petricoin, E. F., 3rd.; Liotta, L. A. *Annu. Rev. Med.* **2004**, *55*, 97–112.
49. Petricoin, E. F.; Belluco, C.; Araujo, R. P.; Liotta, L. A. *Nat. Rev. Cancer* **2006**, *6*, 961–967.
50. Hartwell, L.; Makoff, D.; Paulovich, A.; Ramsey, S.; Swisher, E. *Nat. Biotechnol.* **2006**, *24*, 905–908.
51. Dunn, G. P.; Bruce, A. T.; Ikeda, H.; Old, L. J.; Schreiber, R. D. *Nat. Immunol.* **2002**, *3*, 991–998.
52. Zitvogel, L.; Tesniere, A.; Kroemer, G. *Nat. Rev. Immunol.* **2006**, *6*, 715–727.
53. Aldea, M.; Clofent, J.; Núñez de Arenas, C.; Chamorro, M.; Velasco, M.; Berrendero, J. R.; Navarro, C.; Cuezva, J. M. *Cancer Lett.* **2011**, *311*, 210–218.

54. Sugihara, Y.; Taniguchi, H.; Kushima, R.; Tsuda, H.; Kubota, D.; Ichikawa, H.; Sakamoto, K.; Nakamura, Y.; Tomonaga, T.; Fujita, S.; Kondo, T. *J. Proteomics* **2012**, *75*, 5342–5355.
55. Pierobon, M.; Calvert, V.; Belluco, C.; Garaci, E.; Deng, J.; Lise, M.; Nitti, D.; Mammano, E.; Marchi, F. D.; Liotta, L.; Petricoin, E. *Clin. Colorectal Cancer* **2009**, *8*, 110–117.
56. Heinemann, V.; Douillard, J. Y.; Ducreux, M.; Peeters, M. *Cancer Treat. Rev.* **2013**, *39*, 592–601.
57. Papale, M.; Di Paolo, S.; Magistroni, R.; Lamacchia, O.; Di Palma, A. M.; De Mattia, A.; Rocchetti, M. T.; Furci, L.; Pasquali, S.; De Cosmo, S.; Cignarelli, M.; Gesualdo, L. *Diabetes Care* **2010**, *33*, 2409–2415.
58. Overgaard, A. J.; Hansen, H. G.; Lajer, M.; Pedersen, L.; Tarnow, L.; Rossing, P.; McGuire, J. N.; Pociot, F. *Proteome Sci.* **2010**, *8*, 4.
59. Zhang, R.; Barker, L.; Pinchev, D.; Marshall, J.; Rasamoelisolo, M.; Smith, C.; Kupchak, P.; Kireeva, I.; Ingratta, L.; Jackowski, G. *Proteomics* **2004**, *4*, 244–256.
60. Fabre, S.; Dupuy, A. M.; Dossat, N.; Guisset, C.; Cohen, J. D.; Cristol, J. P.; Daures, J. P.; Jorgensen, C. *Clin. Exp. Immunol.* **2008**, *153*, 188–195.
61. Mateos, J.; Lourido, L.; Fernández-Puente, P.; Calamia, V.; Fernández-López, C.; Oreiro, N.; Ruiz-Romero, C.; Blanco, F. J. *J. Proteomics* **2012**, *75*, 2869–2878.
62. Zajac, A.; Song, A. D.; Qian, W.; Zhukov, T. *Colloids Surf. B Biointerfaces* **2007**, *58*, 309–314.
63. Ghazani, A. A.; Lee, J. A.; Klostranec, J.; Xiang, Q.; Dacosta, R. S.; Wilson, B. C.; Tsao, M. S.; Chan, W. C. *Nano Lett.* **2006**, *6*, 2881–2886.
64. Liang, R. Q.; Tan, C. Y.; Ruan, K. C. *J. Immunol. Methods* **2004**, *285*, 157–163.
65. Borroto, A.; Arellano, I.; Dopfer, E. P.; Prouza, M.; Suchànek, M.; Fuentes, M.; Orfao, A.; Schamel, W. W.; Alarcón, B. *J. Immunol.* **2013**, *190*, 1103–1112.
66. Yue, M.; Stachowiak, J. C.; Lin, H.; Datar, R.; Cote, R.; Majumdar, A. *Nano Lett.* **2008**, *8*, 520–524.
67. Mukhopadhyay, R.; Sumbayev, V. V.; Lorentzen, M.; Kjems, J.; Andreasen, P. A.; Besenbacher, F. *Nano Lett.* **2005**, *5*, 2385–2388.
68. Joo, J.-Y.; Lim, H. B. *J. Anal. At. Spectrom.* **2012**, *27*, 1069–1073.
69. Seuma, J.; Bunch, J.; Cox, C.; McLeod, C.; Bell, J.; Murray, C. *Proteomics* **2008**, *8*, 3775–3784.
70. Quinn, Z. A.; Baranov, V. I.; Tanner, S. D.; Wrana, J. L. *J. Anal. At. Spectrom.* **2002**, *17*, 892–896.

Chapter 7

Mass Spectrometry Imaging in Proteomics and Metabolomics

Benjamin Balluff, Ricardo J. Carreira and Liam A. McDonnell
Center for Proteomics and Metabolomics, Leiden University Medical Center, Leiden, The Netherlands

Chapter Outline

1 THE NEED FOR IMAGING-BASED PROTEOMICS AND METABOLOMICS

Proteins are the essential building blocks of an organism and regulate most of the intra- and intercellular processes. A great number of substrates and products involved in these biochemical reactions are known as metabolites. Metabolites are usually small biochemicals, such as sugars, nucleotides, amines, amino acids, and even lipids. Thus, a metabolite is defined as a small molecule with a molecular weight up to 1000 Da *(1)*. Similar to genomics actions, both proteomics and metabolomics aim for the large-scale investigation of the proteome and metabolome, respectively. However, in comparison to genomics, proteomics and metabolomics technologies have to cope with different challenges, such as the huge dynamic range in protein and metabolite quantities (22 proteins make up 99% of the protein mass in human plasma) and its enormous molecular complexity (10^7 modified proteins) *(2)*.

Comprehensive Analytical Chemistry, Vol. 63. http://dx.doi.org/10.1016/B978-0-444-62651-6.00007-6

This puts high demands on parallelization (multiplexing) and detection sensitivity, especially since it is not possible to amplify the sample material, like in genomics.

Mass spectrometry (MS), a very sensitive and untargeted technology, has become one of the most important analytical approaches in the life sciences. It allows the unlabeled and multiplexed analysis of a broad variety of molecules, ranging from small molecules to proteins, and their modifications in biological samples. Mass spectrometry coupled with chromatographic separation (LC–MS) is currently the work horse in most proteomics and metabolomics laboratories. LC–MS-based approaches are now well established and allow for the large-scale and quantitative proteomic and metabolic profiling of biological samples. In proteomics experiments, these technologies enable the measurement and identification of peptides at a rate of many thousands of sequences per day with femtomole sensitivity *(3)*.

Yet, LC–MS-based approaches have a major downside when it comes to analyzing solid tissue samples. Since sample preparation requires homogenization of the tissue and extraction of the biomolecules, the information about their localization in the tissue is lost. But knowing the localization of molecules in human, animal, or plant tissues is indispensable to understanding the biochemical, physiological, or pathological processes. This is because the molecular composition may vary according to cell type or the disease state of the organism. As tissues are highly complex systems of morphological heterogeneity, a differentiated tissue analysis can only be performed by morphology- or microscopy-based techniques *(4)*.

Non-MS-based methods that account for important spatial information when studying proteins or metabolites in tissues are available. Proteins can be visualized within their morphological context in tissue sections by using antibody-based techniques such as immunofluorescence or immunohistochemistry, which are the gold standard for diagnostic purposes in a clinical pathology setting. Visualization of metabolites *in situ* can be achieved through *in vivo* magnetic resonance spectroscopy, positron emission tomography, or radiolabeling of specific substances. However, these methods require prior knowledge and labeling of the compound to be studied, and therefore cannot be used for the untargeted investigation of samples.

Mass spectrometry imaging (MSI) circumvents these issues by combining MS with microscopy *(5)*. Spatially resolved mass spectrometric experiments directly on the surface of tissue sections enable the reconstruction of images and therefore the localization of hundreds of molecules within their histological context without any labeling. MSI has therefore received considerable interest as an imaging technique for the *ex vivo* molecular interrogation of animal, human, or plant tissue sections *(6)*.

Technological developments in *in situ* ionization techniques have extended the range of molecular classes that can be detected, from elements and small molecules, to macromolecules such as proteins. Advances in the user-friendliness of instrumentation and software have brought this technology even to non-MS-oriented life science and biomedical laboratories, where it has proved its value by providing new insight into the tissues' molecular composition and biochemical processes, and identified novel biomarker candidates for tissue-based diseases such as cancer *(7–9)*.

The content of this chapter covers the basics of MSI with regard to its technological principle, its variants, and important aspects of sample preparation. Due to the special nature of the data, explanations on data analysis strategies and interpretation are also provided. Further experimental implications, such as subsequent protein and metabolite identification strategies or sample staining and coregistration, are discussed as well. In order to make this chapter more complete, the results of several studies that have used MSI for the analysis of proteins and metabolites, including drugs, in tissues are summarized. Finally, recent efforts to improve MSI are addressed, as well as the future direction of a technology in constant development.

2 MASS SPECTROMETRY IMAGING

MSI allows mass spectrometric experiments to be performed on organic or inorganic surfaces while preserving the information of the localization of the measured molecules. The most common method for MSI is to use a focused ionization beam to analyze a small, localized region of the sample. The resulting mass spectrum is stored along with the spatial coordinates of the measured spot. By analyzing an array of positions across the entire sample, images can then be reconstructed by plotting the intensity of a selected mass signal as a function of its position. As MSI works by performing repeated single MS measurements directly on surfaces, the knowledge of the basic concepts of MS are essential for understanding the technique.

2.1 Mass Spectrometry

In mass spectrometry, ionized sample molecules are separated and measured on the basis of their mass (m) to charge (z) ratio, abbreviated m/z. For their detection, the sample molecules travel through three distinct parts of a mass spectrometer. In the first part, the ion source, the molecules are transported into the gas phase and ionized. In the second step, the generated ions are separated in the mass analyzer under high-vacuum according to their m/z value before they hit the ion detector, in a third step. While this classic three-component setup is common for most instruments, mass spectrometers can be distinguished by the choice of method for each one of the three parts. With several different ionization, separation, and detection methods available, the setup of the mass spectrometer has to be carefully chosen according to the aim of the study and the molecular classes to be analyzed. In the following section, we briefly introduce the different molecular classes and address the chemical properties that influence the choice of the instrumental MS setup and sample preparation protocol. We also explain the different and most popular methods for ionization and separation in MSI, which define the abilities and limitations of the experiment.

2.1.1 *Molecular Classes*

Based on their chemical, physical, and biological properties, such as molar mass or functional role in a cell, and also based on the different technical requirements for MS, four main molecular classes are usually distinguished: metabolites, lipids, peptides, and proteins (Table 1) *(10)*. Other molecular groups, such as oligosaccharides and oligonucleotides, are infrequently the focus of MSI studies.

Metabolites are small molecules that participate as substrates or products in metabolic reactions essential for the normal function of a cell. This molecular class comprises a wide range of compounds, from amino acids to lipids, organic acids, and nucleotides *(11)*. The wide range of concentration and different chemical properties make the analysis of these compounds a challenging task. Usually, high-resolution mass spectrometers and chemical derivatization strategies are necessary to resolve isobaric interferences, increase ionization efficiency, and overcome chemical background effects from the matrix-assisted laser desorption/ionization (MALDI) matrix or the tissue matrix itself *(12)*.

TABLE 1 Molecular Classes Usually Covered by MSI

Molecular Class	Mass Range (Da)	Properties	MS Instrument Requirements
Metabolites	<1000	– Wide range of chemical properties and concentrations – Isobaric interferences – High ion density	– High mass resolution – High mass accuracy
Lipids	200–1000	– Structural diversity – High mass defect – High abundance in tissue – Prone to ionization	– High mass resolution – High mass accuracy
Peptides	1000–4000	– Ion suppression – High dynamic range	– Sensitivity – MS/MS – High mass resolution
Proteins	2500–30,000	– High dynamic range – Low ionization efficiency	– Sensitivity

Lipids have been the focus of many MSI studies due to their important role as structural components of cells, signaling molecules, and involvement in disease-related metabolic pathways *(13)*. Their high abundance in tissue and their physical and chemical properties facilitate ionization and MS analysis *(6)*. Lipids are usually detected in positive or negative mode within a mass range of 200–1000 Da.

Historically, most MSI experiments have focused on protein and peptide analysis. However, protein analysis still suffers from low sensitivity, ion suppression issues, and difficulties in identification. This can be partially overcome by performing on-tissue enzymatic digestion with specific enzymes, usually trypsin, to generate peptides. Peptides have higher ionization efficiency and high-resolution measurements or MS/MS experiments can be performed for accurate identification. The drawback is that important information, such as posttranslational modifications, is lost *(3)*.

2.1.2 Techniques for the Ionization of Molecules

The measurability of molecules by MSI is enabled through the local desorption and ionization of the molecules from a surface. In theory, all types of molecules that can undergo these two chemical processes can be measured. Many techniques have been developed or adapted to achieve desorption and ionization of molecules from surfaces, but three different desorption/ionization techniques made their way to commercially available products. The acronyms of these technologies are desorption electrospray ionization (DESI), MALDI, and secondary ion mass spectrometry (SIMS). The principles of these three methods and a comparison of their possibilities and limitations are outlined throughout this section and summarized in Figure 1 and Table 2, respectively.

2.1.2.1 Secondary Ion Mass Spectrometry

SIMS is the oldest of all existing MSI technologies as it was already used in the late 1960s to collect some of the first ion images of monolayers *(18)*. Since then, SIMS has been widely used in material science to study hard surfaces like semiconductors or rocks *(19)*.

The ionization of molecules from a surface is achieved by directing a primary ion beam onto the surface of an object (Figure 1B). Upon impact secondary ions are ejected and transferred under high vacuum to the mass analyzer. The use of a high-energy primary ion beam results in extensive in-source fragmentation that hampers the analysis of compounds larger than 1000 Da. In exchange for the limited mass detection range, SIMS allows the acquisition of high spatial resolution images with pixel sizes in the submicron range and as small as 50 nm *(20)*.

In a biological setting, SIMS has been mainly used for analyzing small molecules, lipids, or inorganic elements in tissue sections on a cellular or

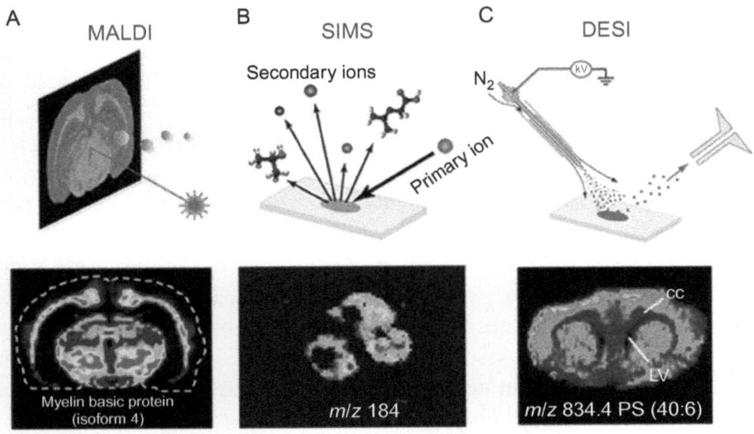

FIGURE 1 Schematic illustrations of MSI ionization techniques. (A) MALDI, (B) SIMS, and (C) DESI and ion maps of (A) myelin basic protein (isoform 4) on sectioned rat brain by MALDI imaging *(14)*, (B) phosphocholine head group on three PC12 cells by SIMS imaging *(15)*, and (C) phosphatidylserine (PS 40:6) on sectioned rat brain by DESI imaging *(16). Reprinted with permission from Liu and Ouyang* (17). *Copyright 2013 by Springer Science and Business Media.* (For color version of this figure, the reader is referred to the online version of this chapter.)

TABLE 2 Commonly Used Ionization Techniques in MSI

	Molecular Classes	Advantages	Disadvantages
MALDI	Proteins, peptides, lipids, metabolites	– Versatile method – High spatial resolution (10–20 μm)	– Molecular coverage and spatial resolution depend on matrix
DESI	Metabolites, lipids	– No sample preparation – On-the-fly derivatization	– Molecular coverage depends on spray composition – Low spatial resolution (100–200 μm)
SIMS	Inorganic ions, metabolites, lipids	– Very high spatial resolution (single cell or subcellular level)	– Limited molecular information – Very high surface sensitivity

subcellular level (Figure 1B). A drawback of SIMS is that an increase in the secondary ion yield also increases the risk of damaging the tissue. With the aim to increase sensitivity in the higher mass domain, research has been performed on the optimal composition of the primary ion beam while maintaining high spatial resolution and minimizing tissue damage. State-of-the-art primary ion beams now use polyatomic ion beams made of gold, bismuth, or C_{60}^{+} ions *(20)*. Although a SIMS experiment requires no prior sample preparation, the sensitivity for higher masses can be increased by coating the sample with a MALDI-like matrix or a noble metal such as gold or silver before analysis. Both matrix-enhanced SIMS (ME-SIMS) and metal-assisted SIMS increase the molecular signals by increasing the amount of surface sputtering that occurs following impact of the focused ion beam on the tissue. This provides increased sensitivities for molecules up to 5000 Da *(21)*.

Another interesting feature of SIMS is the ability to generate depth profiles of the studied surface, hence limited three-dimensional imaging *(22)*. This is achieved by sputtering monolayers repeatedly off the surface. However, as the sputtering needs very high ion doses, such an analysis is usually limited to atoms or very small fragments.

2.1.2.2 DESI

SIMS is considered a "hard ionization" technique due to the intrinsic heavy fragmentation of larger molecules *(23)*. The invention of "soft ionization" techniques such as electrospray ionization (ESI) and MALDI allowed the study of higher mass molecules because of their gentle ionization mechanism and decreased degree of fragmentation *(3)*. For years, both techniques have been successfully used to characterize the protein content of liquid-based extracts. Since the late 1990s and the beginning of the new century, adaptations of these methodologies have been made in order to enable spatially resolved desorption and ionization of larger molecules from biological surfaces.

One of these recent imaging implementations is based on the very well established ESI technique, called DESI. Unlike SIMS, where an ion beam hits the surface of the sample, DESI directs the electrospray to the sample's surface (Figure 1C). Similar to ESI, a high voltage is applied to the nebulizer capillary, which, together with a nebulizing gas and a solvent, creates an electro-aerosol consisting of high-velocity charged microdroplets *(24)*. Upon impact on the surface, the microdroplets desorb the analytes and gas-phase ions are produced by mechanisms similar to ESI, resulting in multiple charged ions *(25)*.

While this method has the advantage of not requiring any sample preparation, there are two major limitations for its use in MSI. First, most of the results reported cover only abundant and active molecules in tissue sections, especially lipids (Figure 1C). Another drawback of DESI is the spatial resolution achieved, approximately 200 μm, a value that is far higher than that obtained by SIMS *(6)*.

However, this is a recent technology and further advances are expected to overcome these limitations. These developments may concern the chemical composition of the solvent or the introduction of sample preparation steps prior to DESI measurements.

2.1.2.3 MALDI

The most popular "soft ionization" technique for MSI is MALDI. Instead of an ion or aerosol beam, a pulsed and focused laser beam is fired against the sample surface in order to initiate the desorption and ionization process (Figure 1A). However, in contrast to the former ionization methods, MALDI requires the sample to be covered with a matrix compound, usually a small organic and aromatic acid. Different matrices enable the ionization/desorption of different types of biomolecules, as summarized in Table 3.

The matrix is usually mixed with acidified organic solvents, such as methanol or acetonitrile, and applied atop and more seldom beneath the tissue sections. During crystallization the analytes are extracted from the tissue and incorporated into the matrix crystals. When the laser beam hits the tissue surface, matrix molecules are excited by the laser energy, resulting in a transition of the embedded sample analytes into the gas phase. During this process, ionization of the analyte molecules (M) takes place, which typically results in singly charged molecules $(M+H)^+$, leading to m/z signals at the nominal mass of the molecule +1 Da *(9)*. The ionization process is mostly performed under vacuum and the resulting gas-phase ions are then transferred to the mass analyzer.

As the matrix absorbs most of the energy, less fragmentation of the analyte compounds occurs, which makes MALDI especially suitable for the ionization of intact large biomolecules such as peptides or proteins (Figure 1A). Although the sensitivity and m/z resolving power decreases with increasing mass, currently most experiments report protein ions <30,000 Da; measurements of proteins with molecular masses up to 100,000 Da have been reported in which special tissue preparation protocols or adapted instrumental setups have been used *(26,27)*. The sensitivity of the mass spectrometer also depends on the selected pixel size of the imaging experiment and the sample amount available at each measurement spot *(20)*. In general, the smaller the spot, the less material to analyze, the higher the demands on the sensitivity of the machine.

TABLE 3 Common MALDI Matrices for MSI

Metabolites	Lipids	Peptides	Proteins
9-Aminoacridine (9-AA)	2,5-Dihydroxybenzoic acid (DHB), 9-AA	α-Cyano-4-hydroxycinnamic acid (CHCA), DHB	CHCA, sinapinic acid (SA)

The spatial resolution offered in MALDI-MSI, also termed MALDI imaging, depends on the laser focus diameter and the average crystal size of the matrix. Recent commercial instrumentation offers spatial resolutions down to 20 μm per pixel.

All methods have their different strengths and limitations, which are summarized in Table 2. When compared to SIMS and DESI, MALDI-MSI provides a good tradeoff between spatial resolution and sensitivity for equally small and higher mass molecules. In addition, as the MALDI instrumentation is commercially available at an affordable cost, this is the most popular MSI method. Consequently, the focus of this chapter is on the MALDI-MSI technique.

2.1.3 Mass Analyzers

Since its introduction many mass analyzers have been used for MSI experiments *(24)*. The vast majority of SIMS and MALDI imaging measurements are currently performed using time-of-flight (ToF) mass analyzers. In a ToF system, all ions are accelerated through a defined ion-acceleration field. The ions then enter a field-free region with a kinetic energy defined by the ion's charge state and the acceleration field strength. In SIMS and MALDI, the majority of ions are singly charged; therefore, all ions enter the field-free region with the same kinetic energy. The velocity of the ions, and thus their ToF through the field-free region, is determined by the strength of the acceleration field and their mass. The ions are separated by the dispersion in their ToF and detected by a microchannel plate or similar detector that provides high sensitivity and high time-resolved detection capabilities.

ToF systems are well suited to the pulsed nature of SIMS and MALDI measurements, and combine many characteristics that are advantageous to imaging experiments. High detection efficiency, parallel detection, and a wide *m/z* detection range have made them ubiquitous for MALDI and SIMS-MSI. Recent improvements in measurement speed have particularly benefited MSI experiments. Modern commercially available MALDI-ToF instruments are equipped with solid-state lasers that allow acquisitions at up to 2 kHz. Such speed is a prerequisite for the development of higher spatial resolution analysis and three-dimensional imaging or for the analysis of large animal/patient series of tissues.

ToF measurements can be performed using a linear or reflectron mass analyzer. The former comprises a basic ToF layout, whereas the latter uses an ion mirror to time-focus the ions onto the detector. The linear detector is used for intact protein analysis owing to its ability to detect a very wide mass range, up to 100,000 Da *(26)*. The reflectron geometry offers higher mass resolution capabilities, but has a more limited mass range, typically below 5000 Da.

MS-based proteomics and metabolomics have led to the development of a wide range of instruments focused on the mass range <5000 Da. This includes

TABLE 4 Mass Analyzer Technologies

	Molecular Classes	Advantages	Disadvantages
ToF	Proteins, peptides, lipids, metabolites	No upper mass limit; high sensitivity	Low mass resolution
FT-ICR	Metabolites, lipids, peptides	Highest mass resolution; accurate mass	Expensive acquisition and maintenance
FT-Orbitrap	Metabolites, lipids, peptides	Very high mass resolution; accurate mass; leading MS/MS capabilities	Imaging sources provided by third-party companies
QQQ	Pharmaceuticals and metabolites	Multiple reaction monitoring: mode for high sensitivity and specificity	Only targeted analysis
Q-ToF	Metabolites, lipids, peptides	Accurate mass; MS/MS; combination with ion mobility	Limited availability and support

ToF analyzers as well as hybrid instruments that combine two or more mass-resolving components, such as quadrupole-ToF (Q-ToF), ion-mobility ToF, and ion-trap-ToF, as well as the high-resolving Fourier transform (FT) analyzer Orbitrap and ion cyclotron resonance (ICR). For targeted analysis, a multiple-reaction monitoring instrument based on triple-quadrupole technologies (QQQ) has provided unrivaled sensitivity for MSI of pharmaceuticals, yet its targeted nature renders it unsuitable for discovery-based research.

High-quality MSI experiments can be performed with almost all modern mass spectrometers provided the tissues have been prepared accordingly and the experiment is well suited to the capabilities of the instrument. The overview provided in Table 4 has been included as a guide to which applications are suited to the different mass analyzers. New developments in mass spectrometers will undoubtedly offer further improvements in sensitivity, speed, mass resolution, spatial resolution, and identification capabilities. Currently, the main factors limiting the broad application of MSI are sample preparation and the efficient analysis of the large, high-dimensionality datasets.

2.2 Sample Preparation

The quality of the results obtained in a MSI experiment are intrinsically related to the sample preparation methodology, which in turn is highly dependent on the molecules to be analyzed and the objective of the study. Briefly, sample preparation can be divided into four major steps:

(i) *Storage*. Tissue samples can be stored for at least 1 year below −80 °C without compromising quality. However, sample degradation starts from the moment the tissue is harvested, affecting the true spatial distribution and abundance of molecules *(28)*. Different methods are available to prevent degradation, from the classical formalin fixation to snap-freezing in dry ice or liquid nitrogen, or recent heat-stabilization technologies that denature proteins at 95 °C, stopping enzymatic activity before storage at −80 °C *(29)*.

(ii) *Sectioning/mounting*. MSI is usually performed on thin tissue sections between 5 and 20 μm. To minimize sample degradation, fresh frozen tissues are sectioned in a cryostat microtome at temperatures between −30 and −12 °C *(10)*. Embedding media such as carboxymethylcellulose or gelatin may be used for fragile samples, but polymer solutions such as the optimal cutting temperature solution should be avoided due to their detrimental interference with MSI experiments. After sectioning, the tissue sections are mounted on conductive glass slides or metal targets cooled to the same temperature as the sample.

(iii) *Tissue treatment*. Tissues may be washed with a set of organic solvents and/or aqueous buffers to remove cofactors. Usually, washing tissue sections has a detrimental effect on the analysis of small molecules such as metabolites and neuropeptides. Yet, when analyzing proteins, washing the tissue may reduce the ion suppression effect and poor crystallization of matrix caused by endogenous cofactors such as lipids and salts *(10)*. On-tissue enzymatic digestion with a specific protease, usually trypsin, may be performed to better analyze and identify proteins through their proteolytic peptides *(30)*. Chemical derivatization may also be used to overcome ion suppression effects and poor ionization efficiency in small molecule analysis *(31)*.

(iv) *Matrix application* (only for MALDI-MSI or ME-SIMS). The choice of matrix, the reproducibility of the application method, and the matrix crystal size are key aspects that directly influence the spatial resolution, sensitivity, and quality of the MSI data *(31)*. As already discussed, the choice of a suitable MALDI matrix depends on the analyte of interest (Table 3). The matrix can be applied manually using a TLC sprayer or automatically with spraying or spotting devices (Table 5) *(32)*. During matrix application, the analytes are extracted from the tissue and incorporated into the resulting matrix crystals. Usually, proteins and peptides require a wet matrix application method to increase the extraction yield. This can be achieved by lowering the percentage of organic solvent in the matrix solution. However, this must be performed carefully to avoid delocalization of the analytes. For lipids and small

TABLE 5 Matrix Deposition Methods

MALDI	Reproducibility	Resolution (μm)	Control of Reagent Volume Deposited
Manual nebulizing	Poor	≥5	Poor
Robotic nebulizing	Good	≥20	Good
Robotic spotting	Excellent	≥150	Excellent
Sublimation	Good	~2	Poor

molecules, which can be extracted more easily and do not need incorporation into the matrix crystals for MALDI MS analysis, dryer matrix application methods such as sublimation may be used.

2.3 Data Processing in MSI

The objective of preprocessing MSI data is the reduction of experimental variance caused by instrumental and chemical noise and variations in sample preparation *(33)*. The preprocessing pipeline typically consists of baseline subtraction, smoothing, normalization, peak picking, and realignment of the datasets *(12,18)*. The importance of preprocessing for the quality of the data and image is illustrated in Figure 2.

Standard baseline subtraction and smoothing methods are applied to each individual MALDI spectrum after measurement or on-the-fly during acquisition in order to reduce the influence of chemical noise (Figure 2D) *(35)*. While this is performed independently on each mass spectrum, other processing steps such as normalization, recalibration, and peak picking have to account for the large number of individual spectra in MSI data.

Normalization is a very important step, as it aims to reduce experimental variance. Normalization is most often performed by dividing each spectrum by a normalization factor (Figure 2G). The most popular normalization factor is calculated as the total ion count (TIC), which is the sum of all ion intensities in a spectrum. Several studies discovered that in MSI the assumptions for TIC applicability hold true only for very homogeneous tissues. In heterogeneous samples, more robust normalization factors based on the median or the TIC with exclusion of very localized mass signals have been proposed *(35–37)*.

Although the inclusion of noise was found to be important for the correct normalization of MSI datasets *(35,36)*, the full data load hampers subsequent statistical analyses *(38)*. The removal of noise signals not only reduces the data load significantly but also aids the semantic interpretation of mass

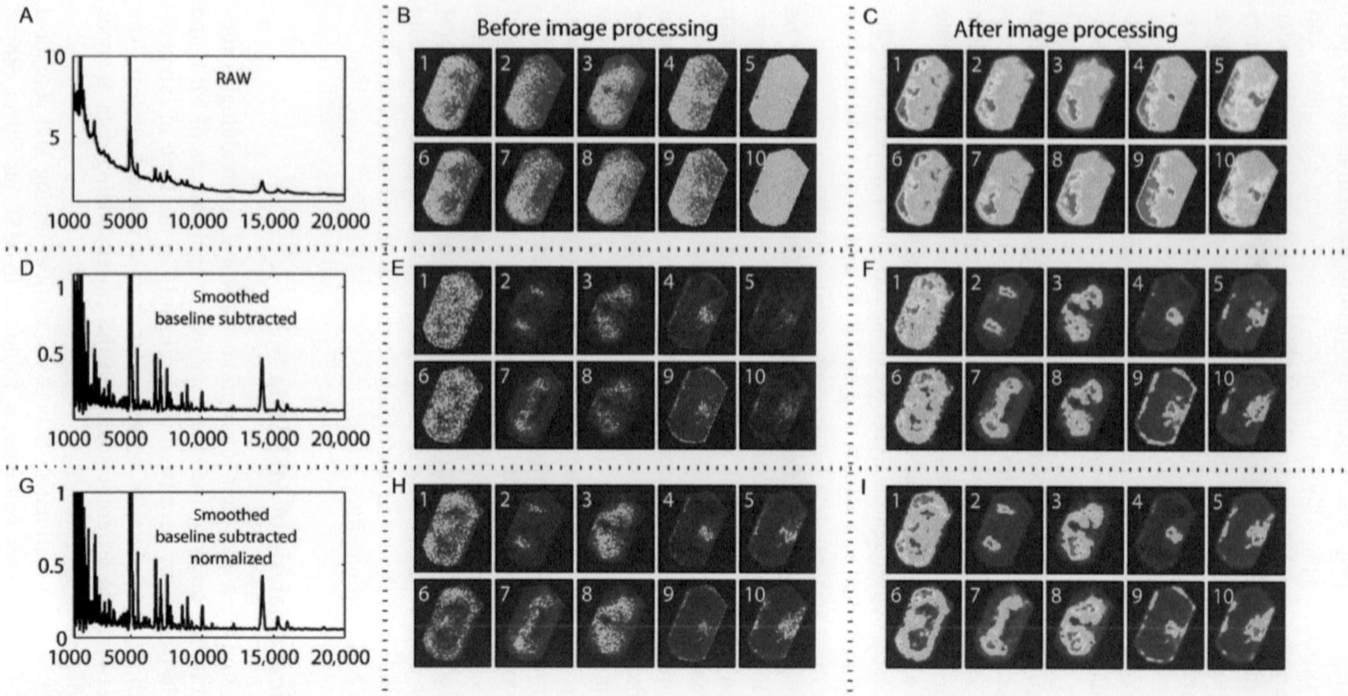

FIGURE 2 Effect of mass spectral and image processing. MALDI imaging analysis of a rat brain tissue section showing the positive effects of mass spectral (A, B, D, E, G, and H) and image processing (C, F, and I) on the total spectra and images of masses (here, 10 example visualizations). *Reprinted with permission from McDonnell* et al. (34). *Copyright 2008 by the American Chemical Society*. (For color version of this figure, the reader is referred to the online version of this chapter.)

spectra. This step is called peak picking, or feature extraction, and is performed after normalization. Although standard peak-picking algorithms can be used, the large amount of spectra in a MSI dataset (usually on the order of 10^3 to 10^4) makes peak picking computationally very intensive. Heuristics have been proposed: the extraction of the features from either a representative subset of spectra *(39)* or from a pooled spectrum based on the mean or maximum values of each mass channel *(38)*.

Finally, mass spectra undergo recalibration and alignment in order to facilitate correct quantitative comparisons of peak intensities. This step is necessary to correct for experimentally derived mass shifts *(33)*. In MSI, mass accuracy deviations are usually corrected using calibrant peaks from standard mixtures that are spotted next to the sample. This recalibration approach corrects for mass shifts to detect the real masses of the molecules. To enable a reliable comparison of spectra, it has been shown to be sufficient to calibrate the spectra on a subset of common endogenous *m/z* signals *(35)*.

Once processed, subsequent analyses may include visual inspection of the distributions of individual *m/z* signals (Figure 1B, E, and H) or statistical evaluations of the mass spectral data between or within samples.

For the visual inspection, additional image processing algorithms may be employed to increase visualization quality (Figure 2C, F, and I). These approaches can use the chemical information in the dataset *(40)*, or can enhance the image quality by regular filtering and smoothing methods *(34)*.

Several commercial and noncommercial software packages implement most of the above preprocessing and image processing techniques, among them the commercial flex-suite from Bruker and SCiLS Lab software (http://scils.de), and the noncommercial Biomap (http://www.maldi-msi.org) or MSI-Reader *(41)*. These programs typically offer the coregistration of additional images to the MSI data. The optical images are typically histological images of the very same tissue analyzed by MSI or of an adjacent tissue section. Knowledge of the underlying morphological constitution is of utmost importance for the interpretation and investigation of the acquired MSI data *(42)*. An annotation of the tissue by a pathologist has therefore become best practice in the MSI community.

2.4 Statistical Analysis of MSI Data

Data analysis strategies and their appropriate statistical methodologies ultimately depend on the goals of the study. As recent applications of MSI, and MALDI-MSI in particular, have been predominantly in the field of biomedical research, this chapter focuses on examples from a clinical context. The relationship between the different goals and the corresponding data analysis strategies are described in Table 6.

In principle, we can distinguish between statistical testing and statistical classification of objects (see columns in Table 6). In MSI, an object can be

TABLE 6 Goals and Methods in MSI Data Analysis

Focus on Spectra	Statistical Testing	Supervised Statistical Classification	Unsupervised Statistical Classification
Within one sample	Finding cell-type specific signals via previous virtual microdissection (Figure 3A)	Spatially resolved annotation of tissue regions after training (e.g., discovery of potentially metastatic cells)	Annotation of tissue based on molecular information by clustering or projection algorithms (Figure 4B)
Between samples	Finding group-associated signals/signatures (e.g., healthy vs. diseased) (Figure 3B)	Predictive classification of samples with unknown status by a trained classifier with samples of known status (e.g., responder, poor survivor)	*Ad hoc* clustering of samples/patients with known or unknown status into different groups (Figure 3B)

either a single mass spectrum within a sample or a representative mass spectrum of a sample (e.g., the mean spectrum of a specific region within a sample) (see rows in Table 6).

Statistical testing is used to find mass signals whose intensities are significantly correlated with a given biological, morphological or clinical endpoint (e.g., tumor vs. nontumor, inflamed vs. noninflamed, or patient disease stage). The first step is to extract the MS data from histologically relevant regions of the sample(s) via so-called virtual microdissection, which is the manual definition of regions of interest within a sample. This results in cell-type-specific molecular MS profiles. The signals of these profiles are then tested individually for statistical correlation with the endpoint or grouping of the data. For comparisons within a sample, the grouping can be the affiliation of the spectra to a certain cell type. An example for this is shown by a study on Barrett's cancer, where MALDI-MSI compared protein profiles from normal, tumor-precursor, and cancer cells (Figure 3A) *(43)*. In the same study, comparisons between patients have been made by using the tumor cell-specific protein profiles of patients with and without metastasis in order to find mass signals (features) for metastatic tumor (Figure 3B) *(43)*.

Testing for such discriminating features can be done by sophisticated statistical techniques for high-dimensionality datasets, such as the significance analysis of microarrays, or classical statistical tests, such as the ANOVA or Kruskal–Wallis tests, corrected by multiple testing methods. If single mass signals are found to correlate with the endpoint of interest, they can be used as a starting point for further investigations. This may be a visual reinspection

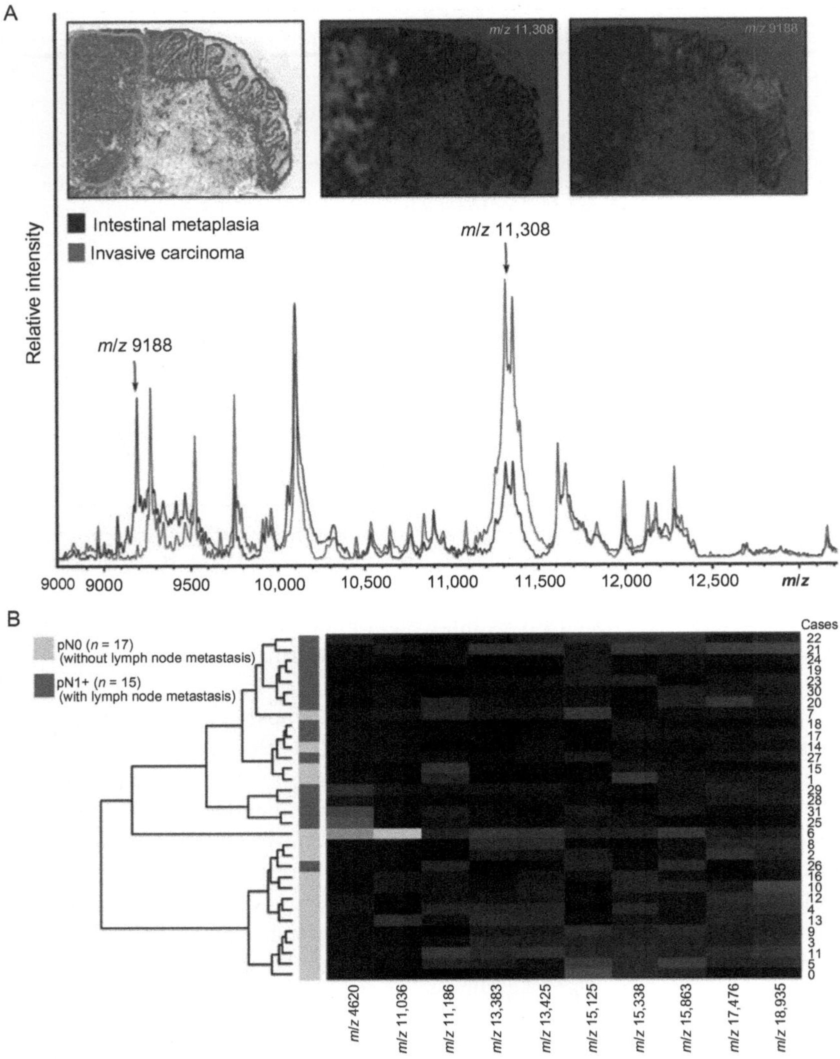

FIGURE 3 Statistical testing and clustering in Barrett's cancer. (A) MALDI imaging reveals cell type-specific profiles for invasive carcinoma (red) and intestinal metaplasia (green) within an individual patient's tissue. Two differentially expressed *m/z* species are marked by arrows: *m/z* 11,308 visualizes for cancer cells, *m/z* 9188 for tumor precursor cells. (B) Unsupervised hierarchical clustering of 32 tumor tissues with (pN1+) and without (pN0) regional lymph node metastasis based on 11 differentially expressed *m/z* signals. *Adapted with permission from Ref.* (43). *Copyright 2012 by Elsevier.* (For interpretation of the references to color in this figure legend, the reader is referred to the online version of this chapter.)

of the distribution of the mass signals in MSI datasets, or an attempt to reveal the molecular identity of the mass signals (see Section 2.5).

The identification of the mass signals is not mandatory if the intention is to use the mass profiles to predict the status of an unknown object. In this case, the discriminating mass signals are combined to train a statistical classifier, known as supervised learning. In MSI, popular multivariate mathematical models to build such classifiers are support vector machines (SVM), random forest, linear discriminant analysis, and artificial neuronal networks. When loaded with input data of an unknown object, the trained classifier can then be used to determine the status of the object. Remember, an object can be a single spectrum or a representative spectrum of a sample/patient. Consequently, classifiers in MSI have been used to classify patients or annotate tissues based on a per-pixel classification. While the first approach usually intends to classify patients according to their disease state, the annotation of tissues may help to detect properties of the cells beyond histology. This has been shown in a study by Willems *et al.* who used a SVM classifier to detect high-grade like tumor cells in intermediate-grade-looking myxofibrosarcoma *(45)*.

Multivariate methods can also be used to investigate the relationships between objects without *a priori* knowledge of a grouping or associations between objects, or unsupervised classification. These methods can be divided into clustering and projection methods. While clustering methods, such as *k*-means or hierarchical clustering, merge objects according to their mass spectral similarity into defined clusters, projection methods, such as the principal component analysis or non-negative matrix factorization, reduce the given multivariate information to a few features while maintaining the information content. Both approaches have been mainly used on a pixel-based level to annotate tissues based on their mass spectral information (Figure 4B) *(44,46)*. An interesting application of this approach is the investigation of histologically challenging tissues. Some studies have used these methods in order to assess the molecular heterogeneity within histologically homogeneous tumors *(47,48)*.

Unsupervised classification or projection can also be used for assessing the relationships between samples. In several studies, it has been shown that cell-type-specific MS profiles can be used to merge samples based on their multivariate molecular information to identify novel patient subgroupings with different clinical behavior (Figure 3B) *(49,50)*.

For a more detailed description on statistical strategies in MSI and explanation of the methods, we strongly recommend the comprehensive paper by Jones *et al.* *(51)*.

2.5 Strategies for the Identification of *m/z* Signals

Statistical analysis can point out the importance of single *m/z* signals, but certainly one of the greatest challenges in MSI is still the disclosure of the

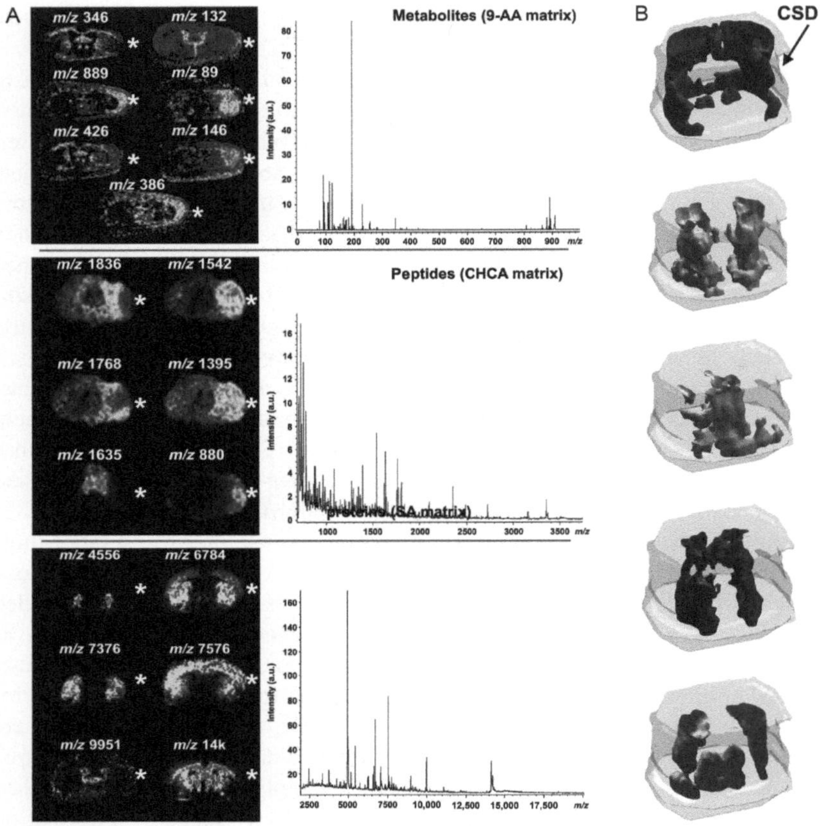

FIGURE 4 Two- and three-dimensional imaging of molecular changes following cortical spreading depression (CSD). (A) MALDI imaging of metabolites (top), peptides (middle), and proteins (bottom) after CSD in wild-type mice, where * indicates the CSD side. Contralateral differences are evident in both the metabolite and peptide datasets. (B) Unsupervised projection using non-negative matrix factorization of the three-dimensional protein MSI dataset of a mouse brain following CSD. *Adapted with permission from Ref.* (44), *Copyright 2012 by Elsevier*. (For color version of this figure, the reader is referred to the online version of this chapter.)

molecular identity of these *m/z* signals. The knowledge of the identity is important as it enables the biological interpretation of the results to further investigate or validate the role of the molecules in the given context beyond the scope of MS.

In standard MS proteomics, well-established and efficient protocols allow for the *en masse* identification of proteins from tissue extracts. A key step within this process is the initial separation and purification of the molecules before submitting them to the mass spectrometer. However, in MSI studies, a general identification strategy for proteins, peptides, and metabolites is still missing due to several reasons. First, in comparison to LC–MS-based approaches, no preceding separation and purification steps are performed on the tissue. Second, most of the preparation protocols from LC–MS are not directly applicable to tissue imaging.

Nevertheless, several groups have managed to establish different strategies in order to identify several *m/z* signals in their studies. Two approaches can be distinguished based on the involvement or exclusion of on-tissue enzymatic digestion.

Identifications from on-tissue without enzymatic digest have been reported *(52,53)*. However, the approach is limited to low mass proteins due to the fact that ion isolation and collision-induced dissociation, as used in commercial MSI instruments, allow reliable and accurate selection and fragmentation of masses up to 4000 Da.

Using enzymatic on-tissue digestion overcomes, this problem as large proteins are cleaved into proteolytic fragments, many of which are structurally informative and lie in the mass range between 1000 and 4000 Da. As the enzyme used for digestion cleaves only before or after specific residues, proteins can be identified based on their peptide sequence by searching against established protein databases. Despite the absence of an explicit purification and separation step, several groups have reported successful identification of *m/z* signals directly from tissue *(50,54)*. The use of an additional separation technique in the mass spectrometer, such as ion mobility for the separation of isobaric peptides, has been shown to increase the number of peptide and protein identifications *(55)*.

Important advances in the number of identified proteins have been made by correlating MSI data with LC–MS analysis of tissue extracts using consecutive tissue sections. The idea is to create a reference database of proteins that can be found in the tissue section *(56)*. Using these databases and on-tissue digestion, several groups were able to identify significantly more peptides by either accurate mass matching of the peptides or querying of MS/MS fragment spectra against this reduced, tissue-specific database *(57,58)*. A recent approach, known as local tissue microextraction, has been reported for the *en masse* identification of peptides or proteins *(59,60)*. Instead of taking extracts from the whole tissue section and submitting it to LC–MS analysis, microextracts were taken from tissue regions of interest previously defined by histology or MSI.

The identification of metabolites in MSI faces similar challenges to the ones described above for protein identification. As in conventional proteomics, nonimaging metabolomics usually couples a preceding separation step to the mass spectrometric analysis. Since this step is missing in MSI, high demands are placed on the mass resolving power of the mass spectrometer, especially in the measurement of small molecules, to resolve the peaks from metabolites with similar masses. Consequently, the method of choice for performing MSI on metabolites relies on ultra-high-resolution instrumentation, such as the FT-ICR *(6)*. The identification, however, can only be successful if a match can be found in a metabolite database. Yet, mass spectral libraries for metabolomics data are still in their infancy, often limiting the identification of a small molecule of interest *(61)*. With increasing interest in metabolomics, a growth in database entries will certainly follow.

3 APPLICATIONS OF MSI

3.1 Applications in Disease Pathology

All presented studies have been performed retrospectively on patient or animal tissues with the aim of identifying molecular signals or signatures for a disease or disease status. While DESI and SIMS focus mainly on the detection of lipids, MALDI imaging can provide molecular signatures ranging from metabolites to proteins.

The application areas of MALDI imaging are very diverse and embrace neurodegenerative and neuronal diseases such as migraine and Parkinson's disease *(44,62)*, skin diseases *(63)*, autoimmune diseases *(64)*, amyloidosis *(65)*, gastrointestinal diseases such as Crohn's disease *(4)*, and osteoarthritis *(66)*. However, the main application field of MALDI imaging is oncology with the aim to discover protein biomarkers for diagnostic and prognostic purposes in all kinds of tumor types, including brain, lung, breast, gastrointestinal, renal, ovarian, skin, and prostate cancers *(7,9)*. The rationale for investigating proteins is that their validity and applicability can be quickly evaluated using established techniques of clinical practice such as immunohistochemistry.

During these studies, MALDI imaging has also disclosed its potential as a diagnostic tool because of its ability to retrieve clinically meaningful and important information in a high-throughput manner, even from minute amounts of tissues, like biopsies. This has been shown by Kim *et al.* who found a diagnostic protein signature obtained from 106 stomach biopsies to identify patients with early-stage gastric cancer and for distinguishing advanced- from early-stage patients, which are potential candidates for endoscopic treatment *(67)*. Similarly, Aichler *et al.* found proteomic changes in pretherapeutic esophageal adenocarcinoma biopsies to be associated with the response of the patients to chemotherapy *(68)*. While this signature has in the first place a predictive potential, the researchers could also elucidate in subsequent *in vitro* experiments the biological role of one of the detected proteins, the mitochondrial protein COX7A2, for chemosensitivity. In other studies, the specific analysis of histologically defined regions also revealed proteomic signatures for both prognosis and recurrence of cancer patients. For instance, a study on 69 melanoma patients found protein signals for the stratification of patients according to their survival expectancy and time of disease recurrence *(69)*.

The successful application of MALDI imaging in clinical proteomics is now being extended to lipids and metabolites in order to make similar discoveries. Some first publications in lipid analysis indicate that similar information can be obtained, although the number of involved patients is still very low *(45,70,71)*.

DESI, in contrast, and despite being a relatively new technique, has already delivered encouraging results in oncological research of lipids in the past few years. In a couple of sample-rich studies, lipid profiles were found to distinguish between diseased and healthy tissues from a variety of different cancers such as human prostate, colorectal, renal cell, bladder, and testis tumors *(72–76)*. Moreover, Eberlin and coworkers reported the use of DESI lipid profiles for the determination of type, grade, and cell concentration in human brain tumors *(77,78)*. DESI has not only been used in an oncological context. It was also used to image the lipid composition of arterial plaques *(79)*.

A similar study was performed using ToF-SIMS, which provides a higher spatial resolution together with the possibility to also investigate inorganic ions *(80)*. Disease applications of ToF-SIMS include the acquisition of high-resolution images of lipid and element distributions in retinas *(81)*, osteoarthritic cartilage *(82)*, and liver diseases *(83)*. SIMS imaging in cancer research has been less frequently used.

Imaging of metabolites is a rapidly growing area of research owing to the absence of alternative methods, and that altered metabolism is implicated in many diseases including neurodegeneration and cancer. The MALDI matrix 9-aminoacridine enables metabolites to be analyzed with even attomole sensitivity as well as image their distributions *(84)*. MALDI imaging has been used to investigate the spatiotemporal behavior of metabolites in the central metabolic pathway regulated by an ischemia reperfusion *(85)* and in the CA3 layer of the mouse hippocampus following ainite-induced seizures *(86)*.

Another recent example confirms the power of MALDI imaging to image molecular changes inaccessible to established methods based on microdissection and tissue extracts. Cortical spreading depression (CSD) is the electrophysiological equivalent of migraine with aura. The lack of clear histopathological features following migraine had severely impeded the characterization of the spatial extent of the molecular changes. MALDI-MSI combined with multivariate analysis could detect extensive disturbances following chemically induced CSD (Figure 4) *(44)*.

3.2 Applications in Drug Imaging

The assessment of the distribution of a drug and its metabolites in tissues can be of high interest for the evaluation of its pharmaceutical properties *(87)*. In comparison to traditional drug imaging modalities, such as autoradiography, MSI offers the label-free and simultaneous imaging of one or several drugs and their metabolites in tissue *(23)*. In consequence, MSI has been rapidly taken up by pharmaceutical companies as part of their lead discovery program and is most often used to highlight drug accumulation in whole-body tissue sections. MSI is now widely used for localizing pharmaceutical compounds in tissues; among them are cancer drugs, such as tyrosine kinase inhibitors (-ibs), psychoactive drugs, drugs for lung diseases, and others. (A full summary is provided in the review of Prideaux and Stoeckli *(8)*.)

One of the practical advantages of MSI is the ability to acquire distribution images much faster than traditional methods *(23)*. A DESI experiment reported only 80 min for the measurement of a whole-body section of a mouse *(88)*. The fact that DESI does not need any previous sample preparation encouraged other groups to use DESI to locate small molecular compounds in single organs or whole-body sections *(89,90)*.

A problem faced by all MSI approaches is the separation of the target molecule from endogenous mass signals. The solutions proposed rely on the measurement of specific fragments of the target compound *(88)*, or on the use of ultra-high-resolving MS *(87)*. Another important issue, which is a basic requirement in pharmacological research, is the absolute quantification of the compounds. Although full quantification is still under development in the field of MSI, initial studies have reported good correlation between the drug signals detected by MSI with those obtained by parallel quantitative LC–MS measurements *(91)*.

4 FUTURE DEVELOPMENTS

MSI complements existing methods for tissue analyses by combining the advantages of microscopic-based techniques with modern MS. Three different ionization methods have enabled a wide range of biological and biomedical research applications, each one offering different advantages, such as high spatial resolution (SIMS), high sensitivity (MALDI), or the measurement of pristine samples (DESI). All methods have proven useful for the acquisition of cell-type-specific molecular profiles or to investigate tissues by enabling the imaging of previously intractable molecules (metabolites, lipids, or drugs).

Several studies have combined results from different MSI approaches or to integrate them with other, non-MS-based imaging techniques. These multimodal studies have reported the successful combination of DESI-MALDI, SIMS-MALDI, and MALDI-MSI/MRI results *(92–94)*. Three-dimensional analysis of tissues or similar samples (cell cultures or microbial colonies) is also getting increased attention (Figure 4B) *(95)*.

In spite of all advantages of MSI, the identification of the *m/z* signals in protein, lipid, or metabolite studies still constitutes a bottleneck. Although several approaches have been proposed (see Section 2.5), an unambiguous identification can only be performed directly by on-tissue analysis. The ability to do so will strongly depend on further advances in MS instrumentation, especially in terms of sensitivity, mass accuracy, and mass resolution coupled with top-down capability for on-tissue analysis. Important future technological developments of instruments will also enable the extension of the mass range for the measurement of several molecular classes at the same time and at higher spatial resolutions.

However, technological advances have to go hand in hand with novel sample preparation protocols. With regard to metabolite imaging, great care needs to be taken in sample preparation, owing to the high solubility of metabolites and their rapid alteration following sacrifice/sampling. Therefore, protocols are being developed that deactivate enzymes present in tissues after collection to minimize the impact of rapid postmortem changes.

As already discussed, new protocols for absolute quantification of *m/z* signals is among the most recent and important efforts in MSI. This will not only help in controlling the reproducibility of experiments, but more importantly, it will enable the absolute comparability between experiments to be assessed. When achieved, this will pave the way for clinical multicenter studies.

We hope that we could familiarize the reader with the technological principle and the possibilities and limitations of MSI, in order to be able to critically judge, but also acknowledge the results and studies performed with this new Omics approach. And maybe this chapter has encouraged the reader to get interested in using this method for his or her own research.

REFERENCES

1. Samuelsson, L. M.; Larsson, D. G. *Mol. Biosyst.* **2008**, *4*, 974–979.
2. Anderson, N. L.; Anderson, N. G. *Mol. Cell. Proteomics* **2002**, *1*, 845–867.
3. Mallick, P.; Kuster, B. *Nat. Biotechnol.* **2010**, *28*, 695–709.
4. Balluff, B.; Rauser, S.; Ebert, M. P.; Siveke, J. T.; Höfler, H.; Walch, A. *Gastroenterology* **2012**, *143*, 544–549.
5. Walch, A.; Rauser, S.; Deininger, S. O.; Hofler, H. *Histochem. Cell Biol.* **2008**, *130*, 421–434.
6. Miura, D.; Fujimura, Y.; Wariishi, H. *J. Proteomics* **2012**, *75*, 5052–5060.
7. McDonnell, L. A.; Corthals, G. L.; Willems, S. M.; van Remoortere, A.; van Zeijl, R. J.; Deelder, A. M. *J. Proteomics* **2010**, *73*, 1921–1944.
8. Prideaux, B.; Stoeckli, M. *J. Proteomics* **2012**, *75*, 4999–5013.
9. Schwamborn, K.; Caprioli, R. M. *Nat. Rev. Cancer* **2010**, *10*, 639–646.
10. Norris, J. L.; Caprioli, R. M. *Chem. Rev.* **2013**, *113*, 2309–2342.
11. Dettmer, K.; Aronov, P. A.; Hammock, B. D. *Mass Spectrom. Rev.* **2007**, *26*, 51–78.
12. Balluff, B.; Schone, C.; Hofler, H.; Walch, A. *Histochem. Cell Biol.* **2011**, *136*, 227–244.
13. Gode, D.; Volmer, D. A. *Analyst* **2013**, *138*, 1289–1315.
14. Seeley, E. H.; Caprioli, R. M. *Proc. Natl. Acad. Sci. U.S.A.* **2008**, *105*, 18126–18131.
15. Lanekoff, I.; Kurczy, M. E.; Hill, R.; Fletcher, J. S.; Vickerman, J. C.; Winograd, N.; Sjövall, P.; Ewing, A. G. *Anal. Chem.* **2010**, *82*, 6652–6659.
16. Wiseman, J. M.; Ifa, D. R.; Song, Q.; Cooks, R. G. *Angew. Chem. Int. Ed. Engl.* **2006**, *45*, 7188–7192.
17. Liu, J.; Ouyang, Z. *Anal. Bioanal. Chem.* **2013**, *405*, 5645–5653.
18. Watrous, J. D.; Alexandrov, T.; Dorrestein, P. C. *J. Mass Spectrom.* **2011**, *46*, 209–222.

19. McPhail, D. S. *J. Mater. Sci.* **2006**, *41*, 873–903.
20. McDonnell, L. A.; Heeren, R. M. *Mass Spectrom. Rev.* **2007**, *26*, 606–643.
21. Chughtai, K.; Heeren, R. M. *Chem. Rev.* **2010**, *110*, 3237–3277.
22. Robinson, M. A.; Graham, D. J.; Castner, D. G. *Anal. Chem.* **2012**, *84*, 4880–4885.
23. Solon, E. G.; Schweitzer, A.; Stoeckli, M.; Prideaux, B. *AAPS J.* **2010**, *12*, 11–26.
24. Pol, J.; Strohalm, M.; Havlicek, V.; Volny, M. *Histochem. Cell Biol.* **2010**, *134*, 423–443.
25. Douglass, K. A.; Venter, A. R. *J. Mass Spectrom.* **2013**, *48*, 553–560.
26. Mainini, V.; Bovo, G.; Chinello, C.; Gianazza, E.; Grasso, M.; Cattoretti, G.; Magni, F. *Mol. Biosyst.* **2013**, *9*, 1101–1107.
27. van Remoortere, A.; van Zeijl, R. J.; van den Oever, N.; Franck, J.; Longuespée, R.; Wisztorski, M.; Salzet, M.; Deelder, A. M.; Fournier, I.; McDonnell, L. A. *J. Am. Soc. Mass Spectrom.* **2010**, *21*, 1922–1929.
28. Skold, K.; Svensson, M.; Norrman, M.; Sjögren, B.; Svenningsson, P.; Andrén, P. E. *Proteomics* **2007**, *7*, 4445–4456.
29. Svensson, M.; Boren, M.; Skold, K.; Fälth, M.; Sjögren, B.; Andersson, M.; Svenningsson, P.; Andren, P. E. *J. Proteome Res.* **2009**, *8*, 974–981.
30. Groseclose, M. R.; Andersson, M.; Hardesty, W. M.; Caprioli, R. M. *J. Mass Spectrom.* **2007**, *42*, 254–262.
31. Goodwin, R. J. *J. Proteomics* **2012**, *75*, 4893–4911.
32. Gustafsson, J. O.; Oehler, M. K.; Ruszkiewicz, A.; McColl, S. R.; Hoffmann, P. *Int. J. Mol. Sci.* **2011**, *12*, 773–794.
33. Hilario, M.; Kalousis, A.; Pellegrini, C.; Muller, M. *Mass Spectrom. Rev.* **2006**, *25*, 409–449.
34. McDonnell, L. A.; van Remoortere, A.; van Zeijl, R. J.; Deelder, A. M. *J. Proteome Res.* **2008**, *7*, 3619–3627.
35. Norris, J. L.; Cornett, D. S.; Mobley, J. A.; Andersson, M.; Seeley, E. H.; Chaurand, P.; Caprioli, R. M. *Int. J. Mass Spectrom.* **2007**, *260*, 212–221.
36. Deininger, S. O.; Cornett, D. S.; Paape, R.; Becker, M.; Pineau, C.; Rauser, S.; Walch, A.; Wolski, E. *Anal. Bioanal. Chem.* **2011**, *401*, 167–181.
37. Fonville, J. M.; Carter, C.; Cloarec, O.; Nicholson, J. K.; Lindon, J. C.; Bunch, J.; Holmes, E. *Anal. Chem.* **2012**, *84*, 1310–1319.
38. McDonnell, L. A.; van Remoortere, A.; de Velde, N.; van Zeijl, R. J.; Deelder, A. M. *J. Am. Soc. Mass Spectrom.* **2010**, *21*, 1969–1978.
39. Alexandrov, T. *BMC Bioinformatics* **2012**, *13*(Suppl 16), S11.
40. Alexandrov, T.; Becker, M.; Deininger, S. O.; Ernst, G.; Wehder, L.; Grasmair, M.; von Eggeling, F.; Thiele, H.; Maass, P. *J. Proteome Res.* **2010**, *9*, 6535–6546.
41. Robichaud, G.; Garrard, K. P.; Barry, J. A.; Muddiman, D. C. *J. Am. Soc. Mass Spectrom.* **2013**, *24*, 718–721.
42. Rauser, S.; Deininger, S. O.; Suckau, D.; Hofler, H.; Walch, A. *Expert Rev. Proteomics* **2010**, *7*, 927–941.
43. Elsner, M.; Rauser, S.; Maier, S.; Schöne, C.; Balluff, B.; Meding, S.; Jung, G.; Nipp, M.; Sarioglu, H.; Maccarrone, G.; Aichler, M.; Feuchtinger, A.; Langer, R.; Jütting, U.; Feith, M.; Küster, B.; Ueffing, M.; Zitzelsberger, H.; Höfler, H.; Walch, A. *J. Proteomics* **2012**, *75*, 4693–4704.
44. Jones, E. A.; Shyti, R.; van Zeijl, R. J.; van Heiningen, S. H.; Ferrari, M. D.; Deelder, A. M.; Tolner, E. A.; van den Maagdenberg, A. M.; McDonnell, L. A. *J. Proteomics* **2012**, *75*, 5027–5035.
45. Willems, S. M.; van Remoortere, A.; van Zeijl, R.; Deelder, A. M.; McDonnell, L. A.; Hogendoorn, P. C. *J. Pathol.* **2010**, *222*, 400–409.

46. Fonville, J. M.; Carter, C. L.; Pizarro, L.; Steven, R. T.; Palmer, A. D.; Griffiths, R. L.; Lalor, P. F.; Lindon, J. C.; Nicholson, J. K.; Holmes, E.; Bunch, J. *Anal. Chem.* **2013**, *85*, 1415–1423.
47. Deininger, S. O.; Ebert, M. P.; Futterer, A.; Gerhard, M.; Rocken, C. *J. Proteome Res.* **2008**, *7*, 5230–5236.
48. Jones, E. A.; van Remoortere, A.; van Zeijl, R. J.; Hogendoorn, P. C.; Bovée, J. V.; Deelder, A. M.; McDonnell, L. A. *PLoS One* **2011**, *6*, e24913.
49. Balluff, B.; Rauser, S.; Meding, S.; Elsner, M.; Schöne, C.; Feuchtinger, A.; Schuhmacher, C.; Novotny, A.; Jütting, U.; Maccarrone, G.; Sarioglu, H.; Ueffing, M.; Braselmann, H.; Zitzelsberger, H.; Schmid, R. M.; Höfler, H.; Ebert, M. P.; Walch, A. *Am. J. Pathol.* **2011**, *179*, 2720–2729.
50. Djidja, M. C.; Claude, E.; Snel, M. F.; Francese, S.; Scriven, P.; Carolan, V.; Clench, M. R. *Anal. Bioanal. Chem.* **2010**, *397*, 587–601.
51. Jones, E. A.; Deininger, S. O.; Hogendoorn, P. C.; Deelder, A. M.; McDonnell, L. A. *J. Proteomics* **2012**, *75*, 4962–4989.
52. Debois, D.; Bertrand, V.; Quinton, L.; De Pauw-Gillet, M. C.; De Pauw, E. *Anal. Chem.* **2010**, *82*, 4036–4045.
53. Minerva, L.; Clerens, S.; Baggerman, G.; Arckens, L. *Proteomics* **2008**, *8*, 3763–3774.
54. Groseclose, M. R.; Massion, P. P.; Chaurand, P.; Caprioli, R. M. *Proteomics* **2008**, *8*, 3715–3724.
55. Stauber, J.; MacAleese, L.; Franck, J.; Claude, E.; Snel, M.; Kaletas, B. K.; Wiel, I. M.; Wisztorski, M.; Fournier, I.; Heeren, R. M. *J. Am. Soc. Mass Spectrom.* **2010**, *21*, 338–347.
56. Meding, S.; Martin, K.; Gustafsson, O. J.; Eddes, J. S.; Hack, S.; Oehler, M. K.; Hoffmann, P. *J. Proteome Res.* **2013**, *12*, 308–315.
57. Gustafsson, J. O.; Eddes, J. S.; Meding, S.; Koudelka, T.; Oehler, M. K.; McColl, S. R.; Hoffmann, P. *J. Proteomics* **2012**, *75*, 5093–5105.
58. Schober, Y.; Schramm, T.; Spengler, B.; Rompp, A. *Rapid Commun. Mass Spectrom.* **2011**, *25*, 2475–2483.
59. Quanico, J.; Franck, J.; Dauly, C.; Strupat, K.; Dupuy, J.; Day, R.; Salzet, M.; Fournier, I.; Wisztorski, M. *J. Proteomics* **2013**, *79*, 200–218.
60. Schey, K. L.; Anderson, D. M.; Rose, K. L. *Anal. Chem.* **2013**, *85*, 6767–6774.
61. Want, E. J.; Cravatt, B. F.; Siuzdak, G. *Chembiochem* **2005**, *6*, 1941–1951.
62. Stauber, J.; Lemaire, R.; Franck, J.; Bonnel, D.; Croix, D.; Day, R.; Wisztorski, M.; Fournier, I.; Salzet, M. *J. Proteome Res.* **2008**, *7*, 969–978.
63. Lazova, R.; Seeley, E. H.; Keenan, M.; Gueorguieva, R.; Caprioli, R. M. *Am. J. Dermatopathol.* **2012**, *34*, 82–90.
64. Bowlus, C. L.; Seeley, E. H.; Roder, J.; Caprioli, R. M.; Gershwin, M. *Cell. Mol. Immunol.* **2011**, *8*, 237–242.
65. Nakanishi, T.; Ito, M.; Nirasawa, T.; Tsuji, M.; Takubo, T. *Clin. Biochem.* **2013**, *46*, 1595–1600.
66. Cillero-Pastor, B.; Eijkel, G. B.; Kiss, A.; Blanco, F. J.; Heeren, R. M. *Arthritis Rheum.* **2013**, *65*, 710–720.
67. Kim, H. K.; Reyzer, M. L.; Choi, I. J.; Kim, C. G.; Kim, H. S.; Oshima, A.; Chertov, O.; Colantonio, S.; Fisher, R. J.; Allen, J. L.; Caprioli, R. M.; Green, J. E. *J. Proteome Res.* **2010**, *9*, 4123–4130.
68. Aichler, M.; Elsner, M.; Ludyga, N.; Feuchtinger, A.; Zangen, V.; Maier, S. K.; Balluff, B.; Schöne, C.; Hierber, L.; Braselmann, H.; Meding, S.; Rauser, S.; Zischka, H.; Aubele, M.; Schmitt, M.; Feith, M.; Hauck, S. M.; Ueffing, M.; Langer, R.; Kuster, B.; Zitzelsberger, H.; Höfler, H.; Walch, A. K. *J. Pathol.* **2013**, *230*, 410–419.

69. Hardesty, W. M.; Kelley, M. C.; Mi, D.; Low, R. L.; Caprioli, R. M. *J. Proteomics* **2011**, *74*, 1002–1014.
70. Thomas, A.; Charbonneau, J. L.; Fournaise, E.; Chaurand, P. *Anal. Chem.* **2012**, *84*, 2048–2054.
71. Uchiyama, Y.; Hayasaka, T.; Masaki, N.; Watanabe, Y.; Masumoto, K.; Nagata, T.; Katou, F.; Setou, M. *Anal. Bioanal. Chem.* **2013**, http://dx.doi.org/10.1007/s00216-013-7062-3.
72. Dill, A. L.; Eberlin, L. S.; Costa, A. B.; Zheng, C.; Ifa, D. R.; Cheng, L.; Masterson, T. A.; Koch, M. O.; Vitek, O.; Cooks, R. G. *Chemistry* **2011**, *17*, 2897–2902.
73. Dill, A. L.; Eberlin, L. S.; Zheng, C.; Costa, A. B.; Ifa, D. R.; Cheng, L.; Masterson, T. A.; Koch, M. O.; Vitek, O.; Cooks, R. G. *Anal. Bioanal. Chem.* **2010**, *398*, 2969–2978.
74. Eberlin, L. S.; Dill, A. L.; Costa, A. B.; Ifa, D. R.; Cheng, L.; Masterson, T.; Koch, M.; Ratliff, T. L.; Cooks, R. G. *Anal. Chem.* **2010**, *82*, 3430–3434.
75. Gerbig, S.; Golf, O.; Balog, J.; Denes, J.; Baranyai, Z.; Zarand, A.; Raso, E.; Timar, J.; Takats, Z. *Anal. Bioanal. Chem.* **2012**, *403*, 2315–2325.
76. Masterson, T. A.; Dill, A. L.; Eberlin, L. S.; Mattarozzi, M.; Cheng, L.; Beck, S. D.; Bianchi, F.; Cooks, R. G. *J. Am. Soc. Mass Spectrom.* **2011**, *22*, 1326–1333.
77. Eberlin, L. S.; Norton, I.; Dill, A. L.; Golby, A. J.; Ligon, K. L.; Santagata, S.; Cooks, R. G.; Agar, N. Y. *Cancer Res.* **2012**, *72*, 645–654.
78. Eberlin, L. S.; Norton, I.; Orringer, D.; Dunn, I. F.; Liu, X.; Ide, J. L.; Jarmusch, A. K.; Ligon, K. L.; Jolesz, F. A.; Golby, A. J.; Santagata, S.; Agar, N. Y.; Cooks, R. G. *Proc. Natl. Acad. Sci. U.S.A.* **2013**, *110*, 1611–1616.
79. Manicke, N. E.; Nefliu, M.; Wu, C.; Woods, J. W.; Reiser, V.; Hendrickson, R. C.; Cooks, R. G. *Anal. Chem.* **2009**, *81*, 8702–8707.
80. Malmberg, P.; Borner, K.; Chen, Y.; Friberg, P.; Hagenhoff, B.; Månsson, J. E.; Nygren, H. *Biochim. Biophys. Acta* **2007**, *1771*, 185–195.
81. Kim, J. H.; Kim, J. H.; Ahn, B. J.; Park, J. H.; Shon, H. K.; Yu, Y. S.; Moon, D. W.; Lee, T. G.; Kim, K. W. *Biophys. J.* **2008**, *94*, 4095–4102.
82. Cillero-Pastor, B.; Eijkel, G.; Kiss, A.; Blanco, F. J.; Heeren, R. M. *Anal. Chem.* **2012**, *84*, 8909–8916.
83. Debois, D.; Bralet, M. P.; Le Naour, F.; Brunelle, A.; Laprevote, O. *Anal. Chem.* **2009**, *81*, 2823–2831.
84. Edwards, J. L.; Kennedy, R. T. *Anal. Chem.* **2005**, *77*, 2201–2209.
85. Miura, D.; Fujimura, Y.; Yamato, M.; Hyodo, F.; Utsumi, H.; Tachibana, H.; Wariishi, H. *Anal. Chem.* **2010**, *82*, 9789–9796.
86. Sugiura, Y.; Taguchi, R.; Setou, M. *PLoS One* **2011**, *6*, e17952.
87. Castellino, S.; Groseclose, M. R.; Wagner, D. *Bioanalysis* **2011**, *3*, 2427–2441.
88. Kertesz, V.; Van Berkel, G. J.; Vavrek, M.; Koeplinger, K. A.; Schneider, B. B.; Covey, T. R. *Anal. Chem.* **2008**, *80*, 5168–5177.
89. Lanekoff, I.; Thomas, M.; Carson, J. P.; Smith, J. N.; Timchalk, C.; Laskin, J. *Anal. Chem.* **2013**, *85*, 882–889.
90. Wiseman, J. M.; Ifa, D. R.; Zhu, Y.; Kissinger, C. B.; Manicke, N. E.; Kissinger, P. T.; Cooks, R. G. *Proc. Natl. Acad. Sci. U.S.A.* **2008**, *105*, 18120–18125.
91. Koeniger, S. L.; Talaty, N.; Luo, Y.; Ready, D.; Voorbach, M.; Seifert, T.; Cepa, S.; Fagerland, J. A.; Bouska, J.; Buck, W.; Johnson, R. W.; Spanton, S. *Rapid Commun. Mass Spectrom.* **2011**, *25*, 503–510.
92. Amstalden van Hove, E. R.; Blackwell, T. R.; Klinkert, I.; Eijkel, G. B.; Heeren, R. M.; Glunde, K. *Cancer Res.* **2010**, *70*, 9012–9021.

93. Eberlin, L. S.; Liu, X.; Ferreira, C. R.; Santagata, S.; Agar, N. Y.; Cooks, R. G. *Anal. Chem.* **2011**, *83*, 8366–8371.
94. Sinha, T. K.; Khatib-Shahidi, S.; Yankeelov, T. E.; Mapara, K.; Ehtesham, M.; Cornett, D. S.; Dawant, B. M.; Caprioli, R. M.; Gore, J. C. *Nat. Methods* **2008**, *5*, 57–59.
95. Seeley, E. H.; Caprioli, R. M. *Anal. Chem.* **2012**, *84*, 2105–2110.

Chapter 8

Advances in NMR-Based Metabolomics

G.A. Nagana Gowda* and Daniel Raftery*,†
*Department of Anesthesiology and Pain Medicine, Northwest Metabolomics Research Center, University of Washington, Seattle, Washington, USA
†Public Health Sciences Division, Fred Hutchinson Cancer Research Center, Seattle, Washington, USA

Chapter Outline

1 INTRODUCTION

Metabolites, biologically relevant compounds with mass less than 1000 Da, are the downstream products of cellular functions and represent virtually every metabolic function in living systems. Altered metabolic levels often represent defects in the functions of specific genes, transcripts, or enzymes. Hence, measurement of metabolite concentrations using biofluids or tissue specimens offers many avenues, for example, for diagnosing and managing diseases; assessing

Comprehensive Analytical Chemistry, Vol. 63. http://dx.doi.org/10.1016/B978-0-444-62651-6.00008-8

disease risk factors associated with drugs, toxins, and the environment; developing personalized treatment options, as well as applications in plant, food, and other sciences. Due to their ability to analyze a large number of metabolites in a single measurement, nuclear magnetic resonance (NMR) spectroscopy and mass spectrometry (MS) represent powerful analytical tools that are widely used in the metabolomics field. The two methods are quite complementary. NMR detects metabolites through specific atomic nuclei such as ^{1}H, ^{13}C, ^{31}P, or ^{15}N, or their combinations. It detects metabolites noninvasively and the detected signal intensity/area varies linearly with concentration across all detected metabolites. The number of NMR-detectable metabolites in a typical biological sample ranges from approximately 20 to 60, though recent advances have pushed this limit to over 100, as will be discussed. In MS, metabolites are detected based on the mass-to-charge ratio (m/z) of the molecular ion or fragment ion. MS offers high sensitivity, which is about three orders of magnitude higher than NMR, and can detect several hundred metabolites routinely. However, reproducibility can be an issue in MS analysis. Owing to the complexity of biological samples, the NMR or MS data are typically very complex, and hence multivariate methods are often needed to analyze the obtained data. Numerous statistical methods including unsupervised and supervised methods are routinely used to help interpret complex metabolite data and identify distinguishing metabolites or spectral features, as will be discussed. Although numerous applications have been developed over the past few years in the field of metabolomics, this review focuses primarily on the development of new methods, as well as a brief description of some important applications. The reader is also referred to a number of excellent reviews that focus on a broader range of applications in metabolomics *(1–9)*.

2 NMR METHODS

Numerous benefits offered by NMR methods far outweigh its somewhat low sensitivity, making it a unique tool for many applications in metabolomics. Importantly, NMR spectroscopy is extremely reproducible and quantitative; a single internal standard can be used for absolute metabolite quantitation since signals in NMR have the same sensitivity, independent of the properties of the metabolite. Second, a combination of several techniques within NMR spectroscopy enables the unambiguous identification of structures for unknown metabolites, which is a very important capability since a large number (often a majority) of metabolites detected in complex biological samples are still unknown. Third, NMR can easily analyze intact biofluids and tissues with no need for sample separation or preparation. This is important considering the fact that the analytical variability linked to the uncontrollable factors associated with sample preparation is a sizable challenge in metabolomics. Finally, metabolic flux analysis is fast emerging as an important route to tracing contributions of metabolites to specific metabolic pathways *(10–12)*; the ability of NMR to unambiguously identify specific isotopomers is unique and promises identification of metabolic pathways and new molecular targets for drug development *(13)*.

A large variety of biological specimens including blood, urine, bile, cerebrospinal fluid, jejunal aspirate, tissue, cells, and cell extracts are employed to profile metabolites and identify biomarkers associated with various diseases and other biological states. However, to date, the most widely used biological specimens are blood and urine; this is because urine and blood can be obtained by noninvasive or minimally invasive methods. In addition, metabolites in blood and urine can potentially report on the pathological status of virtually all cells in the human body. Other biological specimens that are also useful for identifying biomarkers for diseases, including tissue, bile, and cerebrospinal fluid, are typically more difficult to collect; others such as saliva, tear fluid, lung aspirate, etc., have not been as well studied.

NMR provides a unique and routine method to profile metabolites from the intact tissue sample using high-resolution magic angle spinning (HR-MAS) techniques. Due to the association or close proximity of tissue with disease pathologies, analysis of tissue metabolic profiles represents a highly sensitive method for disease detection. Accordingly, numerous studies have achieved high classification accuracy for distinguishing cancer and control tissue using HR-MAS NMR *(14–17)*. Analysis of tissue specimens is also of high interest because of the ability to collect genomic information on the same sample. The latest technological advancements in HR-MAS NMR enable metabolic profiling of intact tissue using as little as a few nanograms of tissue *(18)*. Highly resolved, liquid-like spectra obtainable for intact tissue offer opportunities for disease diagnosis and prognosis *(14)*. Since virtually no sample preparation is required and metabolite data can be obtained within a few minutes, metabolic profiling of biopsied tissue is promising for translational applications *(16)*.

A large number of investigations have also focused on cell lines or animal models for investigations of disease pathogenesis. While cell studies provide the means for understanding pathogenesis under controlled conditions, studies using animal models enable understanding diseases in a more realistic tissue that are still under controlled conditions. Studies using mouse or rat models, for example, can provide reasonable surrogates for many human diseases.

2.1 One-Dimensional NMR Methods

While many NMR active nuclei such as ^{1}H, ^{13}C, ^{31}P, and ^{15}N have been used to analyze metabolites *(19–21)*, ^{1}H NMR analysis is the most widely used in the field because of the ubiquitous nature of ^{1}H and its high NMR sensitivity. Furthermore, ease of analysis and high-throughput capabilities make one-dimensional (1D) NMR experiments, including the 1D NOESY (nuclear Overhauser enhancement spectroscopy) and CPMG (Carr–Purcell–Meiboom–Gill) pulse sequences with water signal suppression (often using presaturation) *(22)*, the most commonly used. The 1D NOESY is useful for biofluids such as urine and extracted metabolite mixtures that are devoid of macromolecules. The CPMG experiment is often used for samples such as blood serum/plasma that have high concentrations of macromolecules such as proteins and lipids. The CPMG sequence attenuates macromolecular signals and provides high-resolution spectra for small-molecule metabolites. One-dimensional NMR pulse sequences that filter small molecules based on differences in their diffusion coefficients are also used, albeit sparingly, for profiling small-molecule metabolites *(23)*.

2.2 Two-Dimensional NMR Methods

To date, ^{1}H NMR-based metabolomics studies have clearly made a major impact on the metabolomics field. However, ^{1}H 1D NMR is not ideal because it suffers from peak overlap due primarily to the high number of ^{1}H peaks per metabolite, as well as the large dynamic range of metabolite concentrations. These factors limit the number of detectable and quantifiable metabolites to approximately 60 or less, except in certain circumstances. Two-dimensional NMR (2D NMR) experiments significantly improve spectral resolution and hence offer avenues to alleviate the peak overlap problem significantly. From the comparison of results obtained by 1D and 2D NMR, it was shown that 2D spectra allowed better characterization of statistically relevant changes in the low-abundance metabolites *(24)*. However, the use of 2D NMR for metabolomics applications has met with some challenges that are primarily due to longer data acquisition times, larger data sizes, and less convenience for data analysis relative to 1D NMR. Although commonly used ^{1}H–^{1}H 2D experiments such as correlation spectroscopy (COSY) and total correlation spectroscopy (TOCSY) provide well-resolved peaks compared to 1D NMR, many of the cross peaks observed for the same metabolite prove redundant and limit the enhancement of the number of metabolites detected compared to 1D NMR. For heteronuclear 2D experiments such as heteronuclear single quantum coherence spectroscopy (HSQC) and heteronuclear multiple quantum correlation spectroscopy (HMQC) the impact of peak redundancy for a given metabolite is not as large due to the broad chemical shift range (e.g., >200 ppm for ^{13}C). However, the low natural abundance of the heteronuclei contributes to increased experimental time and renders such experiments less attractive, especially for high-throughput applications. In view of such limitations, 2D NMR has been used mainly for the still important tasks of validating peak assignments or identifying the structures of unknown metabolites found after performing 1D NMR-based metabolomics studies. Considering that the advantages of 2D NMR generally outweigh the limitations, it is somewhat surprising that relatively few studies have used the various available 2D NMR experiments. Nevertheless, 2D *J*-resolved spectroscopy, COSY,

TOCSY, HSQC, and HMQC experiments have been applied in a number of metabolomics studies *(24–31)*.

2.3 Isotope-Enhanced NMR Methods

Considering the limited sensitivity of NMR and highly complex spectra of biological samples, isotope labeling promises new avenues in metabolomics through the combination of selectivity, sensitivity, and resolution. To date, isotope labeling using nuclei such as ^{13}C, ^{15}N, ^{2}H, and ^{31}P, either *in vivo* or *ex vivo*, has dramatically improved the ability to identify accurately and quantify low-concentration metabolites in biological samples. Thus far, *in vivo* isotope labeling has been used primarily to target altered metabolic pathways, to better understand the composition and origins of various metabolic products from different pathways, and to measure their fluxes *(12,13,32–36)*. In contrast, *ex vivo* isotope labeling has explored the possibility of enhancing the pool-quantifiable metabolites *(19–21,37,38)*.

2.3.1 In Vivo *Isotope-Enhanced Methods*

The information obtained by *in vivo* isotope labeling is generally complementary to that obtained from global metabolite profiling. While global metabolic profiling provides homeostatic levels of metabolites in health or disease states, isotope labeling offers the ability to trace metabolic pathways and measure fluxes connected to specific substrates and thereby promises important clues to disease mechanisms. NMR spectroscopy is especially powerful as it allows, apart from identification and quantification of metabolites, determination of positional isotopomer distributions derived from precursors enriched with stable isotopes. For example, ^{1}H–^{13}C HSQC or even 1D NMR analysis of cells supplied with ^{13}C-labeled glucose can be used to measure and quantify the downstream metabolic products of glycolysis. Lactate is one of these major downstream products in cancer cells; the lactate pool in ^{13}C-glucose-fed cells consists of two major forms, one labeled with the ^{13}C isotope due to the glycolysis of ^{13}C-glucose, and the other normal (unlabeled) lactate arising from pathways unconnected to ^{13}C-glucose metabolism. NMR detects and distinguishes both forms of lactate, and provides rates at which ^{13}C-glucose is consumed and its products formed, all in a single experiment *(10,39)*. The identification of metabolic products of ^{13}C-labeled glutamine in the malignant transformation of cancer cells was also investigated using NMR *(13)*. It is well known that mammalian cells depend on the catabolism of glucose and glutamine for their viability and growth *(40,41)*. However, it was only known that cancer cells depend on high rates of glucose uptake and metabolism to maintain their viability. Recently, it has been observed that a high rate of glutamine metabolism is also exhibited in cancer cells *(42)*. Through isotope-labeled studies, it was shown that the transcriptional regulatory oncogene

Myc is involved in the catabolism of glutamine, leading to glutamine addiction of the tumor cells; such studies promise novel avenues for establishing enzymatic targets for prevention of the growth of tumor cells *(13)*. While cell studies provide the means for understanding metabolic pathways under controlled conditions, *in vivo* labeling studies using animal models or even human subjects enable an understanding of pathogenesis directly in the relevant tissues *(32–34,43)*.

In vivo isotope labeling has also gained interest for enhancing the sensitivity of metabolite detection using heteronuclear NMR experiments *(44,45)*. Here, unlike specific substrate-based labeling for pathway and flux analysis, metabolites are nonselectively labeled with ^{13}C and/or ^{15}N isotopes, for example, by growing cells in media supplemented with labeled substrates. Two-dimensional NMR experiments involving the labeled isotope then provide rich spectra for labeled metabolites. Using such an approach, nearly 500 peaks were detected by an ^{1}H–^{13}C HSQC experiment in *Bombyx mori* and T87 cultured cells *(45)*. Importantly, uniform isotope enrichment of metabolites enables acquiring homonuclear experiments involving heteronuclei, which provide new avenues for metabolite identification. For example, in ^{13}C-labeled *Escherichia coli* 112 carbon backbone topologies were determined based on a 2D ^{13}C–^{13}C constant time TOCSY NMR experiment *(46)*. Many organisms including bacteria, yeast, and plants can be fully ^{13}C-labeled, and information on metabolites can be derived using unique ^{13}C–^{13}C 2D experiments. The quantitation of metabolites using such 2D experiments may be achieved with no need for internal standards *(47)*.

2.3.2 Ex Vivo *Isotope-Enhanced Methods*

Metabolite profiling based on a specific class of metabolites promises a number of additional opportunities to unravel the complexity of biological mixtures and detect a larger pool of quantifiable metabolites. Considering that biological samples such as blood contain over 5000 metabolites, and that detection of all these metabolites in a single measurement is out of reach for current technology, stable isotope tagging of metabolites based on a specific functional group provides new avenues to enhance the resolution and sensitivity of NMR experiments *(19–21,37,38)*. Heteronuclei such as ^{13}C, ^{15}N, and ^{31}P have been used to tag metabolites as they offer benefits in terms of sensitivity, due to their higher isotopic abundance, and resolution, due to their wide chemical shift dispersion compared to ^{1}H. Detection using ^{1}H-X 2D NMR methods such as HSQC, further adds to the sensitivity and resolution. For example, tagging amine functional groups with ^{13}C isotope-based acetylation or formylation and detection using ^{1}H–^{13}C HSQC 2D experiments provide a useful approach for the analysis of the amine-containing metabolites in complex mixtures *(20,37)*. Similarly, reacting carboxyl functional groups with the ^{15}N isotope-containing ethanolamine tag and detection using

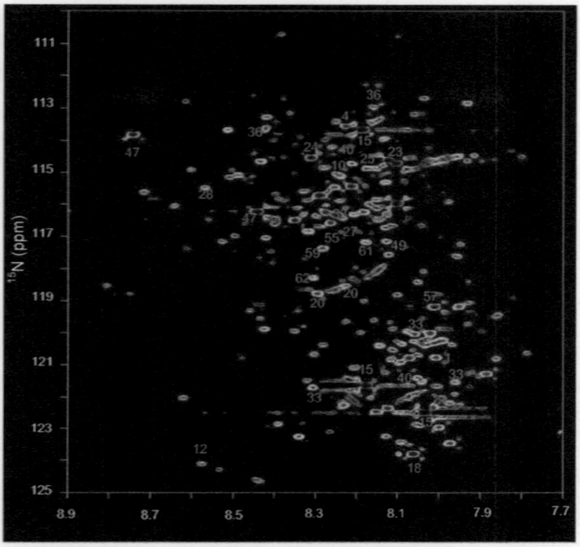

FIGURE 1 Detection of nearly 200 carboxyl-containing metabolites in urine by 2D $^1H–^{15}N$ HSQC NMR after tagging with ^{15}N ethanolamine. *Reproduced with permission from Ref.* (21). (For color version of this figure, the reader is referred to the online version of this chapter.)

$^1H–^{15}N$ 2D NMR provides opportunities to profile carboxyl-containing metabolites *(21)*. Generally, a single peak, devoid of multiplicity, is observed for each tagged metabolite and effective filtering of nontagged metabolites significantly adds to the sensitivity and background suppression, thus enabling detection of a large number of metabolites with improved resolution (Figure 1).

Using this isotope labeling approach, metabolites in human plasma have been quantified *(38)* and altered metabolites have been detected in urine that are due to metabolic disorders such as tyrosinemia type II, argininosuccinic aciduria, homocystinuria, and phenylketonuria *(20)*. Most recently, a smart isotope tag, ^{15}N-cholamine, has been developed for effective detection of the same metabolites using both NMR and MS methods. This approach maximizes the combined strengths of two powerful analytical techniques for a variety of metabolomics applications. ^{15}N-cholamine possesses dual characteristics: an NMR-sensitive heteronuclear isotope with good chemical shift dispersion and a permanent charge that improves MS sensitivity *(48)*.

3 MICRO-COIL NMR

Advances in micro-coil probes offer additional benefits in sensitivity for NMR detection of metabolites and are especially useful for mass limited samples *(49–55)*. Apart from the small coil volume, their solenoidal geometry provides signal enhancement compared to the Helmholtz geometry coils typically employed in conventional, large sample-volume probes *(56)*. Micro-coil probes with a variety of coil sizes and with single- or dual-volume cells have been developed and their performance tested for metabolite analysis *(51,52,57,58)*. Using susceptibility-matched plugs and zero-susceptibility wire, volume efficiency and improved magnetic field homogeneity have been achieved *(52,53)*. Numerous applications that deal with small amounts of sample benefit from signal enhancement by micro-coil probes. For example, the analysis of metabolite fractions eluted from liquid chromatography (LC) columns is one such area of application. Online detection and analysis of compounds using LC-micro-coil-NMR was demonstrated using a combination of 1D NMR and statistical correlation spectroscopy (*vida infra*) *(59)*. Parallel detection by NMR and MS by splitting the chromatography fractions, which offers avenues to exploit the combined strength of the two methods for identification of unknown metabolites, has also been demonstrated *(60,61)*. Considering that the LC fraction obtained from a single chromatographic run is often insufficient for micro-coil NMR detection, methods that incorporate online preconcentration into LC-micro-coil NMR have been developed and applied for the analysis of drug metabolites in biofluids *(62,63)*. Chromotagraphic separation and offline detection using micro-coil NMR has also been demonstrated for confirming the structural identity of plasma metabolites characterized by electrocoulometric array analysis and MS *(55)*. Using a Hadamard-based method, the real-time acquisition of 2D NMR spectra was demonstrated using LC-micro-coil NMR *(64)*. Micro-coil probes integrated with automation accessories are now available commercially, which enable high-throughput analysis and are suitable for routine analysis of small-volume samples for metabolomics applications. Investigations have been made based on such advances; for example, a study recently demonstrated the detection of ovarian cancer based on analysis using 20-μL serum samples from patients with ovarian cancer and matched controls *(65)*.

A recent investigation evaluated the micro-coil NMR method for metabolomics applications by comparing the performance of a 10-μL micro-coil NMR probe with a conventional 5 mm NMR probe emphasizing the analysis of commonly used biofluids, serum, and urine *(54)*. As anticipated, concentration of the samples does improve the sensitivity of detection; however, such an improvement does not follow anticipated linear increase and is dependent on both metabolite and the sample matrix. For example, metabolites with limited solubility, and sample matrices with high salt or protein concentration will adversely impact both relative and absolute concentrations of metabolites upon concentrating samples. Furthermore, the adverse effect of concentration is more pronounced for serum than urine, presumably due to the high protein concentration in serum. Thus, while undoubtedly micro-coil NMR brings superior sensitivity for mass limited samples, care should be exercised while concentrating samples for enhancing sensitivity using micro-coil NMR detection.

4 FAST NMR METHODS

The widespread interest in achieving high-throughput 2D NMR experiments has resulted in a number of new developments in the field. For example, nonlinear sampling and forward maximum entropy reconstruction offer significant reduction in data acquisition times *(66,67)*. High-resolution ^{1}H–^{13}C 2D HSQC spectra were collected for metabolite mixtures using this approach, and a reduction in acquisition time by nearly an order of magnitude was achieved *(66)*. A method that combines nonlinear sampling, forward maximum entropy reconstruction, and *J*-compensated ^{1}H–^{13}C HSQC spectra was shown to achieve a 22-fold reduction in time without the loss of any quantitative information. The method was demonstrated in an analysis of urinary metabolites *(68)*. The SOFAST (band-selective optimized flip angle short transient) approach enables 2D data acquisition in a few seconds through the enhancement of the steady-state magnetization achieved by combining an accelerated T_1 relaxation time *(69)* and optimized flip angle *(70)*. By obtaining SOFAST-^{1}H–^{15}N HMQC within 10–15 s, it was shown for the first time that metabolism in living cells could be monitored in real time *(71)*. Covariance NMR spectroscopy, introduced by Bruschweiler and coworkers, provides high-resolution 2D NMR spectra with minimal data points needed in the t_1 dimension, which significantly enhances the speed of data acquisition needed for high-throughput studies *(72)*. Faster still is the single-scan acquisition method proposed by Frydman that has opened avenues for the *subsecond* acquisition of 2D NMR spectra *(73,74)*. The utility of this very fast 2D NMR method for quantitative analysis of mixtures was evaluated for *J*-resolved and TOCSY experiments; both methods are shown to provide reliable quantitative data, which potentially promise several applications in metabolomics *(75)*.

5 HYPERPOLARIZATION IN NMR

Methods that temporarily boost nuclear polarization can achieve superb sensitivity enhancement to enable real-time monitoring of metabolism *in vivo*. Three major approaches for hyperpolarization—optical pumping of ^{3}He or ^{129}Xe, parahydrogen-induced polarization (PHIP), and dynamic nuclear polarization (DNP)—have demonstrated their utility in a variety of biological applications. The latter two approaches have been particularly successful for hyperpolarizing other nuclei such as ^{13}C or ^{15}N in various substrates for detection of downstream products, and show significant promise for metabolomics applications. The dynamics of metabolic processes *in vivo* can be visualized by hyperpolarization of substrates in aqueous solution followed by infusion, metabolic transformation, and then detection *in vivo*. The metabolic products that remain highly polarized are detected with high sensitivity. Among the hyperpolarization methods that use parahydrogen, the PASADENA (parahydrogen and synthesis allow dramatically enhanced nuclear alignment) is most commonly used *(76,77)*. An advantage of PHIP is that it offers a fast preparation step, on the order of 1 min; however, until recently, this method was restricted to unsaturated molecular precursors, which are necessary for molecular addition of parahydrogen and hence important precursors such as ^{13}C-pyruvate are not amenable to hyperpolarization by this method. Focused on alleviating such limitations, hyperpolarized ^{13}C-lactate was, however, obtained starting from potassium ^{13}C-phosphoenolpyruvate; ^{13}C-lactate is metabolically linked to ^{13}C-pyruvate *(78)*. Recently, it was shown that a temporary association of a substrate and parahydrogen via a transition metal center (without H transfer) enables spontaneous polarization transfer to substrate's nuclei such as ^{1}H, ^{13}C, ^{31}P and ^{19}F, resulting in signal enhancement of up to 800 times *(79)*.

DNP uses paramagnetic centers to transfer polarization from electron spins to neighboring spins of the substrate and is the most widely used hyperpolarization method *(80–82)*. Unlike PHIP, DNP can be used to hyperpolarize many different metabolites, with a wide range of molecular structure. The dissolution DNP approach—wherein hyperpolarization occurs on a solid sample that is dissolved in an appropriate solvent, and that sample is liquified, transported, and injected into a high-resolution NMR spectrometer—has opened new avenues for real-time metabolism studies *(83)*. Numerous efforts have been focused on improving the dissolution DNP method that have enabled, for example, the enhancement of the lifetime of polarized nuclei through deuteration *(84)*. Two-dimensional spectra can be obtained in a single scan by merging with ultrafast acquisition techniques *(81)*, while multiscan correlation spectroscopy has been performed by combining DNP with a flow NMR probe and Hadamard reconstruction *(85)*. Numerous studies have applied the dissolution DNP approach for the metabolism-based studies using a wide range of substrates *(86)*. For example, using hyperpolarization, the performance of ^{13}C-ascorbic and ^{13}C-dehydroascorbic acid, the reduced and oxidized forms of vitamin C, respectively, were evaluated as probes of the tumor redox state *(87)*. While no detectable oxidation was observed for ascorbic acid, reduction of dehydroascorbic acid was detected, consistent with the reduced state of the microenvironment in tumors. Separately, preclinical studies have detected tumor response to drug treatment by measuring the changes in tumor metabolism using hyperpolarized ^{13}C-pyruvate/$^{13}C_2$-fumarate *(88,89)*. Two major drawbacks of DNP are the long hyperpolarization preparation time, on the order of 1 h or more, and the need for an external (expensive) polarizer.

6 STATISTICAL AND DATA ANALYSIS METHODS

The data obtained from NMR spectroscopy are generally complex, and thus multivariate statistical methods are often used to extract the maximum information from the large datasets. Of the multivariate methods, the most frequently used are principal component analysis (PCA), logistic regression, and partial least-squares discriminant analysis (PLS-DA). PCA is often used as a starting point in data analysis. It is an unsupervised method, meaning that it clusters samples without the prior knowledge of the sample class, solely based on the variance of the signals in the metabolite profiles *(90,91)*. Recently, software has been developed to facilitate the quantification of clustering information present in the score plots derived from PCA and PLS-DA through dendrograms *(92,93)*. Supervised methods such as PLS-DA offer the construction of predictive models based on the regression of the data matrix X against matrix Y, which contains class information (e.g., case vs. control) for each sample *(94)*. A number of approaches are used for the selection of the highly ranked metabolites that contribute most to the classification of patients and controls, and the most widely used among these approaches is logistic regression. The validity of such statistical models is then tested using a number of approaches. The most convenient and often-used method is leave-one-out cross-validation analysis, where one sample is removed for model building, then used to test the model; this process is cycled such that each sample is left out one time and model accuracy can be computed. Similarly, leaving 1/3rd, 1/5th, or 1/7th *(95)* of the samples are commonly used variants of this approach. Monte Carlo cross validation (MCCV) *(96)* is often used to further evaluate the robustness and accuracy of the PLS-DA model and has been recently applied to the analysis of NMR-based metabolomics *(97,98)*. Several hundred iterations are typically performed for MCCV by randomly dividing the whole data into training and validation sets for each iteration; a PLS-DA model is developed and cross-validated using the training and validated using the test set. Often a permuted set of data labels is used to compare the results of the average PLS-DA performance. Another approach divides samples into three groups, that is, 1/3rd for model building, 1/3rd for internal testing, and the last 1/3rd for validating the model. This approach can be challenging as it requires a relatively large sample set. The third and best approach is to use an entirely independent set of samples for validating predication models. Validation using independent samples provides the most stringent test, as it helps to alleviate influences of numerous confounding factors. However, obtaining such samples has often proved difficult and continues to be a major challenge.

A number of reviews describe the various steps involved in NMR data analysis *(2,99–101)*. In general, two major approaches are used for statistical analysis of NMR data: a chemometric approach as well as a quantitative metabolomics approach *(102,103)*. The chemometric approach uses the complex NMR spectra directly for analysis after subjecting the data to preprocessing steps such as baseline correction, peak alignment, solvents peak removal, and normalization. Often binning and data scaling are used to account for small peak shifts and better emphasize the low-intensity peaks, respectively. Subsequently, metabolite features that distinguish sample classes are identified and their metabolite identities are established. While this direct approach is relatively straightforward, and can be mostly automated, it has the potential for errors due to imperfect spectral baselines and peak alignments, as well as strong and uneven residual solvent backgrounds. Metabolite peaks in NMR spectra are sensitive to sample conditions and the effects are more pronounced for samples such as urine, which often exhibits varied pH and ionic strength across samples *(104,105)*. Numerous data preprocessing options are currently available, each with a specific advantage *(2,99–101,106–108)*. However, considering the complexity of spectra combined with significant spectral variability across different types of biological samples, it is often challenging to choose the best combination of preprocessing steps without trying several. For example, binning often used in chemometrics is easy to implement, but it restricts resolution and makes it challenging to find the variation of individual metabolites as they often get integrated into a single bin. Moreover, binning noise can adversely affect the outcome of analysis *(109)*. Many approaches including Bayesian parametric modeling that utilizes the free induction decay signal to improve resolution and information recovery *(110)* have been reported to potentially alleviate such problems. However, the combined effect of peak shifts and baseline distortions is still not a completely solved problem and can affect the outcome of the analysis deleteriously.

Quantitative metabolomics, on the other hand, can be described as a targeted approach focused on the analysis of specific metabolite species. In this method, multivariate statistical analysis follows metabolite identification and quantitation. Because of the reliable peak identification and measurement of metabolite integrals, quantitative metabolomics promises greater insights into the dynamics and fluxes of metabolites, as well as robust statistical models for distinguishing classes with better classification accuracy. A major requirement for quantitative metabolomics is good-quality spectral analysis to provide reliable peak assignments and metabolite identification.

7 SPECTRAL ASSIGNMENT AND METABOLITE QUANTITATION METHODS

Considering the high complexity of NMR spectra, the need to develop and improve spectral analysis methods for high-throughput metabolomics applications has received increased attention. A number of investigations are focused on the development of methods for efficient analysis of complex spectra including metabolite identification and quantitation. A relatively large number of published articles as well as several large and publicly available metabolite databases provide a wealth of information for identifying peaks in 1D or 2D spectra of virtually any biological sample *(111–114)*. However, performing manual peak assignment is very tedious, and in addition, requires a significant level of NMR expertise. Significant advances in this area have been made that aid in the analysis of spectra.

7.1 Automation

Automation of peak assignment and quantitation represents an important step, particularly for the analysis of large sample sets with high throughput. A number of different approaches are being applied for improved metabolite identification and quantitation, and these often use various levels of automation. Chenomx provides a semiautomated platform for peak identification and metabolite quantitation using 1D NMR spectra *(115,116)*. Here, metabolites are identified by modeling the spectral features using spectra of pure compounds from pH-dependent chemical shift databases. Metabolite concentrations are then obtained by comparing modeled metabolite peak integrals with that of an internal reference compound such as DSS (4,4-dimethyl-4-silapentane-1-sulfonic acid). The Chenomx algorithm requires manual selection of appropriate metabolites from the library and manual feedback during peak fitting. AMIX, another commercially available software package, provides a number of options for both spectral and statistical analysis using 1D or 2D NMR spectra (http://www.bruker.com), as well as the option to access a very large spectral library. Peak annotation and quantitation of a large number of spectra using a multi-integration tool is especially attractive for high-throughput targeted analysis. An algorithm known as AutoFit, which performs automated spectral deconvolution of 1D NMR spectra, is reported *(117)*. It uses a reference compound library to reconstruct experimental spectra and takes into account a number of fit parameters including peak position, intensity, and peak line width from the experimental spectrum. A Bayesian-fitting algorithm that was recently proposed works fully in an automated mode *(118)*. However, this approach is currently limited to the analysis of relatively small spectral regions due to computational requirements. Another automated approach for annotation and quantification of metabolites in ^{1}H NMR spectra was recently reported *(119)*. This method is based on the generalized fuzzy Hough transform for peak alignment and on PCA for peak selection and quantitation.

Efforts focused on the automated analysis of 2D NMR spectra of complex metabolite mixtures are still somewhat limited. Considering the importance of peak alignment in 2D spectra, a statistical method was developed to align NMR peaks from the same metabolites across multiple 2D NMR spectra *(120)*. A semiautomated method for metabolite identification from 2D TOCSY and HSQC spectra was also developed *(103)*. The software tool, MetaboMiner, makes use of a spectral reference library extracted from publicly available databases to automatically match peaks and identify compounds.

7.2 Analysis of Heteronuclear 2D Spectra

In general, 2D NMR experiments offer significant improvement in spectral resolution due to wider peak dispersion in the 2D plane, facilitating reliable peak assignment and accurate quantitation of metabolites. Heteronuclear 2D experiments are particularly attractive due to the additional resolving power arising from the wider chemical shift range of the heteronuclei such as ^{13}C and ^{15}N. Among the heteronuclear 2D techniques, HSQC has been utilized the most often for metabolite quantitation. However, a challenge in using HSQC spectra for absolute quantitation is that the integral of each 2D NMR peak depends on a number of parameters such as radio frequency pulse power, interpulse delays, the magnitude of heteronuclear spin–spin couplings, and longitudinal (T_1) and transverse (T_2) relaxation times. Hence a single internal or external reference is not sufficient to obtain relative or absolute metabolite concentrations using the 2D peak integrals. Numerous investigations have focused on alleviating this challenge. In one such investigation, a calibration curve for each metabolite was generated from the spectra of standard compound mixtures, and the absolute metabolite concentrations in the test spectra were then obtained by comparing each peak integral with the corresponding calibration curve *(121)*. Another method involved applying a correction factor to the 2D peak integrals calculated from solving the Bloch equations and using product operator formalism to determine metabolite concentrations from 2D ^{1}H–^{13}C HSQC spectra *(122)*. This approach needed no calibration curves or internal references. Calculation of the correction factors involved utilization of T_1 and T_2 relaxation parameters, ^{1}H–^{13}C heteronuclear *J*-coupling, and delays used in the 2D pulse sequence. More recently, a "zero-time" method that requires neither calibration curves nor correction factors for each metabolite was reported *(123,124)*. In this method, a series of HSQC spectra was acquired with incremented evolution times and, from these, a time-zero 2D ^{1}H–^{13}C HSQC spectrum (HSQC$_0$) was generated by extrapolating the peak intensity back to zero time. The peak intensities in the HSQC$_0$ spectrum thus generated were shown to be proportional to the concentrations of individual metabolites. However, this method requires multiple 2D spectra. Separately, an algorithm for deconvoluting 2D peaks using a fast maximum likelihood reconstruction approach was developed and demonstrated for the analysis of metabolites using ^{1}H–^{13}C 2D spectra *(125)*.

8 STOCSY AND RANSY METHODS

Correlation methods provide many opportunities to help identify metabolites in NMR spectra by relating and/or isolating peaks from the same metabolites. A number of papers describe the application of correlation to NMR data. The method, popularly known as *STOCSY* (statistical total correlation spectroscopy), generates correlation coefficients between every pair of 1D NMR peaks. Metabolites are identified based on the set of peaks with high correlations *(126–129)*. Generally, in STOCSY, the correlation analysis is derived from all the 1D NMR spectra from different samples in a dataset; the 2D map of correlations obtained enables easy visualization of all correlations for any peak in 1D NMR spectra. This approach is highly useful for identifying metabolites from overlapped NMR spectra. While STOCSY enables the identification of peaks from the same metabolite, it also provides some connectivity information between peaks from different metabolites that are related biologically and thus have an inherent biological correlation. These additional correlations, while beneficial for biological understanding, can potentially interfere in metabolite identification in complex spectra. SHY (statistical heterospectroscopy), an extension of the STOCSY approach, was developed to correlate data from two different analytical platforms, NMR and MS *(130,131)*. The general principle of data analysis through correlation has been broadly applied by the Nicholson group, and the current state of statistical spectroscopy research and its growing applications have been recently reviewed highlighting the potential benefits of the approach for medicine and systems biology *(132)*.

A new method, RANSY (ratio analysis of NMR spectroscopy), facilitates the isolation of peaks from the same metabolite and thereby improves compound identification using NMR spectroscopy *(133)*. RANSY identifies all the peaks of a specific metabolite on the basis of peak height or integral ratios, and works on the principle that the intensity ratios between NMR peaks from the same metabolite are fixed. When the peak ratios are divided by their coefficients of variation derived from a set of NMR spectra, the individual RANSY spectrum is generated. In contrast, peak ratios derived from unrelated metabolites will typically show large standard deviations, and thus have small RANSY values, except in those rare cases where the metabolites of interest are highly correlated. RANSY was shown to work well for peak identification using both 1D and 2D NMR spectra (Figure 2) and compares favorably to correlation methods. The RANSY concept was recently extended to identify metabolites using mass spectral data *(134)*. Here, the new approach, termed RAMSY (ratio analysis of mass spectrometry), was shown to better identify and isolate mass fragment peaks for the same metabolite using a single mass chromatogram (i.e., multiple mass spectra within a single chromatographic peak). The method potentially promises significant improvement in unknown metabolite identification in MS-based metabolomics, and can be applied to heterospectroscopic datasets as well.

9 APPLICATIONS

NMR spectroscopy-based methods have been applied to numerous areas including biomedical science, toxicology and drug development, food and beverages, environmental science, and more. Such applications have grown exponentially in the past decade, producing more than 1300 papers in 2012 alone, according to a PubMed search. Considering the urgent need to develop better and simpler methods for early disease detection, diagnosis, and therapy monitoring for numerous fatal diseases, investigations in biomedical science

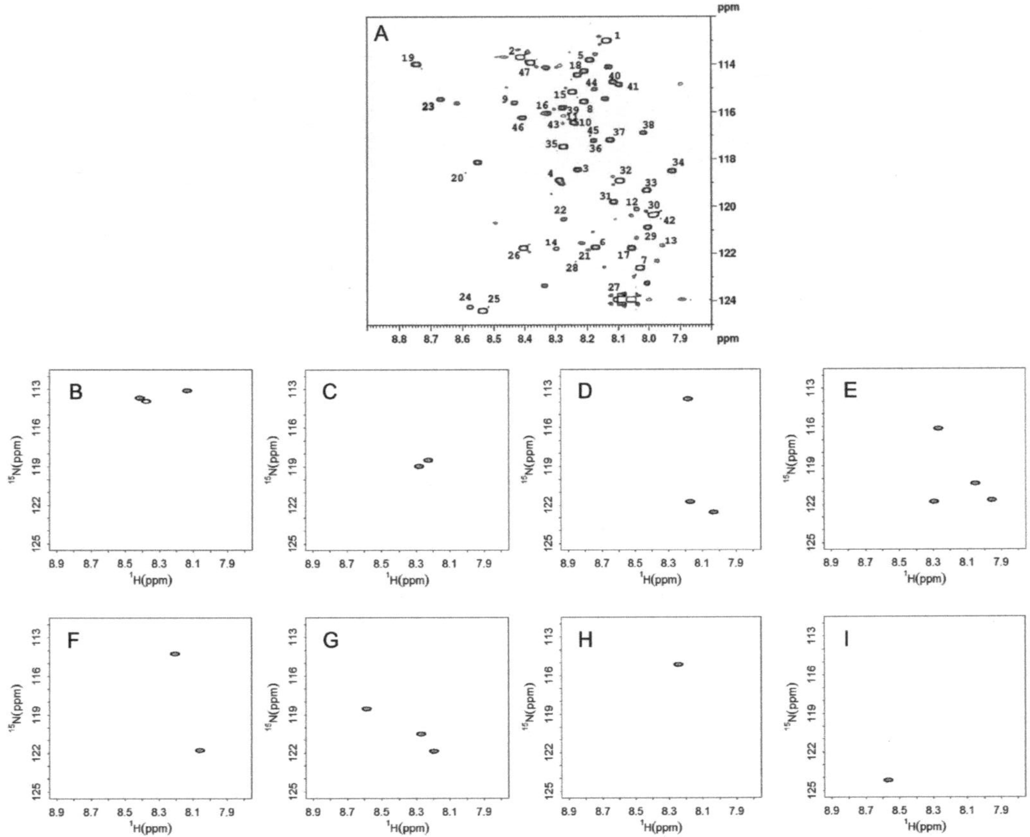

FIGURE 2 (A) ^{1}H–^{15}N HSQC 2D NMR spectrum of a standard mixture containing 33 carboxyl group containing metabolites tagged with ^{15}N label. Multiple peaks for the same metabolites are identified by RANSY. Examples of metabolites detected by RANSY: (B) lactic acid (peaks 1, 2, and 47); (C) glucuronic acid (peaks 3 and 4); (D) citric acid (peaks 5–7); (E) isocitric acid (peaks 11–14); (F) malic acid (peaks 17 and 18); (G) *cis*-aconitic acid (peaks 20–22); (H) leucine (peak 15); and (I) betaine (peak 24). Individual HSQC spectra B–I were generated by thresholding the RANSY ratio of peaks above a value, $R \geq 6.5$. *Reproduced with permission from Ref.* (133).

have been the major focus of metabolomics research to date. In fact, since the beginning of the field, over 90% of the metabolomics investigations have focused on biomedical science. These studies have generated a large body of literature regarding the altered metabolism of pathogeneses in numerous human diseases, and thereby generated much greater insight. A large range of diseases including common diseases such as diabetes, cardiovascular disease, cancer, and neurological diseases, as well as inborn metabolic disorders, have been the subject of numerous investigations. The most widely explored disease is cancer, as virtually all types of cancers are asymptomatic until a late stage, at which time treatment options become limited and the survival rate is poor.

Broadly, NMR-based metabolomics studies in the area of biomedicine have focused on two major applications: (1) understanding disease pathogenesis based on altered metabolic profiles and (2) the discovery of biomarkers for early disease diagnostics. Investigations focused on biomarker discovery have identified numerous metabolite panels that are of value for diagnostic applications. Major efforts have also focused on the translation of findings for clinical applications through commercial ventures *(135)*. Such efforts will benefit from the ability of advanced NMR methods to provide insights into the altered metabolites using small volumes of intact biofluids or tissue with high throughput.

NMR-based methods have also been extensively used in the area of toxicology and drug development. A wealth of information about markers of toxicity has been derived based on the metabolic profiling of tissue or biofluids from animal models obtained after toxin administration *(136)*. The effect of a range of drugs or toxins on the liver and kidney has been explored and several toxicity markers were derived. Drug development includes the assessment of toxicity, and thus NMR-based metabolomics studies have often been focused on drug safety and the effect of drugs on basic metabolism *(137–139)*. Based on the need to develop metabolomics methods and databases for preclinical toxicological screening of drugs, the COMET (Consortium for Metabonomic Toxicology), comprising academia and pharmaceutical industries, extensively evaluated the blood and urine of rodents treated with various toxins using ^{1}H NMR spectroscopy and multivariate statistics *(9,140,141)*. Using a combination of baseline metabolic profiles and histopathological examinations post-dose of commonly used drugs such as acetaminophen, efforts have been made to understand the extent of liver damage; such studies can form a basis for more personalized treatment *(142)*. While drug toxicity studies have benefited significantly from NMR-based metabolomics, other areas are also increasingly exploiting NMR methods. For example, a recent study used NMR to demonstrate that metabolic responses were a sensitive measure of food toxicity *(143)*. Rapid growth in the area of nanoparticles has also witnessed a number of NMR-based metabolomics studies focused on evaluating nanoparticle toxicity *(144,145)*.

Food and beverages comprise another important area explored using NMR-based metabolomics. A variety of food species including fruits, oils, spices, vegetables, and honey have been explored *(5,146–148)*. A number of applications including food-quality testing, storage assessment, shelf-life, and post-harvesting conditions are being explored. Based on metabolic profiles, discrimination between wild and farmed or genetically modified varieties of food has been made and seasonal variation in metabolic profiles has been studied *(147,148)*. Numerous studies have utilized NMR-based metabolomics to explore the association of wine metabolites with environmental and fermentative factors; to classify wines according to variety, region, and vintage; to test adulteration; and to better understand fermentation behavior *(149–152)*. Beers produced at different geographical locations under different fermentation conditions have also been assessed using NMR-based metabolic profiling approaches *(153–155)*.

The interaction of living organisms with their environment is being explored using metabolomics with NMR spectroscopy as one of the major analytical tools *(7)*. Some studies in this area have been focused on the identification of altered metabolites due to a variety of environmental stresses in a number of species including fresh water and marine fish and aquatic and terrestrial invertebrates. For example, a study focused on understanding the elevated levels of carbon dioxide in seawater has used ^{1}H NMR to identify altered metabolites in a marine species in water with different levels of carbon dioxide *(156)*. Separately, with the goal of establishing reliable metabolite indicators of mental stress to aquatic organisms, ^{1}H NMR-based metabolomics studies were made for *Daphnia magna* neonates after exposing them to sublethal concentrations of arsenic, copper, or lithium *(157)*. The toxic effects of pesticides and their degradation products and toxic soil contaminants such as phenanthrene on earthworms were studied using ^{1}H and ^{13}C 1D and 2D NMR-based metabolomics *(158,159)*. Many such investigations have identified a variety of endogenous metabolites as potential biomarkers of exposure to toxic contaminants and hence environmental metabolomics promises a simple method to assess the contaminant bioavailability in the environment. Moreover, exposure to environmental toxins such as cadmium and tobacco in human populations has also been studied using NMR-based metabolomics, and the results demonstrate the utility of such studies for developing risk assessment models as guides to prevent or at least reduce environment- or lifestyle-related diseases *(160)*.

10 CONCLUSIONS

NMR spectroscopy, with its array of unique characteristics and capabilities, is playing a key role in the field of metabolomics. Specifically, the high reproducibility, excellent quantitation and ability to analyze intact samples offer numerous opportunities to understand systems biology, discover biomarkers, and translate laboratory findings on health and diseases to clinical applications. The advances in NMR are multifaceted, with a significant number of investigations focused on improving the sensitivity, resolution, experimental time, and detection of a number of quantifiable metabolites and metabolite identification. Many studies have focused on a very broad range of applications with important implications for a number of disciplines. Clearly, the growth has been tremendous in both basic and applied areas. Numerous approaches including isotope labeling, nonlinear sampling and related methods, covariance NMR, STOCSY and RANSY, and single-scan acquisition and hyperpolarization, used individually or in combination, promise high-throughput as well as real-time studies, greater insights into biological pathways, and disease diagnostics. In sum, the future of NMR-based metabolomics continues to be very promising and exciting.

ACKNOWLEDGMENT

Financial support from the National Institute of General Medical Sciences, National Institutes of Health (Grant No. 2R01GM085291) is gratefully acknowledged.

DISCLOSURE OF POTENTIAL CONFLICT OF INTEREST

Daniel Raftery reports holding equity and an executive role in Matrix-Bio, Inc.

REFERENCES

1. Nicholson, J. K.; Holmes, E.; Kinross, J. M.; Darzi, A. W.; Takats, Z.; Lindon, J. C. *Nature* **2012**, *491*, 384–392.
2. Nagana Gowda, G. A.; Zhang, S.; Gu, H.; Asiago, V.; Shanaiah, N.; Raftery, D. *Expert. Rev. Mol. Diagn.* **2008**, *8*, 617–633.
3. Gebregiworgis, T.; Powers, R. *Comb. Chem. High Throughput Screen.* **2012**, *15*, 595–610.
4. Griffin, J. L.; Atherton, H.; Shockcor, J. P.; Atzori, L. *Nat. Rev. Cardiol.* **2011**, *8*, 630–643.
5. Wishart, D. S. *Trends Food Sci. Technol.* **2008**, *19*, 482–493.
6. Serkova, N. J.; Spratlin, J. L.; Eckhardt, S. G. *Curr. Opin. Mol. Ther.* **2007**, *9*, 572–585.
7. Viant, M. R. *Mol. Biosyst.* **2008**, *4*, 980–986.
8. Lindon, J. C.; Holmes, E.; Nicholson, J. K. *FEBS J.* **2007**, *274*, 1140–1151.
9. Lindon, J. C.; Keun, H. C.; Ebbels, T. M.; Pearce, J. M.; Holmes, E.; Nicholson, J. K. *Pharmacogenomics* **2005**, *6*, 691–699.
10. Lane, A. N.; Fan, T. W. *Metabolomics* **2007**, *3*, 79–86.
11. Lane, A. N.; Fan, T. W.; Higashi, R. M.; Tan, J.; Bousamra, M.; Miller, D. M. *Exp. Mol. Pathol.* **2009**, *86*, 165–173.
12. Fan, T. W.; Lane, A. N. *J. Biomol. NMR* **2011**, *49*, 267–280.
13. Wise, D. R.; DeBerardinis, R. J.; Mancuso, A.; Sayed, N.; Zhang, X. Y.; Pfeiffer, H. K.; Nissim, I.; Daikhin, E.; Yudkoff, M.; McMahon, S. B.; Thompson, C. B. *Proc. Natl. Acad. Sci. U.S.A.* **2008**, *105*, 18782–18787.

14. Jiménez, B.; Mirnezami, R.; Kinross, J.; Cloarec, O.; Keun, H. C.; Holmes, E.; Goldin, R. D.; Ziprin, P.; Darzi, A.; Nicholson, J. K. *J. Proteome Res.* **2013**, *12*, 959–968.
15. Chan, E. C.; Koh, P. K.; Mal, M.; Cheah, P. Y.; Eu, K. W.; Backshall, A.; Cavill, R.; Nicholson, J. K.; Keun, H. C. *J. Proteome Res.* **2009**, *8*, 352–361.
16. Yakoub, D.; Keun, H. C.; Goldin, R.; Hanna, G. B. *Cancer Res.* **2010**, *70*, 9129–9136.
17. Somashekar, B. S.; Kamarajan, P.; Danciu, T.; Kapila, Y. L.; Chinnaiyan, A. M.; Rajendiran, T. M.; Ramamoorthy, A. *J. Proteome Res.* **2011**, *10*, 5232–5241.
18. Wong, A.; Jiménez, B.; Li, X.; Holmes, E.; Nicholson, J. K.; Lindon, J. C.; Sakellariou, D. *Anal. Chem.* **2012**, *84*, 3843–3848.
19. DeSilva, M. A.; Shanaiah, N.; Nagana Gowda, G. A.; Rosa-Pérez, K.; Hanson, B. A.; Raftery, D. *Magn. Reson. Chem.* **2009**, *47*(Suppl. 1), S74–S80.
20. Shanaiah, N.; Desilva, A.; Nagana Gowda, G. A.; Raftery, M. A.; Hainline, B. E.; Raftery, D. *Proc. Natl. Acad. Sci. U.S.A.* **2007**, *104*, 11540–11544.
21. Ye, T.; Mo, H.; Shanaiah, N.; Nagana Gowda, G. A.; Zhang, S.; Raftery, D. *Anal. Chem.* **2009**, *81*, 4882–4888.
22. Nicholson, J. K.; Foxall, P. J. D.; Spraul, M.; Farrant, R. D.; Lindon, J. C. *Anal. Chem.* **1995**, *67*, 793–811.
23. Smith, L. M.; Maher, A. D.; Cloarec, O.; Rantalainen, M.; Tang, H.; Elliott, P.; Stamler, J.; Lindon, J. C.; Holmes, E.; Nicholson, J. K. *Anal. Chem.* **2007**, *79*, 5682–5689.
24. Van, Q. N.; Issaq, H. J.; Jiang, Q.; Li, Q.; Muschik, G. M.; Waybright, T. J.; Lou, H.; Dean, M.; Uitto, J.; Veenstra, T. D. *J. Proteome Res.* **2008**, *7*(2), 630–639.
25. Dumas, M. E.; Canlet, C.; André, F.; Vercauteren, J.; Paris, A. *Anal. Chem.* **2002**, *74*, 2261–2273.
26. Tang, H. R.; Wang, Y.; Nicholson, J. K.; Lindon, J. C. *Anal. Biochem.* **2004**, *325*, 260–272.
27. Xi, Y.; de Ropp, J. S.; Viant, M. R.; Woodruff, D. L.; Yu, P. *Metabolomics* **2006**, *2*, 221–233.
28. Fan, T. W. M.; Bandura, L. L.; Higashi, R. M.; Lane, A. N. *Metabolomics* **2005**, *1*, 325–339.
29. Ludwig, C.; Viant, M. R. *Phytochem. Anal.* **2010**, *21*, 22–32.
30. Parsons, H. M.; Ludwig, C.; Gunther, U. L.; Viant, M. R. *BMC Bioinforma.* **2007**, *8*, 234.
31. Adosraku, R. K.; Choi, G. T.; Constantinou-Kokotos, V.; Anderson, M. M.; Gibbons, W. A. *J. Lipid Res.* **1994**, *35*, 1925–1931.
32. Fan, T. W.; Lane, A. N.; Higashi, R. M.; Farag, M. A.; Gao, H.; Bousamra, M.; Miller, D. M. *Mol. Cancer* **2009**, *8*, 41.
33. Fan, T. W.; Lane, A. N.; Higashi, R. M.; Yan, J. *Metabolomics* **2011**, *7*, 257–769.
34. Locasale, J. W.; Grassian, A. R.; Melman, T.; Lyssiotis, C. A.; Mattaini, K. R.; Bass, A. J.; Heffron, G.; Metallo, C. M.; Muranen, T.; Sharfi, H.; Sasaki, A. T.; Anastasiou, D.; Mullarky, E.; Vokes, N. I.; Sasaki, M.; Beroukhim, R.; Stephanopoulos, G.; Ligon, A. H.; Meyerson, M.; Richardson, A. L.; Chin, L.; Wagner, G.; Asara, J. M.; Brugge, J. S.; Cantley, L. C.; Vander Heiden, M. G. *Nat. Genet.* **2011**, *43*, 869–874.
35. Burgess, S. C.; Babcock, E. E.; Jeffrey, F. M.; Sherry, A. D.; Malloy, C. R. *FEBS Lett.* **2001**, *505*, 163–167.
36. Perdigoto, R.; Furtado, A. L.; Porto, A.; Rodrigues, T. B.; Geraldes, C. F.; Jones, J. G. *Biochim. Biophys. Acta* **2003**, *1637*, 156–163.
37. Ye, T.; Zhang, S.; Mo, H.; Tayyari, F.; Nagana Gowda, G. A.; Raftery, D. *Anal. Chem.* **2010**, *82*, 2303–2309.
38. Nagana Gowda, G. A.; Tayyari, F.; Ye, T.; Suryani, Y.; Wei, S.; Shanaiah, N.; Raftery, D. *Anal. Chem.* **2010**, *82*, 8983–8990.
39. Lloyd, S. G.; Zeng, H.; Wang, P.; Chatham, J. C. *Magn. Reson. Med.* **2004**, *51*, 1279–1282.
40. Coles, N. W.; Johnstone, R. M. *Biochem. J.* **1962**, *83*, 284–291.

41. Eagle, H. *Science* **1955**, *122*, 501–514.
42. DeBerardinis, R. J.; Mancuso, A.; Daikhin, E.; Nissim, I.; Yudkoff, M.; Wehrli, S.; Thompson, C. B. *Proc. Natl. Acad. Sci. U.S.A.* **2007**, *104*, 19345–19350.
43. Lane, A. N.; Fan, T. W.; Bousamra, M., II.; Higashi, R. M.; Yan, J.; Miller, D. M. *OMICS* **2011**, *15*, 173–182.
44. Zhang, F.; Bruschweiler-Li, L.; Brüschweiler, R. *J. Magn. Reson.* **2012**, *225*, 10–13.
45. Chikayama, E.; Suto, M.; Nishihara, T.; Shinozaki, K.; Kikuchi, J. *PLoS One* **2008**, *3*, e3805.
46. Bingol, K.; Zhang, F.; Bruschweiler-Li, L.; Brüschweiler, R. *J. Am. Chem. Soc.* **2012**, *134*, 9006–9011.
47. Bingol, K.; Zhang, F.; Bruschweiler-Li, L.; Brüschweiler, R. *Anal. Chem.* **2013**, *85*, 6414–6420.
48. Tayyari, F.; Nagana Gowda, G. A.; Gu, H.; Raftery, D. *Anal. Chem.* **2013**, *85*, 8715–8721.
49. Lacey, M. E.; Subramanian, R.; Olson, D. L.; Webb, A. G.; Sweedler, J. V. *Chem. Rev.* **1999**, *99*, 3133–3152.
50. Olson, D. L.; Peck, T. L.; Webb, A. G.; Magin, R. L.; Sweedler, J. V. *Science* **1995**, *270*, 1967–1970.
51. Kc, R.; Henry, I. D.; Park, G. H. J.; Raftery, D. *J. Magn. Reson.* **2009**, *197*, 186–192.
52. Kc, R.; Gowda, Y. N.; Djukovic, D.; Henry, I. D.; Park, G. H. J.; Raftery, D. *J. Magn. Reson.* **2010**, *205*, 63–68.
53. Kc, R.; Henry, I. D.; Park, G. H. J.; Aghdasi, A.; Raftery, D. *Concepts Magn. Reson. Part B: Magn. Reson. Eng.* **2010**, *37B*, 13–19.
54. Grimes, J. H.; O'Connell, T. M. *J. Biomol. NMR* **2011**, *49*, 297–305.
55. Bird, S. S.; Sheldon, D. P.; Gathungu, R. M.; Vouros, P.; Kautz, R.; Matson, W. R.; Kristal, B. S. *Anal. Chem.* **2012**, *84*, 9889–9898.
56. Hoult, D. I.; Richards, R. E. *J. Magn. Reson.* **1976**, *24*, 71–85.
57. Henry, D.; Park, G. H. J.; Kc, R.; Tobias, B.; Raftery, D. *Concepts Magn. Reson. Part B: Magn. Reson. Eng.* **2008**, *33B*, 1–8.
58. Bergeron, S. J.; Henry, I. D.; Santini, R. E.; Aghdasi, A.; Raftery, D. *Magn. Reson. Chem.* **2008**, *46*, 925–929.
59. Cloarec, O.; Campbell, A.; Tseng, L. H.; Braumann, U.; Spraul, M.; Scarfe, G.; Weaver, R.; Nicholson, J. K. *Anal. Chem.* **2007**, *79*, 3304–3311.
60. Lindon, J. C.; Nicholson, J. K.; Wilson, I. D. *J. Chromatogr. B* **2000**, *748*, 233–258.
61. Lenz, E. M.; Wilson, I. D. *J. Proteome Res.* **2007**, *6*, 443–458.
62. Djukovic, D.; Liu, S.; Henry, I.; Tobias, B.; Raftery, D. *Anal. Chem.* **2006**, *78*, 7154–7160.
63. Djukovic, D.; Appiah-Amponsah, E.; Shanaiah, N.; Nagana Gowda, G. A.; Henry, I.; Everly, M.; Tobias, B.; Raftery, D. *J. Pharm. Biomed. Anal.* **2008**, *47*, 328–334.
64. Zhou, Z.; Lan, W.; Zhang, W.; Zhang, X.; Xia, S.; Zhu, H.; Ye, C.; Liu, M. *J. Chromatogr. A* **2007**, *1154*, 464–468.
65. Garcia, E.; Andrews, C.; Hua, J.; Kim, H. L.; Sukumaran, D. K.; Szyperski, T.; Odunsi, K. *J. Proteome Res.* **2011**, *10*, 1765–1771.
66. Hyberts, S. G.; Heffron, G. J.; Tarragona, N. G.; Solanky, K.; Edmonds, K. A.; Luithardt, H.; Fejzo, J.; Chorev, M.; Aktas, H.; Colson, K.; Falchuk, K. H.; Halperin, J. A.; Wagner, G. *J. Am. Chem. Soc.* **2007**, *129*, 5108–5116.
67. Hyberts, S. G.; Arthanari, H.; Wagner, G. *Top. Curr. Chem.* **2012**, *316*, 125–148.
68. Rai, R. K.; Sinha, N. *Anal. Chem.* **2012**, *84*, 10005–10011.
69. Pervushin, K.; Vögeli, B.; Eletsky, A. *J. Am. Chem. Soc.* **2002**, *124*, 12898–12902.
70. Ernst, R. R.; Bodenhausen, G.; Wokaun, A. *Principles of Nuclear Magnetic Resonance in One and Two Dimensions;* Oxford University Press: Oxford, 1987.

71. Motta, A.; Paris, D.; Melck, D. *Anal. Chem.* **2010**, *82*, 2405–2411.
72. Bruschweiler, R.; Zhang, F. *J. Chem. Phys.* **2004**, *120*, 5253–5260.
73. Frydman, L.; Scherf, T.; Lupulescu, A. *Proc. Natl. Acad. Sci. U.S.A.* **2002**, *99*, 15858–15862.
74. Frydman, L.; Lupulescu, A.; Scherf, T. *J. Am. Chem. Soc.* **2003**, *125*, 9204–9217.
75. Giraudeau, P.; Remaud, G. S.; Akoka, S. *Anal. Chem.* **2009**, *81*, 479–484.
76. Bhattacharya, P.; Chekmenev, E. Y.; Perman, W. H.; Harris, K. C.; Lin, A. P.; Norton, V. A.; Tan, C. T.; Ross, B. D.; Weitekamp, D. P. *J. Magn. Reson.* **2007**, *186*, 150–155.
77. Chekmenev, E. Y.; Norton, V. A.; Weitekamp, D. P.; Bhattacharya, P. *J. Am. Chem. Soc.* **2009**, *131*, 3164–3165.
78. Shchepin, R. V.; Coffey, A. M.; Waddell, K. W.; Chekmenev, E. Y. *J. Am. Chem. Soc.* **2012**, *134*, 3957–3960.
79. Adams, R. W.; Aguilar, J. A.; Atkinson, K. D.; Cowley, M. J.; Elliott, P. I.; Duckett, S. B.; Green, G. G.; Khazal, I. G.; López-Serrano, J.; Williamson, D. C. *Science* **2009**, *323*, 1708–1711.
80. Frydman, L.; Blazina, D. *Nat. Phys.* **2007**, *3*, 415–419.
81. Mishkovsky, M.; Frydman, L. *ChemPhysChem* **2008**, *9*, 2340–2348.
82. Saunders, M. G.; Ludwig, C.; Gunther, U. L. *J. Am. Chem. Soc.* **2008**, *130*, 6914–6915.
83. Ardenkjaer-Larsen, J. H.; Fridlund, B.; Gram, A.; Hansson, G.; Hansson, L.; Lerche, M. H.; Servin, R.; Thaning, M.; Golman, K. *Proc. Natl. Acad. Sci. U.S.A.* **2003**, *100*, 10158–10163.
84. Sarkar, R.; Comment, A.; Vasos, P. R.; Jannin, S.; Gruetter, R.; Bodenhausen, G.; Hall, H.; Kirik, D.; Denisov, V. P. *J. Am. Chem. Soc.* **2009**, *131*, 16014–16015.
85. Chen, H. Y.; Hilty, C. *Anal. Chem.* **2013**, *85*, 7385–7390.
86. Kurhanewicz, J.; Vigneron, D. B.; Brindle, K.; Chekmenev, E. Y.; Comment, A.; Cunningham, C. H.; Deberardinis, R. J.; Green, G. G.; Leach, M. O.; Rajan, S. S.; Rizi, R. R.; Ross, B. D.; Warren, W. S.; Malloy, C. R. *Neoplasia* **2011**, *13*, 81–97.
87. Bohndiek, S. E.; Kettunen, M. I.; Hu, D.; Kennedy, B. W. C.; Boren, J.; Gallagher, F. A.; Brindle, K. M. *J. Am. Chem. Soc.* **2011**, *133*, 11795–11801.
88. Bohndiek, S. E.; Kettunen, M. I.; Hu, D. E.; Witney, T. H.; Kennedy, B. W.; Gallagher, F. A.; Brindle, K. M. *Mol. Cancer Ther.* **2010**, *9*, 3278–3288.
89. Day, S. E.; Kettunen, M. I.; Gallagher, F. A.; Hu, D. E.; Lerche, M.; Wolber, J.; Golman, K.; Ardenkjaer Larsen, J. H.; Brindle, K. M. *Nat. Med.* **2007**, *13*, 1382–1387.
90. Holmes, E.; Antti, H. *Analyst* **2002**, *127*, 1549–1557.
91. Griffin, J. L. *Philos. Trans. R. Soc. Lond. B Biol. Sci.* **2004**, *359*, 857–871.
92. Werth, M. T.; Halouska, S.; Shortridge, M. D.; Zhang, B.; Powers, R. *Anal. Biochem.* **2010**, *399*, 58–63.
93. Worley, B.; Halouska, S.; Powers, R. *Anal. Biochem.* **2013**, *433*, 102–104.
94. Barker, M.; Rayens, W. *J. Chemom.* **2003**, *17*, 166–173.
95. Heather, L. C.; Wang, X.; West, J. A.; Griffin, J. L. *J. Mol. Cell. Cardiol.* **2013**, *55*, 2–11.
96. Girard, D. A. *Numer. Math.* **1989**, *56*, 1–23.
97. Rocha, C. M.; Carrola, J.; Barros, A. S.; Gil, A. M.; Goodfellow, B. J.; Carreira, I. M.; Bernardo, J.; Gomes, A.; Sousa, V.; Carvalho, L.; Duarte, I. F. *J. Proteome Res.* **2011**, *10*, 4314–4324.
98. Wei, S.; Suryani, Y.; Nagana Gowda, G. A.; Skill, N.; Maluccio, M.; Raftery, D. *Metabolites* **2012**, *2*, 701–716.
99. Ye, T.; Zhang, S.; Nagana Gowda, G. A.; Raftery, D. Nuclear magnetic resonance and statistical analysis. In: *Encyclopedia of Analytical Chemistry;* Meyers, R. A., Ed.; John Wiley: Hoboken, NJ, 2010.

100. Smolinska, A.; Blanchet, L.; Buydens, L. M.; Wijmenga, S. S. *Anal. Chim. Acta.* **2012**, *750*, 82–97.
101. Zhang, A. H.; Sun, H.; Qiu, S.; Wang, X. J. *Magn. Reson. Chem.* **2013**, *51*, 549–556.
102. Djukovic, D.; Nagana Gowda, G. A.; Raftery, D. Mass Spectrometry and NMR Spectroscopye Based Quantitative Metabolomics. In *Proteomic and Metabolomic Approaches to Biomarker Discovery*, 1st ed.; Isssaq, H. J.; Veenstra, T. D., Eds.; Academic Press: San Diego, CA, USA, 2013; pp. 279–298 (chapter 18).
103. Xia, J.; Bjorndahl, T. C.; Tang, P.; Wishart, D. S. *BMC Bioinforma.* **2008**, *9*, 507.
104. Lauridsen, M.; Hansen, S. H.; Jaroszewski, J. W.; Cornett, C. *Anal. Chem.* **2007**, *79*, 1181–1186.
105. Asiago, V.; Nagana Gowda, G. A.; Zhang, S.; Shanaiah, N.; Clark, J.; Raftery, D. *Metabolomics* **2008**, *3*, 328–336.
106. Gartland, K. P.; Beddell, C. R.; Lindon, J. C.; Nicholson, J. K. *Mol. Pharmacol.* **1991**, *39*, 629–642.
107. Anthony, M. L.; Sweatman, B. C.; Beddell, C. R.; Lindon, J. C.; Nicholson, J. K. *Mol. Pharmacol.* **1994**, *46*, 199–211.
108. MacKinnon, N.; Ge, W.; Khan, A. P.; Somashekar, B. S.; Tripathi, P.; Siddiqui, J.; Wei, J. T.; Chinnaiyan, A. M.; Rajendiran, T. M.; Ramamoorthy, A. *Anal. Chem.* **2012**, *84*, 5372–5379.
109. Halouska, S.; Powers, R. *J. Magn. Reson.* **2006**, *178*, 88–95.
110. Rubtsov, D. V.; Waterman, C.; Currie, R. A.; Waterfield, C.; Salazar, J. D.; Wright, J.; Griffin, J. L. *Anal. Chem.* **2010**, *82*, 4479–4485.
111. Wishart, D. S.; Jewison, T.; Guo, A. C.; Wilson, M.; Knox, C.; Liu, Y.; Djoumbou, Y.; Mandal, R.; Aziat, F.; Dong, E.; Bouatra, S.; Sinelnikov, I.; Arndt, D.; Xia, J.; Liu, P.; Yallou, F.; Bjorndahl, T.; Perez-Pineiro, R.; Eisner, R.; Allen, F.; Neveu, V.; Greiner, R.; Scalbert, A. *Nucleic Acids Res.* **2013**, *41*(Database issue), D801–D807.
112. Seavey, B. R.; Farr, E. A.; Westler, W. M.; Markley, J. L. *J. Biomol. NMR* **1991**, *1*, 217–236.
113. Cui, Q.; Lewis, I. A.; Hegeman, A. D.; Anderson, M. E.; Li, J.; Schulte, C. F.; Westler, W. M.; Eghbalnia, H. R.; Sussman, M. R.; Markley, J. L. *Nat. Biotechnol.* **2008**, *26*, 162–164.
114. Lundberg, P.; Vogel, T.; Malusek, A.; Lundauist, P.-O.; Cohen, L.; Dahlqvist, O. *MDL—The Magnetic Resonance Metabolomics Database (mdl.imv.liu.se);* ESMRMB: Basel, Switzerland, 2005.
115. Weljie, A. M.; Newton, J.; Mercier, P.; Carlson, E.; Slupsky, C. M. *Anal. Chem.* **2006**, *78*, 4430–4442.
116. Chenomx NMR Suite Chenomc Inc., Edmonton, AB, Canada. http://www.chenomx.com, 2010.
117. Mercier, P.; Lewis, M. J.; Chang, D.; Baker, D.; Wishart, D. S. *J. Biomol. NMR* **2011**, *49*, 307–323.
118. Zheng, C.; Zhang, S.; Ragg, S.; Raftery, D.; Vitek, O. *Bioinformatics* **2011**, *27*, 1637–1644.
119. Alm, E.; Slagbrand, T.; Aberg, K. M.; Wahlström, E.; Gustafsson, I.; Lindberg, J. *Anal. Bioanal. Chem.* **2012**, *403*, 443–455.
120. Zheng, M.; Lu, P.; Liu, Y.; Pease, J.; Usuka, J.; Liao, G.; Peltz, G. *Bioinformatics* **2007**, *23*, 2926–2933.
121. Lewis, I. A.; Schommer, S. C.; Hodis, B.; Robb, K. A.; Tonelli, M.; Westler, W. M.; Sussman, M. R.; Markley, J. L. *Anal. Chem.* **2007**, *79*, 9385–9390.
122. Rai, R. K.; Tripathi, P.; Sinha, N. *Anal. Chem.* **2009**, *81*, 10232–10238.
123. Hu, K.; Westler, W. M.; Markley, J. L. *J. Am. Chem. Soc.* **2011**, *133*, 1662–1665.
124. Hu, K.; Ellinger, J. J.; Chylla, R. A.; Markley, J. L. *Anal. Chem.* **2011**, *83*, 9352–9360.

125. Chylla, R. A.; Hu, K.; Ellinger, J. J.; Markley, J. L. *Anal. Chem.* **2011**, *83*, 4871–4880.
126. Cloarec, O.; Dumas, M. E.; Craig, A.; Barton, R. H.; Trygg, J.; Hudson, J.; Blancher, C.; Gauguier, D.; Lindon, J. C.; Holmes, E.; Nicholson, J. K. *Anal. Chem.* **2005**, *77*, 1282–1289.
127. Blaise, B. J.; Navratil, V.; Emsley, L.; Toulhoat, P. *J. Proteome Res.* **2011**, *10*, 4342–4348.
128. Holmes, E.; Cloarec, O.; Nicholson, J. K. *J. Proteome Res.* **2006**, *5*, 1313–1320.
129. Alves, C.; Rantalainen, M.; Holmes, E.; Nicholson, J. K.; Ebbels, T. M. D. *Anal. Chem.* **2009**, *81*, 2075–2084.
130. Crockford, D. J.; Holmes, E.; Lindon, J. C.; Plumb, R. S.; Zirah, S.; Bruce, S. J.; Rainville, P.; Stumpf, C. L.; Nicholson, J. K. *Anal. Chem.* **2006**, *78*, 363–371.
131. Crockford, D. J.; Maher, A. D.; Ahmadi, K. R.; Barrett, A.; Plumb, R. S.; Wilson, I. D.; Nicholson, J. K. *Anal. Chem.* **2008**, *80*, 6835–6844.
132. Robinette, S. L.; Lindon, J. C.; Nicholson, J. K. *Anal. Chem.* **2013**, *85*, 5297–5303.
133. Wei, S.; Zhang, J.; Liu, L.; Ye, T.; Nagana Gowda, G. A.; Tayyari, F.; Raftery, D. *Anal. Chem.* **2011**, *83*, 7616–7623.
134. Gu, H.; Nagana Gowda, G. A.; Neto, F.C.; Opp, M.R.; Raftery, D. *Anal. Chem.* **2013**, *85*, 10771–10779.
135. Nagana Gowda, G. A.; Raftery, D. *Curr. Metabolomics* **2013**, *1*, 227–240.
136. Coen, M.; Holmes, E.; Lindon, J. C.; Nicholson, J. K. *Chem. Res. Toxicol.* **2008**, *21*, 9–27.
137. Waters, N. J. *Curr. Drug Metab.* **2010**, *11*, 686–692.
138. Coen, M. *Toxicology* **2010**, *278*, 326–340.
139. Powers, R. *Magn. Reson. Chem.* **2009**, *47*(Suppl. 1), S2–S11.
140. Lindon, J. C.; Nicholson, J. K.; Holmes, E.; Antti, H.; Bollard, M. E.; Keun, H.; Beckonert, O.; Ebbels, T. M.; Reily, M. D.; Robertson, D.; Stevens, G. J.; Luke, P.; Breau, A. P.; Cantor, G. H.; Bible, R. H.; Niederhauser, U.; Senn, H.; Schlotterbeck, G.; Sidelmann, U. G.; Laursen, S. M.; Tymiak, A.; Car, B. D.; Lehman-McKeeman, L.; Colet, J. M.; Loukaci, A.; Thomas, C. *Toxicol. Appl. Pharmacol.* **2003**, *187*, 137–146.
141. Ebbels, T. M.; Keun, H. C.; Beckonert, O. P.; Bollard, M. E.; Lindon, J. C.; Holmes, E.; Nicholson, J. K. *J. Proteome Res.* **2007**, *6*, 4407–4422.
142. Clayton, T. A.; Lindon, J. C.; Cloarec, O.; Antti, H.; Charuel, C.; Hanton, G.; Provost, J. P.; Le Net, J. L.; Baker, D.; Walley, R. J.; Everett, J. R.; Nicholson, J. K. *Nature* **2006**, *440*, 1073–1077.
143. Jones, O. A.; Murfitt, S.; Svendsen, C.; Turk, A.; Turk, H.; Spurgeon, D. J.; Walker, L. A.; Shore, R. F.; Long, S. M.; Griffin, J. L. *Food Chem. Toxicol.* **2013**, *59*, 438–445.
144. Schnackenberg, L. K.; Sun, J.; Beger, R. D. *Methods Mol. Biol.* **2012**, *926*, 141–156.
145. Duarte, I. F. *J. Control. Release* **2011**, *153*, 34–39.
146. Sandusky, P.; Raftery, D. *Anal. Chem.* **2005**, *77*, 7717–7723.
147. Shepherd, L. V.; Fraser, P.; Stewart, D. *Bioanalysis* **2011**, *3*, 1143–1159.
148. Mannina, L.; Sobolev, A. P.; Capitani, D. *Electrophoresis* **2012**, *33*, 2290–2313.
149. Hong, Y. S. *Magn. Reson. Chem.* **2011**, *49*(Suppl. 1), S13–S21.
150. Viggiani, L.; Morelli, M. A. C. *J. Agric. Food Chem.* **2008**, *56*, 8273–8279.
151. Anastasiadi, M.; Zira, A.; Magiatis, P.; Haroutounian, S. A.; Skaltsounis, A. L.; Mikros, E. *J. Agric. Food Chem.* **2009**, *57*, 11067–11074.
152. Son, H. S.; Kim, K. M.; Van den Berg, F.; Hwang, G. S.; Park, W. M.; Lee, C. H.; Hong, Y. S. *J. Agric. Food Chem.* **2008**, *56*, 8007–8016.
153. Duarte, I. F.; Barros, A.; Belton, P. S.; Righelato, R.; Spraul, M.; Humpfer, E.; Gil, A. M. *J. Agric. Food Chem.* **2002**, *50*, 2475–2481.
154. Duarte, I. F.; Barros, A.; Almeida, C.; Spraul, M.; Gil, A. M. *J. Agric. Food Chem.* **2004**, *52*, 1031–1038.

155. Almeida, C.; Duarte, I. F.; Barros, A.; Rodrigues, J.; Spraul, M.; Gil, A. M. *J. Agric. Food Chem.* **2006**, *54*, 700–706.
156. Hammer, K. M.; Pedersen, S. A.; Størseth, T. R. *Comp. Biochem. Physiol. Part D Genomics Proteomics* **2012**, *7*(3), 292–302.
157. Nagato, E. G.; D'eon, J. C.; Lankadurai, B. P.; Poirier, D. G.; Reiner, E. J.; Simpson, A. J.; Simpson, M. J. *Chemosphere* **2013**, *93*, 331–337.
158. Yuk, J.; Simpson, M. J.; Simpson, A. J. *Environ. Pollut.* **2013**, *175*, 35–44.
159. Brown, S. A.; McKelvie, J. R.; Simpson, A. J.; Simpson, M. J. *Environ. Pollut.* **2010**, *158*, 2117–2123.
160. Ellis, J. K.; Athersuch, T. J.; Thomas, L. D.; Teichert, F.; Pérez-Trujillo, M.; Svendsen, C.; Spurgeon, D. J.; Singh, R.; Järup, L.; Bundy, J. G.; Keun, H. C. *BMC Med.* **2012**, *10*, 61.

Chapter 9

The Role of Mass Spectrometry in Nontargeted Metabolomics

Helen G. Gika[*], **Ian D. Wilson**[†] **and Georgios A. Theodoridis**[‡]
[*]*Department of Chemical Engineering, Aristotle University Thessaloniki, Thessaloniki, Greece*
[†]*Department of Surgery and Cancer, Faculty of Medicine, Imperial College, South Kensington, London, United Kingdom*
[‡]*Department of Chemistry, Aristotle University Thessaloniki, Thessaloniki, Greece*

Chapter Outline

1 INTRODUCTION

The metabolome is defined as the total content of small molecules in a biofluid, a cell, an organ, or an organism, and metabolomics represents the study of the metabolome: the analytical investigation to identify the myriad of small molecules and provide their (relative) concentrations but also the correlation of their concentration patterns with pathophysiological conditions, genetic variation, or other factors or properties of the analyzed samples (see also metabonomics) *(1–3)*. To perform meaningful metabolomics, high-level analytical chemistry is generally combined with advanced statistical analysis, biochemistry, and applied sciences such as the agricultural, environmental, or life sciences *(4,5)*. This multidisciplinary research clearly requires the coordination and cooperation of researchers from different fields and

Comprehensive Analytical Chemistry, Vol. 63. http://dx.doi.org/10.1016/B978-0-444-62651-6.00009-X

expertise: analytical chemists, bioinformaticians, chemometricians, biochemists, physicians, plant scientists, and so forth. To stand any reasonable chance of success, careful preparations and thorough experimental planning are needed before experiments are undertaken, covering all aspects of the organization of the study including sample collection, storage and handling, analysis, and data mining. Analysis to obtain holistic metabolic profiles is not a trivial undertaking and should employ the best analytical platforms available. Data mining requires deep understanding of statistics and proficiency in the handling of corresponding software, which often may require the development of custom scripts. In order to improve the knowledge of the various biochemical pathways and fluxes, careful data scrutiny should be applied. When a potential biomarker(s) is identified, its biochemical role should be understood so that it can be incorporated within the biochemical context of the condition being studied (many metabolic changes observed result from "collateral damage" to the system rather than being specific to the particular disease under investigation). Subsequently, the correlation of the various different biomarkers identified in the metabolic profiles as being specific to the condition under study into biochemical networks and fluxes requires excellent knowledge of biochemistry as well as significant effort. Finally, the combination of different Omics data provides a systems biology approach, and there are an increasing number of studies where metabolomics has been combined with one or more of the other Omics fields: genomics/genetics, transcriptomics, and proteomics (e.g., *(6–8)*). In the best-case scenario, the findings of the global metabolic profiling phase of the investigation should be confirmed using a targeted analytical method(s) developed specifically for the newly found markers. New samples should be analyzed and results should be confirmed to verify that findings were not simply the result of some confounding factor (e.g., stress or poor diet) and that the markers are specific. Figure 1 shows a general roadmap of the metabolomics analytical journey from holistic profiling to validated biomarker profile (hypothesis-free to hypothesis-proved).

Although the first applications of metabolomics originated in plant/microbial sciences, the field of animal/human biology now offers the major application area. The truth is, however, that in the early years of the development of global metabolic profiling, the term devised by scientists working on animal biology was that of metabonomics, and this was initially performed using high field ^{1}H NMR spectroscopy *(2)*. The two terms (metabolomics, metabonomics) are now often used interchangeably, but the initial definitions were quite different (Ref. *(2)* provides a perspective on the development of the field and its terminology). Whatever the term used, metabolomics/metabonomics provides a powerful methodology for patient stratification (disease diagnosis/prognosis), drug toxicity assessment, and evaluation of human exposure because the obtained metabolite profiles offer a close to real-time snapshot of the ongoing biochemistry of the system under study. The fact that

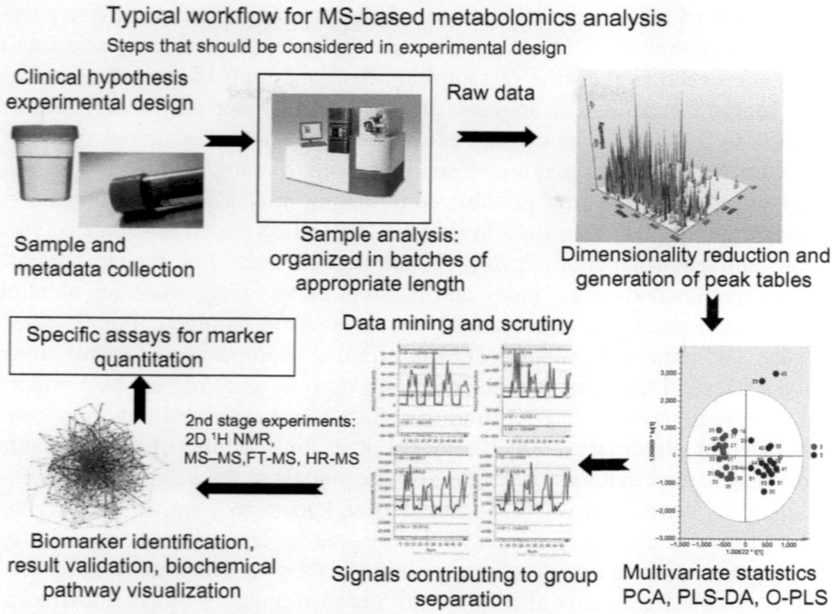

FIGURE 1 Schematic workflow for metabolomics analysis. (For color version of this figure, the reader is referred to the online version of this chapter.)

metabolites represent endpoints of the various biochemical pathways, compared with the upstream genes or proteins, gives hope that the metabolite markers giving rise to the disease-specific metabotype can be directly linked with physiology and the phenotype of the subject under investigation.

2 STUDY DESIGN

Metabolomics is basically a hypothesis-free journey to map the metabolic content. In reality, this represents a journey to the unknown and as such it should be taken very seriously. Preparations and experimental design should include various parameters: the number of subjects likely to be required to achieve statistical significance, the number of samples to be analyzed, the possible confounding factors (age, gender, etc.) that would divide the cohorts into subgroups (often detrimental to statistical power analysis and the validity of results), the different conditions to be assayed, preanalytical procedures such as collection of metadata, sampling and sample handling, the selection of method(s) for analysis, the strategy and the software tools for data treatment, and finally, the selection of parameters to be correlated and associated. Investigations should be carried out according to agreed protocols (and animal or human/patient studies require ethical clearance and in the case of the latter appropriate informed consent) that should be followed to avoid the introduction of biases. Investigators should be trained to the specific requirements of metabolomics investigations, for example, they should understand that factors such as the time of sampling, the conditions of growth of plants (e.g., sunlight or airing) or animals, diet, age, and gender play a very important role in the metabolite content of samples and such intersubject variation has been shown to be a major confounding issue in many studies. In general, information on a number of possible confounding factors such as age, gender, animal strain or plant species/cultivar type, diet, stress conditions, diurnal variation, and environmental factors should be collected. For human studies, aspects such as body mass index and habitual factors (e.g., smoking, alcohol use, or pharmaceutical intervention) should also be recorded. For diseased humans or patients under treatment, finding appropriate controls may prove difficult. Details such as the materials used in sampling are also important: different syringes or vials used to collect blood or other fluids may contain different plasticizers, which may result in the leaching of contaminants into the specimen that then generate massive signals in the subsequent chemical analysis. With regard to blood, it has been known by clinical chemists for many years that plasma and serum are not equal specimens (and this has also been shown for metabolomics/metabonomics data *(9)*). The time from collection to sample freezing is also important because ongoing biochemical reactions can alter the sample metabolome. All these issues should be made clear to the technical personnel and should be a matter for prestudy investigation as part of the design of an experiment, and then a standard operating procedure (SOP) to be used in the study itself should be standardized and specified (examples of such SOPs can be found on the HUSERMET Web site at www.husermet.org).

3 SAMPLE PREPARATION

Sample collection and preparation form distinct and important steps in metabolomics studies. In fact, sample preparation is one of the most error-prone and time-consuming steps of chemical analysis *(10)*. Despite some developments in automation and robotics, many aspects of sample preparation are still performed largely manually, relying on the work of a number of people. Sample preparation aims to prepare the sample for the subsequent analysis and this in most cases is for targeted analysis, meaning that the analyte(s) of interest are concentrated, purified, and transferred in a form suitable for introduction into the analytical instrument. The majority of the developments in sample preparation were thus designed with the aim of removing potential interferences. In Omics studies, and in metabolomics in particular, a new concept should be applied: sample preparation should treat all the low-molecular-mass metabolites that form the sample components equally, and avoid altering the sample's composition even if there is a need for the removal of macromolecular interferences. This is not always realistic, but, even so, sample preparation should be as minimal as possible. As a result, extraction is not universally applied to samples and, when done, it is as generic as possible: the aim is to extract as many metabolites in an unbiased fashion as possible. In fact, in the majority of the published works, the key criterion for method selection was merely the number of features detected *(11,12)* (which, as explained in Section 6, is not equivalent to the number of true metabolites, but only indicative of the latter).

The methods applied in sample preparation are defined by the type of sample and the analytical method selected. So, for example, derivatization is practically a prerequisite in gas chromatography–mass spectrometry (GC–MS) metabolomics unless analysis is confined to only volatile analytes such as in breath analysis or the true-volatiles bouquet contained in wine *(13)*. A dual derivatization scheme is typically applied in GC–MS: methoxyamination is applied first to convert carbonyl groups; next, silylation by MSTFA or BSTFA is applied in order to mask active hydrogens such as those present in OH or NH_2 groups. The conditions used during the derivatization reactions may prove critical for the result of the process: temperature, volume ratio, drying of the sample, and reaction time may alter the efficiency of the derivatization reaction. When dealing with biological fluids it is very important to ensure that the sample is thoroughly dried before derivatization, otherwise the reproducibility of the reaction process is doubtful *(14)*. In LC–MS, the following issues should be considered in sample preparation: (1) particulate matter and macromolecules should be removed, (2) the mass spectrometer (MS) ion source should be regularly cleaned to maintain performance, and (3) matrix effects should be monitored. ^{1}H NMR spectroscopic analysis imposes much lower requirements on the analyst with regard to sample preparation and is typically limited to the removal of particulates, the addition of D_2O, and one sample volume of an appropriate buffer. Sample buffering is very important as the resonance of protons located on molecules at positions near ionizable groups may vary with sample pH, causing problems with subsequent data alignment.

It should be understood that whatever the sample preparation protocol selected, the metabolic content will be modified to an extent. It is now recognized that the assessment of the extent of metabolome modification due to sample preparation is important and a number of publications have dealt with the effects of such manipulations when applied to different samples of human or plant origin *(12,15–19)*. If such measures are not taken, a possibility exists to mistake biases introduced because of the sample preparation method for trends related to the study (as opposed to real effects originating from the standpoint of pathophysiology, biology, or investigator intervention).

With regard to the type of samples analyzed, a number of considerations impose important rules. First of all, good knowledge of the topic of the study and the nature of the specimen is important. For example, it is important to understand that blood is a sample that while in the body is a subject of homeostasis. Urine is collected in the kidney and the concentrations of metabolites excreted are not regulated by such mechanisms. Plant samples or fruit are characterized by huge differences in the concentration range of certain metabolites, as the concentration of some saccharides may be up to seven orders of magnitude higher than the concentration of secondary metabolites *(20)*. The content of lipids or salts of some samples may have an effect on the transportation of particular metabolites to the final analytical stage. Such consideration should be taken seriously; it is thus advised to embark on metabolic profiling of a species or organ only after accumulating knowledge and experience on its specificities. When dealing with solid samples (e.g., fruit or tissue), one should remember that the solvent for metabolite recovery and method selected for extraction should solubilize the metabolites and recover them from a solid network that may show strong affinity to some metabolites. Tissues can be cryolyzed prior to extraction with a combination of different solvents. If the tissue analysis is performed on a specific region of the intact tissue information on the relative content of the markers at the focal point of interest can be obtained (e.g., on a cancerous tumor or on a point of inflammation). The presence of blood in the organs to be analyzed will, in the end, result in the acquisition of a mixed metabolic profile (blood and tissue); such concerns have been in the interest of researchers a protocol for the metabolomics analysis of tissue have been published recently *(21)*.

It is now accepted that the choice and the practice of sample preparation in metabolite profiling can have a major effect on the outcome of the study. We can expect that further research on this topic, documentation, and standardization in this area will continue to provide helpful information and inform analytical practice promoting the harmonization of the practices needed for field consolidation.

4 ANALYTICAL STRATEGIES

Metabolomics and metabonomics are realized using information-rich spectroscopic techniques: NMR spectroscopy and MS, the latter mostly in combination with chromatographic separation (liquid or gas chromatography) *(5,22)*. Other techniques were also tested (e.g., FTIR, NIR, and electrochemical analysis), but with little success *(4)*. Figure 2 provides a graphical representation of the number or publications reported since 2000 as found by Scopus, searching for the terms NMR, MS, or chromatography and metabolomics or metabonomics in the abstract/title/keywords. It is clear that the field is growing and, in particular, chromatography and MS are driving this growth.

NMR spectroscopy is in many respects an ideal analytical technology for metabolic profiling because it is equally sensitive to all metabolites, it can provide quantitative results, direct analysis, metabolite identification, and broad metabolite class coverage. However, NMR spectroscopy gives information only for the abundant metabolites (due to the method's relatively poor

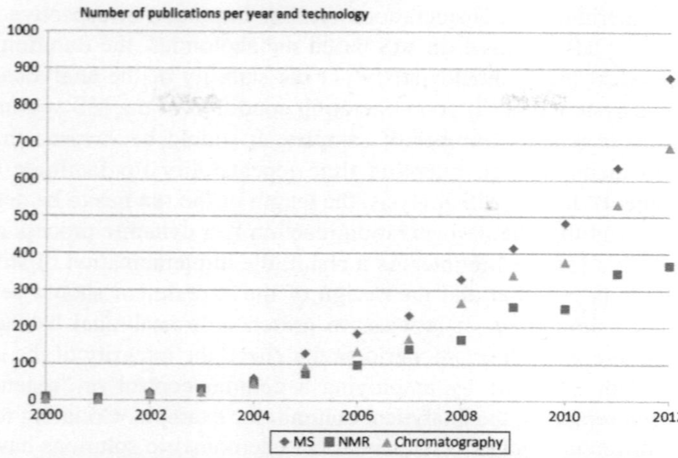

FIGURE 2 Plot of the number of publications on the field of metabolomics/metabonomics and the following analytical technologies: (1) mass spectrometry (blue diamond), (2) chromatography (green triangle), and (3) NMR spectroscopy (red box). Based on search in Scopus. (For interpretation of the references to color in this figure legend, the reader is referred to the online version of this chapter.)

sensitivity with practical analysis times) and requires a major investment in instrument acquisition and highly trained personnel to operate the instrument and perform data processing and treatment. NMR spectroscopy has been, and continues to be, widely used in clinical and toxicity studies as it offers a global perspective of the major constituents of the analyzed sample. Sample preparation is in general minimal and analysis for urine may require 5–10 min and for blood plasma/serum 15–20 min (assuming a modern 600 MHz instrument), as the presence of proteins in the latter requires the application of a variety of different pulse sequences (e.g., the CPMG pulse sequence) to extract the maximum amount of information from the sample *(23)*.

MS-based metabolomics is largely an LC–MS or GC–MS activity and details on the application of these technologies are given later in Sections 5.1 and 5.2. In comparison to LC and GC, little work has been published on CE-MS-based metabolomics *(24,25)*. A major issue is that CE is less robust, mainly due to capillary inner wall modification after the analysis of a number of real samples, as well as the difficulty to maintain stability in the CE electrical circuit. Despite this limitation, the work of the Soga group *(24)* has generated CE-MS methodologies that are used routinely in Human Metabolome Technologies, one of the largest companies in the field of metabolomics, which is based solely on CE-MS technology.

Lately, supercritical fluid chromatography (SFC) is coming up again and is also applied in the field of metabolomics. SFC-MS has found use especially in the analysis of lipids. SFC offers unique characteristics in between those of GC and HPLC. Recent reviews describe the utility of SFC-MS in lipidomics *(26,27)*.

Some experimental considerations remain important irrespective of the technological platform used. In MS-based metabolomics, the duration of the whole analytical batch should satisfy (1) the stability of the analytical chromatographic system and (2) good operating condition of the MS system, after the injection of a large number of samples. It should be stressed that it is essential to maintain high retention time repeatability (to facilitate correct peak alignment). In GC–MS analysis, the length of the run has to be designed, keeping in mind that the derivatization reaction is a dynamic process and the products have a limited lifetime. As a result, the implementation of sufficient cleanup steps is essential and the design of the experiment should seriously consider such limitations. It is wiser to preschedule analytical batches of a logical length (e.g., 24 h or so), periodically check the integrity of the system, and, if needed, intervene by employing a column control or "regeneration step" or even replacing the analytical column, for example. Collating together data from different batches is not trivial but chemometric solutions have been suggested toward these ends *(28)*.

The test samples should always be analyzed in random order to avoid the introduction of unwanted biases and time trends. A way to monitor such unwanted effects is to utilize quality control (QC) samples as a means to effectively examine system performance. QC samples are often made by pooling together subaliquots of biological test samples *(29–31)*. This way a representative bulk sample is generated that should contain all metabolites present in the test samples. QC samples are typically analyzed through the analytical batch and data from the QC injections are scrutinized as a separate dataset by both multivariate analysis but also as typical LC–MS data. Data should pass certain criteria to ensure adequate quality of the dataset, that is, the number of zero values (which should be less than 40%), the CV% of the peak areas (should be less than 30% for a trustworthy peak), the number of peaks that pass the 30% CV filter (which should be higher than 70% in the dataset), the repeatability of retention times and peak areas, and so forth *(29)*.

Apart from monitoring for time trends, QC samples can also be used to correct them; for example, QC data were used to correct for retention time drifts *(32)*, but also to help fuse the data from different analytical batches (e.g., different time periods of data). A number of QC injections are often implemented at the beginning of the analytical run in order to "condition" the system *(5,30,31)*.

Overall, it is well understood that one method cannot cover the whole metabolome, so the current state-of-the-art demonstrates the need for the application of a number of analytical techniques such as reversed-phase liquid chromatography-mass spectrometry ((RP)LC–MS) untargeted as a general profiling tool, another RPLC–MS method to profile the lipids (on the C30 or C18 column), GC–MS to profile the content of volatiles, and one or two LC–MS or CE–MS methods (targeted and untargeted) to profile the polar metabolites (see Section 5.1).

5 MS-BASED UNTARGETED METABOLOMICS

Although the essence of Omics technologies is basically that of a hypothesis-free study, lately targeted metabolomics is becoming a pragmatic approach for a number of researchers *(33–36)*. The difference between the two strategies is basically that in targeted methods, MS/MS is applied to detect a predefined set of metabolites while in untargeted methods full-scan MS mode is employed to detect signals from any analyte potentially present in the sample. A number of arguments can be used to advocate for the use of one versus the other mode and, ultimately, when the metabolome has been fully mapped, the whole metabolic profiling exercise will become one enormous targeted method. However, here, we describe only untargeted strategies; readers interested in targeted modes are referred to Ref. *(37)* for a description of recent developments in this area.

Untargeted LC–MS-based metabolomics in practice refers to LC–MS analysis that utilizes high-resolution mass spectrometry and generates large datasets that are subsequently subjected to multivariate statistical analysis to reduce data complexity and dimensionality and reveal salient trends.

GC–MS analysis is also practically untargeted: in most cases, the mass spectrometer is a single quadrupole, while time of flight (TOF) mass spectrometers are also used *(13)*. Either mass spectrometer typically operates in full-scan mode, thus detecting peaks in an unbiased way. In post-run data acquisition, the analyst may select from two alternative approaches: either to first identify as many peaks as possible and then to try to look for differences in their concentration patterns, or to subject the data to data mining and statistical analysis and then try to identify only the differentiating peaks. The approach selected relies largely on the researcher. The robustness of GC–MS as an analytical platform enables the creation of retention time and spectral libraries and databases such as those from NIST or Wiley and smaller, more dedicated libraries such as the Agilent-Fiehn metabolomics database or the Pfleger–Maurer–Weber-Drugs-and-Pesticides-Library for toxicology.

LC–MS is by its nature less robust than GC–MS, and the comparison of single-MS spectra is impractical *(38)*. Commercial spectral libraries are not available so in LC–MS untargeted mode, the data directs the analysis and the course of investigation: initially, the data should be evaluated and scrutinized with predefined acceptance criteria as suggested previously; if trustworthy, then further investigation is applied to find differentiating signals between the various groups. If such signals are found (hopefully corresponding to characteristic "markers"), extensive work may then be required to identify potential markers.

In the following paragraphs, important analytical issues with regard to the application of MS-based analytical metabolomics are highlighted.

5.1 LC Opportunities and Issues

In metabolomics, LC–MS analysis should be performed using the highest chromatographic and MS resolution. As a result, use of the most efficient chromatographic material is highly recommended: At present, sub-2-μm particles are the gold standard and ultra (high) performance liquid chromatography (U(H)PLC) is the LC mode of choice *(22)*. In comparison to HPLC, U(H)PLC offers the following advantages: (1) higher chromatographic efficiency leading to greater peak capacity, (2) higher sensitivity, and (3) lower occurrence of matrix effects (ionization suppression).

RPLC-MS analysis represents the key method in contemporary bioanalysis and is by far the most advanced mode of LC. Thus, the stationary phases, analytical protocols, and methods are much more advanced and better documented in comparison to any other LC mode. A large number of chemistries are available as stationary phases with different geometrical characteristics of columns and particles. Hence, RPLC is practically unchallenged despite major limitations with regard to the analysis of polar or ionic species. This category includes molecules of high significance in different biochemical pathways such as amines, nucleotides, organic acids, amino acids, peptides, and carbohydrates. The more hydrophilic representatives of all of these classes are hardly retained in RPLC and are eluted typically with, or near, the solvent front. Hence, new separation modes are needed to improve the retention of such polar metabolites. As such, hydrophilic interaction chromatography (HILIC) is promising, and appears complementary to RPLC. HILIC–MS(/MS) has been used with success for the analysis of various specimens (urine, blood, plant extracts, etc.) in different application fields (from biomarker discovery to food science) *(39)*.

With regard to chromatographic conditions, since metabolomics typically aims to achieve the widest metabolome coverage, gradient elution is used in order to efficiently separate as many analytes as possible. The selection of the mobile phase depends on the LC mode to be applied (RP or HILIC) and the type of specimen to be analyzed. Basic chromatographic rules should be applied. For example, specimens that are largely aqueous are better analyzed with aqueous mobile phases. For urine analysis by RPLC, the gradient program should start from about 100% water *(30)*. HILIC can also be used to increase metabolome coverage in urine profiling *(40)*. In contrast, in the case of the analysis of rather lipophilic samples (e.g., extracts of tissues or blood plasma with organic solvents such as chloroform), the initial conditions in RPLC can be set at around 40–50% organic modifier in the mobile phase and thorough column washing steps are required *(32,41,42)*. Gradient elution times from 10 to 30 min are preferred as a compromise between analytical performance and throughput. Following sample analysis the efficient removal of interferences from the LC column requires extensive washing of the LC column with a variety of solvents with different eluotropic strengths. Alcohols work better for the analysis of blood samples, as they are very efficient in removing lipids from the column by disrupting hydrogen bonds formed between the stationary phase and the lipids *(32)*. Gradient elution in HILIC has to follow certain constraints: the increase in the water content of the mobile phase should be gradual and the final amount not exceed 50% (v/v). Achieving satisfactory analytical repeatability in HILIC-based separations requires that re-equilibration times of 10 min or so be applied *(39)*.

5.2 GC Opportunities and Issues

The application of GC–MS in metabolomics is not trivial. Metabolites generally must be derivatized to form more volatile molecules. A critical point in this process is that these reactions represent ongoing processes and need time to reach a plateau. Hence, ample time should be allowed before subjecting trimethylsilyl (TMS)-derivatized samples to GC–MS analysis *(43)*. On the other hand, as indicated above, the stability of the derivatives themselves is also limited. Hence, it is not practical to prepare large numbers of samples in one step and then leave them in the autosampler for extended periods before analysis, as the potential risk for derivative decomposition to occur cannot easily be compensated for.

Researchers should therefore study the performance of the derivatization reactions used with their samples to avoid the possibility that biases are introduced due to the derivatization reactions (QC samples of the type employed in LC–MS can provide much useful data in this respect) *(44)*. Another issue is the need for extensive column cleanup. In most cases, injection occurs in splitless mode. Introduction of excess derivatization reagents may damage the injection liner and the analytical column. Frequent replacement of the liner is thus necessary. Cleanup of the column can be done by back-flush (when this feature is available) or by application of high temperatures (bake-out).

The chief advantage of the application of GC–MS in metabolomics is the utilization of spectral libraries. In contrast to other MS modes, GC–MS can utilize spectral libraries compiled by the application of electron impact (EI) ionization. As a result, the potential for direct identification of unknown peaks (providing that they are in the database) is very good. A further advantage is the very high chromatographic resolution, which allows for the generation of retention indices calculated relating the retention times of analyzed compounds to the retention times of standards (*n*-alkanes or fatty acid methyl esters). Despite the utility of these tools, peak annotation is not a trivial effort. Metabolites often occur in more than one form; an amino acid may be derivatized in one or both functional groups (silylation of one or more active moieties). These forms will elute with different retention times and each peak will then produce a number of fragments in EI-MS. Utilization of libraries greatly assists in the identification of the peaks: spectral libraries contain data for both derivatives and natural forms of the analytes and can recognize a single or double TMS derivative. Furthermore, software tools allow for the deconvolution of complex signals: Automated Mass Spectral Deconvolution and Identification System (AMDIS; free software from NIST) or particular instrument vendor software can perform spectral deconvolution and library searching against a selected database. Automatic deconvolution should, however, always be used with caution and manual inspection of results is highly advised. Recent work resulted in the development of a bespoke Matlab script that works in parallel to AMDIS, allowing rapid manual inspection of chromatographic peaks across all samples and integration of peak areas across all samples *(45)*. The field of GC–MS library searching for compound identification has been reviewed by Halket *et al. (46)*. Finally, a recent version of the software XCMS (see data treatment Section 6) is optimized for the treatment of GC–MS data.

5.3 MS Opportunities and Issues

As indicated earlier, MS is the leading technology in the life sciences; as a result, Omics research relies heavily on MS-generated analytical results.

In LC–MS analysis, the choice of the ionization mode has a significant effect on the analytical outcome and thus the obtained metabolic profile. For example, polar molecules are efficiently ionized in electrospray whereas more apolar molecules are better ionized in atmospheric pressure chemical ionization. Acids, carbohydrates, and phenolic substances may give higher signals in negative ionization, whereas nitrogen containing molecules typically give stronger signals in positive ionization. Lately increased awareness has been focused on laser desorption ionization techniques such as matrix-assisted laser desorption ionization *(47,48)* or desorption/extractive electrospray ionization *(49)*. These latter techniques can provide information on the localization of metabolites (e.g., the distribution of certain metabolites in key lesions of a specimen (an atheroma plaque, the skin of fruits, or the bark of plants)).

With regard to mass analyzers, the current strategy is that metabolomics should employ the highest resolution instrument that is available. In GC–MS, the majority of the published work is performed on nominal mass resolution using single quadrupole instruments. GC–TOF instruments are more expensive, and while they offer higher spectral resolution they have lower dynamic ranges. In U(H)PLC–MS, the majority of the research work is realized via TOF-MS instruments *(5,22)*. U(H)PLC–TOF-MS provides good chromatographic resolution and fast data acquisition. TOF-MS offers high mass accuracy and detection sensitivity. Higher MS resolution can be achieved on FT-MS instruments (Orbitrap and Ion Cyclotron Resonance); however, the latter are relatively slow scanning instruments and the maximum MS resolution requires a duty cycle that may not be compatible with fast and sharp U(H)PLC peaks (peak widths as narrow as 2 s). The latest Orbitrap-MS designs do offer higher scan and acquisition speeds and when combined with ion trap offers many advantages for metabolite identification: such hybrid analyzers may perform a number of MS experiments in a single analysis and combine MS^n with high mass accuracy data. High mass accuracy data is often deemed necessary in metabolite identification in order to reach the correct atomic composition of an unknown. The combination of high mass accuracy MS data with isotope ratios, MS/MS experiments, and library searching are the major MS-based analytical tools to be applied for metabolite identification.

All such developments offer exciting opportunities, but a major obstacle in LC–MS-based metabolomics is the lack of standardization. In contrast to NMR or GC–MS, the application and the development of the field of LC–MS cannot rely on databases due to a number of important issues: (1) correlation of LC–MS data or comparison of data obtained in different MS instruments is not possible, (2) adduct formation cannot be controlled, and (3) interlaboratory comparisons of full-scan MS data becomes very difficult.

In sum, it becomes clear that the development of MS methods for untargeted metabolomics is complex. Methods should be optimized to each type of specimen (urine, plasma, tissue extracts, etc.) and each application and its results need to be carefully monitored.

6 DATA ANALYSIS

In MS-based metabolomics, the mass spectrometer operates in full-scan mode; in LC–MS metabolomics it takes typically about 50–1200 amu to cover the width of masses expected for small molecules; in GC–MS, from 10 to 500 Da, a pragmatic limit for mass measurable by GC. Typical raw data files can be comprehended as a series of full-scan mass spectra joined together. The data content in these files represents a very small portion of their final size; one can understand each spectrum as an excess of zero values or signals close to the detection threshold with only a small number of useful signals (which correspond to the real peaks). The systematic noise that is recorded should be filtered during the data treatment process. In metabolomics, we need to search through hundreds of such bulky data files (typical size of each raw data file is at least a GB), so we need specially developed data treatment tools. Different software packages have been developed; these utilize a series of processes: noise filtering, baseline correction, centering, normalization, peak picking, peak integration, and alignment *(50,51)*. The overall goal of this process is to detect all the peaks of a chromatographic analysis to produce a two-dimensional peak table where samples represent columns and peaks represent rows while the signal intensities are reported in the table cells (Figure 3). Peak-picking software programs are freely available, such as MZmine, MetAlign, or XCMS, or instrument manufacturer-supplied proprietary software. In the case of the former, the data is imported as a universal format such as netCDF or mzXML, so data should be first "translated" using small programs provided by all instrument manufacturers. Some researchers have developed tailor-made scripts (in Matlab or R language) to customize peak picking, data mining, and result visualization to fit the exact need of

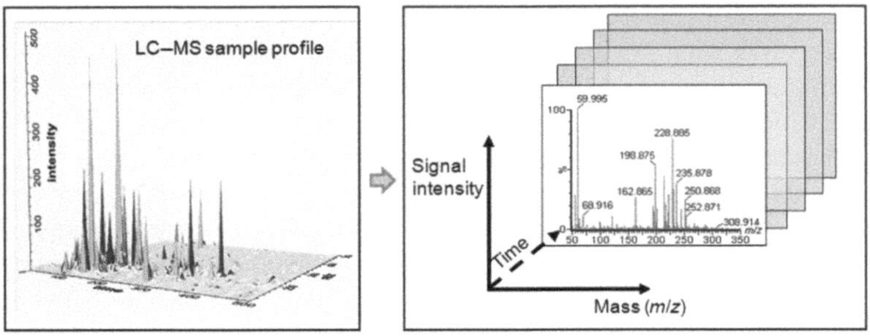

Peak table

Samples	Variables (RT_mass)						
	0.99_105.5	1.06_119.0	1.29_138.6	2.93_198.2	3.12_201.5	3.03_236.2	...
1	105.13	15.35	128.32	12.52	55.63	42.02	
2	130.15	19.52	126.55	13.36	62.36	40.63	
3	50.19	16.25	128.05	36.38	56.36	41.26	
4	120.50	22.36	110.66	16.98	45.45	41.35	
....							

FIGURE 3 Schematic representation of the process of peak picking and alignment algorithm. A three-dimensional chromatogram is divided into panes by dividing retention time into spectral measurements. Each spectrum is seen as a two-dimensional pane with mass (*m*/*z*) in the *X* axis and signal intensity on the *Y* axis. Signals are collected in tables for each sample; detected ions are reported as features. Peak alignment fuses the tables of different samples into a large peak table, where each row represents a sample and each column a feature (ion). Peak tables are then subjected to multivariate analysis and data mining treatment.

analysis (i.e., customize the algorithm to the actual peak width) as GC and LC (especially HILIC) provide very different peak widths.

As indicated, most MS manufacturers have designed their own metabolomic software, which is often provided with the MS operating software. These programs are usually not as flexible as the open-source software as they do not permit modifications to operating algorithms and parameter optimization can also be limited. Commercial software can also perform multivariate statistical analysis such as principal component analysis. Recent versions can also be directly linked to public databases for Web-based searches toward marker identification.

Peak tables are generated so that the data can be subjected to statistical analysis, univariate, and mainly multivariate analysis. However, before this step is undertaken, it is good practice to carefully scrutinize the peak tables manually. For example, a large number of features (e.g., higher than 10,000) does not mean an analysis of high quality but instead may indicate failed peak alignment. Features detected are not equivalent to metabolites present in the sample, as one metabolite may give rise to numerous features in the form of isotope peak adducts, dimers, or fragments. In GC–MS, EI may produce a number of fragments per molecule. To overcome such issues, one can use spectral libraries (GC–MS) or adduct-finding scripts and apply deisotoping of LC–MS data.

Possible error sources in data analysis include (1) varying mass accuracy often observed at the peak inflection points and/or in peaks of low intensity, (2) failed peak alignment, and (3) variations in chromatography. Correcting or controlling peak alignment is not trivial. The selection of peak picking and alignment parameters (mass and retention time window, mass tolerance, etc.) are best performed after the examination of a number of peaks that are widespread in the different samples in the *m/z* and retention time axes. Such peaks should be examined in a number of different analytical injections (potentially all samples) to obtain a confident measure of the most appropriate parameters (mass or retention time windows and tolerances) to be applied. After peak extraction, thorough scrutiny of the peak table is mandatory.

Most groups have devised their own rules to check for data quality. Different researchers follow all, some, or none of these rules. This is indicative of the lack of standardization in the field and the need for harmonization. A list of such rules is (1) the number of zero values in a dataset or a sample should not exceed a certain threshold (e.g., 40%); (2) the 80% rule dictates that features found in less than 20% of the analyzed samples are removed as being noise or irrelevant peaks *(52)*; (3) QCs are used to monitor the quality of data and stability of the analytical system (Figure 4); peak tables generated by QC analyses have to pass predefined criteria: retention time and signal intensity should be repeatable, for example, by showing a coefficient of variation less than 30% (Figure 4B and C) for the majority of the detected peaks *(30)*; (4) features should be considered as potential markers only if several different tests both multivariate (S-plots, loading plots, VIP value/plot) and univariate (ANOVA, *t*-test with $p<0.05$) signify them as important (Figure 4f and G); (5) such markers should be proven stable and trustworthy in the QC samples.

Another mode of data scrutiny is by using "conventional" analysis, processing a number of selected ions as extracted ion chromatograms. Ions should be selected along the retention time and mass range to check for the performance of the analytical system along the run. For peaks highlighted by statistical analysis as potential markers, the stability of retention time, mass detected, peak shape, and peak area in the QC samples should be evaluated with predefined acceptance criteria (e.g., retention time RSD $<3\%$). Figure 4 provides an overview of (some of) the steps taken for data scrutiny of MS-generated metabolomics data.

Data analysis is one of the most time-consuming steps of any metabolomics investigation. Extreme caution is needed in order to avoid unnecessary mistakes such as overfitting; nonvalidated statistical data analysis should be performed by expert chemometricians.

7 IDENTIFICATION OF BIOMARKERS AND BIOCHEMICAL PATHWAY VISUALIZATION

Biomarker identification represents a major bottleneck in nontargeted MS-based metabolomics and in particular in LC–MS studies. This happens

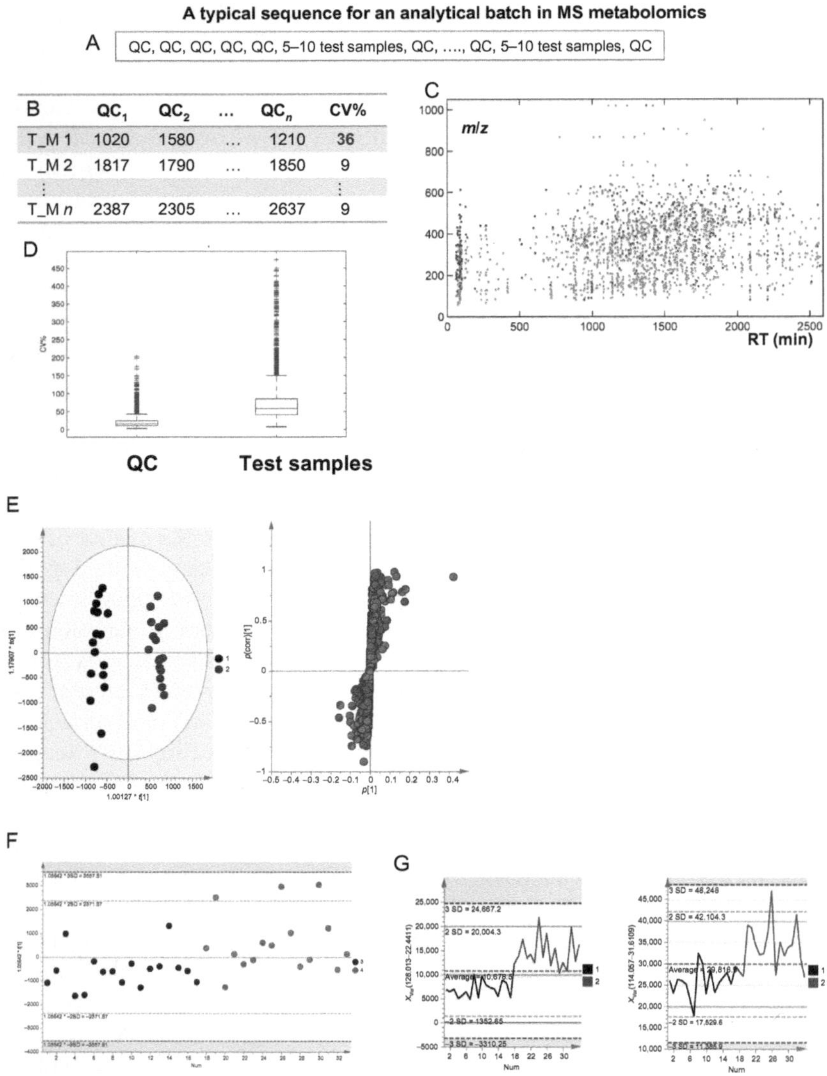

FIGURE 4 Schematic representation of the steps taken for data scrutiny. (A) Illustration of the batch sequence showing QC samples analyzed in the beginning and then sporadically through the batch. Test samples should be randomized in sets of 5–10 samples. (B) A peak table from the QC data is examined for peak area variability (filter CV < 30%). Variables that show CV > 30% are excluded. (C) Variability of features is examined over the RT and *m/z* axis. Red dots represent features with CV > 30% in the QCs, green dots are features that show CV < 30%. This examination could reveal potential analytical pitfalls, for example, in the case where a load of irreproducible features are concentrated in a specific area. (D) Test samples show much higher variability than the QC samples as shown in the corresponding box plots. (E) When the QC data shows

due to a number of issues: (1) mass spectra show limited repeatability between experiments and instruments (with the exception of EI spectra), (2) spectral libraries in LC–MS have not proven robust (so far), and (3) Web-based resources and data repositories for LC–MS metabolomics are not yet very useful, although steps are being taken in that direction *(53)*.

Fragmentation of the work is a major problem for the development of the field. Researchers working in different institutes may expend large amounts of time in identifying the same metabolites. Metabolite identification is often based on lab-based spectral databases, with searches either performed manually or via tailor-made software scripts that search through high mass accuracy data to correlate these with isotope ratios and MS/MS data. As a rule, researchers search in general mass spectral databanks such as METLIN (http://metlin.scripps.edu/), specialized spectral databanks, such as lipidmaps (http://www.lipidmaps.org/), and biochemical or other databases, such as HMDB (http://www.hmdb.ca/) and Chemspider (http://www.chemspider.com/). A shortlist is generated and putative identities should be further reduced by applying MS data rules or the analysis of authentic standards spiked in the sample. A recent review has covered the field of marker identification based on MS data in detail *(54)*.

When a biomarker is finally unambiguously identified, the molecule has to be incorporated into a biochemical pathway. The proposed molecular identities will have to make sense and the trends found (increase or decrease) will have to finally generate new knowledge that would provide a biochemical hypothesis. This is no easy task, even in the case where only a few tens of molecules are involved. Incorporating all of these into biochemical paths needs excellent knowledge of biochemistry and in particular of the biochemistry of the studied system. Informatics tools have been developed recently to assist in this task. These integrate existing public knowledge of the biochemical pathways (e.g., by searching in KEGG, HMDB, or lipid maps) and known gene expression profiles to assist in building associations between detected ions. Existing tools such as MassTrix use only the MS data (retention time data is not yet incorporated) to link to KEGG and annotate metabolites and pathways *(55)*. Obviously, the topic is open for the development of innovative

acceptable repeatability, the whole peak table is processed further and test samples are subjected to multivariate analysis with the aim to reach group differentiation (scores plot on the left, where two groups are shown in red and black). The ions (features) responsible for this separation are found in loadings or the *S* plot as the one given on the right. (F) PLS-DA models should be thoroughly validated with permutation tests and control charts. (G) Ions found to contribute to group clustering are further evaluated with univariate tests (*t*-tests) and visualization plots such as the trends plots shown here. The two features depicted show higher signals in the red group of subfigure (E). (For interpretation of the references to color in this figure legend, the reader is referred to the online version of this chapter.)

tools that could investigate the suitability of the different databases (KEGG, Biocyc, Metabolights, Reacotme, MGI Genome, etc.) in linking the different molecules together and lately a number of research groups have focused their interest to achieve such ends. There then has to be careful consideration of what information the biomarkers are providing, whether they are specific to the disease/condition or are the result of more general effects (collateral damage) and should therefore not be considered further.

8 SYNOPSIS

In this chapter, we have attempted to outline the reasons why MS is applied in metabolic profiling, the numerous existing limitations and strategies to overcome the latter, and the huge potential that the field provides to research. It is obvious that MS (and LC–MS in particular) currently represents the major technological platform for metabolic profiling. However, a number of issues remain. The biggest problem is, in our opinion, that LC–MS-based methods have not yet reached the required level of maturity for the easy application of untargeted metabolomics profiling. With all of these problems, one might wonder if it is indeed useful to work on this field and use this technology. Why should one use MS in metabolomics when the lab next door may reach dissimilar findings working on the same samples? The answer to this question is that it is exactly because of such issues that we need to work further and invest in the field to raise the level of the technology and reach the necessary robustness and standardization. The potential of metabolomics as a whole is such that it will, like genomics and proteomics, remain in the limelight for many years; maybe the name will change, maybe robotics will also be part of it (especially in large-scale works) but it is hard to imagine that there will not be a need for the analysis of small molecules in biological fluids, plants, foods, or environmental samples. Despite its downsides, MS is and will remain a key technology in this pursuit for two main reasons: sensitivity and specificity. Ultimately, standardized analytical protocols will be applied to the metabolic profiling of a variety of different types of samples on a routine basis: these will assist the traceability and determination of the origin of wines, cheeses, or other high-quality products; help clinicians in decision making with regard to drug efficacy (response to therapy) and disease prognosis; be used for patient stratification; and effectively diagnose disease with the help of specific molecular biomarkers or achieve early diagnosis (before the onset of disease). In clinical practice, the discovery of effective biomarkers may have immense impact in patient healthcare quality. Many current clinical laboratory tests are relatively nonspecific (e.g., they may measure liver function but not reveal the underlying cause of liver failure). Results provided by assays have "gray zones," suffer from cross reactivity (possibility for false-positives), and often are only of semiquantitative nature. MS-based analysis may instead provide concrete, mechanism-based answers: absolute identities and absolute quantification.

We are now at the start of a long journey in the development of the necessary methodologies that should result in the harmonization of the methods and the consolidation of the field, finally leading to the routine application of metabolic profiling in the clinic, in the environment, and in the food industry. In the last decade, metabolic profiling has advanced from a small interdisciplinary research field located between analytical chemistry plant biotechnology and the life sciences to become a major research theme *(56)* with specialized conferences, journals, and workshops, a field that amasses large investments from the major instrument manufacturers, the industry and the research funding bodies. Although technologies continue to improve at an astonishing rate, our opinion is that the next key breakthroughs have to come from the informatics sector. The data that is already in place is immense in size and improving the MS resolution to 350,000 or higher is not seen as the key to success. The key point is to learn how to extract only useful information from these oceans of data. Along this line, informatics tools are needed to unite several necessary data scrutiny operations in one software program and to establish platform software tools that can look through data, through different datasets, and through different experiments (e.g., fuse NMR with LC–MS data or HILIC with RPLC data, etc.).

Metabolomics is now an integral part of most major systems biology grants; we can thus expect more coordinated large-scale efforts and projects of high-quality research where metabolite profiling data will be fused with that from other Omics platforms. Most importantly, standardization and establishment of robust and reliable cross-platform metabolomics databases are absolutely necessary. Such initiatives *(57)* are needed to promote the topic and finally assess its utility in enhancing the quality of human life.

REFERENCES

1. Hall, R.; Beale, M.; Fiehn, O.; Hardy, N.; Sumner, L.; Bino, R. *Plant Cell* **2002**, *14*, 1437–1440.
2. Nicholson, J. K.; Lindon, J. C. *Nature* **2008**, *455*, 1054–1056.
3. Nicholson, J. K.; Wilson, I. D. *Nat. Rev. Drug Discov.* **2003**, *2*, 668–676.
4. Lenz, E. M.; Wilson, I. D. *J. Proteome Res.* **2007**, *6*, 443–458.
5. Theodoridis, G.; Gika, H. G.; Wilson, I. D. *Mass Spectrom. Rev.* **2011**, *30*, 884–906.
6. Craig, A.; Sidaway, J.; Holmes, E.; Orton, T.; Jackson, D.; Rowlinson, R.; Nickson, J.; Tonge, R.; Wilson, I.; Nicholson, J. *J. Proteome Res.* **2006**, *5*, 1586–1601.
7. Fiehn, O. *Int. J. Genomics* **2001**, *2*, 155–168.
8. Gavaghan, C. L.; Wilson, I. D.; Nicholson, J. K. *FEBS Lett.* **2002**, *530*, 191–196.
9. Liu, L.; Aa, J.; Wang, G.; Yan, B.; Zhang, Y.; Wang, X.; Zhao, C.; Cao, B.; Shi, J.; Li, M.; Zheng, T.; Zheng, Y.; Hao, G.; Zhou, F.; Sun, J.; Wu, Z. *Anal. Biochem.* **2010**, *406*, 105–112.
10. Theodoridis, G.; Papadoyannis, I. N. *Mikrochim. Acta* **2001**, *136*, 199–204.
11. Bruce, S. J.; Tavazzi, I.; Parisod, V.; Rezzi, S.; Kochhar, S.; Guy, P. A. *Anal. Chem.* **2009**, *81*, 3285–3296.
12. Theodoridis, G.; Gika, H.; Franceschi, P.; Caputi, L.; Arapitsas, P.; Scholz, M.; Masuero, D.; Wehrens, R.; Vrhovsek, U.; Mattivi, F. *Metabolomics* **2012**, *8*, 175–185.

13. Aprea, E.; Gika, H.; Carlin, S.; Theodoridis, G.; Vrhovsek, U.; Mattivi, F. *J. Chromatogr. A* **2011**, *1218*, 4517–4524.
14. Dunn, W. B.; Broadhurst, D.; Begley, P.; Zelena, E.; Francis-McIntyre, S.; Anderson, N.; Brown, M.; Knowles, J. D.; Halsall, A.; Haselden, J. N.; Nicholls, A. W.; Wilson, I. D.; Kell, D. B.; Goodacre, R.; HUSERMET Consortium. *Nat. Protoc.* **2011**, *6*, 1060–1083.
15. Trygg, A. J.; Gullberg, J.; Johansson, A. I.; Jonsson, P.; Antti, H.; Marklund, S. L.; Moritz, T. *Anal. Chem.* **2005**, *77*, 8086–8094.
16. Agnolet, S.; Jaroszewski, J.; Verpoorte, R.; Staerk, D. *Metabolomics* **2010**, *6*, 292–302.
17. del C. Alvarez, M.; Donarski, J. A.; Elliott, M.; Charlton, A. J. *Metabolomics* **2010**, *6*, 541–549.
18. Anastasiadi, M.; Zira, A.; Magiatis, P.; Haroutounian, S.; Skaltsounis, A.; Mikros, E. *J. Agric. Food Chem.* **2009**, *57*, 11067–11074.
19. Bruce, S.; Jonsson, P.; Antti, H.; Cloarec, O.; Trygg, J.; Marklund, S.; Moritz, T. *Anal. Biochem.* **2008**, *372*, 237–249.
20. Gika, H. G.; Theodoridis, G. A.; Vrhovsek, U.; Mattivi, F. *J. Chromatogr. A* **2012**, *1259*, 121–127.
21. Want, E. J.; Masson, P.; Michopoulos, F.; Wilson, I. D.; Theodoridis, G.; Plumb, R. S.; Shockcor, J.; Loftus, N.; Holmes, E.; Nicholson, J. K. *Nat. Protoc.* **2013**, *8*, 17–32.
22. Theodoridis, G. A.; Gika, H. G.; Want, E. J.; Wilson, I. D. *Anal. Chim. Acta* **2012**, *711*, 7–16.
23. Beckonert, O.; Keun, H. C.; Ebbels, T. M. D.; Bundy, J.; Holmes, E.; Lindon, J. C.; Nicholson, J. K. *Nat. Protoc.* **2007**, *2*, 2692–2703.
24. Hirayama, A.; Kami, K.; Sugimoto, M.; Sugawara, M.; Toki, N.; Onozuka, H.; Kinoshita, T.; Saito, N.; Ochiai, A.; Tomita, M.; Esumi, H.; Soga, T. *Cancer Res.* **2009**, *69*, 4918–4925.
25. Ramautar, R.; Somsen, G. W.; de Jong, G. J. *Electrophoresis* **2009**, *30*, 276–291.
26. Bamba, T.; Lee, J. W.; Matsubara, A.; Fukusaki, E. *J. Chromatogr. A* **2012**, *1250*, 212–219.
27. Matsubara, A.; Fukusaki, E.; Bamba, T. *Bioanalysis* **2010**, *2*, 27–34.
28. Draisma, H. H. M.; Reijmers, T. H.; van der Kloet, F.; Bobeldijk-Pastorova, I.; Spies-Faber,-E.; Vogels, J. T. W. E.; Meulman, J. J.; Boomsma, D. I.; van der Greef, J.; Hankemeier, T. *Anal. Chem.* **2010**, *82*, 1039–1046.
29. Gika, H. G.; Theodoridis, G. A.; Earll, M.; Wilson, I. D. *Bioanalysis* **2012**, *4*, 2239–2247.
30. Gika, H. G.; Theodoridis, G. A.; Wingate, J. E.; Wilson, I. D. *J. Proteome Res.* **2007**, *6*, 3291–3303.
31. Lai, L.; Michopoulos, F.; Gika, H.; Theodoridis, G.; Wilkinson, R. W.; Odedra, R.; Wingate, J.; Bonner, R.; Tate, S.; Wilson, I. D. *Mol. Biosyst.* **2009**, *6*, 108–120.
32. Zelena, E.; Dunn, W. B.; Broadhurst, D.; Francis-McIntyre, S.; Carroll, K. M.; Begley, P.; O'Hagan, S.; Knowles, J. D.; Halsall, A.; Wilson, I. D.; Kell, D. B. *Anal. Chem.* **2009**, *81*, 1357–1364.
33. Christians, U.; Klawitter, J.; Hornberger, A.; Klawitter, J. *Curr. Pharm. Biotechnol.* **2011**, *12*, 1053–1066.
34. Lu, W.; Clasquin, M. F.; Melamud, E.; Amador-Noguez, D.; Caudy, A. A.; Rabinowitz, J. D. *Anal. Chem.* **2010**, *82*, 3212–3221.
35. Lu, W.; Bennett, B. D.; Rabinowitz, J. D. *J. Chromatogr. B* **2008**, *871*, 236–242.
36. Roberts, L. D.; Souza, A. L.; Gerszten, R. E.; Clish, C. B. *Curr. Protoc. Mol. Biol.* **2012**, *30*, 1–24.
37. Griffiths, W. J.; Koal, T.; Wang, Y.; Kohl, M.; Enot, D. P.; Deigner, H.-P. *Angew. Chem. Int. Ed. Engl.* **2010**, *49*, 5426–5445.
38. Gika, H. G.; Theodoridis, G. A.; Earll, M.; Snyder, R. W.; Sumner, S. J.; Wilson, I. D. *Anal. Chem.* **2010**, *82*, 8226–8234.

39. Spagou, K.; Tsoukali, H.; Raikos, N.; Gika, H.; Wilson, I. D.; Theodoridis, G. *J. Sep. Sci.* **2010**, *33*, 716–727.
40. Gika, H. G.; Theodoridis, G. A.; Wilson, I. D. *J. Sep. Sci.* **2008**, *31*, 1598–1608.
41. Michopoulos, F.; Lai, L.; Gika, H.; Theodoridis, G.; Wilson, I. *J. Proteome Res.* **2009**, *8*, 2114–2121.
42. Want, E. J.; Wilson, I. D.; Gika, H.; Theodoridis, G.; Plumb, R. S.; Shockcor, J.; Holmes, E.; Nicholson, J. K. *Nat. Protoc.* **2010**, *5*, 1005–1018.
43. Kanani, H.; Chrysanthopoulos, P. K.; Klapa, M. I. *J. Chromatogr. B* **2008**, *871*, 191–201.
44. Sangster, T.; Major, H.; Plumb, R.; Wilson, A. J.; Wilson, I. D. *Analyst* **2006**, *131*, 1075–1078.
45. Behrends, V.; Tredwell, G. D.; Bundy, J. G. *Anal. Biochem.* **2011**, *415*, 206–208.
46. Halket, J. M.; Waterman, D.; Przyborowska, A. M.; Patel, R. K. P.; Fraser, P. D.; Bramley, P. M. *J. Exp. Bot.* **2005**, *56*, 219–243.
47. Amantonico, A.; Urban, P. L.; Fagerer, S. R.; Balabin, R. M.; Zenobi, R. *Anal. Chem.* **2010**, *82*, 7394–7400.
48. Shroff, R.; Rulíšek, L.; Doubský, J.; Svatoš, A. *Proc. Natl. Acad. Sci. U.S.A.* **2009**, *106*, 10092–10096.
49. Li, X.; Hu, B.; Ding, J.; Chen, H. *Nat. Protoc.* **2011**, *6*, 1010–1025.
50. Katajamaa, M.; Oresic, M. *J. Chromatogr. A* **2007**, *1158*, 318–328.
51. Sugimoto, M.; Kawakami, M.; Robert, M.; Soga, T.; Tomita, M. *Curr. Bioinform.* **2012**, *7*, 96–108.
52. Bijlsma, S.; Bobeldijk, I.; Verheij, E. R.; Ramaker, R.; Kochhar, S.; Macdonald, I. A.; van Ommen, B.; Smilde, A. K. *Anal. Chem.* **2006**, *78*, 567–574.
53. Tohge, T.; Fernie, A. R. *Phytochemistry* **2009**, *70*, 450–456.
54. Werner, E.; Heilier, J.-F.; Ducruix, C.; Ezan, E.; Junot, C.; Tabet, J.-C. *J. Chromatogr. B* **2008**, *871*, 143–163.
55. Wägele, B.; Witting, M.; Schmitt-Kopplin, P.; Suhre, K. *PLoS One* **2012**, *7*, e39860.
56. Wishart, D. S.; Knox, C.; Guo, A. C.; Eisner, R.; Young, N.; Gautam, B.; Hau, D. D.; Psychogios, N.; Dong, E.; Bouatra, S.; Mandal, R.; Sinelnikov, I.; Xia, J.; Jia, L.; Cruz, J. A.; Lim, E.; Sobsey, C. A.; Shrivastava, S.; Huang, P.; Liu, P.; Fang, L.; Peng, J.; Fradette, R.; Cheng, D.; Tzur, D.; Clements, M.; Lewis, A.; De Souza, A.; Zuniga, A.; Dawe, M.; Xiong, Y.; Clive, D.; Greiner, R.; Nazyrova, A.; Shaykhutdinov, R.; Li, L.; Vogel, H. J.; Forsythe, I. *Nucleic Acids Res.* **2009**, *37*, D603–D610.
57. Steinbeck, C.; Conesa, P.; Haug, K.; Mahendraker, T.; Williams, M.; Maguire, E.; Rocca-Serra, P.; Sansone, S.-A.; Salek, R. M.; Griffin, J. L. *Metabolomics* **2012**, *8*, 757–760.

Chapter 10

Direct Mass Spectrometry-Based Approaches in Metabolomics

Clara Ibáñez, Virginia García-Cañas, Alberto Valdés and Carolina Simó
Laboratory of Foodomics, Institute of Food Science Research (CIAL), CSIC. Nicolás Cabrera 9, Madrid, Spain

Chapter Outline

1 INTRODUCTION

The general approach in metabolomics is the analysis of as many low-molecular-weight compounds as possible in a given sample to obtain maximal biochemical information. The reality is that considering the number of different types of metabolites in a single living system (lipids, carbohydrates, and many other small compounds, such as amino acids, organic acids, nucleic acids, fatty acids, phytochemicals, minerals, etc.), actual analytical technologies can still only capture a part of the metabolome. Two analytical platforms are by far the most predominantly used for metabolomic analyses: *mass spectrometry* (MS) and nuclear *magnetic resonance* (NMR) technologies. Currently MS-based approaches yield a higher sensitivity than NMR when analyzing minimal amounts of complex mixtures. In particular, the use of high- and ultra-high-resolution mass spectrometers greatly improves analytical performance and offers the best combination of selectivity and sensitivity. "Conventional" methods for analyzing metabolites by MS usually involve a physical/chemical purification/fractionation prior to MS analysis. Thus, to avoid sample matrix effects hyphenation of high-resolution separation

Comprehensive Analytical Chemistry, Vol. 63. http://dx.doi.org/10.1016/B978-0-444-62651-6.00010-6

techniques and MS is usually carried out at the expense of less high-throughput performance.

There is a clear need for more rapid, high-throughput MS approaches for metabolomics studies. In the quest to achieve the maximum high-throughput production of metabolite information in the largest possible number of samples, an extensive array of direct ionization or desorption/ionization techniques have been developed and combined. Using these approaches, any chromatographic or electrophoretic step prior to MS detection is avoided and thus direct analysis of samples (processed or not) is carried out. Sample introduction and ionization system used prior to MS analysis will always cause discrimination of specific metabolite classes. Thus, the choice of the most adequate methodology will require careful consideration, taking into account the goal of the metabolomic work. When direct MS analysis approaches are used, ionization suppression and its effect on sensitivity are not negligible since the presence of multiple chemical species and other matrix components will have considerable impact on the ionization of metabolites. Moreover, the overlapping of MS signals of isobaric species will be a common drawback of direct ionization technologies. Despite these limitations the major potential of this approach is its high-throughput character, especially when the number of samples to be analyzed is high. Thus, by using a variety of sample introduction/desorption/ionization techniques, metabolome screening in complex samples can be obtained in a few seconds by direct MS analysis. Moreover, through direct MS analysis the chemical composition within the spatial context of biological samples is also possible. Most MS-based approaches to analyze tissues require certain sample preparation that leads to destruction of histology structures. MS imaging (MSI) allows the capability to capture the chemical composition within the spatial context of biological tissues. In this chapter, an overview of the main desorption/ionization techniques coupled with MS applied to direct metabolite profiling or fingerprinting is presented.

2 MATRIX-ASSISTED AND MATRIX-FREE LASER DESORPTION/IONIZATION MS

Matrix-assisted laser desorption/ionization (MALDI) development has largely focused on high-molecular-weight polymers and biopolymers. With the development of new-generation time-of-flight (TOF) mass analyzers with remarkable improvements in mass resolution, the interest in application of MALDI-MS to small molecules has been recently renewed *(1,2)*. Using MALDI, sample is spotted on a metal plate with a solid or liquid matrix, and is co-crystallized with a highly UV-absorbing substance, which is generally a low-molecular-weight compound. When compared with direct infusion MS-based approaches, MALDI-MS has some advantages, especially its high tolerance toward salts and buffers and, in addition, the amount of sample consumed during analysis is very small. Application of MALDI-MS to small molecules typically involves a target approach *(1,2)*, and global metabolite analysis is relatively recent. One of the main limitations of MALDI-MS in metabolite analysis is signal suppression due to matrix background ions (in the low-mass range, $m/z < 700$) from conventional matrices such as 2,5-dihydroxybenzoic acid (DHB) and α-cyano-4-hydroxycinnamic acid (CHCA)), which interferes with the analysis of the metabolites. The use of a fluorophenyl porphyrin as matrix, instead of a standard low-molecular-weight matrix, did not produce ions in the low-mass region (100–500 Da) *(3)*. Using this matrix, determination of the fatty acid composition in a variety of vegetable oils was carried out by MALDI-TOF MS with minimum sample treatment *(4)*. To improve the analysis of low-mass molecules, Guo *et al. (5)* developed a novel approach to suppress the production of matrix-related background ions (from the common matrix CHCA) in MALDI by adding the surfactant cetrimonium bromide. As a result of the CHCA-related ion background suppression, very clean mass spectra were routinely obtained in the low-mass range. The use of small, nonpolar polymers (oligomers) based on oligothiophene or oligobenzodioxin as matrices allowed the analysis of model small-molecular-weight compounds *(6)*. It was suggested that the mechanisms for forming positive ions was based on charge transfer, rather than on proton transfer. In a different work, the use of ionic liquid-based matrices *(7)* or other matrix-assisted compounds such as 9-aminoacridine *(8,9)* have been presented for high-throughput metabolomics applications. These strategies are in continuous development for a more amenable analysis of small molecules minimizing matrix interferences.

Significant effort has also been made to develop laser desorption/ionization (LDI) techniques that can be performed without matrix, allowing placement of the sample directly onto a surface. Sample preparation is simplified in this case, and inhomogeneous cocrystallization processes of the sample and matrix are avoided. One of the first matrix-free LDI-MS methods was introduced in 1999 by Siuzdak through the development of LDI on porous silicon (desorption ionization on porous silicon (DIOS)) *(10)*. The porous silicon substrate is easily obtained by electrochemical anodization of crystalline silicon in hydrofluoric acid-based solutions. This strategy was presented as an encouraging matrix-free strategy to counteract these interferences. Both unoxidized *(11)* and oxidized *(12)* porous silicon surfaces showed to be a successful and simple method for high-throughput analysis of metabolites. However, this approach still has some limitations, for example, due to the infiltration of analytes into the pores of the silicon, the control of the position and size of the crystallized analytes is difficult, thus resulting in a less effective energy transfer when the analyte is not in the silicon surface. *Nanostructure-initiator mass spectrometry* (NIMS) was introduced as an alternative matrix-free approach to DIOS to produce low-background and high-sensitivity MS measurements *(13)*. NIMS uses initiator molecules trapped in nanostructured surfaces or "clathrates" to release and ionize intact molecules adsorbed on the surface. A list of compounds used as initiators for NIMS can be found elsewhere *(14)*. When the surface is heated with a laser (or ion beam), the initiator violently erupts from the pore, triggering the desorption/ionization of the analyte. In contrast to MALDI, these compounds do not absorb ultraviolet energy, analytes do not cocrystallize with the initiator, and most of them do not ionize. NIMS-MS has been used for direct biofluid analysis (blood and urine) *(13)*. Nanostructure-assisted LDI (NALDI) is a patented matrix-free technology (NALDI^{TM} chip, from Bruker Daltonics). The nanomaterial on the target absorbs the laser energy and allows for the desorption/ionization process of the analyte *(15)*. As a result of the absence of the matrix, mass spectra present very low chemical background. Disposable nanostructured target plates are commercially available, and they have been used in a variety of small molecule analysis, such as phospholipids *(16)*. Many other matrix-free methods have been developed for laser desorption/ionization of small molecules, with potential applications in metabolomics. A variety of surface properties have been studied for matrix-free ionization of small molecules, for example, diamond nanowires *(17)*, sol-gel-derived silver-nanoparticles-impregnated thin biofilm *(18)*, nanoporous gallium nitride-silver nanoparticles *(19)*, metal oxide surfaces *(20)*, and nanofilament silicon *(21)*. Comprehensive reviews of innovative technologies using energy-absorbing materials for matrix-free LDI-MS have been recently published *(22–24)*.

3 DIRECT-INFUSION MS

Among atmospheric pressure ionization (API) techniques, both electrospray ionization (ESI) and atmospheric pressure chemical ionization (APCI) ion sources are surely the most employed techniques in MS analysis using direct infusion approaches. When using direct infusion approaches, sample typically needs to be treated to dissolve the compounds of interest in the appropriate solvent. By using direct-infusion MS-based approaches, sample preparation can be considered a "bottleneck." In this sense, automated multiwell devices for metabolite purification provide the highest throughput in sample preparation. When large-scale multibatch experiments are designed, robust workflows have to be developed to minimize experiment analytical variation *(25)*. Lin *et al.* compared the classification and biomarker discovery capacities of direct-infusion ESI-MS and liquid chromatography (LC)–MS *(26)*. For that purpose, serum samples from patients with kidney cancer and healthy controls were analyzed by both analytical techniques. It was observed that direct-infusion ESI-MS had comparable classification and prediction capabilities to LC–MS but consumed only ~5% of the analysis time. In contrast, biomarker discovery of LC–MS (48 variables) was better than that of direct-infusion ESI-MS (23 variables).

By using the direct-infusion approach based on ESI-MS, multidimensional MS-based shotgun lipidomics (MDMS-SL) has also demonstrated to be a successful innovative approach in nontargeted analysis of lipids *(27)*. Using this approach, a two-dimensional mass spectrum is constructed. The first dimension is composed by the molecular ions in *m/z* values, while the second dimension is comprised of the mass corresponding to the neutrally lost fragments or the monitored fragment ions in *m/z* values (Figure 1). The cross

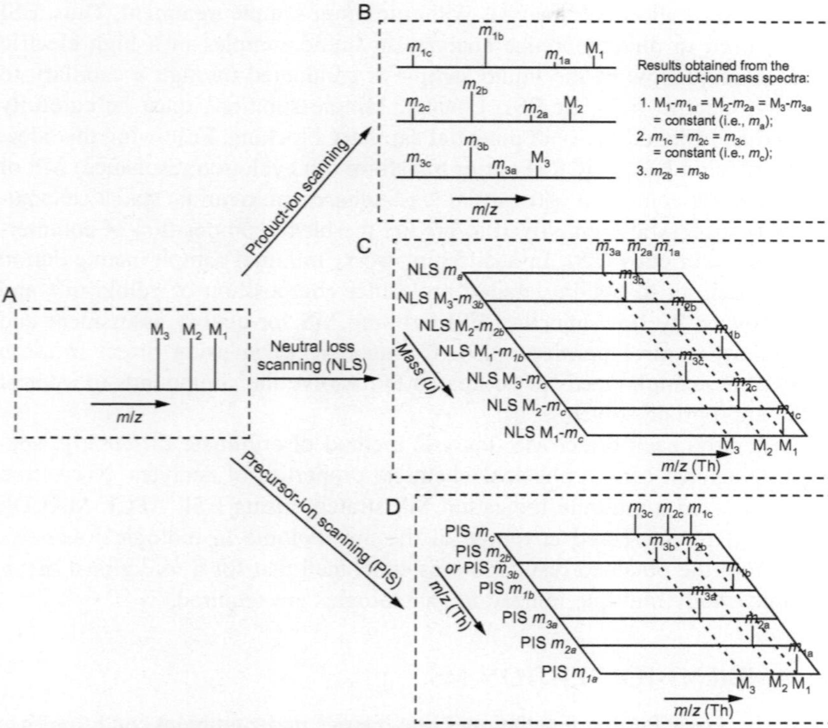

FIGURE 1 Schematic illustration of the interrelationship among the MS/MS techniques for the analysis of individual molecular species of a class of interest. We only illustrate the analysis of three species (M1, M2, and M3) of a class for simplicity, whereas there exist up to hundreds of individual molecular species within a class. We assume that this class of lipid species, similar to a class of phospholipids or sphingolipids, possesses one common neutral-loss fragment with mass of m_a (i.e., M1-m_{1a}=M2-m_{2a}=M3-m_{3a}=m_a (a constant)), one common fragment ion at *m/z* m_c (i.e., m_{1c}=m_{2c}=m_{3c}=m_c), and a specific ion to individual species at *m/z* m_{1b}, m_{2b}, and m_{3b}, respectively, which might not be identical to each other. Specifically, the common neutral fragment and the common fragment ion both result from the head group of the class; the individual species-specific ions represent the fatty acyl moieties of the species. Thus, the residual part of each individual species can be derived from these fragments in combination with the *m/z* of each molecule ion. (A) A simplified full-mass scan. (B) The product-ion analysis of these molecule ions. (C) Scanning of the individual neutral-loss fragment between a specific molecule ion and its individual fragment ion. (D) Scanning of each individual fragment ion. It should be emphasized that, although the analyses of fragments with either neutral-loss scanning (NLS) or precursor-ion scanning (PIS) are much more complicated than those in product-ion scanning in this simplified case, the analyses by NLS or PIS are much simpler than that with product-ion scanning. *Reproduced from Ref.* (27).

peaks of a given primary molecular ion in the first dimension with the second dimension represent the fragments of a given molecular ion. The major difficulty of this approach is the accurate interpretation of spectra. On the other hand, the main drawback of this methodology is ion suppression that can be partially avoided with exhaustive sample purification.

Direct-infusion MS is a very interesting approach, especially when sample characteristics allow MS analysis with minimum sample treatment. Thus, ESI can be used to directly ionize analytes in liquid samples in a high electric field. A small flow of the liquid sample is conducted through a capillary to the high electric field for ESI. Usually, sample solutions must be carefully cleaned and filtered to avoid potential capillary blocking. Following this idea, direct-infusion ESI-FTICR (Fourier transform ion cyclotron resonance) MS of a coffee drink combined with partial least-squares multivariate statistical analysis was successfully employed to predict the blend composition of commercial coffee varieties *(28)*. In a different work, minimal sample manipulation was carried out to obtain detailed molecular composition of edible oils and fats analyzed by flow-injection ESI-Orbitrap MS for quality assessment and authenticity control purposes *(29)*. Commonly, when using direct infusion approaches, sample needs to be treated to dissolve the compounds of interest in the appropriate solvent.

Generally, each direct MS analysis method discriminate differently, specifically against certain physical–chemical properties of analytes. Nordstrom *et al.* followed a multiple ionization MS strategy using ESI, APCI, MALDI, and DIOS for increased coverage of the metabolome in biological samples *(30)*. From the obtained results, it was concluded that for a true global metabolomics study multiple ionization technologies are required.

4 AMBIENT-IONIZATION MS

Recently, a new family of techniques that operate under ambient conditions has emerged with the attempt to minimize the need for sample preparation and separation (purification/fractionation) prior to MS analysis by combining the sampling and ionization/desorption processes into a single step. In this sense, since the early 2000s the highest efforts are aimed at the performance of these "ambient" ionization techniques. Ambient ionization refers to a variety of combinations of sample introduction systems, and desorption and ionization methods that allow direct analysis of sample surfaces in open-air conditions with little or no sample pretreatment and, in most cases through noninvasive procedures *(31–34)*. For this reason, MALDI and "traditional" API techniques such as ESI and APCI are not considered to belong to this group since they usually still require extensive sample preparation and/or vacuum conditions.

The development of *ambient MS* was initiated with the introduction of *desorption electrospray ionization* (DESI) by Cooks in 2004 *(35)*. Since then, a variety of possibilities combining different desorption and ionization methods have been developed. Almost 30 ambient sampling/ionization approaches were involved in MS analysis in the 2-year period 2009–2010, as reviewed by Harris *et al.* *(36)*. Among them, DESI and *direct analysis in real time* (DART) were the two most prevalent techniques.

DESI is derived from traditional ESI and, as described by the developers, DESI shares the advantages of the matrix-free DIOS and the advantageous production of multiply charged ions of ESI. DESI-MS has demonstrated its usefulness in high-throughput differential metabolomics of biological samples with minimal sample preparation *(37)*. Using the DESI approach, electrosprayed aqueous droplets are directed at a surface of interest in air and act as projectiles desorbing ions from the surface as a result of electrostatic and pneumatic forces. As can be seen in Figure 2, electrical charge applied to the solution produces charged droplets, which in aqueous solutions lead to an excess of hydronium ions (H_3O^+) or hydroxide ions (OH^-), and hence to protonated or deprotonated analytes, which are observable in the positive-ion and negative-ion modes, respectively *(38)*.

DESI-MS allowed differentiation between diseased (lung cancer) and healthy mice urine samples *(37)*. DESI-MS has also been demonstrated to be a promising tool in food safety control. Thus, successful analysis of a group of agrochemicals (insecticides, herbicides, and fungicides) was carried out, spotting fruit and vegetable extracts onto a conventional smooth poly(tetrafluoroethylene) surface *(39)*. However, the real potential of this approach was demonstrated by the direct DESI-MS/MS analysis of fruit peels from market samples without any further sample treatment *(39)*. DESI-MS also allowed the rapid analysis of sulfur volatiles in several onion varieties to distinguish phenotypes (tearless and normal) by simply scratching leaves and recording the extractable ions for <0.5 min *(40)*. In this field, DESI-MS

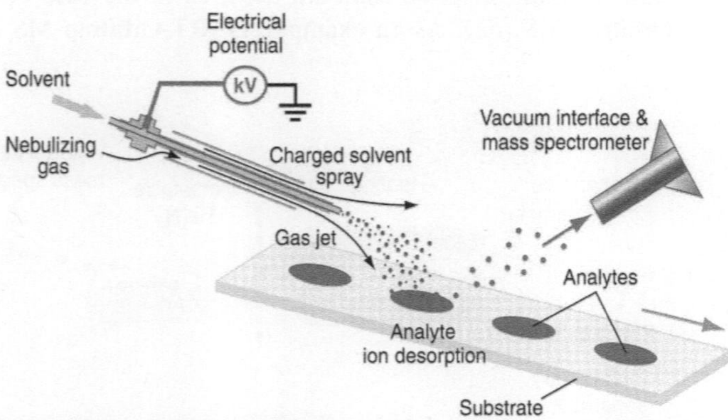

FIGURE 2 Schematic showing the DESI analyses for ambient high-throughput MS of unprepared samples. *Reproduced from Ref.* (38). (For color version of this figure, the reader is referred to the online version of this chapter.)

implemented in portable instruments is being performed in Cooks's laboratory for a rapid, *in situ*, direct qualitative and quantitative (ultra)trace analysis of agrochemicals in foodstuffs.

DESI was followed by DART in 2005 *(41)*. DART can be considered an API-related technique and is based on the thermo-desorption of condensed-phase analytes by a (distal) plasma discharge in a heated gas stream (helium or nitrogen). Metastable atoms generated from gas interact with ambient molecules, such as water, to create gas-phase ionic reagents that in turn react and ionize analyte on a surface or present as a vapor in the atmosphere (Figure 3) *(38)*. DART-MS has been applied to the analysis of small molecules in numerous types of samples without prior preparation *(42)*. DART is capable of analyzing low- to high-polarity compounds (up to 1 kDa) in both negative-ion and positive-ion modes; however, it is not really suitable for ionic compounds. Analysis of small molecules in plasma samples without sample preparation has been demonstrated by DART-MS *(43)*. Even living organisms can be subjected to DART-MS analysis *(44)*. Although more slowly than DESI-MS, DART-MS is now beginning to deliver its potential in metabolomics. DART-MS proved to be a powerful analytical technique for rapid metabolic fingerprints of human serum *(45,46)*. Metabolic fingerprints obtained by DART-MS of tomato and pepper have been recently reported for classification purposes (crops grown under organic vs. conventional conditions) *(47)*. Monitoring tea fermentation/maturation of tea through a nontargeted metabolite analysis approach was also possible by using DART-MS *(48)*. Hajslova *et al.* have critically reviewed DART-MS applications for food quality, safety, and authentication purposes *(49)*. As an example of its utility in food safety, DART-MS permitted the measurement of xenobiotics in real time on the fruit peel *(50)*. As occurred with other novel ambient desorption ionization techniques, maximal performance is achieved when ultra-high-resolution mass spectrometers are used, as in the case of FTICR MS *(51)* or Orbitrap MS *(52)*. As an example, DART-Orbitrap MS showed

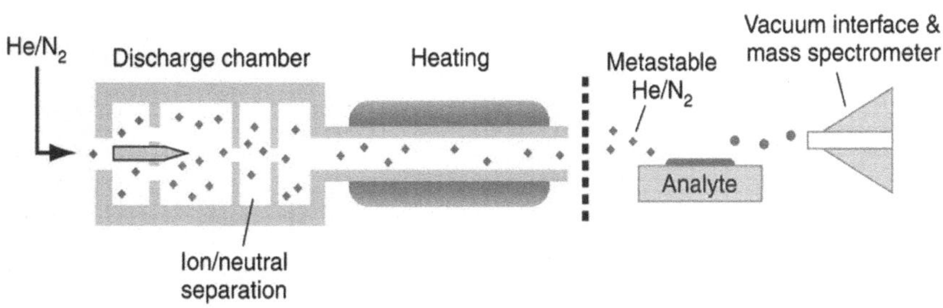

FIGURE 3 Schematic showing the DART analyses for ambient high-throughput MS of unprepared samples. *Reproduced from Ref.* (38). (For color version of this figure, the reader is referred to the online version of this chapter.)

its potential in the metabolic profiling of flavonoids and other phenolic compounds in propolis *(53)*.

Paper spray ionization is another recently developed ionization method for a fast direct MS analysis of complex mixtures on a paper substrate in open environment conditions *(54)*. It shares characteristics of ESI and ambient ionization methods. Sample is loaded by dropping or by wiping the surface of interest onto a paper of triangle shape. The electrospray is induced from the sharp tip of the triangular paper wetted with a small amount of sample by applying a high voltage (about 5 kV). The geometry of the paper substrate, the onset voltage for spray, and the sample load have been investigated for their effects on the ionization efficiency. The capabilities of paper spray ionization have been mostly demonstrated for the direct analysis of biological samples in drug monitoring applications *(55)*, thus the measurement of therapeutic drugs and their metabolites in dried blood spots. The paper spray ionization method has also been used as a direct sampling ionization method for MS analysis of additives in foods *(56)*. Thus, a piece of paper wetted with methanol was used to wipe a 10-cm^2 area on the peel of a lemon. Although the identification of particular compounds from food surfaces showed the potential of this approach, a further development should be carried out for its implementation in nontargeted metabolomics applications. Of particular interest was the modification based on paper spray ionization made by its own developers *(57,58)*, in which a fresh triangular piece of onion and spinach leaves served as both sample and substrate (Figure 4). Using this original approach, stress-induced changes in glucosinolates could be followed on the minute timescale.

5 IMAGING MS

The use of imaging mass spectrometry (IMS) for surface-based analysis of metabolites in tissue sections/surfaces is a remarkable novel approach *(59)*, particularly in metabolomics *(60)*. Recent advancements in the field of IMS have specifically been reviewed and discussed, showing the great potential of this technique in small molecule analysis *(61,62)*. Through a computer-controlled *xy* stage to the ionization source, the surface of the sample is typically scanned with a local desorbing and ionizing probe, and the generated ions from the surface are analyzed by MS *(63)*. In Figure 5, basics of IMS are represented *(64)*. Thus, two- and three-dimensional constructions of chemical abundance of metabolites from cells or tissues allow a deeper knowledge concerning the spatial organization of metabolic processes, cell-to-cell communication, molecular transport, and so on. Since altered chemical/molecular distributions are diagnostic for diseases, a direct examination of biological processes will result in a better understanding of the pathophysiology. The application of IMS to drug discovery/development is particularly attractive because it provides the opportunity to detect the localization of a certain drug

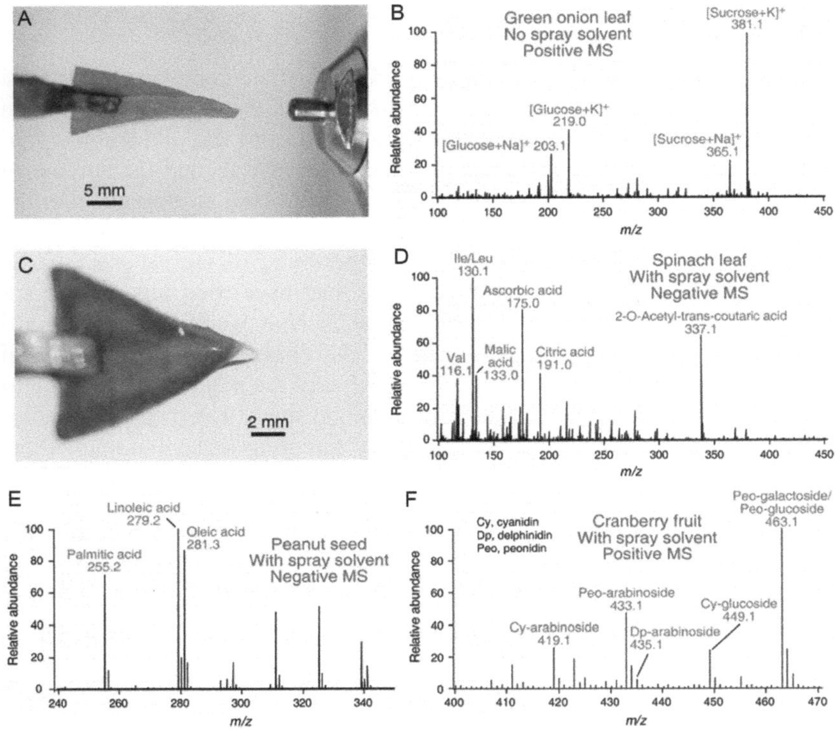

FIGURE 4 (A) Photograph of leaf spray ionization of green onion leaf cut to a point and held by a high-voltage connector in front of the atmospheric inlet of a mass spectrometer. (B) Leaf spray spectrum acquired from green onion leaf in positive ion mode, showing sucrose and glucose ions. (C) Photograph of leaf spray ionization of spinach leaf in negative-ion mode. The spinach leaf was cut into a triangle and methanol was applied on the leaf to achieve leaf spray ionization. (D) Leaf spray spectrum acquired from spinach leaf, showing amino acids and organic acids. (E) Leaf spray spectrum acquired from peanut seed in negative-ion mode, showing three fatty acids. (F) Leaf spray spectrum acquired from cranberry fruit in positive ion mode, showing a series of phytochemicals. Assignments given are based on exact mass and/or MS/MS data. *Reproduced from Ref.* (57). (For interpretation of the references to color in this figure legend, the reader is referred to the online version of this chapter.)

and its metabolites, as well as detect metabolite changes as a result of drug administration. Although the majority of the IMS experiments are based on imaging animal tissue sections or small tumor biopsies, IMS is starting to be applied to three-dimensional cell and tissue culture systems. Typically, samples from biological sources such as a biopsy, postmortem organ, and tumor section, as well as plants and single cells, are the object of an imaging study. In IMS rigorous sample preparation is vital to achieve the most accurate, reproducible, and validated data as possible *(65)*. In most cases, dissection of the sample is generally followed by freezing the tissue (in liquid

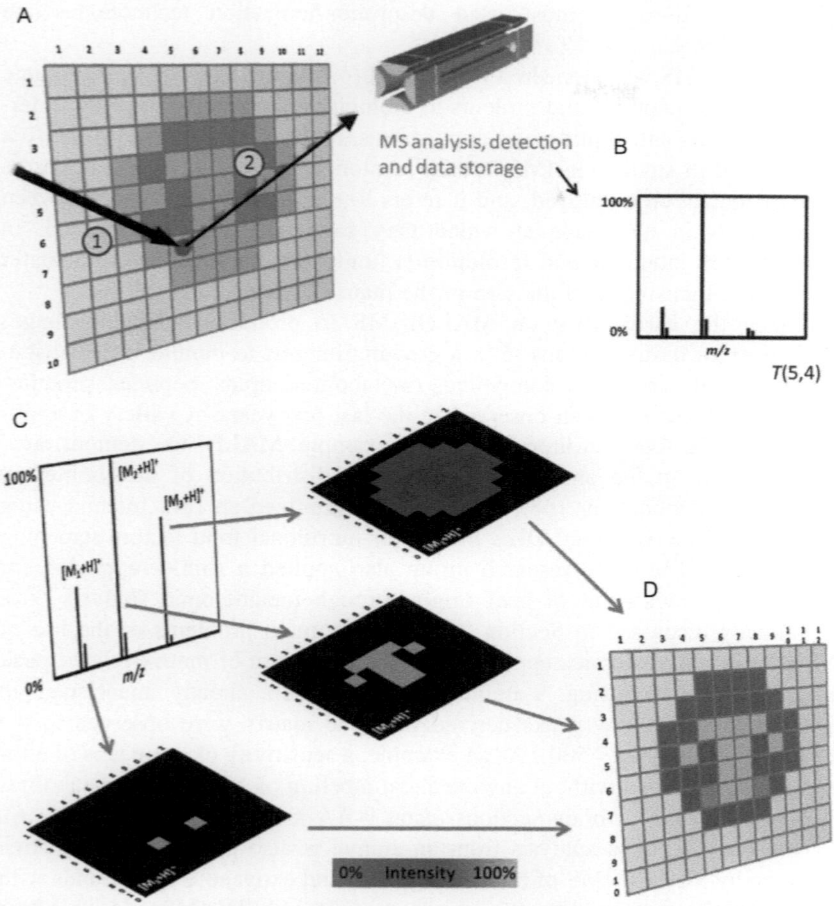

FIGURE 5 MSI imaging concepts and methods. (A) In a typical MSI experiment, the total area is subdivided (conceptually) into pixels that are individually inspected. (B) For each pixel, a single mass spectrum or the average of several mass spectra is collected and stored together with its spatial coordinates. (C) After the entire surface is scanned, an average mass spectrum can be created. The distribution of specific ions can be visualized by the creation of chemical images where the color scale (false color) represents the normalized intensity of particular ions. Each pixel from the image is associated with the original mass spectrum/mass spectra acquired at the specific point. The numbers 1 and 2 on panel A represent the steps desorption and ionization process. (D) The aim of imaging is to display the distribution of chemicals across a surface. *Reproduced from Ref.* (64). (For interpretation of the references to color in this figure legend, the reader is referred to the online version of this chapter.)

nitrogen or isopentane) and storing it at −112°F (−80°C) until use. Different types of desorption/ionization sources are currently in use or being developed to be used in IMS *(66)*. Each technique has advantages over the alternatives in terms of sensitivity, selectivity, spatial resolution, or sample preparation.

Brief descriptions of most used desorption/ionization technologies are described below.

MALDI-IMS was pioneered by Caprioli *(67)* in the late 1990s to generate ion images of peptides and proteins in biological samples. Without considering proteomics and peptidomics applications, lipids have been one of the first targets in IMS studies. In IMS, spatial resolution is usually dependent on the type of ionization employed, and it refers to the minimum distance between two objects in an image at which they can be distinctly discerned. In MALDI-IMS lateral spatial resolution is limited by the laser beam diameter (shape and focusing) and the size of the matrix crystal.

Since the early work on MALDI-IMS to produce molecular images directly from tissue sections *(67)*, a growing interest to monitor the distribution of a wide variety of compounds (metabolites, lipids, peptides, proteins, and xenobiotics) has been observed in the last few years. A variety of applications can be found in literature. As an example, MALDI has demonstrated its versatility in the analysis of the spatial distribution of metabolites in plant-based applications *(68)*. Zaima *et al.* optimized an IMS method using MALDI and conventional DHB matrix for nutritional food factors screening in rice *(69)*. The same research group also applied a similar approach for authenticity assessment of beef origin through metabolomic analysis *(70)*. As already discussed in Section 2, one of the main problems of the use of MALDI in metabolomics application is the formation of matrix-related peak interferences. By using 9-aminoacridine (9-AA), already mentioned in Section 2, only a few peaks derived from the matrix were observed in the low-mass range ($m/z \sim 500$). As an example, a sensitivity of some tens of attomoles per pixel and without any chemical labeling of 13 primary metabolites was obtained on rat brain sections using 9-AA as a matrix *(71)* (Figure 6). Whole-body sections analysis from an animal is also possible for label-free tracking by MALDI-IMS of both endogenous and exogenous compounds with spatial resolution and molecular specificity *(72)*. MALDI-IMS using 9-AA as a matrix has also been proved to be sensitive enough for the detection of metabolites and for two-dimensional imaging with single-cell sensitivity *(73)*. MALDI-MSI involving both CHCA and 9-AA was used to analyze the distribution of metabolites (amino acids, sugars, phosphorylated metabolites) in wheat seeds at different stages of development and under temperature stress *(74)*.

In addition to MALDI, matrix-free LDI techniques have been proved to be useful in IMS studies of small molecules *(62)*. Thus, NIMS has also been proved to be a highly sensitive matrix-free method (in which functionalized surfaces are used to absorb the laser, eliminating the need for matrices) for tissue imaging in metabolomics *(75,76)*. It has been described that NIMS surface can be easily treated or modified with different chemical initiators. As a result, distinct metabolite profiles from the same biological sample can be obtained. For instance, coating NIMS surface with cationization agents

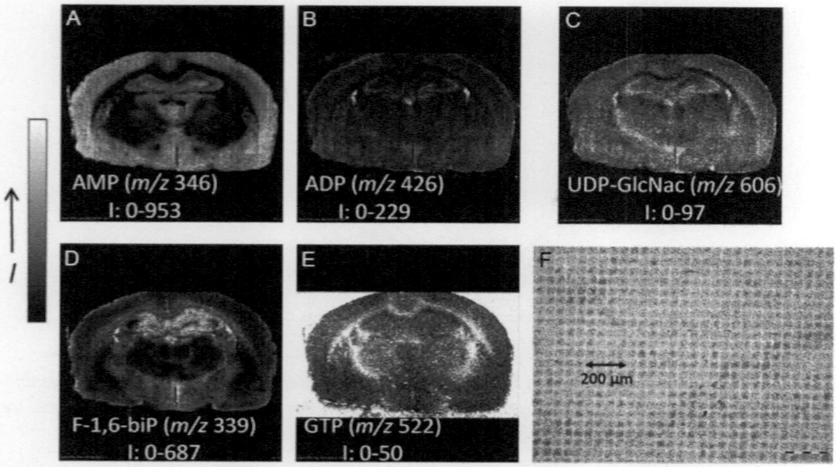

FIGURE 6 MALDI chemical images of (A) AMP, (B) ADP, (C) UDP-GlcNac, (D) F-1,6-biP, and (E) GTP acquired in the negative-ion mode from a rat brain section deposited on a stainless steel plate, after deposition of a homogeneous layer of 9-AA. Field of view: $8.3 \times 8.3\ mm^2$, pixel size 50 μm. The values of intensity (*I*) indicated under each image correspond to the minimal and maximal intensity in a pixel. (F) Optical image of a brain tissue section after 9-AA deposition and analysis by MALDI imaging with a 50 μm pixel size. *Reproduced from Ref.* (71). (For color version of this figure, the reader is referred to the online version of this chapter.)

($AgNO_3$) permitted the acquisition of images of brain sterol localization in a mouse model. Abnormal cholesterol biosynthesis in pathological brain tissues could be identified following this approach *(77)*. Sturm *et al.* compared images obtained by MS by using both NIMS and MALDI for neuropeptide and lipid imaging in a crustacean model organism (*Cancer borealis*) brain *(78)*. Similar lipid profiles were obtained using both strategies; however, MALDI-IMS gave better performance in the neuropeptide imaging experiments than NIMS. Other interesting applications have been published using other matrix-free methodologies, such as NALDI. As an example, NALDI-MS images of tumors through lipid analysis were demonstrated to be an encouraging technology for biomarker discovery *(79)*.

MALDI-IMS is still limited to about 20 μm spatial resolution (with 10–50 μm the most commonly achievable spatial resolution at this time), typically obtained with commercial ion sources *(80)*. Recently, Zavalin *et al.* demonstrated that 5 μm spatial resolution can be achieved for MALDI-IMS instruments by spatial filtration of the laser beam by using a 25 μm ceramic pinhole filter *(81)*. *Secondary ion mass spectrometry* (SIMS), developed in the 1960s *(82)*, offers a complementary or alternative method to MALDI-IMS and other matrix-free LDI-MS methods for the acquisition of higher spatial resolution images. In SIMS, a focused high-energy primary ion beam (such as Cs^+, Au^{3+}, Bi_5^+, or C_{60}^+) is used to directly bombard the sample

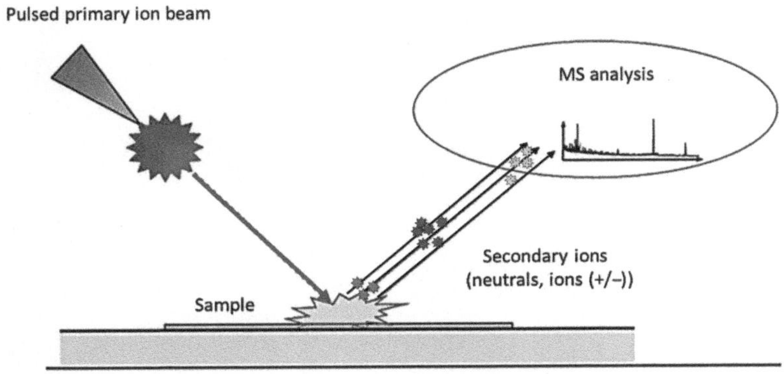

FIGURE 7 Schematic showing the SIMS analyses for IMS. *Reproduced from http://www.geobiologie.uni-goettingen.de/people/vthiel/tof_sims/index_e.shtml.* (For color version of this figure, the reader is referred to the online version of this chapter.)

surface. The primary ions transfer their energy to molecules on the surface, resulting in the desorption of ionized molecules (secondary ions), which are then analyzed by MS (Figure 7). Secondary ions are either positive or negative, depending on the primary ions' identity. Due to the use of primary ion beam, images can be acquired at high lateral resolution ($\geq$50 nm) *(83)*. In contrast to MALDI, SIMS produces higher fragmentation of desorbed ions that makes desorption of large intact molecules very difficult. Due to high fragmentation rate and low ionization efficiency, the size of biological molecules detected by SIMS analysis is limited ($\sim$2 kDa). The different types of ion beams can be used to increase the intact ion yield of larger or more labile compounds. As pointed by Fletcher *et al.*, sensitive SIMS–IMS still remains the requirement to significantly increase the secondary ion yields *(84)*. The use of SIMS in lipid MSI has been recently evaluated and successfully applied to elucidating a number of biological processes *(85)*. SIMS offers particularly powerful capabilities in the single-cell MSI area *(86)*. TOF MS are most commonly coupled with SIMS and MALDI sources for surface imaging. But the ultra-high mass-resolving power capabilities of FTICR MS can provide better mass identification capabilities of secondary ions with high specificity in tissue imaging *(87)*. However, analysis time continues to be lengthy, and lower spatial resolution capabilities were obtained when compared to SIMS-TOF IMS platforms.

In addition to laser beam- or ion beam-based methods, other matrix-free-based methods using a gas/liquid jetstream can also be used for desorption and ionization of compounds from the sample surface. This is the case of the DESI method under ambient conditions. DESI has very recently started to be explored for the analysis of small compounds imaging on intact surfaces *(88)*. In contrast to MALDI and SIMS in which the sample must be confined (in most cases) in a high-vacuum region of the instrument, in DESI–IMS the tissue surface is maintained at atmospheric pressure in open air and probed with a focused spray of charged microdroplets of a polar solvent. Lateral spatial resolution provided by DESI is typically 250 μm *(89)*, lower than MALDI or SIMS. In contrast, DESI usually requires less sample preparation. DESI–MSI has become an attractive tool for discovering the distribution of secondary metabolites in plants *(90)* and lipids in a variety of tissue samples *(91)*. Recently, Li *et al. (92)* explored the capabilities of IMS using DESI for the study of secondary metabolites in barley leaf surface. Although direct DESI analysis of the untreated leaves was not possible, it was certainly possible by stripping and analyzing the epidermis from the leaf. Thus, a number of hydroxynitrile glucosides from three different cultivars of barley were successfully identified and imaged throughout the leaves' surfaces. DESI–IMS has been recently applied for a better understanding of the molecular signatures of plant surfaces' thin-layer chromatographic imprints of leaves/petals of several plants *(93)*. In Figure 8, a particular example of the images of selected ions from petals and leaved TLC imprints is shown.

Laser ablation electrospray ionization (LAESI) is especially designed for biological samples containing water. Using this technique, a focused mid-IR laser excites the OH vibrations in a sample's water molecules. Phase explosion causes a rapid microscale ablation, ejecting a mixture of molecules, clusters, and particulate matter from the sample surface. Variations in the water content

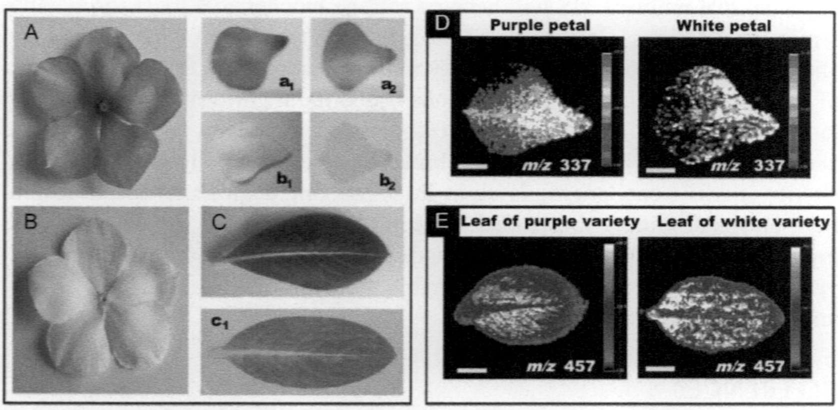

FIGURE 8 Photographs of flowers, petals, and a leaf of Madagascar periwinkle *C. roseus* and their TLC imprints. Images of (A) pink flower, (a1) single petal of a pink flower, and (a2) TLC-imprint of a pink petal. Images B, b1, and b2 correspond to the same data for a white flower. Images C and c1 correspond to a leaf and its imprint. Imprints do not correspond to the same petals or leaf whose photographs are shown. Images D and E correspond to one of the DESI MS images collected from petal and leaf showing the difference in spatial distribution between purple and white varieties of periwinkle, using the ion at *m/z* 337 and 457, respectively. Scale bars of both the images in D and E are the same (5 mm). *Reproduced from* (93). (For interpretation of the references to color in this figure legend, the reader is referred to the online version of this chapter.)

or tensile strength of tissues can affect the spatial resolution. LAESI was successfully applied to three-dimensional imaging MS of metabolites in leaf tissues, obtaining specific secondary metabolite accumulation patterns that correlate with the biochemical roles of these chemical species in plant defense and photosynthesis *(94)*. The feasibility of metabolite imaging using LAESI on chemically untreated sections of brain tissue at atmospheric pressure has also been demonstrated *(95)*. Among all mentioned methodologies, only DESI and LAESI techniques operate exclusively under atmospheric pressure, and sample treatment is minimum, which makes them suitable approaches for screening purposes (analysis of large sample sets). Both DESI and LAESI are comparable in terms of lateral spatial resolution (300–400 μm) *(96)*, and thus, further improvements are needed in this matter. Other less common ambient desorption/ionization techniques allow IMS to be performed under atmospheric pressure on untreated samples outside the MS. The latest developments and applications in this field have been reviewed by Wu *et al.* *(64)*.

6 CONCLUSIONS

A variety of novel direct MS-based approaches with promising utility in metabolomics have been introduced in this chapter. Among them, ambient MS is a very active area of research to perform high-throughput analysis, and it is expected that ambient MS will become routine biochemical tools for metabolomics applications. On the other hand, IMS is a rapidly growing technology, and although it is still in an early stage in the metabolomics field, the potential of metabolite imaging in tissue sections is enormous. IMS offers complementary information to conventional metabolomics. However, the simultaneous and spatially resolved detection of a broad range of metabolites with high sensitivity is still a challenging issue in IMS. Imaging acquisition speed, resolution, and data-mining tools need to be further developed. As the IMS field grows, new instrumentation and methods, such as improved lasers, matrix/solvent combinations, and advanced imaging software, will maximize sensitivity and identification capabilities.

REFERENCES

1. Cohen, L. H.; Gusev, A. I. *Anal. Bioanal. Chem.* **2002**, *373*, 571–586.
2. van Kampen, J. J.; Burgers, P. C.; de Groot, R.; Gruters, R. A.; Luider, T. M. *Mass Spectrom. Rev.* **2011**, *30*, 101–120.
3. Ayorinde, F. O.; Hambright, P.; Porter, T. N.; Keith, Q. L., Jr. *Rapid Commun. Mass Spectrom.* **1999**, *13*, 2474–2479.
4. Ayorinde, F. O.; Garvin, K.; Saeed, K. *Rapid Commun. Mass Spectrom.* **2000**, *14*, 608–615.
5. Guo, Z.; Zhang, Q.; Zou, H.; Guo, B.; Ni, J. *Anal. Chem.* **2002**, *74*, 1637–1641.
6. Woldegiorgis, A.; von Kieseritzky, F.; Dahlstedt, E.; Hellberg, J.; Brinck, T.; Roeraade, J. *Rapid Commun. Mass Spectrom.* **2004**, *18*, 841–852.

7. Vaidyanathan, S.; Gaskell, S.; Goodacre, R. *Rapid Commun. Mass Spectrom.* **2006**, *20*, 1192–1198.
8. Edwards, J. L.; Kennedy, R. T. *Anal. Chem.* **2005**, *77*, 2201–2209.
9. Miura, D.; Fujimura, Y.; Tachibana, H.; Wariishi, H. *Anal. Chem.* **2010**, *82*, 498–504.
10. Wei, J.; Buriak, J. M.; Siuzdak, G. *Nature* **1999**, *399*, 243–246.
11. Vaidyanathan, S.; Jones, D.; Broadhurst, D. I.; Ellis, J.; Jenkins, T.; Dunn, W. B.; Hayes, A.; Burton, N.; Oliver, S. G.; Kell, D. B. *Metabolomics* **2005**, *1*, 243–250.
12. Vaidyanathan, S.; Jones, D.; Ellis, J.; Jenkins, T.; Chong, C.; Anderson, M.; Goodacre, R. *Rapid Commun. Mass Spectrom.* **2007**, *21*, 2157–2166.
13. Northen, T. R.; Yanes, O.; Northen, M. T.; Marrinucci, D.; Uritboonthai, W.; Apon, J.; Golledge, S. L.; Nordström, A.; Siuzdak, G. *Nature* **2007**, *449*, 1033–1036.
14. Woo, H. K.; Northen, T. R.; Yanes, O.; Siuzdak, G. *Nat. Protoc.* **2008**, *3*, 1341–1349.
15. Daniels, R. H.; Dikler, S.; Li, E.; Stacey, C. *JALA* **2008**, *13*, 314–321.
16. Colantonio, S.; Simpson, J. T.; Fisher, R. J.; Yavlovich, A.; Belanger, J. M.; Puri, A.; Blumenthal, R. *Lipids* **2011**, *46*, 469–477.
17. Coffinier, Y.; Szunerits, S.; Drobecq, H.; Melnyk, O.; Boukherroub, R. *Nanoscale* **2012**, *4*, 231–238.
18. Gamez, R. C.; Castellana, E. T.; Russell, D. H. *Langmuir* **2013**, *29*, 6502–6507.
19. Nie, B.; Duan, B. K.; Bohn, P. W. *ACS Appl. Mater. Interfaces* **2013**, *5*, 6208–6215.
20. McAlpin, C. R.; Voorhees, K. J.; Corpuz, A. R.; Richards, R. M. *Anal. Chem.* **2012**, *84*, 7677–7683.
21. Tsao, C. W.; Devoe, D. L. *Methods Mol. Biol.* **2011**, *790*, 183–189.
22. Dattelbaum, A. M.; Iyer, S. *Expert Rev. Proteomics* **2006**, *3*, 153–161.
23. Rainer, M.; Qureshi, M. N.; Bonn, G. K. *Anal. Bioanal. Chem.* **2011**, *400*, 2281–2288.
24. Urban, P. L.; Amantonico, A.; Zenobi, R. *Mass Spectrom. Rev.* **2011**, *30*, 435–478.
25. Kirwan, J. A.; Broadhurst, D. I.; Davidson, R. L.; Viant, M. R. *Anal. Bioanal. Chem.* **2013**, *405*, 5147–5157.
26. Lin, L.; Yu, Q.; Yan, X.; Hang, W.; Zheng, J.; Xing, J.; Huang, B. *Analyst* **2010**, *135*, 2970–2978.
27. Han, X.; Yang, K.; Gross, R. W. *Mass Spectrom. Rev.* **2012**, *31*, 134–178.
28. Garrett, R.; Vaz, B. G.; Hovell, A. M.; Eberlin, M. N.; Rezende, C. M. *J. Agric. Food Chem.* **2012**, *60*, 4253–4258.
29. Vichi, S.; Cortés-Francisco, N.; Caixach, J. *J. Mass Spectrom.* **2012**, *47*, 1177–1190.
30. Nordstrom, A.; Want, E.; Northen, T.; Lehtio, J.; Siuzdak, G. *Anal. Chem.* **2008**, *80*, 421–429.
31. Huang, M.-Z.; Cheng, S.-C.; Cho, Y.-T.; Shiea, J. *Anal. Chim. Acta.* **2011**, *702*, 1–15.
32. Venter, A.; Nefliu, M.; Cooks, R. G. *Trends Anal. Chem.* **2008**, *27*, 284–290.
33. Ifa, D. R.; Jackson, A. U.; Paglia, G.; Cooks, R. G. *Anal. Bioanal. Chem.* **2009**, *394*, 1995–2008.
34. Weston, D. J. *Analyst* **2010**, *135*, 661–668.
35. Takats, Z.; Wiseman, J. M.; Gologan, B.; Cooks, R. G. *Science* **2004**, *306*, 471–473.
36. Harris, G. A.; Galhena, A. S.; Fernández, F. M. *Anal. Chem.* **2011**, *83*, 4508–4538.
37. Chen, H.; Pan, Z.; Talaty, N.; Raftery, D.; Cooks, R. G. *Rapid Commun. Mass Spectrom.* **2006**, *20*, 1577–1584.
38. Cooks, R. G.; Ouyang, Z.; Takats, Z.; Wiseman, J. M. *Science* **2006**, *311*, 1566–1570.
39. Garcia-Reyes, J. F.; Jackson, A. U.; Molina-Diaz, A.; Cooks, R. G. *Anal. Chem.* **2009**, *81*, 820–829.
40. Joyce, N. I.; Eady, C. C.; Silcock, P.; Perry, N. B.; van Klink, J. W. *J. Agric. Food Chem.* **2013**, *61*, 1449–1456.

41. Cody, R. B.; Laramee, J. A.; Durst, H. D. *Anal. Chem.* **2005**, *77*, 2297–2302.
42. Gross, J. H. *Anal. Bioanal. Chem.* **2013**, *406*, 63–80.
43. Zhao, Y.; Lam, M.; Wu, D.; Mak, R. *Rapid Commun. Mass Spectrom.* **2008**, *22*, 3217–3224.
44. Yew, J. Y.; Cody, R. B.; Kravitz, E. A. *Proc. Natl. Acad. Sci. U.S.A.* **2008**, *105*, 7135–7140.
45. Zhou, M.; McDonald, J. F.; Fernández, F. M. *J. Am. Soc. Mass Spectrom.* **2010**, *21*, 68–75.
46. Jones, C. M.; Fernández, F. M. *Rapid Commun. Mass Spectrom.* **2013**, *27*, 1311–1318.
47. Novotná, H.; Kmiecik, O.; Gałązka, M.; Krtková, V.; Hurajová, A.; Schulzová, V.; Hallmann, E.; Rembiałkowska, E.; Hajšlová, J. *Anal. Control Expo. Risk Assess.* **2012**, *29*, 1335–1346.
48. Fraser, K.; Lane, G. A.; Otter, D. E.; Harrison, S. J.; Quek, S. Y.; Hemar, Y.; Rasmussen, S. *Food Chem.* **2013**, *141*, 2060–2065.
49. Hajslova, J.; Cajka, T.; Vaclavik, L. *Trends Anal. Chem.* **2011**, *30*, 204–218.
50. Farre, M.; Pico, Y.; Barcelo, D. *Anal. Chem.* **2013**, *85*, 2638–2644.
51. Rummel, J. L.; McKenna, A. M.; Marshall, A. G.; Eyler, J. R.; Powell, D. H. *Rapid Commun. Mass Spectrom.* **2010**, *24*, 784–790.
52. Cajka, T.; Riddellova, K.; Zomer, P.; Mol, H.; Hajslova, J. *Food Addit. Contam. Part A Chem. Anal. Control Expo. Risk Assess.* **2011**, *28*, 1372–1382.
53. Chernetsova, E. S.; Bromirski, M.; Scheibner, O.; Morlock, G. E. *Anal. Bioanal. Chem.* **2012**, *403*, 2859–2867.
54. Wang, H.; Liu, J.; Cooks, R. G.; Ouyang, Z. *Angew. Chem. Int. Ed. Engl.* **2010**, *49*, 877–880.
55. Zhang, Z.; Xu, W.; Manicke, N. E.; Cooks, R. G.; Ouyang, Z. *Anal. Chem.* **2012**, *84*, 931–938.
56. Liu, J.; Wang, H.; Manicke, N. E.; Lin, J. M.; Cooks, R. G.; Ouyang, Z. *Anal. Chem.* **2010**, *82*, 2463–2471.
57. Liu, J.; Wang, H.; Cooks, R. G.; Ouyang, Z. *Anal. Chem.* **2011**, *83*, 7608–7613.
58. Zhang, J. I.; Li, X.; Ouyang, Z.; Cooks, R. G. *Analyst* **2012**, *8*, 3091–3098.
59. Ait-Belkacem, R.; Sellami, L.; Villard, C.; DePauw, E.; Calligaris, D.; Lafitte, D. *Trends Biotechnol.* **2012**, *30*, 466–474.
60. Miura, D.; Fujimura, Y.; Wariishi, H. *J. Proteomics* **2012**, *75*, 5052–5060.
61. Svatos, A. *Trends Biotechnol.* **2010**, *28*, 425–434.
62. Greer, T.; Sturm, R.; Li, L. *J. Proteomics* **2011**, *74*, 2617–2631.
63. Trim, P. J.; Djidja, M. C.; Muharib, T.; Cole, L. M.; Flinders, B.; Carolan, V. A.; Francese, S.; Clench, M. R. *J. Proteomics* **2012**, *75*, 4931–4940.
64. Wu, C.; Dill, A. L.; Eberlin, L. S.; Cooks, R. G.; Ifa, D. R. *Mass Spectrom. Rev.* **2013**, *32*, 218–243.
65. Goodwin, R. J. *J. Proteomics* **2012**, *75*, 4893–4911.
66. Watrous, J. D.; Dorrestein, P. C. *Nat. Rev. Microbiol.* **2011**, *9*, 683–694.
67. Caprioli, R. M.; Farmer, T. B.; Gile, J. *Anal. Chem.* **1997**, *69*, 4751–4760.
68. Kaspar, S.; Peukert, M.; Svatos, A.; Matros, A.; Mock, H.-P. *Proteomics* **2011**, *11*, 1810–1850.
69. Zaima, N.; Goto-Inoue, N.; Hayasaka, T.; Setou, M. *Rapid Commun. Mass Spectrom.* **2010**, *24*, 2723–2729.
70. Zaima, N.; Goto-Inoue, N.; Hayasaka, T.; Enomoto, H.; Setou, M. *Anal. Bioanal. Chem.* **2011**, *400*, 1865–1871.
71. Benabdellah, F.; Touboul, D.; Brunelle, A.; Laprevote, O. *Anal. Chem.* **2009**, *81*, 5557–5560.
72. Khatib-Shahidi, S.; Andersson, M.; Herman, J. L.; Gillespie, T. A.; Caprioli, R. M. *Anal. Chem.* **2006**, *78*, 6448–6456.
73. Miura, D.; Fujimura, Y.; Yamato, M.; Hyodo, F.; Utsumi, H.; Tachibana, H.; Wariishi, H. *Anal. Chem.* **2010**, *82*, 9789–9796.

74. Burrell, M. M.; Earnshaw, C. J.; Clench, M. R. *J. Exp. Bot.* **2007**, *58*, 757–763.
75. Calavia, R.; Annanouch, F. E.; Correig, X.; Yanes, O. *J. Proteomics* **2012**, *75*, 5061–5068.
76. Greving, M. P.; Patti, G. J.; Siuzdak, G. *Anal. Chem.* **2011**, *83*, 2–7.
77. Patti, G. J.; Shriver, L. P.; Wassif, C. A.; Woo, H. K.; Uritboonthai, W.; Apon, J.; Manchester, M.; Porter, F. D.; Siuzdak, G. *Neuroscience* **2010**, *170*, 858–864.
78. Sturm, R. M.; Greer, T.; Chen, R.; Hensen, B.; Li, L. *Anal. Methods* **2013**, *5*, 1623–1628.
79. Tata, A.; Fernandes, A. M.; Santos, V. G.; Alberici, R. M.; Araldi, D.; Parada, C. A.; Braguini, W.; Veronez, L.; Silva Bisson, G.; Reis, F. H.; Alberici, L. C.; Eberlin, M. N. *Anal. Chem.* **2012**, *84*, 6341–6345.
80. Lee, Y. J.; Perdian, D. C.; Song, Z. H.; Yeung, E. S.; Nikolau, B. J. *Plant J.* **2012**, *70*, 81–95.
81. Zavalin, A.; Yang, J.; Caprioli, R. *J. Am. Soc. Mass Spectrom.* **2013**, *24*, 1153–1156.
82. Castaing, R.; Slodzian, G. J. *J. Microscopie* **1962**, *1*, 395–399.
83. Klitzing, H. A.; Weber, P. K.; Kraft, M. L. *Methods Mol. Biol.* **2013**, *950*, 483–501.
84. Fletcher, J. S.; Lockyer, N. P.; Vickerman, J. C. *Surf. Interface Anal.* **2011**, *43*, 253–256.
85. Passarelli, M. K.; Winograd, N. *Biochim. Biophys. Acta* **2011**, *1811*, 976–990.
86. Lanni, E. J.; Rubakhin, S. S.; Sweedler, J. V. *J. Proteomics* **2012**, *75*, 5036–5051.
87. Smith, D. F.; Kiss, A.; Leach, F. E.; Robinson, E. W.; Paša-Tolić, L.; Heeren, R. M. A. *Anal. Bioanal. Chem.* **2013**, *405*, 6069–6076.
88. Wiseman, J. M.; Ifa, D. R.; Song, Q.; Cooks, R. G. *Angew. Chem. Int. Ed.* **2006**, *45*, 7188–7192.
89. Ifa, D. R.; Wiseman, J. M.; Song, Q.; Cooks, R. G. *Int. J. Mass Spectrom.* **2007**, *259*, 8–15.
90. Thunig, J.; Hansen, S. H.; Janfelt, C. *Anal. Chem.* **2011**, *83*, 3256–3259.
91. Dill, A. L.; Ifa, D. R.; Manicke, N. E.; Ouyang, Z.; Cooks, R. G. *J. Chromatogr. B* **2009**, *877*, 2883–2889.
92. Li, B.; Bjarnholt, N.; Hansen, S. H.; Janfelt, C. *J. Mass Spectrom.* **2011**, *46*, 1241–1246.
93. Hemalatha, R. G.; Pradeep, T. *J. Agric. Food Chem.* **2013**, *61*, 7477–7487.
94. Nemes, P.; Barton, A. A.; Vertes, A. *Anal. Chem.* **2009**, *81*, 6668–6675.
95. Nemes, P.; Woods, A. S.; Vertes, A. *Anal. Chem.* **2010**, *82*, 982–988.
96. Nemes, P.; Vertes, A. *Anal. Chem.* **2007**, *21*, 8098–8106.

Chapter 11

Functional Glycomics Analysis: Challenges and Methodologies

Nathan W. Stebbins and Ram Sasisekharan
Department of Biological Engineering, Koch Institute for Integrative Cancer Research, Massachusetts Institute of Technology, Cambridge, Massachusetts, USA

Chapter Outline

1 INTRODUCTION

After sequencing the human genome, the most puzzling finding was the small number of protein-encoding genes identified (~30,000). Given the phenotypic complexity of *Homo sapiens* compared to *Drosophila melanogaster*, *Caenorhabditis elegans*, or *Saccharomyces cerevisiae*, initial estimates of the human genome size were two to five times greater than what was ultimately reported by Venter *et al. (1)*. This finding led authors to consider the importance of non-genetically encoded mechanisms in generating the functional diversity of proteins, namely, post-translational modifications (PTMs) *(1)*. The modified central dogma now includes PTMs as key mediators of cellular phenotype and, in this post-genomics era, PTMs are a central focus of investigation. Glycosylation, defined as the covalent attachment of

Comprehensive Analytical Chemistry, Vol. 63. http://dx.doi.org/10.1016/B978-0-444-62651-6.00011-8

a carbohydrate to a glycoconjugate carrier, is the most abundant and diverse PTM observed in nature *(2)*. The attachment of a glycan moiety encodes a vast array of information associated with cellular function. In mammals, glycans are attached to cell surface proteins and lipids, but may also exist as free reducing sugars (e.g., human milk glycans), unconjugated in the extracellular matrix (ECM) (e.g., hyaluronic acid), or attached to intracellular proteins (e.g., *O*-GlcNAcylation) *(2)*. Structurally, glycans are either branched or linear, and further classified by the nature of linkage to the glycoconjugate. Branched glycans can be Asn-linked or Ser/Thr-linked to a glycoprotein, or attached to a glycolipid *(2,3)*. In contrast, linear glycans, which are predominantly glycosaminoglycans (GAGs), are Ser/Thr-linked to a proteoglycan core *(2,4,5)* (Figure 1).

Glycans interact with components of the extracellular milieu, including ECM proteins, growth factors, receptors, enzymes, cells, and pathogens *(2,4)*. As a consequence of these interactions, glycans regulate diverse processes including cell growth *(6,7)*, development *(8,9)*, morphogenesis *(10)*, cell adhesion *(11,12)*, cell–cell interaction *(13)*, immunity *(2,14)*, and host–pathogen interactions *(15,16)*. The glycome is highly dynamic and is remodeled either biosynthetically or postsynthetically during various biological processes or changes in cellular state *(7,17)*. As such, changes in glycan composition or abundance are observed during development, tumorigenesis and metastatic progression *(18–21)*, autoinflammatory disease *(22)*, and in the immune evasion of pathogens *(23)*.

In addition to the regulation of cell physiology in both normal and disease states, glycans exert their function at a molecular level. Glycosylation influences the structural properties of proteins, regulating features such as protein folding, solubility, stability, immunogenicity, and clearance *(24–26)*. These features are of interest within the pharmaceutical industry and have been studied extensively in the context of developing effective biologic therapeutics. For example, monoclonal antibodies contain key *N*-glycosylation sites in the Fc region. Notably, glycosylation in the Fc domain of IgG_1 is required for antibody-dependent cell-mediated cytotoxicity (ADCC) or complement-dependent cytotoxicity (CDC) *(26,27)*. Furthermore, alterations in the structures or composition of Fc glycans can enhance or diminish ADCC/CDC (e.g., hypersialylation or increased core fucosylation results in decreased ADCC/CDC *(25)*).

Clearly, glycosylation plays an integral role in a diverse array of biological and disease processes, yet decoding the function of complex carbohydrates has proved difficult. Key aspects of glycan chemical properties and fundamental modes of action suggest a need for a unique experimental toolkit combine with a systems-level approach for determining biological function. First, glycans possess several unique structural elements (i.e., branching, anomericity, varied monosaccharide linkages) that give rise to diverse chemical structures and a high degree of isomerism *(28–30)*. These features pose significant barriers pertaining to the development of analytical techniques capable of elucidating fine chemical structure. Second, glycan biosynthesis is a

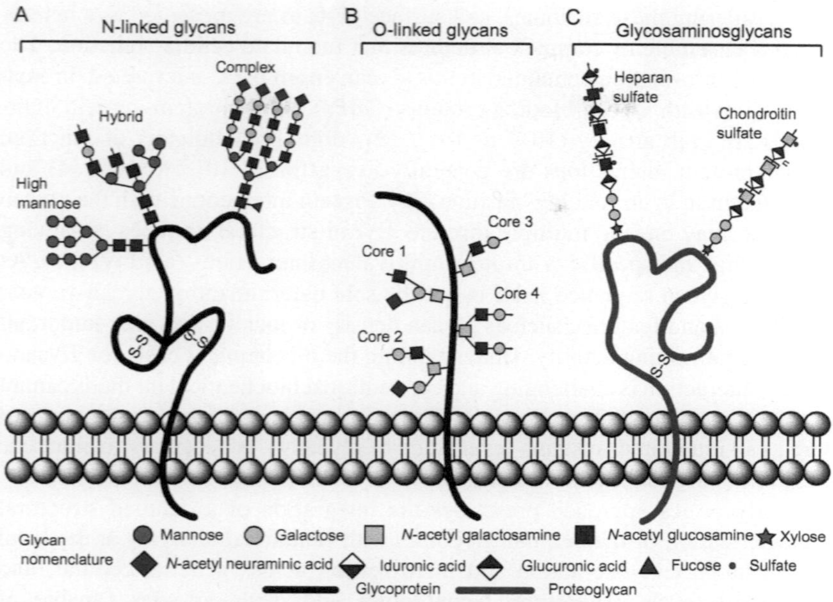

FIGURE 1 Structural diversity of glycans. Glycans are classified based on topology and linkage to the glycoconjugate. They can be branched (N-linked and O-linked glycans) or linear (glycosaminoglycans). Glycan structures are depicted using CFG cartoon notation. (A) N-linked glycans are characterized by the carbohydrate attachment to a protein via the asparagine residue of the polypeptide sequon, N-X-S/T, where X is any amino acid other than proline. All *N*-glycans share a common core structure ($Man_3GlcNAc_2$) and can be categorized into three subtypes based on the nature of the extension of the core: high mannose, complex, and hybrid. (B) Mucin-type O-linked glycans are characterized by the covalent linkage of GalNAc to a serine or threonine, and are classified into eight different core structures (core 1–4 depicted). Unlike N-linked glycans, *O*-glycans do not have a single-peptide consensus sequence and several GalNAc polypeptide transferase (pp-GalNAc-Ts) enzymes utilize unique polypeptide motifs for *O*-GalNAcylation. (C) Glycosaminoglycans (GAGs) are linear polysaccharides composed of repeating hexosamine-uronic acid disaccharide units and are O-linked to a proteoglycan core. Several classes of GAGs exist and are classified based on their disaccharide composition; heparin/heparan sulfate (IdoA/GlcA-GlcN) and chondroitin sulfate (GlcA-GalNAc) are depicted here. (For color version of this figure, the reader is referred to the online version of this chapter.)

non-template driven, non-proofreading process and involves the concerted activity of several hundred enzymes, including glycosyltransferases, glycosidases, and enzymes involved in sugar–nucleotide biosynthesis and transport *(2,4,30)*. Their biosynthetic complexity complicates the use of classical molecular biology or functional genomic strategies for interrogating glycan structure–function relationships. Furthermore, the lack of a template or proofreading mechanism leads to the expression of a heterogeneous, often related, set of glycan structures, and results in microheterogeneity and variation in site occupancy at the level of the glycoconjugate.

Considering these attributes, cell surface glycans are presented as a heterogeneous ensemble of complex structures that modulate cellular function. The functional information contained in this glycan ensemble is interpreted through interactions with glycan-binding proteins (GBPs). While protein–protein interactions are high affinity (10^{-9} to 10^{-12} M) "digital" modulators of function, glycan–protein interactions are generally low affinity (10^{-3} to 10^{-6} M) and affect function in an "analog" fashion *(30)*. Protein interactions with the glycan ensemble may engage multiple discrete glycan structural epitopes, achieving high affinity and specificity through multivalent interactions (avidity) *(31)*. As such, the glycan sequence itself is not the sole determinant of glycan–protein interactions and features, such as glycan density or localization play important roles in determining activity. Understanding the biochemical basis for glycan–protein interaction is challenging and conventional biochemical methods cannot capture such interactions accurately. Therefore, a glycomics approach—defined here as an integrated, systems-level study of glycans—is necessary to elucidate these structure–function relationships.

A glycomics approach prescribes the integration of a detailed structural characterization of the cellular glycome, with functional analyses at the level of glycan–protein interaction and interrogation across genetic, cellular, and organismal levels. Therefore, technologies and methodologies capable of measuring each of these attributes are required. In recent years, there have been significant advances in the development of high-throughput methods for structural glycomics analysis, generation of glycan-related reagents, model systems for functional studies, and databases that contain computational tools and glycan-related data. These advances enable glycobiologists to probe the biological role of glycans in an unprecedented fashion. Focusing largely on branched mammalian glycans, we review the fundamental methods used for structural elucidation of the glycome and tools for functional analysis of glycans across multiple levels. Here, we emphasize the integrated efforts essential to characterize glycan structure and determine structure–function relationships.

2 STRUCTURAL ANALYSIS OF GLYCANS

The characterization of glycan structures is central to uncovering their biological role. However, even with modern analytical tools, the unique chemical complexities of glycans make their analysis challenging. Many monosaccharide building blocks are isobaric stereoisomers (e.g., Glc, Man, and Gal are same MW), glycans can be branched or linear, with varied anomeric configuration (α or β) and diverse linkage types between monosaccharides *(2)*. Furthermore, glycans can be neutral or acidic, variable in polymer size ($n=1-10^3$), and contain modifications like acetylation, phosphorylation, or sulfation *(32)*. These unique features give rise to a large number of isobaric isomers, and a glycan pool consisting of heterogeneous physicochemical properties. Considering these features, no single method is capable of measuring all features of the glycan structures present within a glycan pool. For example, mass-based characterization of glycans is not straightforward, as mass spectrometry (MS) cannot readily distinguish isobaric structures nor provide definitive details on anomericity or glycosidic linkage. Conversely, liquid chromatography (LC) can resolve isomers but cannot provide direct structural information, thus it must be coupled with sequential enzyme digests or complemented by direct mass composition analysis. These examples illustrate the need for multidimensional methods to probe the glycome.

As multimethodological analysis is required, a tradeoff exists between measuring detailed structural information and high-throughput compositional analysis which yields valuable, yet incomplete structural information. For instance, glycosylation changes that occur during cancer are the focus of many diagnostic or clinical biomarker discovery efforts. In these studies, the profiling of alterations in structural features (i.e., changes in core fucosylation or terminal sialylation *(33)*) or quantifying the relative amounts of certain structures (i.e., abundance of fucosylated triantennary glycan in hepatocellular carcinoma *(34)*) are informative. In contrast, studies concerned with *de novo* characterization of novel glycan structure *(35)* or mechanistic determination of glycan–protein interactions require rigorous determination of composition, linkage, and anomericity to determine exact sequence or three-dimensional structural conformation *(36,37)*.

A number of techniques such as MS, high-performance liquid chromatography (HPLC), capillary electrophoresis (CE), nuclear magnetic resonance (NMR), and glycan-binding/-degrading reagents are used together to identify structure. Each method offers a unique and complementary dimension of data, bringing with it a unique set of advantages and limitations (outlined in Table 1). In this section, we examine the fundamentals of structural characterization focused on the methods, limitations, and complementary levels of structural detail gleaned from each method. We discuss the general considerations for glycan analysis, commonly used separation technologies for isomeric resolution, MS-based methods for compositional and detailed structural analysis, and structural motif profiling using lectins reagents.

2.1 General Consideration for Glycan Analysis

The release of the glycan pool from the glycoconjugate carrier is a critical step in structural glycomics characterization. Due to the differences in the monosaccharide linkage to the peptide backbone, N-, O-linked, and GAGs each have unique or preferred methods of release. *N*-glycans share a common core structure ($Man_3GlcNAc_2$) and are released using Peptide-*N*-glycosidase F, an enzyme that cleaves the bond between the peptide asparagine and the *N*-glycan's core GlcNAc *(38)*. Chemical methods for *N*-glycan release such

TABLE 1 Analytical Technologies for Structural Glycomics Analysis

Method	Detection Limit	Structural Data	Strengths	Weaknesses
CE	amol–fmol	Separation, profiling	Rapid separation, isomeric resolution, small sample volume	Limited databases for CE, standards required for structural annotation, limited compatibility with MS
HPLC	fmol	Separation, profiling, sequencing	Isomeric resolution, databases for peak assignment, amenable to online MS detection	Moderate to low throughput, standards or complementary methods (i.e., exoglycosidase or MS) needed for unambiguous characterization
MS	fmol–pmol	Composition, profiling	Potential for automation, databases for assisted or automated annotation	Isomers indistinguishable, separation/derivatization required to prevent ion suppression
MS^n	fmol–pmol	Composition, sequence, linkages	Detailed structural information, databases for assisted analysis	Data profile is complex, significant expertise required, low throughput
NMR	nmol	Complete structure, stereochemistry	Detailed structural information (linkage, anomericity, conformation)	Significant expertise required, low throughput, large amount of pure material required
Lectin arrays Lectin binding	pmol–nmol	Structural motifs	Structural analysis in tissue context, living cells can be screened, no fractionation required	Limited by availability and specificity of characterized lectins, quantification is challenging

as hydrazinolysis and β-elimination can be used but are not favorable as compared to enzymatic release *(39)*. In contrast, Ser/Thr-linked glycans are more diverse in their attachment to the protein, and there is no single enzyme that is capable of releasing all *O*-glycans. For mucin-type *O*-glycans, chemical methods such as hydrazinolysis *(39)* and β-elimination *(40)* are used. Alkaline β-elimination cleaves the *O*-glycosidic bond and releases the *O*-glycans in their reduced alditol form *(40)*.

Glycans lack any natural absorbance and fluorescence properties. As such, the detection of glycans requires derivatization with a label, allowing for detection using multiwavelength UV or laser-induced fluorescent detection (LIF). Labeling glycans with dyes serves to (i) homogenize physicochemical properties of the glycan pool, which enables analysis of neutral and acidic glycans simultaneously, or (ii) confer novel chemical properties, which allows for the use of solid phases that are incompatible with native glycans (e.g., hydrophobic labels enable separation using RP-HPLC) *(41)*. Glycan derivatization is achieved using reductive amination. Many fluorophores have been reported, and the choice of label depends heavily on the separation and detection method. For HPLC analysis, hydrophobic or weakly charged fluorophores, such as 2-aminobenzamide (2-AB) or 2-aminoanthranilic acid, are commonly used, whereas highly charged labels, such as 8-aminopyrene-1,3,6-trisulfonic (APTS) acid, are used for CE analysis *(41–43)*. Glycan labeling strategies result in stoichiometric 1:1 labeling of glycans, which is advantageous for ensuring reproducibility and reliable quantification based on fluorescent intensity. Notably, a large excess of dye is required for the reaction to go to completion, and cleanup steps (i.e., paper chromatography, microscale HILIC SPE, or RP chromatography) are required to remove excess dyes, which can result in reduced sensitivity and accuracy *(41,44)*.

2.2 High-Performance Liquid Chromatography

LC is the most common and versatile technology for glycan characterization. LC is used to profile glycan structures, sequence complex glycan mixtures, and separate glycans prior to downstream analysis (e.g., CE or MS). In general, separation is performed using anion exchange (AEX) or normal-phase HPLC (NP-HPLC); however, a diversity of modes, such as reverse-phase (RP), porous graphitized carbon (PGC), and lectin affinity chromatography have been reported and are gaining popularity *(41,45)*. AEX is used for preparatory fractionation schemes or as the first dimension of a multidimensional analysis. Glycan pools often contain structures with variable number of charged sialic acids, and prefractionation based on charge can lead to more efficient separation using hydrophilic interaction chromatography (HILIC) or CE, and prevent ion suppression during MS of native glycans *(41,46)*.

HILIC is a form of NP-HPLC that is widely used for glycan analysis. Retention is based on the hydrophilicity of the glycan structure, which is dependent on properties such as composition, charge, size, and topology. HILIC analysis of

2-AB-labeled glycans is the most robust and widely used method for glycan profiling and sequencing *(43,47)*. This method (i) enables analysis of neutral and acidic glycans simultaneously, (ii) resolves positional and linkage isomers, (iii) allows quantitation of peaks, (iv) demonstrates reliable migration times, and (v) is amenable to online MS hyphenation. In general, the structural identity of the peaks is deduced by spiking synthetic glycans into a sample, repeating the chromatographic run, and observing changes in peak area of the elution profile. This method of characterization is slow, and limited by synthetic glycans, which are expensive, available only in small quantities, and do not represent the full spectrum of possible structures. To circumvent the use of synthetic glycan standards, a partially hydrolyzed dextran glucose ladder, consisting of a variable number of glucose units (e.g., $n = 1$–20 mer), can be used to define migration times in terms of glucose units (GUs). The use of GU for calibration minimizes run-to-run variability, and allows for standardized reporting data. The GlycoBase (http:/glycobase.nibrt.ie/glycobase/about.action) database contains the GU values for several hundred 2-AB-labeled *N*-glycan structures resolved using HILIC. Comparison of GU values against the database values enables rapid peak identification of 2-AB-labeled *N*-glycan structures *(47,48)*.

Detailed structural characterization and *de novo* peak annotation is possible by integrating LC with sequential enzyme digests for a technique termed exoglycosidase sequencing. Libraries of exoglycosidases, which cleave terminal glycosidic linkages in a specific manner, can be sequentially applied to remove monosaccharides from a glycan structure *(49)*. LC is then performed again, and the GU shifts in the resultant elution profile provide detailed information about the composition, linkage, and position of the cleaved monosaccharide. Automated analysis software, AutoGU, was created to aid in automated peak assignment following exoglycosidase sequencing. The GlycoBase database and AutoGU software have enabled faster more reliable interpretation of HPLC data *(48)*. Paralleling these developments, a 96 well-plate format for sample preparation was created and led to advancements in sample throughput *(47)*. This novel workflow has been employed for several large-scale studies of the human serum *N*-glycome. Such studies have revealed disease-associated glycosylation changes in rheumatoid arthritis patients *(47)*, and correlated variability, heritability, and environmental determinants with variation in *N*-glycome ($n = 1008$) *(46)*. Recently, high-throughput HPLC profiling was integrated with a genome-wide association study to identify SNPs associated with changes to the *N*-glycome *(50)*. This study revealed HNF1α as a master regulator of *N*-glycome fucosylation, and has important implications for the regulation of glycosylation in both physiologic and disease processes.

2.3 Capillary Electrophoresis

CE has proved to be an invaluable tool in glycan analysis for biomedical research, clinical diagnostics, and the biopharmaceutical industry *(51)*. During

CE, a strong electric field is applied to a narrow-bore capillary (50–75 μm inner diameter) to separate molecular constituents based on electrophoretic migration and electro-osmotic flow (EOF) *(52)*. Migration time of the analytes depends on charge, hydrodynamic radius, and, in some modes, interaction with separation solution (micelles, polymers, chiral selectors, etc.). CE's greatest strengths are its small sample consumption (few nL) and its ability to efficiently separate isomers in a rapid manner (timescale: 12–60 min). These features make CE a favorite for glycomics profiling and clinical glyco-biomarker discovery *(53,54)*.

The two most widely used modes of CE analysis are capillary zone electrophoresis (CZE) and capillary gel electrophoresis (CGE). CZE is performed using a bare fused-silica capillary in normal or reverse polarity modes, where an appropriate electrolyte buffer is used to separate analytes into migration "zones" based on the sum of electrophoretic migration and EOF. A major challenge in CZE is the reproducible migration of analytes due to EOF. Reducing the EOF by altering buffer composition (e.g., lowering conductivity) or by coating the inner surface of the capillary through additions of buffer additives (e.g., Dextran-PEG) leads to improved reproducibility. In contrast, CGE completely suppresses EOF and employs a neutral capillary filled with a gel mesh polymer matrix to separate the sample mixture based on size and charge (analogous to polyacrylamide gel electrophoresis (PAGE)). Separation based solely on size can be achieved through the removal of acidic glycans such as sialic acids. CGE has been applied to separate uniformly charged species such as neutral or desialylated *N*-glycans derivatized with charge tags *(55)*. Detection is facilitated using online LIF, and structural assignment of peaks is performed similar to HPLC (i.e., GU standard, exoglycosidase sequencing, spiking synthetic glycans). In contrast to HPLC, the databases and computational tools to aid in peak assignment of CE-LIF resolved glycans are not as well developed, thereby limiting CE's widespread adoption for sequencing.

The GlycoFibroTest best demonstrates the strength of CGE-LIF for clinical biomarker analysis. The test was developed in an effort to create a fibrotic staging system using serum biomarkers as an alternative to an invasive tissue biopsy in patients with fibrotic liver disease *(54,56)*. Rapid sample preparation workflows were developed to decrease analysis time and minimize purification steps. Briefly, using a PCR thermocycler and 96 well-plate, serum samples were denatured, subjected to PNGase F release, derivatized with APTS, and analyzed using a multicapillary gel-based DNA analyzer *(42)*. Glycan mixtures are resolved within 12–25 min and the ratios of key structures (e.g., undergalactosylated biantennary, galactosylated triantennary) can effectively monitor and stage liver fibrosis. This platform for rapid glycomics profiling has been extended to other disease indications, such as diagnosis of hepatocellular carcinoma and differentiating disease-state progression toward esophageal adenocarcinoma (EAC) *(57)*.

2.4 MS for Glycomics

Similar to proteomics and metabolomics, MS has become the primary workhorse for glycomics analyses. Mass-based analysis of glycan is advantageous compared to indirect analysis methods (e.g., HPLC migration time), mainly in terms of the amount of structural information obtained. MS is used for glycan composition profiling, while MS/MS or MS^n can elucidate detailed structural information about position and linkage isomers. Notably, MS has not superseded the need for other methods (e.g., NMR, exoglycosidase treatment, or lectin analysis), and orthogonal means of characterization are employed to further inform and validate structural data obtained using MS. In fact, coupling LC or CE with MS can greatly enhance the information quality and content, and integration across multiple techniques enables more rapid and robust sequencing of glycan structure. In this section, we review the basic methodologies employed for glycomics analyses using MS.

Matrix-assisted laser desorption/ionization (MALDI) and electrospray ionization (ESI) methods are the two primary ionization methods for glycomics analyses *(58)*. As with any other biomolecule the chemical properties of the analyte pool affect ionization efficiency. Isolated glycan pools are composed of neutral and acidic structures that are variable in size and hydrophilicity. The heterogeneous physicochemical properties lead to ion suppression during MS analysis (e.g., in positive mode acidic compounds are suppressed and the reverse occurs in negative mode) *(59)*. Glycan derivatization serves to homogenize the physicochemical properties of the glycan pool and reduce ion suppression. Permethylation is the most common derivatization method for MS and greatly enhances sensitivity stabilize labile acidic glycan moieties, and increase the amount of structural information attained during fragmentation (MS^n) *(60)*. Additionally, stable isotope permethylation can be used for relative quantitation of glycomics samples *(61)*. Reducing end derivatives also enhance ionization efficiency, and similar isotopic labeling strategies have been developed for quantitative analysis *(41)*.

MALDI is widely used for glycan profiling due to the ease of sample preparation, potential for automation, speed of data acquisition, and the simplicity of data interpretation (i.e., only singly charged ions observed) *(62)*. When profiling native glycans, one major limitation of this technique is that the acidic moieties, such as sialic acids, are often lost due to the high degree of vibrational excitation of the ions during ionization. Derivation techniques such as permethylation or the use of alternative matrices have been developed to circumvent such issues. MALDI-TOF-MS is the most frequently used method for profiling permethylated glycans. The major advantage of this method is that both neutral and acidic glycan structures can be detected in positive ion mode, quantitation of glycans is reliable *(63)*, and it is amenable to fragmentation analysis using tandem MS (e.g., TOF/TOF or QIT-TOF). Additionally, automated software such as

Cartoonist was developed for automated or assisted annotation of MALDI-MS spectra of permethylated *N*-glycans using biosynthetic plausibility as a constraint *(64)*. This software further enables rapid and accurate annotation of spectra and minimizes laborious MS/MS fragmentation needed to characterize ambiguous peaks. Because of these features, MALDI is a valuable tool for large-scale diagnostics and biomarker studies and has been applied extensively to analyze the changes in the composition or abundance of *N*- and *O*-glycans present in the serum of cancer patients or on the surface of cancer cell lines *(65)*.

ESI is “softer” than MALDI, generates more stable parent ions, and can be used with native or derivatized glycans. For ESI, preparation of samples is more cumbersome as ESI is sensitive to salts and detergents *(59)*. Additionally, the presence of multiply charged peaks makes spectral interpretation more difficult. One of the major limitations of all MS is its inability to discriminate isobaric structures. Consequently, on- and offline LC has been applied to separate the samples prior to analysis. Perhaps the greatest advantage of ESI-MS is the ability for online interfacing with LC *(66)*. The added dimension of separation enhances sensitivity due to diminished ion suppression, and provides resolution of positional and linkage isomers. Furthermore, retention time data coupled with mass composition can enable detailed structural characterization without the use of MS/MS analysis. HILIC, RP-LC, and more recently PGC have been shown to interface with MS, and have been used for many glycomics applications *(41,58,67)*. One limitation of using LC hyphenation is that it is relatively slow (timescale: hours). CE–MS is an attractive possibility to address this limitation, due to its ability to rapidly separate carbohydrates. Unfortunately, there are some technical hurdles that make CE–MS hyphenation challenging; however, recent developments in CE–ESI-MS interfaces have enabled the resolution of a dextran ladder containing 15 GU in 30 min, and is a promising step forward for rapid CE–MS-based carbohydrate analysis *(68)*.

Tandem MS may be used to elucidate detailed structural information on glycans. Indeed, similar to proteomics peptide fragmentation for sequencing, glycans can be subjected to fragmentation to elucidate fine structural information such as linkage or branching patterns. Fragmentation is typically achieved using collision-induced dissociation, where precursor ions are selected and dissociated using gas collisions. Fragmentation of permethylated glycans is the most well-developed method for structural analysis of glycans. Cross-ring fragmentation of permethylated glycans provides the most information about linkage and branching of glycan structures. In many cases, a single round of fragmentation is sufficient for isomeric resolution; however, multiple rounds of fragmentation can provide a highly informative product ion spectra resulting in more accurate and detailed structural annotation. Similar to Cartoonist, many computations tools (e.g., Glycoworkbench, STAT, and StrOligo) have been developed to aid in the interpretation of MS/MS and MS^n fragmentation spectra *(69–71)*.

2.5 Lectins as Structural Probes

GBPs, such as lectins or glycan-binding antibodies, are valuable tools for glycomics analysis due to their specificity for defined glycan motifs and ability to capture structural information in the context of tissue. Lectins are non-immunological glycan-binding proteins that exhibit specificity toward a particular glycan motif or structural feature (*31,72*). For instance, the *Sambucus nigra* lectin (SNA-I) exhibits specificity for Neu5Ac α2,6-Gal(NAc) and is routinely used for the detection of terminal α2,6-linked sialic acid structures (*73*). Currently, more than 100 lectins have been isolated and characterized. They are derived from a variety of sources, such as natural animal GBPs (e.g., Galectins) or plant sources (e.g., SNA-I from elderberry bark) (*73*). In contrast to lectins, antiglycan antibodies are typically highly specific for a single glycan determinant (e.g., anti-Sialyl Lewis antibody *(74)*) and they have been developed for clinical biomarker detection or therapeutic use *(74,75)*. Experimentally, lectins and glycan-binding antibodies are used as structural probes, whereby binding (or loss of binding) indicates the presence or absence of specific glycan motifs. Compared to analytical methods that provide detailed information on glycan structure, lectin analyses provide an orthogonal measurement of the structural glycome based on broad structural features or motifs. Perhaps one of the greatest strengths of lectins is their ability to measure certain structural features, such as monosaccharide linkage isomers or acidic glycans, much faster and easier than analytical techniques, which require special derivatization, purification, separation, and multiple rounds of MS *(31,76)*. These benefits are contrasted by a limited number of structures recognized by lectins/GBPs *(77)*. Furthermore, definitive knowledge of lectin specificity is also a limitation, as some of the most widely used lectins (e.g., MAL and MAH for detection of α2,3-linked sialic acids) have conflicting reports regarding the structural requirements for their binding *(78)*. Attempts to comprehensively define these specificities are being pursued using frontal affinity chromatography (FAC) *(76)*.

Classically, lectins and glycan-binding reagents have been used for SDS-PAGE—"lectin blotting," flow cytometry analysis *(79)*, enzyme-linked lectin assay, and histopathological analysis *(80)*. Notably, recent developments in protein array technologies have enabled the development of a lectin array for characterizing the structural glycome through motif profiling of cells and glycoproteins *(73,81)*. Lectin arrays do not require laborious glycan release or fractionation method for analysis, so they can be performed in a high-throughput fashion. For lectin array analysis, panels of lectins are printed onto a glass slide or functionalized surface and incubated with fluorescently labeled glycoproteins, cellular micellae preparations, or whole cells *(73,81,82)*. The slide is then washed, imaged using a standard confocal-type scanner, and the fluorescent signal is quantified. Due to the low affinity of GBP–glycan interactions, many binding events are lost during the washing step, thus limiting the overall sensitivity (high pM) of this method. Optimization of evanescent-field fluorescent detection technology has enabled analysis without wash steps, and therefore improved the sensitivity (low pM) and ability to quantify interactions *(83)*. While it is not possible to use lectins to unambiguously determine the glycans structure, data derived from lectin arrays provides a unique glycan motif signature for a biological sample *(73,77,79)*.

Array profiling has been performed using single-color labeling of the sample; however, this only provides qualitative information about the glycome due to the unknown or variable dynamic ranges of printed lectins. The single-color array has been used for glycosylation analysis of glycoprotein samples *(84)* and to determine unique signatures for whole cells at various developmental stages *(79)*. Ratiometric comparative glycomics analysis has been achieved using two-color labeling strategy where glycan isolates from different samples are labeled with a different fluorophore (e.g., Cy3 and Cy5), mixed, and applied to the array *(81)*. The ratiometric lectin array was recently applied to detect changes in glycosylation that occur during progression of Barrett's esophagus (a precancerous state) to EAC. Motif profiling revealed that *Tritiicum vulgare* agglutinin (WGA) binding is lost in dysplastic and neoplastic esophageal tissue compared to healthy tissue *(85)*. WGA is a lectin that binds GlcNAc groups and sialic acids. After confirmation at the level of tissue staining, fluorescent WGA lectin was proposed for use as a negative contrast agent for the rapid endoscopic identification of dysplastic or neoplastic tissue to help guide biopsy *(85)*. Variations on the lectin array have been developed *(86)*. The antibody–lectin sandwich array offers a novel platform for characterizing glycosylation in the context of the native glycoprotein. For an antibody–lectin sandwich array, capture antibodies that bind a glycoprotein (e.g., MUC1 or CEA) are printed onto a glass slide and incubated with a serum or glycoprotein sample. After a target protein is captured, a panel of lectins is used to detect the glycan structures expressed on the captured antigen *(86)*. Together, these techniques illustrate the utility of lectins as structural probes as an orthogonal measurement of the structural glycome across multiple levels, including surface glycans, glycans on specific glycoproteins, and *in vivo* tissue-level analysis of glycan structures.

3 FUNCTIONAL ANALYSIS OF GLYCANS

While characterizing the structural glycome is a critical step toward defining structure–activity relationships, it should be noted that sequence alone is not sufficient to explain glycan function. The greatest challenge in decoding glycan structure–function relationships is understanding how a heterogeneous structural ensemble governs cellular function, as attributes such as density and localization affect glycan–protein interactions. A detailed understanding of the biochemical basis for glycan–protein interactions is paramount in order to bridge the mechanistic gap between structure and function. Additionally, cell or organism model systems with defined functional readouts are needed to measure/perturb glycan-protein interactions and enable extension of biochemical analyses to high-level biological function. In this section, two key approaches toward elucidating function are discussed, namely, the genetics approach to glycomics and methods to study glycan–protein interactions.

3.1 Genetics Approach to Glycomics

Transcriptional analysis of the genes involved in glycan biosynthesis and glycan–protein interactions provide another unique insight toward a global characterization of the glycome. Integration of transcriptomics data with analytical structural analysis validates and complements the structural data obtained and enables inquiry into the complex biosynthetic regulation of glycans *(30,87)*. In the absence of structural data, glycogene expression measurements provide useful clues as to the glycan structures present and guide the level of detail to be targeted during structural analysis *(6,88)*. Collectively, these measurements yield integral information for a systems-level understanding of glycan structure–function relationships.

Advances in microarray technology have enabled systems-level transcriptomics analysis to become common practice. Many databases house gene expression datasets derived from biomedical research model systems or clinical samples (e.g., National Center for Biotechnology Information's GEO or ONCOMINE). However, there are limitations when extracting or measuring glycogene expression data obtained using commercially available arrays. Typically, glycan biosynthesis genes are underrepresented, poorly annotated, or use probe sets that are unable to distinguish glycogene isoforms. Furthermore, glycogene transcript abundance is lower than many downstream signaling transcripts and most commercial arrays lack the sensitivity or dynamic range needed to quantify glycogene expression. Focused glycogene microarrays have been created to address these issues. Using Affymatrix chip technology, the GlycoGene-chip features probe sets for the detection of human and mouse transcripts specific to glycan biosynthesis enzymes, degradation enzymes, and glycan-binding proteins *(89,90)*. The design of the glycogene arrays is amenable to quantifying the low-abundance transcripts that are present in most mammals. The GlycoGene-chip has been used to measure the gene expression of several thousand mouse and human samples, and this data is publicly available through consortium for functional glycomics (CFG). Further strides have been made toward increasing the sensitivity, dynamic range, and accuracy of glycotranscriptomic analyses through the development of a qRT-PCR platform capable of measuring ~700 glycan-related genes *(17,87)*. Data generated using these platforms have been integrated with the MS-based glycan structural glycomics data to reveal correlations that inform the mechanisms of transcriptional regulation of glycan synthesis. For instance, various mouse tissues (liver, kidney, brain, and testis) were profiled using MALDI-MS and the qRT-PCR platform *(87)*. This analysis revealed certain structural features that correlated well with gene expression data (e.g., low *Fut8*, a core α1,6 fucosyltransferase, in liver correlated with the reduction of core fucosylation), whereas other features demonstrated complex biosynthetic regulations (e.g., no change in expression of sialyltransferase in kidneys, although a decrease in sialylated *N*-glycan structures was observed) *(87)*. This is no surprise as several hundred enzymes are responsible for the synthesis of glycan structures.

Computational models have been created in an attempt to study the relationship between glycogene expression and structure, and to predict glycan structures from gene expression data *(91,92)*. The GECS *N*-Glycan Prediction Server, for instance, uses reaction pattern libraries of glycosyltransferase reactions and the co-occurrence frequencies of observed N-linked glycan structures to predict *N*-glycan structures from microarray data (http:/www.genome.jp/tools/gecs/) *(91,93)*. While there are limitations in the accuracy of this tool, in pilot studies using gene expression data from cancer cell lines, the software predicted glycan structures containing Sialyl Lewis^{x}, which agreed with experimental results *(91)*. Undoubtedly, as computational models of glycan biosynthesis are refined, they will be invaluable tools for metabolic engineering, understanding the regulation of glycan diversity, and even aiding in the interpretation of experimentally derived structural data.

Probing the functional role of glycans at a cellular or organismal level through genetic manipulation provides another critical dimension toward understanding structure–function relationships. Numerous knockout animals and cell lines have been created in an effort to understand the role of glycosyltransferases/glycosidases, GBPs, and glycoproteins/proteoglycans. Unlike proteins, single glycan structures cannot simply be "knocked out" or "knocked in" as most functional epitopes involve several glycan biosynthesis genes. Even so, knockout models are valuable tools for understanding biosynthetic regulation of glycan and creating model systems to explore glycan function in the context of higher-order biological processes. Toward this end, many stable glycosylation mutants of cultured cell lines have been developed. The lectin-resistant CHO mutants (Lec mutants), for example, were generated using mutagenesis and selection with cytotoxic plant lectins. This yielded mutants with truncated or altered *N*- and *O*-glycan repertoires, and Lec mutants have been used to determine glycan modifications that influence ADCC, or understand various glycan-mediated signaling processes *(94)*. Additionally, several mouse models have been created in order to study the functional role of glycans or GBP. Recently, a large phenotypic screen of 36 mutant mouse strains deficient in glycosyltransferases of glycan-binding proteins was completed (the data is available through the CFG) *(95)*. Knockout studies often reveal the pleiotropic role of glycans and GBPs and complex biosynthetic regulation of the glycome (e.g., knockout of β4-Gal-T1 showed increases in Galβ1-3GlcNAc linkages and shifts in sialic acid) *(96)*. These models provide an exciting avenue to explore the function of the glycome at a systems level.

3.2 Glycan Array and Synthesis Technologies

The function of the glycome can be viewed in the context of glycan–protein interactions and the signaling consequences thereof. Glycan–protein interactions are critical events in multiple pathologies including tumor metastasis, immune infiltration, and infection of pathogens *(97,98)*. Consequently, determining the specificities and affinities of GBPs can provide valuable mechanistic information needed to relate glycan structure to function. Capturing biologically meaningful glycan-binding events is a key challenge in glycomics. Most GBP–glycan interactions are low affinity and show graded affinity or complex/mixed binding specificities toward cognate glycan "motifs." For instance, the endothelial scavenger receptor C-type lectin is selective for Lewis-type glycan motifs, whereas DC-SIGN binds to both Lewis type and high mannose *(99)*. As such, GBPs should be screened against a large number of glycan structures to comprehensively and quantitatively define their binding preferences. In an effort to systematically measure GBP specificities, the glycan array was developed. In a glycan array, large libraries of oligosaccharides are immobilized on a solid phase (e.g., glass slide or multiwell plate), incubated with varying dilutions of labeled GBP, washed, and the GBP–glycan interactions are subsequently quantified. Characterization of GBP–glycan interactions offers insight into the functional role of the GBP. For instance, the glycan-binding activity of Galectins-1, -2, and -3 were assessed using glycan array to reveal that each galectin differs in their binding to sialylated glycans, and these differences are responsible for mediating their diverse biological roles *(100)*.

In a physiological setting, GBPs achieve their specificity, high affinity, and function through multivalent interactions with structurally related glycans. It is nontrivial to identify the biologically relevant densities of glycan structures and valency in GBP presentation. Thus, many strategies exist to control these variables. For instance, valency may be enhanced synthetically by generating multiprotein complexes (e.g., bead or particle immobilization, or complexing biotinylated GBPs using streptavidin tetramers). Alternatively, whole cells or viruses can be analyzed to explore glycan binding in the context of natural valency. Oligosaccharide density, immobilization chemistry, linker characteristics, and spatial orientation can influence glycan–protein interactions *(97,98)*. Concepts such as density and multivalent presentation can significantly impact the binding specificity and affinity data obtained using the array. Unfortunately, it is not tractable to examine the full experimental space of glycan structure, glycan density, linker choice, and valency in protein presentation simultaneously. Consequently, the insight gleaned from glycan array data should be complemented with additional analyses, including validation of biochemical data with focused techniques (e.g., SPR or FAC), molecular and computational modeling, and functional data at the cellular level.

Capturing the full spectrum of glycan diversity on an array is another major challenge of this platform. Libraries of synthetic oligosaccharide are needed to represent the breadth of glycan structures that compose the cellular glycome. While the number of possible unique glycan structures is vast, even the most comprehensive glycan arrays only represent 500–600 structures *(97)*. Given the (relative) paucity of structures represented on arrays, it is clear that glycan synthesis is a bottleneck. Developments in synthesis hold great potential for advancing glycan array technology and our understanding of functional glycomics. Currently, glycans are derived naturally or synthesized using chemical and/or enzymatic methods. Using natural sources of glycans (i.e., isolated from cells or glycoproteins) is an attractive prospect as natural isolates capture the physiological diversity present in a biologic context *(101)*. In this manner, it is possible to create glycan array using oligosaccharides isolated from cells representing unique disease or cellular states. In practice, using naturally derived glycans is challenging, as it is difficult to isolate large amounts of pure glycans and to unambiguously characterize glycan structure simultaneously. Chemical methods for glycan synthesis face enormous challenges, namely, in yield and achieving the regio- and stereochemistry observed in nature. Recently, effective one-pot and solid-phase methods for combinatorial synthesis of glycans have been developed (reviewed elsewhere *(102–104)*). To date hundreds of oligosaccharides have been synthesized for glycan array, analytical applications, for use as agonists/antagonists of GBP–glycan interactions, and even for carbohydrate-based therapeutics or vaccines.

4 INTEGRATING STRUCTURE AND FUNCTION: A CASE STUDY

As discussed above, key aspects of glycan chemical properties and their fundamental mode of action suggest a need for a glycomics approach to determining biological function. Consequently, data across multiple lines of inquiry (structural, biochemical, and functional) should be integrated to elucidate structure–function relationships of glycans. Thus far we have primarily focused on individual methods that enable a systems-level analysis of glycan structure or function, with little regard to integration of structural and functional data. In an effort to provide tangible insight into the power of an integrated approach, we discuss the interaction of Influenza A virus Hemagglutinin (HA) with sialylated glycan receptors as a case study of a robust paradigm born out of an integrated glycomics approach.

In recent years, there has been substantial concern regarding the prospect of another influenza pandemic. Influenza A viruses are found in wild aquatic birds where they exist commensally without causing disease, making birds the natural reservoir for influenza. As observed in past influenza pandemics (e.g., 1957, 1968, and 2009), genetic reassortment of avian-derived viruses leads to adaptation, enabling efficient transmission between humans. This prospect of "host switching" is particularly concerning in light of highly lethal avian subtype such as *H5N1* (bird flu), which has had several localized outbreaks in humans with a high fatality rate (~60%) *(105,106)*. Fortunately, transmission of avian H5N1 to humans has only been observed in populations that are in direct contact with infected poultry, and H5N1 has not yet demonstrated efficient human-to-human transmission. The lethality of the virus and previous observations of avian to human transmission underscore the need to understand the molecular changes that govern human infectivity and transmissibility of influenza A viruses *(107)*.

The influenza A infection cycle begins with attachment of viral hemagglutinin (HA) to sialylated glycan receptors on host cells. Thus HA–glycan interaction is a critical step for regulating influenza infectivity and selectivity. Previous biochemical analyses identified that recognition of Neu5Ac $\alpha2\rightarrow6$ Galactose (Gal) (hereafter $\alpha2\rightarrow6$) is a key feature of human-adapted HA, whereas avian-adapted HA recognizes Neu5Ac $\alpha2\rightarrow3$ Gal (hereafter $\alpha2\rightarrow3$). In addition to biochemical characterization, structural analysis of cocrystal structures also corroborated the hypothesis that influenza virus host switching (i.e., avian to human adaptation) requires HA binding preferences to switch from $\alpha2\rightarrow3$ glycans found on avian tissue to $\alpha2\rightarrow6$ glycans found on human tissue *(32,107,108)*. Although this hypothesis explained transmission specificity for a number of viruses, confounding cases emerged where HAs exhibited mixed $\alpha2\rightarrow3/\alpha2\rightarrow6$ binding and showed variable transmission efficiencies in a ferret animal model of human transmission *(107,109)*. These results suggested that defining the glycan receptors in terms of HA preferences for $\alpha2\rightarrow3$ or $\alpha2\rightarrow6$ glycosidic linkage alone is not sufficient to explain avian to human adaptation.

In order to better understand avian- and human-adapted HA in the context of HA glycan-binding specificity, complementary methodologies were employed to decipher fine structural properties of sialylated glycan receptors and bridge structural analysis with biochemical binding analysis of HA–glycan interactions (Figure 2). The upper respiratory tract is the primary target for human-to-human transmission of influenza A virus, thus characterizing the diversity and distribution of the sialylated glycan receptors in this tissue was critical to define the virus's target glycan receptor. First, a panel of lectins was applied to stain tissue sections of the upper respiratory tract, and fine structure of *N*-glycans isolated from a human bronchial epithelial cell line was determined and quantified using MALDI-MS/MS coupled with sialidase pretreatment. Using these complementary tools, multiantennary sialylated (predominantly $\alpha2\rightarrow6$ linked) glycans with long branches composed of multiple lactosamine repeats were identified *(15)*. Next, a glycan array platform was used to determine the identity of the structural motifs necessary for human-adapted HA binding. The glycan array was designed to incorporate

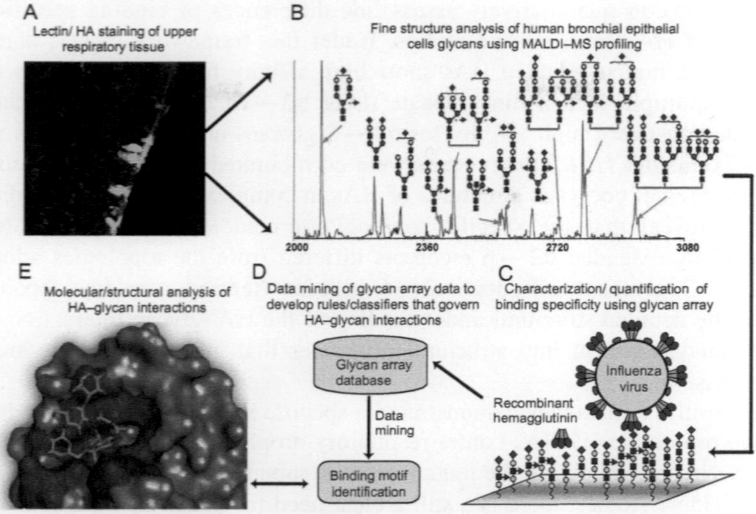

FIGURE 2 Integrated analyses to elucidate the key determinants of HA–glycan interactions that govern influenza A pathogenesis. (A) Lectin staining of the human upper respiratory tissues was used to determine the tissue-level distribution of α2,3 and α2,6 sialylated N- and O-linked glycans. (B) Detailed structural information on the composition of the glycan pool isolated from human bronchial epithelial cells was determined using MALDI-MS in combination with sialidases. (C) Glycan-binding specificity was determined using recombinant HA or whole virus on a glycan array platform. The array platform was designed to incorporate target structures based on their predominant expression in the upper respiratory tract. (D) The HA/virus binding data were mined to obtain rules or classifiers that govern the glycan-binding specificity of human-adapted or wild-type HAs. (E) The rules that were obtained were corroborated using X-ray cocrystal structures of HA–glycan complexes and molecular simulation of HA–glycan interactions. (For color version of this figure, the reader is referred to the online version of this chapter.)

the physiologically relevant glycan structures that were identified in the glycan characterization of upper respiratory tissue. To characterize binding preferences, data-mining tools were applied to glycan array data obtained with influenza A viruses and recombinant HAs from various avian- and human-adapted strains. This informatics-based approach identified patterns of structural features in glycan motifs that differentiated avian- versus human-adapted HA specificity and indicated that extension length on the nonreducing end of sialic acid was important for human-adapted HA binding *(15)*. To validate predictions generated by the data-mining approach, a binding assay was developed to quantitatively measure HA and virus specificity and affinity to key α2→3 and α2→6 glycan structures identified previously with MS and informatics. To accurately capture and compare specificity and affinity of HA–glycan interactions across different HAs, this assay used an antibody-based precomplexing strategy to enhance valency of HA presentation and ensure fixed avidity when comparing HAs *(16)*. This binding assay provided a framework to quantitatively assess fine differences of binding specificity/affinity of HAs from diverse strains. Under this framework, it was demonstrated that human-adapted HAs show high affinity for α2→6 glycans containing multiple lactosamine repeats (long α2→6), whereas avian-adapted HAs demonstrated high affinity for α2→3 glycans and α2→6 glycans with short extensions *(15)*. These results were corroborated with molecular modeling analyses of cocrystal structures of HAs in complex with sialic acid receptors to reveal that the specific topology (termed *umbrella-like* topology) adopted by extended α2→6 receptors differed from the topologies adopted by α2→3 glycans and short α2→6 glycans (termed *cone-like* topology) *(15)*. The detailed structural understanding of the HA–glycan interaction provides unique insight into structural properties that govern human-to-human transmission.

Recently, two studies demonstrated a specific set of "hallmark" mutations in two prototypic H5 HAs confer respiratory droplet viral transmission in ferrets to viruses containing the mutant H5 HA subunit *(110,111)*. In light of the recent H5N1 studies, there is a still a clear need for better characterization of the molecular determinants guiding the avian to human host switching to aid in influenza surveillance and guide vaccine development. Given our previous characterization of the glycan receptor topological determinants of human adaptation of H5 HA, we sought to apply this framework to understand the structural determinants of H5 HAs that govern the quantitative switch in receptor-binding preferences for long α2→6. We developed a structural framework to systematically analyze the receptor-binding site (RBS) of H5 HA from the perspective of the structural topology of its glycan receptor, and the residues and interresidue interactions in the RBS. The RBS of H5 HA was compared to the RBS of its closest phylogenetic neighbor, pandemic H2 HA *(106)*. From this comparison, key molecular features in the RBS that interact with the base region (Neu5Ac α2→6 Gal β1→) and extension region (→4 GlcNAc β1→4 Galβ1→) of the long α2→6 glycan in the open umbrella topology were defined. After the features required for a quantitative switch of binding preferences of H5 HA to human receptors were described, the acquisition of these features over the natural sequence evolution of H5 HA were investigated. A phylogenetic analysis of H5N1 HA1 sequences was constructed and revealed that several clades, namely, a subset of rapidly evolving and currently circulating clades (clades 2.2.1 and 7), have acquired amino acid changes characteristic of one or more key structural features of human adaptation. Based on this analysis, representative strains from the currently circulating clades that have acquired amino acid changes that match one or two of the human-adapted RBS features were recombinantly expressed and used on the glycan array platform to determine binding affinity/specificity toward α2,3- or α2,6-linked receptor. In these strains, mutations were made to confer the additional human-adapted RBS features, and then probed for their changes in binding preferences. From this analysis, we were able to demonstrate that as few as one or two mutations in some of the recent H5 HAs could confer a quantitative switch to preference for human receptors. Additionally, we introduced several of the "hallmark" mutations that lead to aerosol transmission in two prototype H5 HAs into more recent H5 HA strains *(110,111)*. It is important to note that the introduction of each mutation has strain-specific results and, importantly, did not quantitatively switch any of the mutant HAs binding preference to human receptors. This case study exemplifies an elegant application of an integrated glycomics framework used to understand the key molecular changes that govern influenza host switching as well as outline a robust framework for guiding the development of effective influenza surveillance strategies.

The example described previously provides tangible insight into an integrated framework for the structure–function study of glycans. The integrated approach provided a mechanistic bridge between structural analyses of physiologically relevant sialylated glycan receptors on upper airways and the binding specificity of avian- and human-adapted HA in the context of relevant glycan motifs *(107)*. Additionally, application and extension of this integrated framework enabled the study of specific mutations in the RBS that result in a quantitative switch in binding to human receptors in the context of currently evolving strains of H5N1 *(106,112)*. The aforementioned studies have many important implications in influenza surveillance, vaccine development, and pandemic preparedness. More broadly, this example serves to demonstrate (i) how multi-methodological approaches were used to elucidate fine structural details, and (ii) sequence alone does wholly govern glycan–protein interaction as features, such as topology and multivalency are also key determinants of structure–function relationships.

5 CONCLUSIONS

The study of the glycome holds promise to further uncover the regulatory and phenotypic complexities of the cell. However, unique challenges pertain to glycan structural complexity and nuances that govern their biological function make their analysis challenging and underscore the need for an integrated, systems-level approach to decode structure–function relationships. Early efforts in glycomics were focused on relieving technological bottlenecks such as sample throughput and data analysis; however, as technology and tools are no longer a staggering limitation, attention has shifted toward deciphering the mysteries of the glycome. Paralleling technological advancements, databases such as the CFG have played an integral role in advancing the field of glycomics. CFG was a collaborative initiative aimed at generating and distributing reagents and resources, as well as creating a relational database containing multiple diverse datasets (gene expression, GBP array data, tissue histology/mouse models, and glycan structural data). Acknowledging the power of an integrated approach, CFG allows users to explore data from multiple orthogonal glycomics tools in the context of a biological sample or question in an effort to uncover novel paradigms pertaining to structure–function relationships. Indeed, over the past decade, dramatic improvements in technologies and methods for structural and functional analysis have brought this once burgeoning field into maturity.

Challenges lie ahead in terms of the development of robust methods for mechanistically bridging structure–function relationships. Finer determinants that govern glycan–protein interactions, such as conformation and multivalency, cannot be captured by a detailed structural characterization, although they are largely responsible for influencing specificity and affinity. Currently, there are no universal technologies to capture these features and this represents an emerging bottleneck in bridging structure–function relationships. As described in the example of HA-sialylated glycan receptor interactions, bioinformatics and molecular modeling tools were used to capture the nuances of glycan-binding specificities and were key elements in bridging structural and functional data. Thus, robust tools that enable determination of glycan motifs from glycan array data and for molecular modeling of glycan–protein interactions are needed to address the finer mechanistic determinants that govern glycan-binding protein specificity.

ACKNOWLEDGMENTS

This work was funded in part by the National Institutes of Health (R37 GM057073-13 and T32 ES007020-38) and the National Research Foundation-supported Interdisciplinary Research Group in Infectious Diseases of SMART (Singapore MIT Alliance for Research and Technology).

REFERENCES

1. Venter, J. C.; Adams, M. D.; Myers, E. W.; Li, P. W.; Mural, R. J.; Sutton, G. G.; Smith, H. O.; Yandell, M.; Evans, C. A.; Holt, R. A.; Gocayne, J. D.; Amanatides, P.; Ballew, R. M.; Huson, D. H.; Wortman, J. R.; Zhang, Q.; Kodira, C. D.; Zheng, X. H.; Chen, L.; et al. *Science* **2001**, *291*, 1304–1351.
2. Varki, A.; Cummings, R. D.; Esko, J. D.; Freeze, H. H.; Stanley, P.; Bertozzi, C. R.; Hart, G. W.; Etzler, M. E., Eds.; *Essentials of Glycobiology*; Cold Spring Harbor Laboratory Press: Cold Spring Harbor (NY), 2009.
3. Moremen, K. W.; Tiemeyer, M.; Nairn, A. V. *Nat. Rev. Mol. Cell Biol.* **2012**, *13*, 448–462.
4. Sasisekharan, R.; Raman, R.; Prabhakar, V. *Annu. Rev. Biomed. Eng.* **2006**, *8*, 181–231.
5. Sasisekharan, R.; Shriver, Z.; Venkataraman, G.; Narayanasami, U. *Nat. Rev. Cancer* **2002**, *2*, 521–528.
6. Pérez-Garay, M.; Arteta, B.; Pagès, L.; de Llorens, R.; de Bolòs, C.; Vidal-Vanaclocha, F.; Peracaula, R. *PLoS One* **2010**, *5*, e12524.
7. Dai, Y.; Yang, Y.; MacLeod, V.; Yue, X.; Rapraeger, A. C.; Shriver, Z.; Venkataraman, G.; Sasisekharan, R.; Sanderson, R. D. *J. Biol. Chem.* **2005**, *280*, 40066–40073.
8. Hennet, T. *Biochim. Biophys. Acta* **2012**, *1820*, 1306–1317.

9. Holst, C. R.; Bou-Reslan, H.; Gore, B. B.; Wong, K.; Grant, D.; Chalasani, S.; Carano, R. A.; Frantz, G. D.; Tessier-Lavigne, M.; Bolon, B.; French, D. M.; Ashkenazi, A. *PLoS One* **2007**, *2*, e575.
10. Inatani, M.; Irie, F.; Plump, A. S.; Tessier-Lavigne, M.; Yamaguchi, Y. *Science* **2003**, *302*, 1044–1046.
11. Sanderson, R. D. *Semin. Cell Dev. Biol.* **2001**, *12*, 89–98.
12. Zhao, Y.-Y.; Takahashi, M.; Gu, J.-G.; Miyoshi, E.; Matsumoto, A.; Kitazume, S.; Taniguchi, N. *Cancer Sci.* **2008**, *99*, 1304–1310.
13. Haines, N.; Irvine, K. D. *Nat. Rev. Mol. Cell Biol.* **2003**, *4*, 786–797.
14. Kolarich, D.; Lepenies, B.; Seeberger, P. H. *Curr. Opin. Chem. Biol.* **2012**, *16*, 214–220.
15. Chandrasekaran, A.; Srinivasan, A.; Raman, R.; Viswanathan, K.; Raguram, S.; Tumpey, T. M.; Sasisekharan, V.; Sasisekharan, R. *Nat. Biotechnol.* **2008**, *26*, 107–113.
16. Srinivasan, A.; Viswanathan, K.; Raman, R.; Chandrasekaran, A.; Raguram, S.; Tumpey, T. M.; Sasisekharan, V.; Sasisekharan, R. *Proc. Natl. Acad. Sci. U.S.A.* **2008**, *105*, 2800–2805.
17. Nairn, A. V.; Aoki, K.; dela Rosa, M.; Porterfield, M.; Lim, J.-M.; Kulik, M.; Pierce, J. M.; Wells, L.; Dalton, S.; Tiemeyer, M.; Moremen, K. W. *J. Biol. Chem.* **2012**, *287*, 37835–37856.
18. López-Ferrer, A.; Barranco, C.; de Bolós, C. *Am. J. Clin. Pathol.* **2002**, *118*, 749–755.
19. Rosen, S.; Lemjabbar-Alaoui, H. *Expert Opin. Ther. Targets* **2010**, *14*, 935–949.
20. Satomaa, T.; Heiskanen, A.; Leonardsson, I.; Angström, J.; Olonen, A.; Blomqvist, M.; Salovuori, N.; Haglund, C.; Teneberg, S.; Natunen, J.; Carpén, O.; Saarinen, J. *Cancer Res.* **2009**, *69*, 5811–5819.
21. Goetz, J.; Mechref, Y.; Kang, P.; Jeng, M.-H.; Novotny, M. V. *Glycoconj. J.* **2009**, *26*, 117–131.
22. Arnold, J. N.; Saldova, R.; Hamid, U. M. A.; Rudd, P. M. *Proteomics* **2008**, *8*, 3284–3293.
23. Vigerust, D. J.; Shepherd, V. L. *Trends Microbiol.* **2007**, *15*, 211–218.
24. Ghaderi, D.; Taylor, R. E. R.; Padler-Karavani, V.; Diaz, S.; Varki, A. *Nat. Biotechnol.* **2010**, *28*, 863–867.
25. Shire, S. J.; Gombotz, W.; Bechtold-Peters, K.; Andya, J., Eds.; *Current Trends in Monoclonal Antibody Development and Manufacturing*; Springer New York: New York (NY), 2010.
26. Jefferis, R. *Nat. Rev. Drug Discov.* **2009**, *8*, 226–234.
27. Arnold, J. N.; Wormald, M. R.; Sim, R. B.; Rudd, P. M.; Dwek, R. A. *Annu. Rev. Immunol.* **2007**, *25*, 21–50.
28. Laine, R. A. *Glycobiology* **1994**, *4*, 759–767.
29. Mariño, K.; Bones, J.; Kattla, J. J.; Rudd, P. M. *Nat. Chem. Biol.* **2010**, *6*, 713–723.
30. Raman, R.; Raguram, S.; Venkataraman, G.; Paulson, J. C.; Sasisekharan, R. *Nat. Methods* **2005**, *2*, 817–824.
31. Ambrosi, M.; Cameron, N. R.; Davis, B. G. *Org. Biomol. Chem.* **2005**, *3*, 1593–1608.
32. Robinson, L. N.; Artpradit, C.; Raman, R.; Shriver, Z. H.; Ruchirawat, M.; Sasisekharan, R. *Electrophoresis* **2012**, *33*, 797–814.
33. Adamczyk, B.; Tharmalingam, T.; Rudd, P. M. *Biochim. Biophys. Acta* **2012**, *1820*, 1347–1353.
34. Liu, X.-E.; Desmyter, L.; Gao, C.-F.; Laroy, W.; Dewaele, S.; Vanhooren, V.; Wang, L.; Zhuang, H.; Callewaert, N.; Libert, C.; Contreras, R.; Chen, C. *Hepatology* **2007**, *46*, 1426–1435.
35. Guerrini, M.; Zhang, Z.; Shriver, Z.; Naggi, A.; Masuko, S.; Langer, R.; Casu, B.; Linhardt, R. J.; Torri, G.; Sasisekharan, R. *Proc. Natl. Acad. Sci. U.S.A.* **2009**, *106*, 16956–16961.

36. Guglier, S.; Hricovíni, M.; Raman, R.; Polito, L.; Torri, G.; Casu, B.; Sasisekharan, R.; Guerrini, M. *Biochemistry* **2008**, *47*, 13862–13869.
37. Raman, R.; Venkataraman, G.; Ernst, S.; Sasisekharan, V.; Sasisekharan, R. *Proc. Natl. Acad. Sci. U.S.A.* **2003**, *100*, 2357–2362.
38. Waddling, C. A.; Plummer, T. H.; Tarentino, A. L.; Van Roey, P. *Biochemistry* **2000**, *39*, 7878–7885.
39. Patel, T.; Bruce, J.; Merry, A.; Bigge, C.; Wormald, M.; Jaques, A.; Parekh, R. *Biochemistry* **1993**, *32*, 679–693.
40. Aminoff, D.; Gathmann, W. *Anal. Biochem.* **1980**, *53*, 44–53.
41. Ruhaak, L. R.; Zauner, G.; Huhn, C.; Bruggink, C.; Deelder, A. M.; Wuhrer, M. *Anal. Bioanal. Chem.* **2010**, *397*, 3457–3481.
42. Ruhaak, L.; Hennig, R. *J. Proteome Res.* **2010**, *9*, 6655–6664.
43. Zauner, G.; Deelder, A. M.; Wuhrer, M. *Electrophoresis* **2011**, *32*, 3456–3466.
44. Pabst, M.; Kolarich, D.; Pöltl, G.; Dalik, T.; Lubec, G.; Hofinger, A.; Altmann, F. *Anal. Biochem.* **2009**, *384*, 263–273.
45. Mauko, L.; Lacher, N. A.; Pelzing, M.; Nordborg, A.; Haddad, P. R.; Hilder, E. F. *J. Chromatogr. B* **2012**, *911*, 93–104.
46. Knez, A.; Polas, O.; Gornik, O.; Rudan, I.; Campbell, H.; Hayward, C.; Wright, A.; Kolc, I.; Donoghue, N. O.; Bones, J.; Rudd, P. M.; Lauc, G. *J. Proteome Res.* **2009**, *8*, 694–701.
47. Royle, L.; Campbell, M. P.; Radcliffe, C. M.; White, D. M.; Harvey, D. J.; Abrahams, J. L.; Kim, Y.-G.; Henry, G. W.; Shadick, N. A.; Weinblatt, M. E.; Lee, D. M.; Rudd, P. M.; Dwek, R. A. *Anal. Biochem.* **2008**, *376*, 1–12.
48. Campbell, M. P.; Royle, L.; Radcliffe, C. M.; Dwek, R. A.; Rudd, P. M. *Bioinformatics* **2008**, *24*, 1214–1216.
49. von der Lieth, C.-W.; Lütteke, T.; Frank, M., Eds.; *Bioinformatics for Glycobiology and Glycomics*; John Wiley & Sons Ltd.: Chichester (UK), 2009.
50. Lauc, G.; Essafi, A.; Huffman, J. E.; Hayward, C.; Knežević, A.; Kattla, J. J.; Polašek, O.; Gornik, O.; Vitart, V.; Abrahams, J. L.; Pučić, M.; Novokmet, M.; Redžić, I.; Campbell, S.; Wild, S. H.; Borovečki, F.; Wang, W.; Kolčić, I.; Zgaga, L.; Gyllensten, U.; Wilson, J. F.; Wright, A. F.; Hastie, N. D.; Campbell, H.; Rudd, P. M.; Rudan, I. *PLoS Genet.* **2010**, *6*, e1001256.
51. Beck, A., Ed.; *Glycosylation Engineering of Biopharmaceuticals*; Humana Press: Totowa, NJ, 2013.
52. Peterson, J. R.; Mohammad, A., Eds.; *Basic Principles and Modes of Capillary Electrophoresis*; Humana Press: Totowa (NJ), 2001.
53. Vanderschaeghe, D.; Szekrényes, A.; Wenz, C.; Gassmann, M.; Naik, N.; Bynum, M.; Yin, H.; Delanghe, J.; Guttman, A.; Callewaert, N. *Anal. Chem.* **2010**, *82*, 7408–7415.
54. Phillips, T. M.; Kalish, H., Eds.; *Clinical Applications of Capillary Electrophoresis*; Humana Press: Totowa (NJ), 2013.
55. Guttman, A.; Chen, F.; Evangelista, R.; Cooke, N. *Anal. Biochem.* **1996**, *233*, 234–242.
56. Vanderschaeghe, D.; Laroy, W.; Sablon, E.; Halfon, P.; Van Hecke, A.; Delanghe, J.; Callewaert, N. *Mol. Cell. Proteomics* **2009**, *8*, 986–994.
57. Mitra, I.; Zhuang, Z.; Zhang, Y.; Yu, C.-Y.; Hammoud, Z. T.; Tang, H.; Mechref, Y.; Jacobson, S. C. *Anal. Chem.* **2012**, *84*, 3621–3627.
58. Zaia, J. *Mass Spectrom. Rev.* **2004**, *23*, 161–227.
59. Zaia, J. *OMICS* **2010**, *14*, 401–418.
60. Ciucanu, I.; Kerek, F. *Carbohydr. Res.* **1984**, *131*, 209–217.

61. Alvarez-Manilla, G.; Warren, N. L.; Abney, T.; Atwood, J.; Azadi, P.; York, W. S.; Pierce, M.; Orlando, R. *Glycobiology* **2007**, *17*, 677–687.
62. Harvey, D. J. *Mass Spectrom. Rev.* **1999**, *18*, 349–450.
63. Wada, Y.; Azadi, P.; Costello, C. E.; Dell, A.; Dwek, R. A.; Geyer, H.; Geyer, R.; Kakehi, K.; Karlsson, N. G.; Kato, K.; Kawasaki, N.; Khoo, K.-H.; Kim, S.; Kondo, A.; Lattova, E.; Mechref, Y.; Miyoshi, E.; Nakamura, K.; Narimatsu, H.; Novotny, M. V.; Packer, N. H.; Perreault, H.; Peter-Katalinic, J.; Pohlentz, G.; Reinhold, V. N.; Rudd, P. M.; Suzuki, A.; Taniguchi, N. *Glycobiology* **2007**, *17*, 411–422.
64. Goldberg, D.; Sutton-Smith, M.; Paulson, J.; Dell, A. *Proteomics* **2005**, *5*, 865–875.
65. Mechref, Y.; Hu, Y.; Garcia, A.; Hussein, A. *Electrophoresis* **2012**, *33*, 1755–1767.
66. Vanderschaeghe, D.; Festjens, N.; Delanghe, J.; Callewaert, N. *Biol. Chem.* **2010**, *391*, 149–161.
67. Ruhaak, L. R.; Deelder, A. M.; Wuhrer, M. *Anal. Bioanal. Chem.* **2009**, *394*, 163–174.
68. Maxwell, E. J.; Ratnayake, C.; Jayo, R.; Zhong, X.; Chen, D. D. Y. *Electrophoresis* **2011**, *32*, 2161–2166.
69. Ethier, M.; Saba, J. A.; Spearman, M.; Krokhin, O.; Butler, M.; Ens, W.; Standing, K. G.; Perreault, H. *Rapid Commun. Mass Spectrom.* **2003**, *17*, 2713–2720.
70. Ceroni, A.; Maass, K.; Geyer, H. *J. Proteome Res.* **2008**, *7*, 1650–1659.
71. Gaucher, S. P.; Morrow, J.; Leary, J. A. *Anal. Chem.* **2000**, *72*, 2331–2336.
72. Gabius, H.-J.; André, S.; Jiménez-Barbero, J.; Romero, A.; Solís, D. *Trends Biochem. Sci.* **2011**, *36*, 298–313.
73. Hirabayashi, J.; Yamada, M.; Kuno, A.; Tateno, H. *Chem. Soc. Rev.* **2013**, *42*, 4443–4458.
74. Sawada, R.; Sun, S.-M.; Wu, X.; Hong, F.; Ragupathi, G.; Livingston, P. O.; Scholz, W. W. *Clin. Cancer Res.* **2011**, *17*, 1024–1032.
75. Welinder, C.; Baldetorp, B.; Borrebaeck, C.; Fredlund, B.-M.; Jansson, B. *Glycobiology* **2011**, *21*, 1097–1107.
76. Hirabayashi, J. *Glycoconj. J.* **2004**, *21*, 35–40.
77. Rakus, J. F.; Mahal, L. K. *Annu. Rev. Anal. Chem.* **2011**, *4*, 367–392.
78. Geisler, C.; Jarvis, D. L. *Glycobiology* **2011**, *21*, 988–993.
79. Tao, S.-C.; Li, Y.; Zhou, J.; Qian, J.; Schnaar, R. L.; Zhang, Y.; Goldstein, I. J.; Zhu, H.; Schneck, J. P. *Glycobiology* **2008**, *18*, 761–769.
80. Thompson, S. M.; Fernig, D. G.; Jesudason, E. C.; Losty, P. D.; van de Westerlo, E. M. A.; van Kuppevelt, T. H.; Turnbull, J. E. *J. Biol. Chem.* **2009**, *284*, 35621–35631.
81. Pilobello, K. T.; Slawek, D. E.; Mahal, L. K. *Proc. Natl. Acad. Sci. U.S.A.* **2007**, *104*, 11534–11539.
82. Zheng, T.; Peelen, D.; Smith, L. M. *J. Am. Chem. Soc.* **2005**, *127*, 9982–9983.
83. Kuno, A.; Uchiyama, N.; Koseki-Kuno, S. *Nat. Methods* **2005**, *2*, 851–856.
84. Rosenfeld, R.; Bangio, H.; Gerwig, G. J.; Rosenberg, R.; Aloni, R.; Cohen, Y.; Amor, Y.; Plaschkes, I.; Kamerling, J. P.; Maya, R. B.-Y. *J. Biochem. Biophys. Methods* **2007**, *70*, 415–426.
85. Bird-Lieberman, E. L.; Neves, A. A.; Lao-Sirieix, P.; O'Donovan, M.; Novelli, M.; Lovat, L. B.; Eng, W. S.; Mahal, L. K.; Brindle, K. M.; Fitzgerald, R. C. *Nat. Med.* **2012**, *18*, 315–321.
86. Chen, S.; LaRoche, T.; Hamelinck, D. *Nat. Methods* **2007**, *4*, 437–444.
87. Nairn, A. V.; York, W. S.; Harris, K.; Hall, E. M.; Pierce, J. M.; Moremen, K. W. *J. Biol. Chem.* **2008**, *283*, 17298–17313.
88. Potapenko, I. O.; Haakensen, V. D.; Lüders, T.; Helland, A.; Bukholm, I.; Sørlie, T.; Kristensen, V. N.; Lingjaerde, O. C.; Børresen-Dale, A.-L. *Mol. Oncol.* **2010**, *4*, 98–118.

89. Comelli, E. M.; Amado, M.; Head, S. R.; Paulson, J. C. *Biochem. Soc. Symp.* **2002**, (69), 135–142.
90. Comelli, E. M.; Head, S. R.; Gilmartin, T.; Whisenant, T.; Haslam, S. M.; North, S. J.; Wong, N.-K.; Kudo, T.; Narimatsu, H.; Esko, J. D.; Drickamer, K.; Dell, A.; Paulson, J. C. *Glycobiology* **2006**, *16*, 117–131.
91. Kawano, S.; Hashimoto, K.; Miyama, T.; Goto, S.; Kanehisa, M. *Bioinformatics* **2005**, *21*, 3976–3982.
92. Bennun, S. V.; Yarema, K. J.; Betenbaugh, M. J.; Krambeck, F. J. *PLoS Comput. Biol.* **2013**, *9*, e1002813.
93. Suga, A.; Yamanishi, Y.; Hashimoto, K.; Goto, S.; Kanehisa, M. *Genome Inform.* **2007**, *18*, 237–246.
94. Patnaik, S. K.; Stanley, P. *Methods Enzymol.* **2006**, *416*, 159–182.
95. Orr, S. L.; Le, D.; Long, J. M.; Sobieszczuk, P.; Ma, B.; Tian, H.; Fang, X.; Paulson, J. C.; Marth, J. D.; Varki, N. *Glycobiology* **2013**, *23*, 363–380.
96. Kotani, N.; Asano, M.; Iwakura, Y.; Takasaki, S. *Biochem. J.* **2001**, *834*, 827–834.
97. Rillahan, C. D.; Paulson, J. C. *Annu. Rev. Biochem.* **2011**, *80*, 797–823.
98. Park, S.; Gildersleeve, J. C.; Blixt, O.; Shin, I. *Chem. Soc. Rev.* **2012**, *42*, 4310–4326.
99. Taylor, M. E.; Drickamer, K. *Glycobiology* **2009**, *19*, 1155–1162.
100. Stowell, S. R.; Arthur, C. M.; Mehta, P.; Slanina, K. A.; Blixt, O.; Leffler, H.; Smith, D. F.; Cummings, R. D. *J. Biol. Chem.* **2008**, *283*, 10109–10123.
101. Yu, Y.; Mishra, S.; Song, X.; Lasanajak, Y.; Bradley, K. C.; Tappert, M. M.; Air, G. M.; Steinhauer, D. A.; Halder, S.; Cotmore, S.; Tattersall, P.; Agbandje-McKenna, M.; Cummings, R. D.; Smith, D. F. *J. Biol. Chem.* **2012**, *287*, 44784–44799.
102. Seeberger, P. H.; Werz, D. B. *Nature* **2007**, *446*, 1046–1051.
103. Blixt, O.; Razi, N. *Methods Enzymol.* **2006**, *415*, 137–153.
104. Smoot, J. T.; Demchenko, A. V. *Adv. Carbohydr. Chem. Biochem.* **2009**, *62*, 161–250.
105. Neumann, G.; Chen, H.; Gao, G. F.; Shu, Y.; Kawaoka, Y. *Cell Res.* **2010**, *20*, 51–61.
106. Tharakaraman, K.; Raman, R.; Viswanathan, K.; Stebbins, N. W.; Jayaraman, A.; Krishnan, A.; Sasisekharan, V.; Sasisekharan, R. *Cell* **2013**, *153*, 1475–1485.
107. Shriver, Z.; Raman, R.; Viswanathan, K.; Sasisekharan, R. *Chem. Biol.* **2009**, *16*, 803–814.
108. Gamblin, S. J.; Haire, L. F.; Russell, R. J.; Stevens, D. J.; Xiao, B.; Ha, Y.; Vasisht, N.; Steinhauer, D. A.; Daniels, R. S.; Elliot, A.; Wiley, D. C.; Skehel, J. J. *Science* **2004**, *303*, 1838–1842.
109. Tumpey, T. M.; Maines, T. R.; Van Hoeven, N.; Glaser, L.; Solórzano, A.; Pappas, C.; Cox, N. J.; Swayne, D. E.; Palese, P.; Katz, J. M.; García-Sastre, A. *Science* **2007**, *315*, 655–659.
110. Imai, M.; Watanabe, T.; Hatta, M.; Das, S. C.; Ozawa, M.; Shinya, K.; Zhong, G.; Hanson, A.; Katsura, H.; Watanabe, S.; Li, C.; Kawakami, E.; Yamada, S.; Kiso, M.; Suzuki, Y.; Maher, E. A.; Neumann, G.; Kawaoka, Y. *Nature* **2012**, *486*, 420–428.
111. Herfst, S.; Schrauwen, E. J. A.; Linster, M.; Chutinimitkul, S.; de Wit, E.; Munster, V. J.; Sorrell, E. M.; Bestebroer, T. M.; Burke, D. F.; Smith, D. J.; Rimmelzwaan, G. F.; Osterhaus, A. D. M. E.; Fouchier, R. A. M. *Science* **2012**, *336*, 1534–1541.
112. Tharakaraman, K.; Jayaraman, A.; Raman, R.; Viswanathan, K.; Stebbins, N. W.; Johnson, D.; Shriver, Z.; Sasisekharan, V.; Sasisekharan, R. *Cell* **2013**, *153*, 1486–1493.

Chapter 12

Applications of Glycan Microarrays to Functional Glycomics

Ying Yu, Xuezheng Song, David F. Smith and Richard D. Cummings
Department of Biochemistry and the Glycomics Center, Emory University School of Medicine, Atlanta, Georgia, USA

Chapter Outline

1 INTRODUCTION

Glycans are comprised of monosaccharides, which, along with nucleic acids, proteins, and lipids, represent one of the four fundamental macromolecules that constitute all living systems. Cells of all species are coated with complex glycans within glycoproteins, glycolipids, and other macromolecules that are involved in myriad biological processes including molecular recognition, cell signaling, immune response, infection, and inflammation *(1)*. It is estimated that more than 50% of all proteins are glycosylated *(2)*. Aberrant glycosylation is associated with congenital disorders, cancers, and autoimmune and other diseases *(3)*. Analogous to the genome, proteome, and other "omes," the glycome is the entire repertoire of glycans and glycoconjugates in specific cells, tissues, or organisms. Therefore, *glycomics* refers to the systematic analysis of the structure and function of the glycome. However, glycomics has

Comprehensive Analytical Chemistry, Vol. 63. http://dx.doi.org/10.1016/B978-0-444-62651-6.00012-X

lagged significantly behind genomics and proteomics because of its inherent complexities. For example, in contrast to nucleic acids and proteins, which are generated by template-driven biosynthesis, the structure of a glycan is not directly encoded in the genome. Rather, glycans are constructed from monosaccharides through concerted actions of glycosyltransferases (GTs); thus, glycans can express diversity that far exceeds that of the linear biopolymers owing to variations of monosaccharide composition, linkage positions, anomeric configurations, branching patterns, and additional modifications including sulfation and phosphorylation. Beyond the overwhelming complexity of glycan structures, the glycome of cells is highly dynamic and tissue specific due to the influence of other "omes," including the metabolome, along with the physiological state and environment of cells. A single glycosylation site on a polypeptide can possess different glycan structures depending on the type and status of the cell. Such variation in structure along with the often low abundance of some glycans makes characterization of the glycome extremely challenging *(4)*. Furthermore, unlike genomics and proteomics, where a single, high-throughput technique is used for sequencing and synthesis of nucleic acids and proteins, many different approaches are usually required for glycomic analysis of the different glycan and glycoconjugate classes.

Nevertheless, advances in technology have significantly accelerated the progress toward our understanding of the importance of glycosylation in biology and disease. A wide variety of high-resolution and highly sensitive analytic methods rooted in traditional methodology have been utilized in glycomics, including high-resolution mass spectrometry (MS), high-performance liquid chromatography (HPLC), capillary electrophoresis (CE), nuclear magnetic resonance (NMR) spectroscopy, and lectin microarrays *(5)*. MS has become the primary tool for studying glycan structures and profiling the glycomes of various cells/tissues *(6,7)*. It has also led to new insights into the complexity of glycomes in different organisms, as well as discoveries of glycan-based biomarkers in diseases *(8,9)*. However, MS analysis suffers from the difficulty in resolving structures of isobaric variations of glycans, which have identical masses, and potential isomeric aspects of determinants within a single glycan; thus, completely defined structures are rarely reported because of the complexity and limitations of techniques needed for complete structural identification. In addition, even predicted structures are currently derived from tandem MS that requires highly sophisticated equipment and expertise. Lectin microarrays indirectly provide global information about the structural determinants present in samples, but no information on individual glycan structures, and are more useful in globally providing information on potential changes in glycosylation, glycomic profiling, and indirectly identifying potential glycan-related biomarkers *(10,11)*. A major advantage of lectin microarrays is that there is no need for release of glycans from the parent glycoproteins, and whole cell extracts can be used. Lectin microarray analyses, however, are limited by the availability of well-defined and quality-controlled lectins and by the lack of detailed understanding of lectin specificity.

The definition of glycan structures obviously requires multiple analytical tools, but understanding the function of glycans as they participate in diverse recognition systems needs alternative approaches. Glycan microarrays offer an unprecedented opportunity to tackle both of these key problems. In the past two decades, glycan microarray technology has become a powerful tool for rapidly screening the specificity of glycan-binding proteins (GBPs), anti-glycan antibodies, and glycan-processing enzymes *(12)*. In addition to the high-throughput feature, the miniaturization of array format, which requires nanograms of glycans for printing, is particularly valuable when considering the difficulties in obtaining pure glycans and when analyzing minute amounts of recombinant GBPs. Another major advantage is that the glycans within a glycan microarray can be presented as a multivalent display on the solid support so that even a low-affinity binding characteristic of certain protein–glycan interactions is supported. Besides using it as a screening platform, our group has also contributed to development of natural glycan microarrays for deciphering the function and structure of the human glycome *(13)*. In this chapter, we provide an overview of glycan microarray fabrication and applications and describe our efforts on method development that highlight our progress on *functional glycomic* studies using shotgun glycan microarray (SGM) strategies.

2 GENERATION OF GLYCAN MICROARRAYS

2.1 Glycan Sources

The utility of a microarray depends on the source, number, and diversity of the immobilized glycan library, which can be either synthesized by chemical or chemoenzymatic methods or isolated from natural sources. Three chemical methods have been developed for rapidly creating synthetic glycans. The automated solid-phase glycan synthesis mimics the protocol and the machinery for peptide synthesis, involving iterative glycosylation and deprotection steps *(14)*. The programmable reactivity-based one-pot synthesis uses a computer program, Optimer, to identify an appropriate combination of building blocks for any given structure and then assembles all the units in a sequential manner *(15)*. On the other hand, the automated solution-phase oligosaccharide synthesis relies on fluorous tag in conjugation with FSPE (fluorous solid-phase extraction) to easily purify the intermediates while significantly reducing the usage of building blocks *(16)*. These methods were applied in preparation of various glycans in the past decade, including malaria glycosylphosphatidylinositol *(17)*, Lewis blood-group determinants *(18)*, tumor-associated antigens *(19)*, high-mannose glycans *(16)*, and sialosides *(20,21)*. The major advantage of these chemical approaches is the elimination/

simplification of purifying the intermediates, which offers great potential to quickly obtain homogeneous and diverse glycan libraries in large quantity. However, the obtained diversity is limited by the difficulties in the synthesis of some linkages and branched structures, as well as by the accessibility of starting building blocks. A complementary approach to chemical synthesis is the introduction of highly regio- and stereo-specific GTs. Some linkage and anomeric obstacles are easily addressed by the use of enzymes, avoiding the numerous protection and deprotection steps *(22)*. Traditionally, glycosylation is performed in solution and the product is easily purified for use in microarray preparations. An alternative to expand the glycan library with GTs is through on-array glycosylation *(23,24)*, which introduces glycan modifications onto numerous glycans simultaneously. Notably, the application of enzymatic glycosylation is limited by the availability of GTs and the need to purify the products from precursors and intermediates. In addition, while mammalian GTs are often highly specific for the acceptor structures and specific products are easily predicted, nonmammalian GTs may have less specificity toward glycan acceptors and/or unpredictable activities.

In spite of the large amount of effort extended for developing high-throughput glycan synthesis, the major obstacle in glycomic studies remains the inability to access sufficient amounts of defined glycans for functional studies. An alternative source of glycan for building libraries of defined structures is from natural sources, where glycans are isolated from body fluids or liberated from glycoconjugates, tagged with a fluorescent molecule, and separated by HPLC. The natural glycan libraries comprise highly diverse and complex structures, which should represent the glycome of the sources *(12)*. However, the isolation of a glycan library from natural sources is often tedious and generates compounds in only low yields and a variety of methods are required for the different classes of glycans. The release of N-linked glycans from glycoproteins is accomplished with the widely used *N*-glycanases, PNGase F and PNGase A, whereas the enzymatic cleavage of O-linked glycans is more difficult due to the lack of an *O*-glycanase with broad specificity *(25)*; existing *O*-glycanase is limited to the disaccharide core 1 or core 3 *O*-glycans. More commonly, complex *O*-glycans are released through reductive β-elimination in strong base, which yields the chemically inert alditols to prevent degradation (Figure 1), but the reduction of the aldehyde function at the reducing end precludes further fluorescent/functional tagging and HPLC detection. To overcome this problem, alternative procedures, such as nonreductive β-elimination and hydrazinolysis, were developed *(26,27)*. Robust but gentle and efficient enzymatic or chemical methods are urgently needed in order to explore the complex *O*-glycome.

The release of free-reducing glycans from glycosphingolipids (GSLs), like the *O*-glycans, has suffered similar reagent and reaction obstacles. We recently addressed this impediment using a simple nonenzymatic procedure to efficiently release reducing glycans in mild condition (Figure 1) *(28)*.

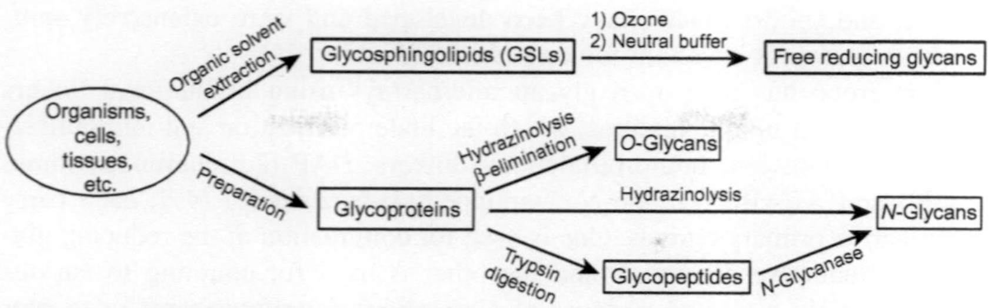

FIGURE 1 Strategies for isolation of glycans from natural sources.

The extracted GSLs were briefly treated by ozone and heated in neutral buffer to release free-reducing glycans that can be permethylated for MS analysis or conjugated to a bifunctional linker for HPLC separation and microarray study. Like other methods based on β-elimination, this method is also not suitable for sphinganine and phytosphingosine GSLs that lack the C=C bond.

Glycosaminoglycans (GAGs) are linear polymeric glycan structures attached either free, as for hyaluronan, or linked to Ser residues within proteoglycans, which represent a major class of glycoconjugates. The major classes of GAGs are hyaluronan, chondroitin sulfate, dermatan sulfate, heparin/heparan sulfate, and keratan sulfate. The synthesis and separation of these classes of glycans are usually very difficult, due to their high negative charges and often very heterogeneous structures. There has been some effort on presenting these glycans on glycan microarrays for their functional screening, but most of the so-called GAG microarrays only incorporate a limited number of synthetic GAG oligosaccharides and/or fractions collected from natural GAGs *(29–33)*, and only limited application of these microarrays to the determination of specific GAG–protein interactions has been reported.

2.2 Glycan Immobilization

The methods for attachment of glycans onto a solid platform broadly fall into two categories: noncovalent and covalent immobilization. Noncovalent immobilization relies on hydrophobic absorption or electrostatic interaction to absorb free or derivatized glycans onto nonfunctionalized or functionalized surfaces. This technique was mainly adopted in some of the early microarray fabrication and is still used for producing arrays of neoglycolipids (NGLs) on nitrocellulose-coated slides *(34)*. For covalent immobilization, the reducing glycans can be printed directly on activated surfaces, but more commonly are synthesized with a reactive functional group at the reducing end. Alternatively, free-reducing glycans are chemically conjugated at the reducing end with reactive tags (functional linkers), such as amines, thiols, and azides, which can then be covalently attached onto solid supports that are functionalized with compatible reactive groups. Varieties of methods for modifying glycans and solid surfaces have been developed and were extensively summarized in several excellent reviews *(12,26,35–37)*.

Our group has developed glycan microarrays using bifunctional linkers because of their unique features, which facilitate purification and immobilization. The fluorescent homo-bifunctional linkers, DAP (2,6-diaminopyridine) *(38,39)* and AEAB (2-amino-*N*-(2-aminoethyl)-benzamide) *(40)*, each carry two reactive primary amines. One is used for conjugation of the reducing glycans via reductive amination and the other is used for coupling to various epoxy- or NHS-activated surfaces of glass slides or microspheres or to proteins using alternative chemistries *(38)*. The hetero-bifunctional linker, PNPA (*p*-nitrophenyl anthranilate), retains a reactive ester after glycan conjugation, which allows direct coupling to protein and other biomolecules or reaction with other bifunctional chemicals *(41)*. The properties of these three linkers are summarized in Table 1. Overall, DAP can be efficiently conjugated to reducing glycans and permits the quantification of printed glycans; the immobilization efficiency of the aryl amine toward large glycans, however, is low compared to reactivity of primary alkyl amines. By contrast, AEAB conjugation is efficient for all types of glycans and the AEAB-tagged glycans immobilize efficiently on NHS-activated solid surfaces. Thus, this has been our method of choice to prepare various glycan microarrays. PNPA derivatives have been used to generate neoglycoproteins and also GSL microarrays after the ester was converted to an alkyl amine and these derivatives are also suitable for solid-support immobilization. Notably, the one-step derivatization with these linkers results in an open-ring structure that potentially leads to the loss of information at the reducing end. Although most of the glycan-binding proteins interact with the determinants at the nonreducing end or the backbone of glycans, this loss of structure at the reducing end may be of particular concern for short glycans, such as mono- or disaccharides or for glycans with unique reducing-end structural features such as core-fucose and xylose, where the reducing end may be part of the recognition motif. To address this issue, we developed a procedure that retains the close-ring configuration *(46)*. The reducing glycan is converted to glycosylamine and then transformed into stable glycosylamide with acryloyl chloride, followed by the introduction of an active aldehyde that is subsequently conjugated to AEAB. The final glycan derivative conserves the closed-ring structure like natural glycans.

A novel strategy for covalent immobilization of underivatized glycans utilizes glass surface functionalized with self-assembled photoactive phthalimide chromophores, which react with the C—H groups of the sugars to form covalent bonds upon UV illumination *(47)*. This photochemical approach does not require glycan modification prior to deposition and it is easy to perform as photons are the only reagent involved. However, the photocoupling may target more than one C—H group on the sugar rings and yield multiple configurations. An alternative approach utilizes a three-dimensional dendrimer

TABLE 1 Comparison of Three Bifunctional Linkers Used for Microarray Preparation

	DAP	AEAB	PNPA
Chemical name	2,6-Diaminopyridine	2-Amino-*N*-(2-aminoethyl)-benzamide	*p*-Nitrophenyl anthranilate
Chemical structure	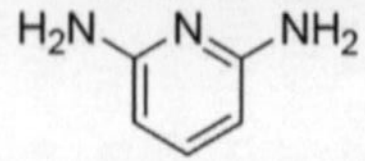	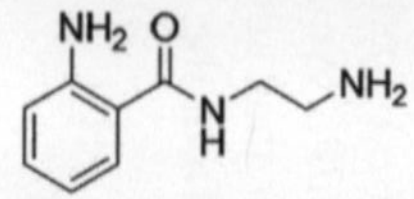	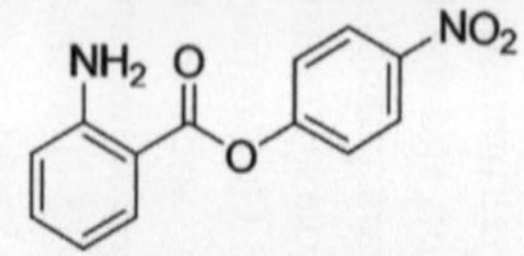
Conjugated glycan	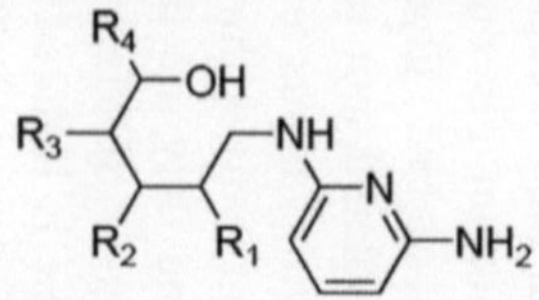	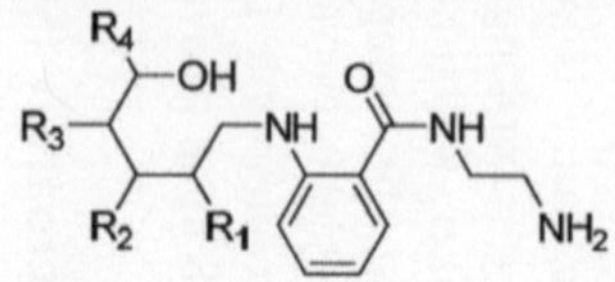	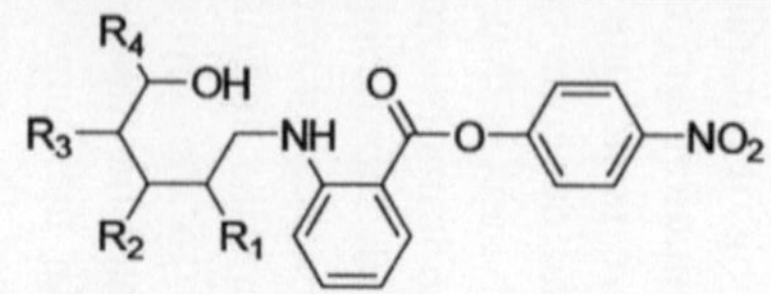
Conjugation efficiency	Modest	High	Modest-high
On-slide quantification	Yes	No	No
Application of conjugated glycans	Coupled to activated glass slides, proteins, and microspheres *(38)*	Used for various glycan microarray preparation *(40,42–44)*	Used to prepare microarrays of neoglycoproteins and glycosphingolipids *(41,45)*

monolayer surface functionalized with terminal aminooxy and hydrazide groups for site-specific coupling of natural glycans, so that the glycan chains are presented in defined orientation and cluster configuration *(48)*. With so many methods available, there is no one completely superior to the others and the choice of array format depends on each application. The advantage of using AEAB-derivatized glycans is that the fluorescent moiety allows purification of the glycans prior to immobilization and the fluorescence detection allows researchers to work with submicrogram quantities of glycans.

Despite the abundant approaches for microarray construction, quantification of the immobilized glycans remains difficult. One option relies on the fluorescence of the linker attached to the glycan, such as DAP; analyses of the attached glycan–DAP conjugates has been performed and reveals the amounts of immobilized glycans are in the range of 5–100 fmol/spot *(39)*. To avoid interfering with the widely used fluorescent-based microarray assays, most of the fluorescent linkers are low molecular weight and excitable at low wavelength (<400 nm), which is not compatible with common scanners. In addition, if a low-wavelength excitation source is available, most glass begins to fluoresce at these shorter wavelengths, making quantification of the glycan on the slide impossible. The DAP linker, which has an optimal excitation of 330 nm, was detected on the slide at the compromised excitation wavelength of 488 nm, but the fluorescence yield was low *(39)*. An interesting alternative approach is the use of a photo-cleavable linker that allows the quantification of immobilized glycans by MS upon cleavage from the solid surface *(49,50)*.

2.3 Shotgun Glycan Microarray (SGM)

To address the challenges of identifying glycans within the glycome that are potentially functional through their recognition by proteins, we developed a general strategy termed "shotgun glycomics." This approach involves the fractionation of fluorescent-tagged glycans from natural sources, and subsequent printing of the isolated, but undefined glycans, as a SGM *(45)*. The SGM may be screened with various reagents, including GBPs, antibodies, pathogens, and so on; those glycans within the SGM that are recognized thus are candidates of interest, and they are then selected for detailed structural analysis *(45)*. The generation of SGMs is based on our strategy for natural glycan arrays *(40)*. A glycome is isolated from a natural source, tagged with a bifunctional linker, subjected to multidimensional HPLC purification, and quantified by fluorescence (Figure 2). The purified tagged glycans constitute a tagged glycan library (TGL) representing a unique glycome or subglycome. Using the TGL as a source of glycans, the corresponding SGM provides a stable platform for studying the interaction of an isolated glycome with proteins and pathogens. The major advantage of this technique is the incorporation of fluorescent tag, which dramatically increases the sensitivity of detection and thus allows for separation and collection of low-abundance glycans from the

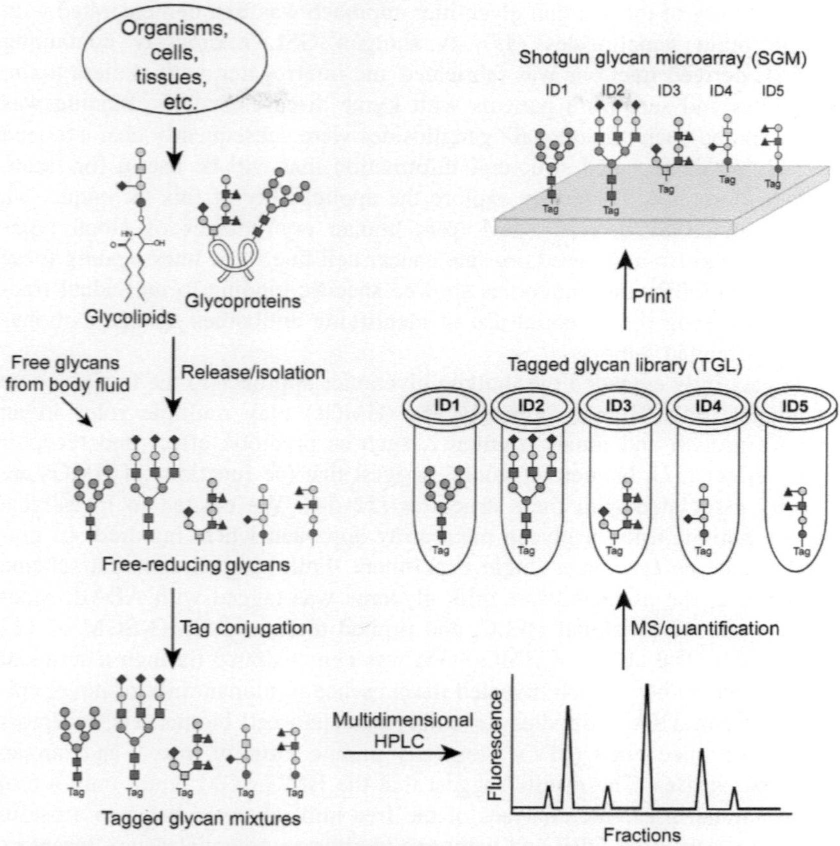

FIGURE 2 Steps in generating a shotgun glycan microarray (SGM). Glycoproteins and glycolipids are extracted from organisms, tissues, cells, or other biological samples. Free-reducing glycans are chemically or enzymatically released and conjugated with a bifunctional linker (tag). Free glycans in body fluid, such as human milk, are directly labeled with a tag. The tagged glycan mixtures are fractionated by multidimensional HPLC and individual fractions are collected to make the tagged glycan library (TGL), in which each glycan is given a unique ID. The TGL is used to generate the shotgun microarray.

complex mixture. Additionally, this strategy offers an opportunity to examine biological activities on individual glycans in a high-throughput manner, and thus to establish the relationships between glycan structures and their potential functions. Although the multi-HPLC fractionation is time-consuming and labor-intensive, a well-constructed TGL, where the composition of each glycan and its potential degree of purity is predicted through MALDI-TOF analysis, becomes a permanent repository of a given glycome and a resource for future studies on that glycome, since the microarray printing process is at a nanogram scale.

The utility of the shotgun glycomics approach was first demonstrated with bovine brain gangliosides *(45)*. A shotgun GSL microarray containing 40 GSL-derived fractions was fabricated and interrogated with cholera toxin, antibodies and sera from patients with Lyme disease. Specific binding was observed and such "functional" gangliosides were subsequently characterized by MS, which revealed structural information that will be useful for future biological studies. To further explore the applicability of this technique, we also created GSL-derived SGM from human erythrocytes of blood types O and A, and from cultured prostate cancer cell line PC3. Interrogating these arrays with GBPs and antibodies showed specific binding to individual fractions, indicating the potential use in identifying antibodies against carbohydrates of human tumors.

We recently extended the shotgun glycomics approach to the free glycome in human milk. Human milk glycans (HMGs) play multiple roles in an infant's growth and innate immunity, such as prebiotic effect and receptor decoy effect *(51)*. Numerous studies suggest that the functions of HMGs are directly associated with their structures *(52–54)*. We elected to investigate these functions using a glycan microarray approach where hundreds of glycans can be assayed in a single experiment. Following the general scheme in Figure 2, the extracted free milk glycome was tagged with AEAB, separated by two-dimensional HPLC, and printed to form a HMG SGM of 127 HMGs *(42)*. The utility of HMG SGM was demonstrated through a series of recognition studies, which revealed the presence in human milk of novel epitopes for anti-TRA antibodies for embryonic stem cell biomarkers, receptors for minute mice virus (MVM; formerly minute virus of mice), and human influenza viruses. The results suggest that the HMG SGM is not only a tool for functional glycomic analyses of the free milk glycome but also a useful source for screening GBP and pathogen binding to potential decoy receptors since this library is comprised of a large number of common and unique structures. We are expanding the HMG library to identify receptors for many other pathogens associated with infant diseases.

Another application of shotgun glycomics was to generate a SGM containing *N*-glycans and glycolipids from different life stages of *Schistosoma mansoni (55)*. Adapted from our approach, enzymatically released glycans were conjugated to 2-aminobenzoic acid (2-AA), fractionated by one-dimensional chromatography, and printed to form a large array containing over 1000 samples. Although no characterization was carried out on the prepared SGM, it was exploited to identify differential glycan recognition profile characteristics for two different age groups.

3 EXAMPLES OF REPORTED GLYCAN MICROARRAYS

Wang *et al.* reported the first glycan microarray in early 2002 *(56)*. Since then, there has been an explosion of publications on methodology developments, which resulted in the production of assorted glycan microarrays. Table 2 provides a list of the complex glycan microarrays reported to date. Some small microarrays (<10 glycans) and arrays containing mostly monosaccharides and disaccharides are not included here *(35)*.

Glycan microarrays can generally be grouped into defined arrays and natural arrays based on the structural information of presented glycans. The majority of the microarrays are composed of structurally defined glycans. The most widely used glycan microarray was developed by the Consortium for Functional Glycomics (CFG) and currently contains over 600 structurally defined glycans, designed to represent terminal sequences found on *N*-glycans, *O*-glycans, and GSLs of mammalian cells (http:/www.functionalglycomics.org). Most of the glycans were generated by chemical or chemoenzymatic synthesis, whereas a small portion were isolated from natural sources and structurally defined. In some cases, the identical glycans are presented on the array with different spacers, which vary from a few carbon atoms to a short peptide. The CFG glycan microarray provides a discovery platform that allows examination of GBP specificities in a high-throughput format and has been available as a free service supported by the National Institute of General Medical Sciences of the National Institutes of Health. Since the initial launch of the printed CFG array in 2005, which replaced the previous ELISA-based array of biotinylated glycans captured on streptavidin-coated microtiter plates, enormous amounts of data have been generated for a broad range of GBPs and pathogens, which assisted the comprehensive understanding of GBP functions. Another such comprehensive glycan microarray is based on neoglycolipid (NGL) technology, where synthetic and natural glycans are conjugated to a hydrophobic lipid via microscale reductive amination or oxime ligation and purified to construct a NGL library, which is printed directly on and absorbed to a solid surface coated with nitrocellulose for recognition studies *(29,75)*. Since its appearance in 2002, the library for the NGL array has expanded to encompass more than 700 defined NGLs from all glycan classes and is continuing to grow in number and structural diversity *(34)*. The NGL microarrays are printed as individual "designer" arrays depending on the application, which has enabled the identification of ligands for important GBPs *(57,76)* and receptors for viruses *(77–79)*.

While the CFG array and NGL array are powerful tools for generally exploring protein–carbohydrate interactions because of their broad coverage of mammalian glycans, there are other defined microarrays that contain a smaller repertoire of glycans that are custom-designed for specific biological applications (Table 2). For example, the GAG arrays were prepared to elucidate the GAG–protein interactions *(63,66)*. The microbial glycan arrays were employed to detect specific antibodies in sera from pathogen-infected patients *(56,72)*. Our laboratory generated a phosphorylated high-mannose glycan microarray to investigate the specificity of cation-independent and -dependent mannose-6-phosphate receptors *(43)*. These custom arrays of defined glycans

TABLE 2 Summary of Complex Glycan Microarrays Reported in the Literature

Year	Array Type	# of Glycans	Glycan Type	Glycan Source	Immobilization Method	Application	References
2002	Defined	48	Microbial polysaccharides	–	Noncovalent	Discovery of cellular marker	*(56)*
From 2002	Defined	100 to ~600	All types of neoglycolipids	Mixed	Noncovalent	Various	*(29,57)*
2002	Defined	22	Polysaccharides, neoglycoproteins	Natural	Noncovalent	Monoclonal antibody (mAb) specificity	*(58)*
2003	Defined	21	*O*-Glycans and HMGs	Synthetic	Biotin-strep	Antibody specificity	*(59)*
From 2004	Defined	200 to ~600	All types of glycans	Mixed	Amine-NHS	Various	*(60)*
2005	Defined	~30	Neoglycoproteins	Synthetic	Covalent	Ab specificity	*(61,62)*
2006	Mixed	160	β-Glucans-NGL and other NGLs	Mixed	Noncovalent	Ligands for β-glucan receptor	*(57)*
2006	Defined	4–12	GAGs	Synthetic	Amine-NHS	Protein interaction	*(63,64)*
2006	Natural	6	GAGs	Natural	Aldehyde-amine	Protein interaction	*(65)*
2006	Defined	4	Chondroitin sulfate	Synthetic	Aminooxy-aldehyde	Growth factors	*(30,66)*
2007	Defined	33	Bacterial glycans	Synthetic	Photocoupling	Pathogen antibodies	*(67)*
2008	Defined	34	Sialoside analogue	Synthetic	Amine-NHS	Siglec	*(68)*

TABLE 2 Summary of Complex Glycan Microarrays Reported in the Literature—Cont'd

Year	Array Type	# of Glycans	Glycan Type	Glycan Source	Immobilization Method	Application	References
2008	Natural	50	Plant cell wall glycans	Natural	Noncovalent	mAb specificity	*(69)*
2008	Defined	7	Parasite glycosylphosphatidylinositol	Synthetic	Thiol-maleimide	Sera	*(70)*
2008	Defined	8	Globo H	Synthetic	Amine-NHS	Cancer patient sera	*(71)*
2008	Defined	15	Salmonella *O*-antigen	Mixed	Amine-NHS	Patient sera	*(72)*
2009	Mixed	>200	*N*-Glycans and HMGs	Natural	AEAB-NHS	Galectins	*(40)*
2009	Defined	27	Sialosides	Synthetic	Amine-NHS	Influenza hemagglutinin	*(73)*
2009	Defined	26	Phosphorylated high-mannose *N*-glycans	Natural + enzymatic	Amine-NHS	P-type lectins	*(43)*
2011	Natural (SGM)	40	Glycosphingolipids	Natural	Amine-NHS	Lyme serum	*(45)*
2011	Defined	80	Sialoside analogue	Mixed	Amine-NHS	Influenza virus	*(44)*
2012	Natural (SGM)	~1000	*S. mansoni* *N*-glycans, glycolipids	Mixed	Amine-epoxy	*S. mansoni*-infected sera	*(55)*
2012	Defined	24	Sialylated poly-*N*-acetyllactosamines	Synthetic	Amine-NHS	Influenza virus	*(74)*
2012	Natural (SGM)	140	Human milk glycans (HMGs)	Natural	Amine-NHS	Anti-TRA mAbs, viruses	*(42)*

yielded valuable data to understand the role of glycan–protein interactions. Defined glycan arrays are used as hypothesis-generating platforms in order to understand GBP function, since knowing the specificity of a GBP can often provide clues to its function. However, it is important to understand that the data from a defined glycan array is not sufficient alone to identify physiological ligands since the printed glycans are not from the biological source being studied and in some cases synthetic glycans may not be found in the cell or organ of interest.

Natural glycan microarrays, on the other hand, are composed of glycans isolated from natural sources and the glycan structures must be defined through detailed analysis. Although some of the defined glycan arrays also include natural glycans, the structures of these glycans are usually fully characterized before immobilization. The natural glycan microarray, such as the SGM, is a unique tool to explore the recognition of glycans derived from a biological source by a large variety of GBPs and thus better to understand the function of a glycome, as discussed in Section 2.3. It should be noted that identification of a candidate glycan on a natural glycan array is still a hypothesis-generating process, and additional experiments are required to support the hypothesis that the glycan is involved in a biologically relevant process.

4 IDENTIFICATION OF VIRUS RECEPTORS WITH GLYCAN MICROARRAYS

The applications of glycan microarrays in numerous systems have been covered in previous reviews *(12,50,80)*. Perhaps one of the most significant applications of glycan microarrays has been the analysis of virus binding to the glycans presented on these microarrays. At the initial stage of infection, most viruses, as well as many other pathogens, use specific host cell surface glycans as recognition and attachment sites. The occurrence of a particular receptor on a restricted set of cell types limits the cell tropism, and the variation of such receptors among species limit the host range. Therefore, identification of virus receptors is necessary for understanding the pathobiology of infection.

4.1 Influenza Virus

Influenza viruses initiate infection by binding to sialic acid of host glycans via viral surface protein hemagglutinin (HA). The receptor specificity varies by species, resulting in different host ranges. It is widely believed that human influenza viruses prefer glycans with terminal α2-6 sialic acid, while avian viruses prefer α2-3 sialic acid linkage *(81)*. This subtle difference in receptor structure appears to be the key determinant of species barrier that prevents avian viruses from infecting humans. However, some early studies showed that the recognition of influenza viruses was not limited to the terminal sialic acid and the inner glycan structures also influenced the HA binding *(82)*. The emergence of glycan microarrays provided an ideal, high-throughput platform to study the fine specificities of influenza viruses.

The CFG glycan microarray, which contains a large collection of structurally defined sialylated glycans, has been a standard platform for screening the specificity of recombinant HA or whole virus isolates *(83)*. So far, a variety of human pandemic, seasonal, and swine strains, as well as avian viruses and parainfluenza viruses, have been evaluated on CFG microarrays (http:/www.functionalglycomics.org). The results not only confirmed the sialic acid linkage preference for human and avian species, but also clearly showed the involvement of underlying structures in HA recognition *(84,85)*. Although all of the glycans bound by influenza viruses were exclusively sialylated, and a few strains, such as Brisbane H1N1, did not discriminate between Siaα2-3- and Siaα2-6-linked sialosides on CFG array, many of the influenza strains were selective for a subset of either Siaα2-3- or Siaα2-6-linked glycans *(85)*. Other than the CFG microarray, custom-designed microarrays, such as NGL microarray *(78)* and sialylated poly-*N*-acetyllactosamine microarray *(74)*, were also used to explore the receptor-binding specificities of HA. In addition, the Centers for Disease Control and Prevention has used a sialoside array that is comprised of sialylated glycans from the CFG microarray to characterize emerging and potentially pandemic influenza viruses *(86–88)*. Recently, a detailed analysis of human H3N2 influenza viruses, isolated from 1968 to 2012, binding to the CFG microarray determined that the agglutinating titer of the viruses corresponded to the avidity of the virus for glycans on the microarray *(89)*. Over the time period of virus evolution, year-to-year patterns of binding varied, and the general glycan structures bound by these viruses drifted from short, branched sialylated structures to those containing longer poly-*N*-acetyllactosamines. Interestingly, all of the viruses spread in seasonal epidemics worldwide, thus suggesting there was no major impact of differences in binding to the glycan array on the ability of the virus to replicate, cause disease, and be transmitted among humans.

A novel sialylated glycan microarray consisting of sialic acids attached in α2,3- and α2,6-linkages to different underlying type 1 (Galβ1-3GlcNAc) and type 2 (Galβ1-4GlcNAc) structures was constructed using a one-pot three-enzyme synthesis *(44)* to generate an array of 80 glycans possessing 16 modified sialic acid derivatives *(90)*. Many of the glycans on this sialylated array were not present on the CFG microarray, providing a complementary assay to systematically study the effects of sialic acid modifications on influenza virus recognition. Of three strains of human parainfluenza viruses assayed on this unique microarray, binding was restricted to Siaα2-3-linked glycans terminating with Neu5Ac or Neu5Gc and some of their derivatives, but not to Kdn and its derivatives. Interestingly, the tested influenza A virus also does not bind glycans terminating in Kdn or Kdn derivatives; however, they equivalently recognized glycans terminated with either α2-6-linked Neu5Ac or its

9-*O*-lactoyl derivative (Neu5Ac9Lt) *(90)*. Since this latter modified sialic acid occurs in significant amounts in humans (i.e., 20% of the sialic acid in normal human serum is Neu5Ac9Lt *(91)*), our results suggested that more attention should probably be given to the role of sialic acid modifications in the pathogenesis of viruses and other microorganisms. In addition, we also examined a few human influenza viruses with our shotgun HMG array and showed that HMGs contained natural glycans capable of binding influenza viruses *(42)*. Consistent with the findings observed from CFG array, the human influenza A viruses have a strong preference toward α2-6-sialylated glycans, but not all α2-6-sialylated glycans were recognized by the viruses, which once again indicates that virus binding is not restricted to the terminal sialic acid moiety. Detailed structural analysis revealed one common motif for the receptors, which is an α2-6 sialic acid linked type 2 chain (Figure 3A).

In a recent study on influenza A virus–glycan interactions using a glycomic approach, *N*-glycan and *O*-glycan composition of human lung and bronchus were analyzed by MS, and found to contain a wide spectrum of Siaα2-3- and Siaα2-6-terminated glycans in both lung and bronchus *(92)*. These results suggest that alveolar tropism of influenza viruses appears to involve both Siaα2-3- and Siaα2-6-terminated glycans, emphasizing the complexity of this system. The information on the predicted structures was then used in combination with influenza A virus binding data from four defined

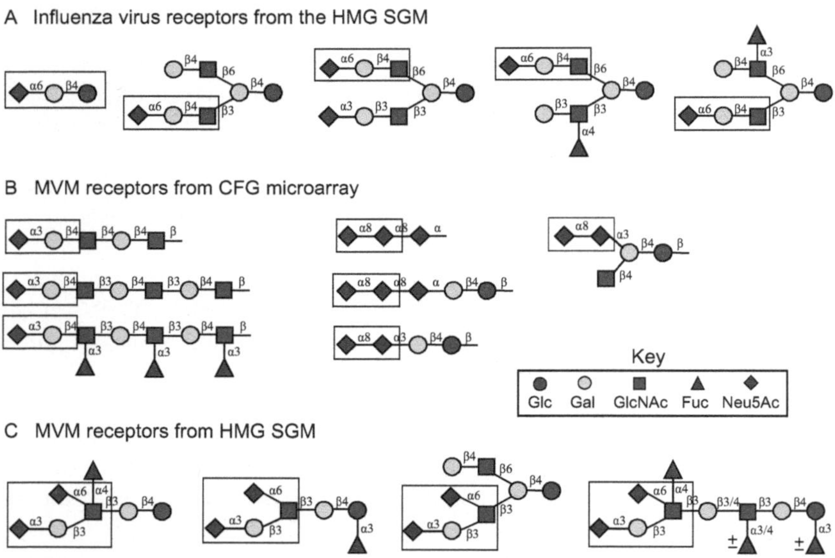

FIGURE 3 Examples of virus receptors identified by glycan microarrays. Examples of influenza virus A glycan receptors identified from HMG shotgun array (A), and MVM receptors identified from CFG microarray (B), and the HMG SGM (C). Common motifs for virus recognition are highlighted by boxes.

glycan microarrays to determine if binding patterns on glycan arrays could predict replication of human, avian, and swine viruses in human *ex vivo* respiratory tract tissues. However, even with the CFG microarray, which was the most comprehensive of the four glycan arrays analyzed, the influenza A binding patterns were not predictive of productive virus replication in human respiratory tract tissue. The study concluded that more comprehensive and focused arrays need to be developed to evaluate potential infectivity of emerging influenza A viruses. While it is tempting to conclude that the solution to this dilemma is simply to synthesize the observed glycans in this glycomics analysis, such an approach may be extremely difficult, if not impossible, due to the lack of detailed structural information and the difficulty of synthesis. Thus, it is worth considering that the shotgun glycomics approach may represent the most efficient alternative means to obtain the glycans from the natural source of tissue or glycoconjugate and possibly discover novel and previously unidentified components that function as viral receptors.

4.2 Minute Virus of Mice

Minute virus of mice (MVM) is a member of the parvovirus family. MVM is a major threat to laboratory animals due to its widespread distribution. The initial attachment of MVMs to host cells requires the recognition and binding to sialic acid *(93)*. Two strains of MVM, the prototype strain (MVMp) and the immunosuppressive strain (MVMi), have served as models for studying the capsid determinant of tissue tropism, pathogenicity, and glycan–receptor interaction. In 2006, screening of MVMp, MVMi, and three mutants on CFG glycan array (Version 3.0), which contained 180 glycans, identified specific binding to several sialylated glycans *(94)*. The three structures bound by all of the strains carried a common α2-3 sialic acid linked type 2 motif, Neu5Acα2-3Galβ1-4GlcNAc (Figure 3B). Furthermore, MVMi also bound to four α2-8-linked multisialylated glycans.

The human milk glycome of free glycans is rich in sialylated glycans. Fifty-seven percent of glycans on the HMG SGM array contain one or two sialic acids *(42)*. While exploring the general application of the HMG SGM array on pathogen binding, we examined several MVMs, including two prototype strains, empty capsid (MVMp-WT) and virus-like particle (MVMp-VLP), and two immunosuppressive strain mutants, MVMi-agD and MVMi-ggA. Interestingly, all of the viruses selectively recognized a small set of sialylated HMGs that carry sialyl α2-3-linked type 1 motif, which is a novel finding compared with the CFG array data. Detailed structural analysis on glycans selected based on their ability to bind virus revealed a common disialyl LNT motif, an α2-3-sialylated type 1 chain with an additional sialic acid in α2-6 linkage to GlcNAc (Neu5Acα2-3Galβ-3(Neu5Acα2-6)GlcNAc) (Figure 3C). Other modifications of glycans containing this motif, such as fucosylation and branching, do not hinder virus recognition. Although there is presently no evidence that such structures occur in the mouse, such a disialyl LNT motif with the unusual Neu5Acα2-6GlcNAc determinant could be one of the recognition determinants for MVM. Because there is very little overlap between the structures found on the CFG array and HMG SGM microarrays, the data from two microarrays provide complementary information and together they clearly indicate that MVMs uniquely prefer multisialylated glycan receptors. However, our results together with the CFG data also showed that sialylation *per se* is insufficient for MVM recognition, since the viruses did not bind to all the multisialylated glycans.

4.3 Rotavirus

Rotavirus (RV) is a major intestinal pathogen that causes severe dehydration and diarrhea in infants and young children, responsible for an estimated 500,000 deaths each year worldwide *(95)*. The initial cell attachment of RVs is mediated by interactions with specific glycans through the VP8* domain of VP4 spike protein *(96)*. However, the identity of such glycans remains unknown and the role of sialic acid is controversial. It was generally believed that the human RVs were "sialidase-insensitive," whereas animal RVs were "sialidase-sensitive." In contrast to this theory, nuclear magnetic resonance and crystallography studies showed that the animal viruses VP8* bound to glycans with terminal sialic acid, whereas some human RV VP8s could interact with internal sialic acid of gangliosides *(97–100)*, raising the possibility that human RVs also recognize sialosides. However, screening on the CFG glycan microarray containing 511 glycans revealed that the recombinant VP8* of a human strain (HAL1166) did not bind to any sialylated glycan with either terminal or internal sialic acid; instead, recombinant VP8* specifically recognized a set of glycans containing blood group A antigen, GalNAcα1-3(Fucα1-2)Galβ1-4GlcNAc *(101)*. Further crystallographic and virus infectivity studies were conducted to demonstrate that A-type histo-blood group antigen is a cell attachment factor for the P[14] RV strain. Recognizing the power of glycan microarray analyses to facilitate identification of RV receptors, the same group sought to determine the binding specificity of neonatal G10P[11] viruses with CFG array *(102)*. Previous epidemiologic surveys found that RV infections in neonates are distinct from those in older children and often associated with unusual strains. Indeed, glycan microarray data showed that the three tested neonatal strains N155, N1509, and B223 all strongly bound to poly-*N*-acetyllactosamine structures, whereas binding to previously known receptors such as GM1 and GD1 was low. Because Galβ1-4GlcNAc is the precursor for H type II blood group antigen, the authors concluded that this binding is biologically relevant for virus infection and conducted hemagglutination assays, binding studies and differential infectivity on CHO cells, which all confirmed the findings of the glycan array. Although the cellular receptor of RVs is still under study, it is becoming clearer that RVs have enormous genetic and strain diversity with respect to their functional receptors.

There have also been studies indicating that HMGs inhibit RV infectivity, resulting in the speculation that HMGs protect breast-fed infants from RV infection *(52,103,104)*. We are currently investigating the RV recognition of human milk free glycome using our HMG SGM. Preliminary data confirmed that the virus recognition is strain-dependent (Yu *et al.*, unpublished data). We are in the process of characterizing the bound HMGs, which are potentially natural receptors for RVs. This study will provide insight into the RV infection in infants and identify relevant HMGs associated with RV binding.

In summary, glycan arrays have offered a unique tool to rapidly identify possible receptors for pathogen attachment. Beyond the three examples discussed here, the receptor specificities of many other viruses, including adeno-associated viruses *(105,106)*, simian virus 40 *(77)*, and bluetongue virus *(107)*, were also analyzed on various glycan microarrays. In all cases, such analyses reveal distinct viral recognition of glycans.

5 CONCLUSIONS

The generation of free glycans, as in human milk and other body fluids, as well as the robust glycosylation of proteins and lipids, represents a tremendous set of chemically distinct macromolecular structures. An understanding of the functional roles of these glycans and their structures is key to solving the molecular mechanisms of many important biological processes and diseases. However, the extreme complexity of glycans has hindered the progress of glycomics for many years. In the past 20 years, the glycomics field has significantly moved forward with the development of high-sensitivity and high-resolution techniques, and each technique has advantages and disadvantages. However, it is not enough to simply define glycan structures; identifying roles of glycans in molecular recognition is paramount. Clearly, meeting the challenges of glycomics requires the combination of several complementary technologies. The glycan microarray technique has revolutionized the approach to investigate protein–glycan interactions and will continue to benefit research in this area. However, even the most comprehensive glycan array administered by the CFG represents only a small portion of the >7000 possible glycan determinants in the human glycome *(108)*. Therefore, an obvious future direction is to expand the number and diversity of glycan arrays and different modes of glycan presentation. It must be emphasized that the results from glycan microarrays provide information on binding specificity, which can be used to guide further investigation. Ideally, additional studies should be done to fully define the biological function of a glycan in a biological process.

ACKNOWLEDGMENTS

Research reported in this chapter was supported by the National Institute of General Medical Sciences of the National Institutes of Health (Grant Nos. R01GM085448, R24GM098791, and U54GM062116). The authors thank Dr. Jamie Heimburg-Molinaro for editing the chapter. The content is solely the responsibility of the authors and does not necessarily represent the official views of the National Institutes of Health.

REFERENCES

1. Paulson, J. C.; Blixt, O.; Collins, B. E. *Nat. Chem. Biol.* **2006**, *2*, 238–248.
2. Apweiler, R.; Hermjakob, H.; Sharon, N. *Biochim. Biophys. Acta* **1999**, *1473*, 4–8.
3. Freeze, H. H. *Nat. Rev. Genet.* **2006**, *7*, 537–551.
4. Robinson, L. N.; Artpradit, C.; Raman, R.; Shriver, Z. H.; Ruchirawat, M.; Sasisekharan, R. *Electrophoresis* **2012**, *33*, 797–814.
5. Krishnamoorthy, L.; Mahal, L. K. *ACS Chem. Biol.* **2009**, *4*, 715–732.
6. Zaia, J. *Chem. Biol.* **2008**, *15*, 881–892.
7. Kolarich, D.; Lepenies, B.; Seeberger, P. H. *Curr. Opin. Chem. Biol.* **2012**, *16*, 214–220.
8. Wada, Y. *Eur. J. Mass Spectrom. (Chichester, England)* **2007**, *13*, 101–103.
9. Mechref, Y.; Hu, Y.; Garcia, A.; Hussein, A. *Electrophoresis* **2012**, *33*, 1755–1767.
10. Gupta, G.; Surolia, A.; Sampathkumar, S. G. *OMICS* **2010**, *14*, 419–436.
11. Hirabayashi, J.; Yamada, M.; Kuno, A.; Tateno, H. *Chem. Soc. Rev.* **2013**, *42*, 4443–4458.
12. Rillahan, C. D.; Paulson, J. C. *Annu. Rev. Biochem.* **2011**, *80*, 797–823.
13. Smith, D. F.; Cummings, R. D. *Mol. Cell. Proteomics* **2013**, *12*, 902–912.
14. Seeberger, P. H. *Chem. Soc. Rev.* **2008**, *37*, 19–28.
15. Lee, J. C.; Greenberg, W. A.; Wong, C. H. *Nat. Protoc.* **2006**, *1*, 3143–3152.
16. Jaipuri, F. A.; Pohl, N. L. *Org. Biomol. Chem.* **2008**, *6*, 2686–2691.
17. Schofield, L.; Hewitt, M. C.; Evans, K.; Siomos, M. A.; Seeberger, P. H. *Nature* **2002**, *418*, 785–789.
18. Routenberg Love, K.; Seeberger, P. H. *Angew. Chem. Int. Ed. Engl.* **2004**, *43*, 602–605.
19. Werz, D. B.; Castagner, B.; Seeberger, P. H. *J. Am. Chem. Soc.* **2007**, *129*, 2770–2771.
20. Wu, C. Y.; Wong, C. H. *Top. Curr. Chem.* **2011**, *301*, 223–252.
21. Esposito, D.; Hurevich, M.; Castagner, B.; Wang, C. C.; Seeberger, P. H. *Beilstein J. Org. Chem.* **2012**, *8*, 1601–1609.
22. Muthana, S.; Cao, H.; Chen, X. *Curr. Opin. Chem. Biol.* **2009**, *13*, 573–581.
23. Voglmeir, J.; Sardzik, R.; Weissenborn, M. J.; Flitsch, S. L. *OMICS* **2010**, *14*, 437–444.
24. Bojarova, P.; Rosencrantz, R. R.; Elling, L.; Kren, V. *Chem. Soc. Rev.* **2013**, *42*, 4774–4797.
25. Rakus, J. F.; Mahal, L. K. *Annu. Rev. Anal. Chem. (Palo Alto, Calif.)* **2011**, *4*, 367–392.
26. Lonardi, E.; Balog, C. I.; Deelder, A. M.; Wuhrer, M. *Expert Rev. Proteomics* **2010**, *7*, 761–774.
27. Kozak, R. P.; Royle, L.; Gardner, R. A.; Fernandes, D. L.; Wuhrer, M. *Anal. Biochem.* **2012**, *423*, 119–128.
28. Song, X.; Smith, D. F.; Cummings, R. D. *Anal. Biochem.* **2012**, *429*, 82–87.
29. Fukui, S.; Feizi, T.; Galustian, C.; Lawson, A. M.; Chai, W. *Nat. Biotechnol.* **2002**, *20*, 1011–1017.
30. Shipp, E. L.; Hsieh-Wilson, L. C. *Chem. Biol.* **2007**, *14*, 195–208.
31. de, P. J. L.; Seeberger, P. H. *Mol. BioSyst.* **2008**, *4*, 707–711.
32. Rogers, C. J.; Clark, P. M.; Tully, S. E.; Abrol, R.; Garcia, K. C.; Goddard, W. A.; Hsieh-Wilson, L. C., III. *Proc. Natl. Acad. Sci. U.S.A.* **2011**, *108*, 9747–9752, S9747/1–S9747/32.

33. Takada, W.; Fukushima, M.; Pothacharoen, P.; Kongtawelert, P.; Sugahara, K. *Anal. Biochem.* **2013**, *435*, 123–130.
34. Liu, Y.; Childs, R. A.; Palma, A. S.; Campanero-Rhodes, M. A.; Stoll, M. S.; Chai, W.; Feizi, T. *Methods Mol. Biol.* **2012**, *808*, 117–136.
35. Culf, A. S.; Cuperlovic-Culf, M.; Ouellette, R. J. *OMICS* **2006**, *10*, 289–310.
36. Park, S.; Gildersleeve, J. C.; Blixt, O.; Shin, I. *Chem. Soc. Rev.* **2012**, *42*, 4310–4326.
37. Oyelaran, O.; Gildersleeve, J. C. *Curr. Opin. Chem. Biol.* **2009**, *13*, 406–413.
38. Xia, B.; Kawar, Z. S.; Ju, T.; Alvarez, R. A.; Sachdev, G. P.; Cummings, R. D. *Nat. Methods* **2005**, *2*, 845–850.
39. Song, X.; Xia, B.; Lasanajak, Y.; Smith, D. F.; Cummings, R. D. *Glycoconj. J.* **2008**, *25*, 15–25.
40. Song, X.; Xia, B.; Stowell, S. R.; Lasanajak, Y.; Smith, D. F.; Cummings, R. D. *Chem. Biol.* **2009**, *16*, 36–47.
41. Luyai, A.; Lasanajak, Y.; Smith, D. F.; Cummings, R. D.; Song, X. *Bioconjug. Chem.* **2009**, *20*, 1618–1624.
42. Yu, Y.; Mishra, S.; Song, X.; Lasanajak, Y.; Bradley, K. C.; Tappert, M. M.; Air, G. M.; Steinhauer, D. A.; Halder, S.; Cotmore, S.; Tattersall, P.; Agbandje-McKenna, M.; Cummings, R. D.; Smith, D. F. *J. Biol. Chem.* **2012**, *287*, 44784–44799.
43. Song, X.; Lasanajak, Y.; Olson, L. J.; Boonen, M.; Dahms, N. M.; Kornfeld, S.; Cummings, R. D.; Smith, D. F. *J. Biol. Chem.* **2009**, *284*, 35201–35214.
44. Yu, H.; Chokhawala, H. A.; Huang, S.; Chen, X. *Nat. Protoc.* **2006**, *1*, 2485–2492.
45. Song, X.; Lasanajak, Y.; Xia, B.; Heimburg-Molinaro, J.; Rhea, J. M.; Ju, H.; Zhao, C.; Molinaro, R. J.; Cummings, R. D.; Smith, D. F. *Nat. Methods* **2011**, *8*, 85–90.
46. Song, X.; Lasanajak, Y.; Xia, B.; Smith, D. F.; Cummings, R. D. *ACS Chem. Biol.* **2009**, *4*, 741–750.
47. Carroll, T.; Wang, D.; Turro, N. J.; Koberstein, J. T. *Langmuir* **2006**, *22*, 2899–2905.
48. Zhou, X.; Turchi, C.; Wang, D. *J. Proteome Res.* **2009**, *8*, 5031–5040.
49. Lee, J. C.; Wu, C. Y.; Apon, J. V.; Siuzdak, G.; Wong, C. H. *Angew. Chem. Int. Ed. Engl.* **2006**, *45*, 2753–2757.
50. Liang, C. H.; Wu, C. Y. *Expert Rev. Proteomics* **2009**, *6*, 631–645.
51. Bode, L. *Nutr. Rev.* **2009**, *67*(Suppl. 2), S183–S191.
52. Newburg, D. S.; Ruiz-Palacios, G. M.; Morrow, A. L. *Annu. Rev. Nutr.* **2005**, *25*, 37–58.
53. Bode, L.; Jantscher-Krenn, E. *Adv. Nutr.* **2012**, *3*, 383S–391S.
54. Kobata, A. *Proc. Jpn. Acad. Ser. B Phys. Biol. Sci.* **2010**, *86*, 731–747.
55. van Diepen, A.; Smit, C. H.; van Egmond, L.; Kabatereine, N. B.; Pinot de Moira, A.; Dunne, D. W.; Hokke, C. H. *PLoS Negl. Trop. Dis.* **2012**, *6*, e1922.
56. Wang, D.; Liu, S.; Trummer, B. J.; Deng, C.; Wang, A. *Nat. Biotechnol.* **2002**, *20*, 275–281.
57. Palma, A. S.; Feizi, T.; Zhang, Y.; Stoll, M. S.; Lawson, A. M.; Diaz-Rodriguez, E.; Campanero-Rhodes, M. A.; Costa, J.; Gordon, S.; Brown, G. D.; Chai, W. *J. Biol. Chem.* **2006**, *281*, 5771–5779.
58. Willats, W. G.; Rasmussen, S. E.; Kristensen, T.; Mikkelsen, J. D.; Knox, J. P. *Proteomics* **2002**, *2*, 1666–1671.
59. Galanina, O. E.; Mecklenburg, M.; Nifantiev, N. E.; Pazynina, G. V.; Bovin, N. V. *Lab Chip* **2003**, *3*, 260–265.
60. Blixt, O.; Head, S.; Mondala, T.; Scanlan, C.; Huflejt, M. E.; Alvarez, R.; Bryan, M. C.; Fazio, F.; Calarese, D.; Stevens, J.; Razi, N.; Stevens, D. J.; Skehel, J. J.; van Die, I.; Burton, D. R.; Wilson, I. A.; Cummings, R.; Bovin, N.; Wong, C. H.; Paulson, J. C. *Proc. Natl. Acad. Sci. U.S.A.* **2004**, *101*, 17033–17038.

61. Manimala, J. C.; Li, Z.; Jain, A.; VedBrat, S.; Gildersleeve, J. C. *ChemBioChem* **2005**, *6*, 2229–2241.
62. Li, Q.; Anver, M. R.; Li, Z.; Butcher, D. O.; Gildersleeve, J. C. *Int. J. Cancer* **2010**, *126*, 459–468.
63. de Paz, J. L.; Noti, C.; Seeberger, P. H. *J. Am. Chem. Soc.* **2006**, *128*, 2766–2767.
64. Noti, C.; de Paz, J. L.; Polito, L.; Seeberger, P. H. *Chemistry* **2006**, *12*, 8664–8686.
65. de Paz, J. L.; Spillmann, D.; Seeberger, P. H. *Chem. Commun. (Camb.)* **2006**, *7*, 3116–3118.
66. Gama, C. I.; Tully, S. E.; Sotogaku, N.; Clark, P. M.; Rawat, M.; Vaidehi, N.; Goddard, W. A., 3rd.; Nishi, A.; Hsieh-Wilson, L. C. *Nat. Chem. Biol.* **2006**, *2*, 467–473.
67. Wang, D.; Carroll, G. T.; Turro, N. J.; Koberstein, J. T.; Kovac, P.; Saksena, R.; Adamo, R.; Herzenberg, L. A.; Herzenberg, L. A.; Steinman, L. *Proteomics* **2007**, *7*, 180–184.
68. Blixt, O.; Han, S.; Liao, L.; Zeng, Y.; Hoffmann, J.; Futakawa, S.; Paulson, J. C. *J. Am. Chem. Soc.* **2008**, *130*, 6680–6681.
69. Moller, I.; Marcus, S. E.; Haeger, A.; Verhertbruggen, Y.; Verhoef, R.; Schols, H.; Ulvskov, P.; Mikkelsen, J. D.; Knox, J. P.; Willats, W. *Glycoconj. J.* **2008**, *25*, 37–48.
70. Kamena, F.; Tamborrini, M.; Liu, X.; Kwon, Y. U.; Thompson, F.; Pluschke, G.; Seeberger, P. H. *Nat. Chem. Biol.* **2008**, *4*, 238–240.
71. Wang, C. C.; Huang, Y. L.; Ren, C. T.; Lin, C. W.; Hung, J. T.; Yu, J. C.; Yu, A. L.; Wu, C. Y.; Wong, C. H. *Proc. Natl. Acad. Sci. U.S.A.* **2008**, *105*, 11661–11666.
72. Blixt, O.; Hoffmann, J.; Svenson, S.; Norberg, T. *Glycoconj. J.* **2008**, *25*, 27–36.
73. Wang, C. C.; Chen, J. R.; Tseng, Y. C.; Hsu, C. H.; Hung, Y. F.; Chen, S. W.; Chen, C. M.; Khoo, K. H.; Cheng, T. J.; Cheng, Y. S.; Jan, J. T.; Wu, C. Y.; Ma, C.; Wong, C. H. *Proc. Natl. Acad. Sci. U.S.A.* **2009**, *106*, 18137–18142.
74. Nycholat, C. M.; McBride, R.; Ekiert, D. C.; Xu, R.; Rangarajan, J.; Peng, W.; Razi, N.; Gilbert, M.; Wakarchuk, W.; Wilson, I. A.; Paulson, J. C. *Angew. Chem. Int. Ed. Engl.* **2012**, *51*, 4860–4863.
75. Feizi, T.; Chai, W. *Nat. Rev. Mol. Cell Biol.* **2004**, *5*, 582–588.
76. Liu, Y.; Palma, A. S.; Feizi, T. *Biol. Chem.* **2009**, *390*, 647–656.
77. Campanero-Rhodes, M. A.; Smith, A.; Chai, W.; Sonnino, S.; Mauri, L.; Childs, R. A.; Zhang, Y.; Ewers, H.; Helenius, A.; Imberty, A.; Feizi, T. *J. Virol.* **2007**, *81*, 12846–12858.
78. Childs, R. A.; Palma, A. S.; Wharton, S.; Matrosovich, T.; Liu, Y.; Chai, W.; Campanero-Rhodes, M. A.; Zhang, Y.; Eickmann, M.; Kiso, M.; Hay, A.; Matrosovich, M.; Feizi, T. *Nat. Biotechnol.* **2009**, *27*, 797–799.
79. Reiss, K.; Stencel, J. E.; Liu, Y.; Blaum, B. S.; Reiter, D. M.; Feizi, T.; Dermody, T. S.; Stehle, T. *PLoS Pathog.* **2012**, *8*, e1003078.
80. Wu, C. Y.; Liang, P. H.; Wong, C. H. *Org. Biomol. Chem.* **2009**, *7*, 2247–2254.
81. Nicholls, J. M.; Chan, R. W.; Russell, R. J.; Air, G. M.; Peiris, J. S. *Trends Microbiol.* **2008**, *16*, 149–157.
82. Gambaryan, A.; Yamnikova, S.; Lvov, D.; Tuzikov, A.; Chinarev, A.; Pazynina, G.; Webster, R.; Matrosovich, M.; Bovin, N. *Virology* **2005**, *334*, 276–283.
83. Heimburg-Molinaro, J.; Tappert, M.; Song, X.; Lasanajak, Y.; Air, G.; Smith, D. F.; Cummings, R. D. *Methods Mol. Biol.* **2012**, *808*, 251–267.
84. Stevens, J.; Blixt, O.; Glaser, L.; Taubenberger, J. K.; Palese, P.; Paulson, J. C.; Wilson, I. A. *J. Mol. Biol.* **2006**, *355*, 1143–1155.
85. Bradley, K. C.; Jones, C. A.; Tompkins, S. M.; Tripp, R. A.; Russell, R. J.; Gramer, M. R.; Heimburg-Molinaro, J.; Smith, D. F.; Cummings, R. D.; Steinhauer, D. A. *Virology* **2011**, *413*, 169–182.

86. Stevens, J.; Blixt, O.; Chen, L. M.; Donis, R. O.; Paulson, J. C.; Wilson, I. A. *J. Mol. Biol.* **2008**, *381*, 1382–1394.
87. Wan, H.; Sorrell, E. M.; Song, H.; Hossain, M. J.; Ramirez-Nieto, G.; Monne, I.; Stevens, J.; Cattoli, G.; Capua, I.; Chen, L. M.; Donis, R. O.; Busch, J.; Paulson, J. C.; Brockwell, C.; Webby, R.; Blanco, J.; Al-Natour, M. Q.; Perez, D. R. *PLoS One* **2008**, *3*, e2923.
88. Yang, H.; Carney, P. J.; Donis, R. O.; Stevens, J. *J. Virol.* **2012**, *86*, 8645–8652.
89. Gulati, S.; Smith, D. F.; Cummings, R. D.; Couch, R. B.; Griesemer, S. B.; St George, K.; Webster, R. G.; Air, G. M. *PLoS One* **2013**, *8*, e66325.
90. Song, X.; Yu, H.; Chen, X.; Lasanajak, Y.; Tappert, M. M.; Air, G. M.; Tiwari, V. K.; Cao, H.; Chokhawala, H. A.; Zheng, H.; Cummings, R. D.; Smith, D. F. *J. Biol. Chem.* **2011**, *286*, 31610–31622.
91. Sillanaukee, P.; Ponnio, M.; Jaaskelainen, I. P. *Eur. J. Clin. Invest.* **1999**, *29*, 413–425.
92. Walther, T.; Karamanska, R.; Chan, R. W.; Chan, M. C.; Jia, N.; Air, G.; Hopton, C.; Wong, M. P.; Dell, A.; Malik Peiris, J. S.; Haslam, S. M.; Nicholls, J. M. *PLoS Pathog.* **2013**, *9*, e1003223.
93. Cotmore, S. F.; Tattersall, P. *Adv. Virus Res.* **2007**, *70*, 183–232.
94. Nam, H. J.; Gurda-Whitaker, B.; Gan, W. Y.; Ilaria, S.; McKenna, R.; Mehta, P.; Alvarez, R. A.; Agbandje-McKenna, M. *J. Biol. Chem.* **2006**, *281*, 25670–25677.
95. Tate, J. E.; Burton, A. H.; Boschi-Pinto, C.; Steele, A. D.; Duque, J.; Parashar, U. D. *Lancet Infect. Dis.* **2012**, *12*, 136–141.
96. Lopez, S.; Arias, C. F. *Curr. Top. Microbiol. Immunol.* **2006**, *309*, 39–66.
97. Blanchard, H.; Yu, X.; Coulson, B. S.; von Itzstein, M. *J. Mol. Biol.* **2007**, *367*, 1215–1226.
98. Haselhorst, T.; Fleming, F. E.; Dyason, J. C.; Hartnell, R. D.; Yu, X.; Holloway, G.; Santegoets, K.; Kiefel, M. J.; Blanchard, H.; Coulson, B. S.; von Itzstein, M. *Nat. Chem. Biol.* **2009**, *5*, 91–93.
99. Yu, X.; Coulson, B. S.; Fleming, F. E.; Dyason, J. C.; von Itzstein, M.; Blanchard, H. *J. Mol. Biol.* **2011**, *413*, 929–939.
100. Yu, X.; Dang, V. T.; Fleming, F. E.; von Itzstein, M.; Coulson, B. S.; Blanchard, H. *J. Virol.* **2012**, *86*, 13456–13466.
101. Hu, L.; Crawford, S. E.; Czako, R.; Cortes-Penfield, N. W.; Smith, D. F.; Le Pendu, J.; Estes, M. K.; Prasad, B. V. *Nature* **2012**, *485*, 256–259.
102. Ramani, S.; Cortes-Penfield, N. W.; Hu, L.; Crawford, S. E.; Czako, R.; Smith, D. F.; Kang, G.; Ramig, R. F.; Le Pendu, J.; Prasad, B. V.; Estes, M. K. *J. Virol.* **2013**, *87*, 7255–7264.
103. Morrow, A. L.; Ruiz-Palacios, G. M.; Jiang, X.; Newburg, D. S. *J. Nutr.* **2005**, *135*, 1304–1307.
104. Hester, S. N.; Chen, X.; Li, M.; Monaco, M. H.; Comstock, S. S.; Kuhlenschmidt, T. B.; Kuhlenschmidt, M. S.; Donovan, S. M. *Br. J. Nutr.* **2013**, *26*, 1–10, [Epub ahead of print].
105. Wu, Z.; Miller, E.; Agbandje-McKenna, M.; Samulski, R. J. *J. Virol.* **2006**, *80*, 9093–9103.
106. Bell, C. L.; Vandenberghe, L. H.; Bell, P.; Limberis, M. P.; Gao, G. P.; Van Vliet, K.; Agbandje-McKenna, M.; Wilson, J. M. *J. Clin. Invest.* **2011**, *121*, 2427–2435.
107. Bhattacharya, B.; Roy, P. *Viruses* **2010**, *2*, 1218–1235.
108. Cummings, R. D. *Mol. BioSyst.* **2009**, *5*, 1087–1104.

Chapter 13

High-Resolution Analytical Tools for Quantitative Peptidomics

Sayani Dasgupta and Lloyd D. Fricker
Department of Molecular Pharmacology, Albert Einstein College of Medicine, Bronx, New York, USA

Chapter Outline

1 INTRODUCTION

Peptidomics is the identification and elucidation of the peptide content or "peptidome" of a living organism, tissue, or cell. Peptides play important roles in regulating biological processes, functioning in cell–cell signaling as peptide hormones and neuropeptides. In addition, peptides have been proposed to function in intracellular regulation of protein–protein interactions. Quantitative peptidomics is a powerful tool to compare peptide profiles between different physiological states. The term "peptidomics" was first used more than 10 years ago *(1,2)*, and the technique has since been applied to the identification and analysis of hundreds of peptides from a variety of tissue samples. Peptidomics has been used to characterize bioactive peptides regulating complex mechanisms such as circadian rhythm, feeding, and behavior in rodents and in lower organisms *(3–7)*. Quantitative peptidomics has been used to measure changes in peptide content of specific brain regions and/or cells *(8–11)* under disease conditions or in response to drug administration *(12,13)*. Quantitative peptidomic approaches have also been used to identify the substrate specificity of known enzymes as well as pathways regulating the synthesis

Comprehensive Analytical Chemistry, Vol. 63. http://dx.doi.org/10.1016/B978-0-444-62651-6.00014-3

and degradation of bioactive peptides *(14–17)*. In addition, peptidomic tools have been applied to the discovery of novel antimicrobial peptides and development of potential biomarkers for various diseases *(18–22)*.

Peptidomics and proteomics share similarities in terms of principle and tools required for identification and analysis, although there are some distinct differences between these techniques. Endogenous peptides are often cleavage products of larger precursor proteins, and the goal of peptidomics is to identify the precise form of each peptide. Thus, samples for peptidomics are not enzymatically digested prior to analysis. The absence of enzymatic processing allows the identification of the peptide in its native form, including posttranslational modifications *(23,24)*. In contrast, the goal of most proteomic studies is to identify proteins present in the sample, without concern for the precise molecular form of each protein. To identify the proteins, they are typically digested with trypsin to produce peptides that can then be sequenced *(25)*. Another difference between the overall goal of proteomics and peptidomics analyses has to do with the tendency for different cleavage products of the same peptide precursor to have distinct physiological functions. For example, limited cleavage of prodynorphin generates peptides with highest affinity for kappa opioid receptors while more complete processing of prodynorphin generates smaller peptides with highest affinity for delta opioid receptors. Thus, it is important to identify all peptides produced from a single protein *(26–28)*. This is not a requirement for proteomic studies, since it is possible to ascertain the identity of a protein from a small number of tryptic peptides. Another difference between proteomics and peptidomics is with sample handling and the issue of protein degradation, which needs to be minimized for peptidomic studies. This is usually not a disadvantage for proteomics since the identification of a protein is not hindered by the breakdown of a small fraction of the protein. However, because bioactive peptides are typically present in tissues at low levels, even a small amount of degradation of highly abundant proteins can overwhelm the signal from the endogenous peptides. Therefore, it is essential to eliminate postmortem protein degradation so that the endogenous bioactive peptides can be detected. To counter the problem of protein degradation, the proteases in the tissue need to be heat-inactivated prior to processing of the tissue for peptidomics *(10,29–31)*.

In this review, we focus on the various techniques for quantification of peptides/proteins and their advantages and disadvantages with respect to their application in peptidomics.

2 ABSOLUTE QUANTIFICATION

Determination of absolute quantities of peptide levels mostly employ the stable isotope dilution (SID) methodology coupled to mass spectrometric analysis. SID involves the use of stable isotope-labeled standards wherein quantification is performed by comparing the mass spectrometric profiles of analytes to their corresponding labeled standards. It is important to choose the internal standard carefully to accurately measure absolute quantities from signal intensities.

Different compounds have variable ionization intensities, which are further affected by factors such as chromatographic retention times and suppression by compounds present in the analyte fluid and sample matrix, so it is essential to use isotope-labeled internal standards, which are physicochemically identical to the molecule of interest rather than structural analogs, to normalize for effects that can lead to erroneous quantification. Isotope-labeled standards also account for any differences arising out of sample preparation and absorptive loss due to selective binding on surfaces during chromatography *(32)*.

Precision and sensitivity of quantitation has been greatly advanced by the use of selected reaction monitoring (SRM). SRM allows for selective monitoring of specific molecular ions with known fragmentation properties. SRM is performed on triple-quadrupole mass spectrometers, where the first mass analyzer selects ions with m/z corresponding to that of the peptide of interest for collision-induced dissociation. Out of the various fragmentation ions produced, only specific ions uniquely derived from the peptide precursor of interest are further selected for detection and measurement in the second analyzer. SRM enables the detection of several precursor-fragment ion pairs within a set of specific retention times. This facilitates the measurement of many target peptides simultaneously, which is known as multiple reaction monitoring (MRM) *(33)*.

A combination of SID and SRM has been applied to the absolute quantification of proteins and posttranslational modifications directly from cell lysates through measuring their tryptic peptide levels. A tryptic digest of a protein mixture is supplemented with a known amount of isotopic-labeled peptide derived from the target protein precursor and subjected to SRM mass spectrometry. Similar retention times and ionization properties enable matching of native peptides to their appropriate labeled standard and quantification from comparing their mass spectral intensities, leading to absolute measurement of endogenous protein levels *(34,35)*. Absolute quantification is limited to the measurement of selected peptides/proteins for which internal standards are available, unlike the relative quantitative peptidomic approaches that can detect unknowns as well as known peptides.

3 RELATIVE QUANTIFICATION

Relative quantification methods provide information on the identity of peptides/proteins in a sample as well as their levels expressed in amounts relative to each other. Some of these methods rely on label-free strategies while others incorporate stable isotopes into one or more of the samples, allowing them to be combined and analyzed together. In both cases, the strength of the signal for each peptide is a reflection of the amount of peptide present in the sample, providing quantitative information.

In the following section, both label-free and isotopic label approaches are described in detail. In both approaches, the point of the experiment is to compare the relative levels of peptides in two or more distinct groups;

Figure 1 shows an example in which an experimental group is compared to a control group. The group can consist of animals, tissues, cells, or other sources of proteins/peptides. Due to normal biological variation, it is important to include replicates of the members of each group. Figure 1 shows four biological replicates for a total of eight samples that need to be compared. In both label-free and isotopic label approaches, the peptides are first extracted from the tissues or cells. In the label-free approach, the samples are directly analyzed by liquid chromatography/mass spectrometry (LC/MS). While this would seem to save time (due to the simpler sample preparation), each biological replicate is usually analyzed in at least three technical replicates, requiring much more time for LC/MS and subsequent analysis than the isotopic label approaches. The technical replicates are first averaged to determine the levels of peptides in each of the biological samples. Then, the individual values for each biological replicate are averaged together and standard error of the mean (or standard deviation) is calculated. Finally, the average of the experimental group is compared to the average of the control group and statistical testing performed to determine if there is a significant difference between any of the peptides in the two groups.

The isotopic approach involves the introduction of stable isotopes so that the peptides in each group can be distinguished by mass. The isotopes chosen are usually ^{15}N, ^{13}C, ^{18}O, or ^{2}H (or a combination of these). In all cases, the isotopes are stable (i.e., nonradioactive) so that they are safe to work with and will not contaminate the mass spectrometer. In some cases, the stable isotope is incorporated into the protein/peptide in cell culture, or even in living animals. In other cases, the isotopic label is added in a chemical reaction once the peptides are extracted from tissue/cells. The example shown in Figure 1 is for a multiplex label, in which eight different isotopic forms can be used. In this example, all eight biological samples can be included in a single LC/MS run, thus saving considerable instrument time compared to the label-free approach. Data analysis for the isotopic label approaches involve averaging levels among the biological replicates and then comparing the experimental and control values, as for the label-free techniques.

3.1 Label-Free Quantification

As the name suggests, label-free quantification does not require the incorporation of an isotopic label into peptides for quantitation of relative concentrations. Each sample is separately prepared, subjected to individual LC/MS runs, and quantification is performed through analysis of their mass spectra. There are two approaches used for label-free quantification. The first approach is based on the assumption that peptide/protein levels follow a linear relationship with area under the curve of the ion spectra. This method involves quantitation of relative levels by comparing the peak intensity of ions from multiple peptides. But experimental variations in sample preparation and

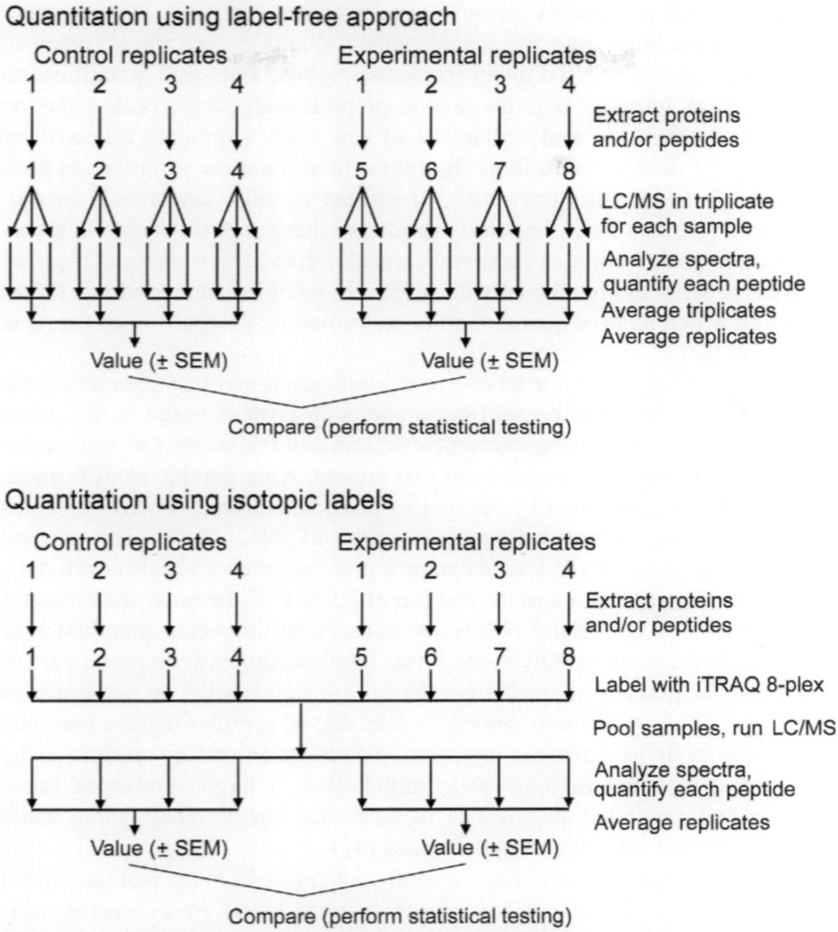

FIGURE 1 Quantitative proteomics/peptidomics using label-free and isotopic labeling approaches. In a typical experiment comparing two conditions (control vs. experimental groups), biological replicates are performed; in this example, four biological replicates are indicated. In both approaches, proteins/peptides are first extracted from the samples. For the label-free approach, the samples are directly analyzed on liquid chromatography/mass spectrometry (LC/MS). However, label-free approaches typically run technical replicates of each sample; three technical replicates are indicated here, resulting in 24 total LC/MS runs. Peptides are quantified by the spectral counts or another method, the technical replicates are averaged, and then the biological replicates are averaged and the standard error of the mean (SEM) or standard deviation calculated. The experimental and control groups are compared using statistical tests. For the isotopic labeling approach, the samples are individually labeled with a distinct chemical tag that differs in mass due to the presence of stable isotopes (^{2}H, ^{13}C, ^{15}N, or others). Following the labeling (and quenching of unreacted labels), the samples are pooled and analyzed on LC/MS. With iTRAQ or other reagents that are available in eight different isotopic forms, all of the biological replicates in this example can be tested in a single LC/MS run. The data are analyzed by comparing the peak intensity of each of the isotopic forms, the biological replicates are averaged together, and control and experimental groups compared, as above.

sample injection procedures can result in run-to-run variations in peak intensities, retention times, and *m/z* values, which greatly reduce the accuracy of data interpretation *(36)*. To avoid imprecise results, label-free quantification based on peak intensity requires several preprocessing steps. Peak detection by baseline correction and smoothing of raw mass spectra is followed by application of data-normalization algorithms to account for variations in peak intensities between different runs. Subsequent to peak detection, consecutively eluting peaks from the same peptide are merged into features in a process termed peak/feature extraction. Features derived from identical peptides in different samples are aligned with respect to retention times and *m/z* ratios, following which relative quantitation is performed by comparison of the peak intensities *(37)*.

The second approach for label-free quantification involves identifying the total number of MS/MS spectra from a particular protein, based on the direct correlation between the abundance of a protein and the number of fragmentation spectra of peptides derived from that protein. After peptide identification, quantification is carried out by calculating peak areas from extracted ion chromatograms of the corresponding peaks in the LC/MS *(38,39)*. This method has its limitations as well; low-abundance proteins often go undetected, deficiencies in ion-trapping capacity and ion efficiency of the mass spectrometer lead to saturation of spectral counts and consequent imprecise quantitation of high-abundance proteins. Often, the linear relationship between protein abundance and the number of spectra is affected by peptides that do not ionize or fragment efficiently. Peptide sharing or detection of peptides that are common to multiple proteins further complicates identification and quantification by the spectral counting method *(40)*. Furthermore, a large number of false-positives are generated in analysis of extracted ion chromatograms while applying spectral counting to peptidomics *(41)*.

Label-free quantitation has several advantages over isotopic label approaches, but also some disadvantages. Isotopic labels are expensive, so if there is no cost for each LC/MS run, label-free quantification is an economically viable option. However, because of the larger number of LC/MS runs that need to be performed with label-free approaches than with isotopic label approaches (especially multiplex labels), the label-free approaches can be considerably more expensive if there is a cost for each LC/MS run. Also, the number of samples that can be compared by labeling techniques are limited by the number of available labels, whereas there is no such restriction in label-free methods, which can accommodate more samples per experiment (although with an increase in the number of LC/MS runs that need to be performed and analyzed in each experiment). Label-free methods have a higher range for quantitation, making them useful to detect large variations in peptide/protein levels. However, the accuracy of label-free approaches is lower than that of isotopic label approaches, making it difficult to detect small changes in peptide levels using the label-free methods. The total number of peptides/proteins identified is typically higher in label-free approaches than with isotopic label approaches because the latter has greater mass spectral complexity due to differently labeled versions of the same peptide *(38,42)*. Overall, there are advantages and disadvantages to each approach, and selection of the optimal approach depends on the scope of the experiment.

3.2 Metabolic Labeling

Incorporation of a stable isotopic tag into proteins/peptides in metabolically active cells was first described to quantify protein abundance in yeast *(43)*. Wild-type and mutant cell populations were grown in media containing the naturally abundant isotopes of nitrogen and enriched in ^{15}N, respectively, followed by trypsin digestion and LC-ESI-MS/MS analysis to identify and quantify relative phosphopeptide levels in both populations *(43)*.

SILAC (stable isotope labeling by amino acids in cell culture) was later described for the *in vivo* incorporation of specific amino acids in mammalian cells. The experimental cell population is grown in media containing an isotopic form of an essential amino acid. For proteomics, the amino acids used for incorporation are usually arginine and lysine because these are cleavage sites for trypsin, thus providing an isotopic tag on the C-terminus of every tryptic fragment. Apart from mass difference, the labeled and unlabeled forms of the same peptide in the experimental and control cell populations are chemically identical, which enables their relative quantitation by comparing the peak intensities through standard mass spectrometric techniques. SILAC does not require any peptide labeling steps and confers the advantage of almost complete incorporation of the labeled amino acids, thus precluding any differences in labeling efficiencies between the two samples. In addition, it is very useful to quantify small changes in protein/peptide levels. But not all cells are suitable for SILAC; the requirement for cells to be grown in dialyzed serum makes it difficult to be adopted for those cell lines, which do not grow well under this condition *(44,45)*.

A distinct disadvantage of SILAC also arises from the metabolic conversion of [$^{13}C_6$]- and [$^{13}C_6$,$^{15}N_4$]-labeled arginine to proline, which leads to inaccurate estimation of isotopically labeled proline containing peptides. A method based on using the [$^{15}N_4$]-arginine in combination with the "light" form of lysine and [$^{13}C_6$,$^{15}N_4$]-arginine with the [$^{13}C_6$,$^{15}N_2$]-lysine allows for internal control of arginine conversion since both [$^{15}N_1$]-proline and [$^{13}C_5$,$^{15}N_1$]-proline will be formed at the same rate under both conditions *(46)*.

Unlike mammalian cells, which cannot synthesize all amino acids, SILAC is not well suited to studying the protein/peptide levels of microorganisms, most of which are prototrophic for all amino acids. Application of SILAC in such cases is mostly restricted to auxotrophic strains, which renders the technique ineffective for the proteome analysis of many commercially important microbes. Native SILAC (nSILAC), a recently developed modification of the existing technique, has been shown to be useful for the analysis of protein/peptide levels in microorganisms. nSILAC relies on the downregulation of lysine biosynthesis enzymes in the presence of exogenous lysine, thus leading to incorporation of isotopically labeled "heavy" lysine present in the media *(47)*.

SILAC is difficult for studies on rodents due to the expense of isotopic label and the need for lengthy exposure times to ensure adequate incorporation of the label into the proteome. Despite the high cost, SILAC had been performed for some specific analyses of protein turnover in rodents *(48,49)*.

Nearly all of the published SILAC studies have been for studies on proteomics, although it is technically possible to use this technique for studies on endogenous peptides. One peptidomic study used SILAC to investigate the production of peptide hormones in cell culture *(50)*.

3.3 Chemical Labeling

Stable isotopic labels can be introduced into reactive groups present in peptides/proteins through chemical modifications. For peptidomics, the most commonly targeted reactive group is the free amine present on the N-terminus of most peptides and also on the side-chain of lysine residues. For proteomics, free amines have been used as well as the side-chain of cysteine residues. In this section, we review different chemical labeling strategies and their applicability in peptidomics.

There are several advantages of chemical labeling over the metabolic labeling approach. First, a wide range of samples can be analyzed, without the need to first culture cells in a special medium. Animal tissues can be used for the analysis, which would be prohibitively expensive using metabolic labeling (and impossible for human tissue). Second, some isotopic tags allow for multiple samples to be compared in the same LC/MS run. This allows for several different replicates to be analyzed at the same time, providing a greater estimate of the reproducibility of the data.

The ideal chemical isotopic tag is one that will react with every protein or peptide present in the sample and be reactive enough that the protein/peptide becomes fully labeled. However, the intrinsic reactivity of the compound should not be so high that it degrades upon storage under proper conditions. Also, the ideal tag will be stable upon mass spectrometry and not readily break apart under standard conditions. The ideal chemical tag will either be commercially available or relatively easy to synthesize, and the price should be reasonable. At a minimum, the tag needs to be available in a heavy and light form, but ideally would be available in a range of masses so that multiple replicates can be combined. It is also important that peptides labeled with the heavy and light forms of the tag coelute from HPLC; otherwise, the quantitation is less accurate when comparing different spectra (due to ion suppression). Some of the isotopic tags come close to the properties of the ideal reagent, although all have their drawbacks (described below).

3.3.1 ICAT

Isotope-coded affinity tags (ICATs) were first used for the quantitative comparison of protein expression in *Saccharomyces cerevisiae*. The three basic components of an ICAT reagent are a thiol-reactive group (usually iodoacetamide) to specifically target cysteine residues present in the peptide/protein, an isotopically "light" or "heavy" linker, and a biotin moiety. The protein samples are treated with the ICAT prior to enzymatic digestion, following which the peptides are purified on an avidin affinity column. ICAT was originally synthesized as a deuterated tag containing none or eight deuteriums and yielded peptide pairs differing by a mass shift of 8 Da on the mass spectrum. One disadvantage is that deuterium interacts with hydrophobic surfaces weakly as compared to hydrogen, causing deuterated peptides to elute earlier on a reverse-phase column (deuterium isotope effects). This difference in retention times of the labeled peptide pairs can lead to inaccurate quantitation *(51)*.

A modified version of the ICAT (clCAT) was developed later, featuring an acid-cleavable biotin tag and substitution of ^{1}H with ^{12}C and ^{2}H and ^{13}C in the isotopically "light" and "heavy" forms, respectively. Peptides thus labeled, coelute during MS, which results in more accurate identification. Removal of the biotin tag after peptide purification leads to relatively cleaner fragmentation spectra, thus further improving the accuracy and rate of identification *(52)*.

Several other versions of the ICAT labeling technique have emerged in recent years to widen its potential (e.g., for the quantification of phosphopeptides). Phosphoprotein isotope-coded affinity tag (PhIAT) is based on the labeling of phosphorylated serine and threonine residues by deuterated and nondeuterated ethanedithiol. Peptide complexes are subsequently enriched on an avidin affinity column and analyzed by mass spectrometry to detect, identify, and quantify differences in phosphopeptide levels *(53,54)*. IGOT (isotope-coded glycosylation-site-specific tagging) combines affinity chromatography and proteolytic labeling (further discussed in this review) for the identification of N-linked glycoproteins. Consequent to isolation of glycopeptides on a lectin affinity column, ^{18}O tag is stably incorporated into glycosylation sites by *N*-glycosidase action. The ^{18}O peptides are further identified by mass spectrometry *(54)*.

ICAT has also been adapted for absolute quantification purposes. The metal-coded affinity tag method introduces different lanthanide ions instead of stable isotope into peptides, which enables their absolute quantification by inductively coupled plasma mass spectrometry (ICP-MS) *(55,56)*. Appending a fluorophore to an ICAT reagent also facilitates absolute quantification by fluorescence detection. Fluorescence isotope-coded affinity tag, as this method has been named, further aids in purification of labeled peptides by immunoprecipitation with anti-fluorescein antibody *(57)*.

Being a relatively rare residue, labeling of cysteines by ICAT reduces complexity in the sample mixture *(51)*, making it a convenient methodology for proteomics. The same property makes it inefficient for peptidomic analyses, since most endogenous peptides do not contain cysteine residues.

3.3.2 iTRAQ/TMT/mTRAQ

Since most peptides contain a free N-terminus and one or more internal lysines, labeling amine groups will permit detection of a large number of peptides, thus making it a useful strategy for peptidomics. Isobaric tags for relative and absolute quantification (iTRAQ) and tandem mass tags (TMTs) use the tandem mass spectra of labels identical in mass and structure (also known as isobars/isotopomers) for the quantification of peptides/proteins. The labels comprise a reporter group, balance group, and an amine-reactive *N*-hydroxysuccinimide group, which derivatizes N-terminal amine and side-chain amines of lysine residues present in peptides/proteins. Differential isotopic composition of carbon, nitrogen, and oxygen atoms keeps the combined mass of the reporter and balance groups constant throughout the isobaric tags. Introduction of isobaric tags produces isotopically labeled peptides with identical mass and chromatographic properties. These peptides coelute in MS, but the reporter group dissociates as a signature ion with distinct *m*/*z* in the MS/MS mode, thus allowing for the identification and subsequent quantification of the parent peptides *(58,59)*.

The first iTRAQ reagents used a fourfold multiplex strategy (reporter ion *m*/*z* from 114 to 117) that allowed multiplexing of experimental samples, a property further enhanced by the development of eight isobaric variants (reporter ion *m*/*z* from 113 to 121) *(60)*. TMTs were initially developed as a 2-plex technique, and have been recently expanded to a 6-plex (reporter ion *m*/*z* from 126–131) and 8-plex version (with two variants of the TMT127 and TMT129 labels) for improved multiplexing ability *(61,62)*. Although iTRAQ 4-plex has been shown to have higher peptide identification rates over iTRAQ 8-plex as well as TMT 6-plex *(63)*, the ability of the latter to facilitate simultaneous analysis of a large number of samples cannot be overlooked, especially while performing high-throughput studies.

Because the iTRAQ/TMT-labeled peptides have the same nominal mass when analyzed by the first stage MS, there is lower mass spectral complexity as well as increased sensitivity. This allows for the identification of a greater number of ions by MS/MS *(58)*. In spite of these advantages, isobaric labeling suffers from certain disadvantages. Peptide identification and quantification require tandem mass spectra of peptides, which are usually fewer in number as compared to the total number of spectra generated during mass spectrometry. Thus, not all peptides that are seen on MS can be quantified. Furthermore, the iTRAQ/TMT reagents are expensive and difficult to synthesize.

There are several examples of peptidomic studies using iTRAQ labeling. One study identified substrates of prolyl peptidases in mouse models, and another explored the effect of diurnal variation on the endogenous peptide composition of human parotid saliva *(64,65)*.

Recently, mTRAQ (MRM tags for relative and absolute quantification) was developed by Applied Biosystems as a proteomic tool to be used specifically in conjunction with MRM. Derived from iTRAQ reagents, these amine-reactive nonisobaric compounds were initially synthesized as light and heavy versions, differing only in their composition of ^{13}C and ^{15}N atoms. Peptides labeled with the light form (mass 141 Da) and heavy form (mass 145 Da, identical to iTRAQ 117) differ by 4 Da. This generates unique MRM transitions (precursor/product pair) for the same peptide labeled with the two mTRAQ reagents, which causes labeled peptide pairs to be monitored independently within a complex mixture. MRM combined with mTRAQ allows for detection of peptides present in low amounts in complex mixtures while avoiding the use of deuterated standards for absolute quantitation. This method has been successfully applied for absolute quantitation in proteomic studies by using one label to tag the synthetic peptide standard and the other to label the protein mixture *(66)*. mTRAQ has also been used for the relative quantification of proteins, particularly biomarkers *(67–69)*. Recent development of a triplex version of mTRAQ (mass 149 Da) permits multiplexing of samples *(70)*. Though mTRAQ has been used to validate methods for enrichment of endogenous peptides in human plasma and their identification *(71)*, it has not been widely used for peptidomics.

3.3.3 TMAB

The trimethylammonium butyrate (TMAB) isotopic tags, which are synthesized as amine-reactive *N*-hydroxysuccinimide derivatives, were first described by Regnier and coworkers *(72)*. The chemical name of the TMAB-*N*-hydroxysuccinimide reagent (TMAB-NHS) is [3-(2,5-dioxopyrrolidin-1-yloxycarbonyl)propyl] trimethylammonium chloride. The label incorporates a quaternary amine so that the labeled peptide retains a positive charge in place of the free amine (unlike acetylation and related labels described in the next section).

TMAB-NHS was initially synthesized in a light and heavy form, containing all hydrogens (D0) and nine deuteriums (D9), respectively. The multiplexing capability of TMAB labels was greatly improved upon synthesis of additional D3, D6, and ^{13}C, D9 (referred to as D12-TMAB) forms (Figure 2). Although TMAB-NHS labels are not commercially available, they are relatively easy to synthesize from inexpensive reagents *(73)*. Peptides labeled with TMAB coelute, presumably due to clustering of deuterium in a polar environment, which reduces its affinity for the reverse-phase column matrix and hence minimizes isotope effects *(72)*. The 3 Da mass difference

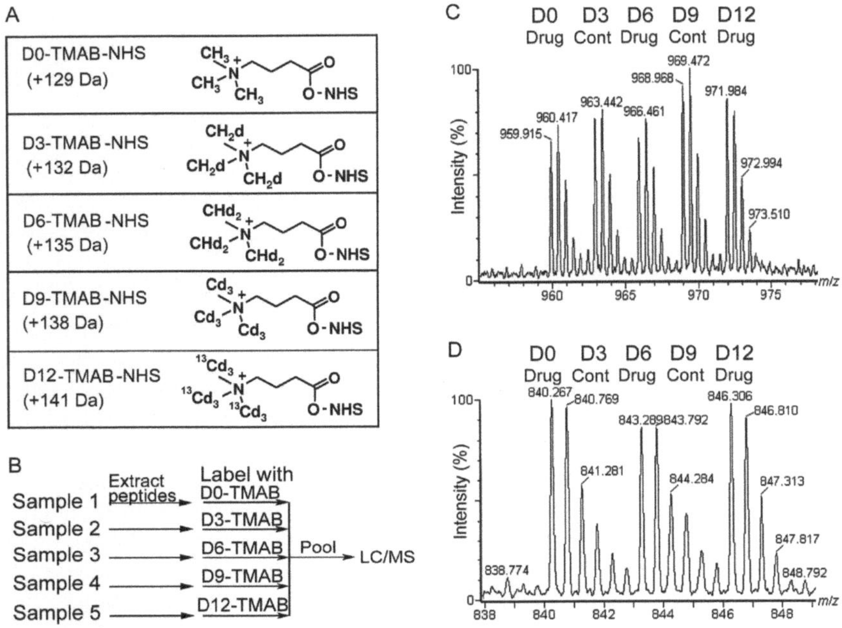

FIGURE 2 Quantitative peptidomics using trimethylammoniumbutyrate (TMAB) labels. (A) Chemical structures of the labeling reagents, which are reactive *N*-hydroxysuccinimide (NHS) esters that combine with free amines to form an amide bond. The methyl groups on the amine contain various ratios of hydrogen/deuterium (^{1}H, or ^{2}H) and ^{12}C/^{13}C. The form containing ^{12}C and only H was initially named H9-TMAB, and renamed D0-TMAB. The form with one deuterium on each of the three methyl groups is D3-TMAB, the form with two deuteriums on each methyl group is D6-TMAB, and the form with three deuteriums on each methyl group is D9-TMAB. The form with nine total deuteriums and the ^{12}C substituted with ^{13}C is named D12-TMAB because the mass is 12 Da heavier than the D0 form. (B) Up to five different samples can be labeled with the various TMAB tags, as shown. (C and D) Representative results from an experiment of HEK293T cells treated for 1 h with 50 nM bortezomib, a proteasome inhibitor used as a drug to treat multiple myeloma and other cancers. In this experiment, two replicates of control cells were labeled with D3 and D9-TMAB, while three replicates of drug-treated cells were labeled with D0, D6, and D12-TMAB. The peptide in panel C was not altered by bortezomib treatment; this peptide was subsequently identified by tandem MS analysis as Ac-AAKVFESIGKFGLALA (an N-terminal fragment of prohibitin) labeled with two TMAB tags. The peptide in panel D was greatly elevated by bortezomib treatment; this peptide was identified as Ac-AGQAFRKFLPLFD (an N-terminal fragment of heat shock 10 kDa protein 1, also known as chaperonin 10) labeled with one TMAB tag.

between the labeled peptides allows most to be sufficiently resolved on MS (Figure 2C), although some peptides are not adequately resolved and can be difficult to quantify (Figure 2D). This problem can be eliminated if only three tags are used: D0, D6, and D12, providing a 6 Da difference between each isotopic form.

Another disadvantage of using TMAB labels is that the isotopic tag disintegrates during collision-induced dissociation in MS/MS mode, causing neutral loss of trimethylamine *(74)*. This can complicate peptide identification if the investigator is not aware of this; however, if a high enough energy is used for collision-induced dissociation, the dissociation of the trimethylamine group is complete and so this is not a major problem for MS/MS identification. The Mascot computer program includes TMAB labels and neutral loss of trimethylamine in their list of modifications (under the name GIST) for the D0, D3, D6, and D9-TMAB tags (although not the D12-TMAB tag).

TMAB labels have been extensively used in peptidomics, especially for the study of neuropeptides. Using mice models deficient in endopeptidases such as prohormone convertase 1/3 (PC1/3) and carboxypeptidase E (CPE), TMAB peptidomics has been useful for exploring the role of these enzymes in neuropeptide biosynthesis as well as the identification and quantitation of several neuropeptides. Numerous peptides derived from secretory pathway proteins (classical neuropeptides) have been identified in the pituitary as well as brain regions of CPE-deficient ($Cpe^{fat/fat}$) and PC1/3-deficient mice *(75–79)*. Since neuropeptides function as cell–cell signaling molecules and regulate a number of biological processes, it is of interest to quantify their levels under specific physiological conditions. Quantitative peptidomics using TMAB labeling has been applied to measuring changes in relative peptide levels in the brain under conditions of food deprivation, exercise, and administration of drugs such as cocaine and morphine *(12,80,81)*.

Differential isotopic labeling with TMAB has been employed for the study of neuropeptides derived from cytosolic proteins, particularly hemopressin, which is a hemoglobin-derived peptide and a known CB1 cannabinoid receptor antagonist. Novel forms of hemopressin with CB1 agonist activity were discovered and their regulation in $Cpe^{fat/fat}$ mice was examined *(82,83)*. Apart from identifying these "nonclassical" neuropeptides *(28)*, TMAB-labeling-based peptidomic approaches have unveiled a possible third category of peptides, which elicit their biological effects within the cell in which they are produced. Many such "intracellular peptides" have been detected and the enzymatic pathway for their synthesis has been elucidated in human cell lines *(84–86)*. In addition, TMAB tags have been used to determine substrate specificity of enzymes *(16)*.

3.3.4 Amine Labeling by Other Reagents

A relatively inexpensive method of labeling peptides is to tag their free amines with isotopically light and heavy forms of commercially available reagents. One of the first reagents to be tested was acetic anhydride, which can be purchased with the normal isotopic distribution of atoms (i.e., containing mostly hydrogen) or with six atoms of deuterium (although only three are incorporated into the labeled peptide). An advantage of these labels is that they are relatively inexpensive and commercially available. A disadvantage is that the heavy and light forms of the labeled peptides generally do not completely coelute from the LC due to deuterium isotope effects, which can result in imprecise quantification *(14,87)*. Another problem is that free amines are turned into amides, and unless there is another positively charged amino acid within the peptide (Arg or His), the peptide will not be positively charged. Because most LC/MS is performed in the positive-ion mode, neutral peptides will not be detected.

Other amine-labeling reagents with similar problems are *N*-acetoxysuccinimide (available with three deuteriums) and propionic anhydride (available with five deuteriums). Both of these labels suffer from isotopic effect of deuterium on chromatographic retention times, similar to that observed with acetic anhydride *(72,88)*. In addition, the free amines of the N-termini and lysine side-chains are covered by the modifying reagent, rendering the peptide neutral if there is no internal arginine or histidine.

Free amines of peptides/proteins can also be labeled with succinic anhydride, available with natural distribution of atoms (i.e., mostly hydrogen and ^{12}C) or with four deuteriums such that succinylation causes a mass difference of 4 Da per tag incorporated into the peptide *(9,89,90)*. In addition, succinic acid with four ^{13}C atoms can be purchased and easily converted into the anhydride. The use of ^{13}C eliminates the problem with non-coelution observed with the deuterated and nondeuterated forms of acetylated peptides. However, the extra expense of the ^{13}C-labeled reagent is not necessary because the deuterated and nondeuterated forms of the succinylated peptides show fairly good coelution. Labeling with succinic anhydride labeling thus provides for more accurate quantitation than acetic anhydride *(72,91)*. Isotopic labeling by succinylation has been used for the identification and quantitation of N-linked glycoproteins as well. After proteolytic removal of nonglycosylated peptides, glycopeptides are isotopically labeled by succinic anhydride. Labeled peptides are then cleaved by peptide-*N*-glycosidase and analyzed by MS *(92)*. One of the main disadvantages of succinic anhydride labeling is that peptides without arginine or histidine residues go undetected in the positive-ion mode, as with acetylated peptides. Furthermore, succinylated peptides have a reduced signal intensity in the mass spectrum compared to acetylated peptides, possibly due to the increased number of acidic groups in the labeled peptide *(74)*.

Another strategy for stable isotope labeling of N-terminus and lysine side-chain amines is reductive amination by nondeuterated and deuterated formaldehyde. Dimethylation of peptides occurs by the formation of a Schiff base, followed by reduction with cyanoborohydride. Peptides thus labeled differ by a mass of 4 Da for each labeled pair. The labeled peak pairs show minimum isotope effects on reverse-phase separation and hence coelute *(93)*. Dimethylation using formaldehyde with the natural isotopic abundance (hydrogen and ^{12}C) versus a heavy form with two deuteriums and one atom of ^{13}C

results in a mass difference of 6 Da between the labeled peptides, which enhances resolution by MS *(94)*. This technique has been adapted for multiplexing in proteomic studies by combining light and heavy forms of formaldehyde with normal (hydrogen) versus deuterated cyanoborohydride *(95)*.

3.4 Proteolytic Labeling

Isotopic labeling of peptides at the carboxy-terminus by ^{18}O occurs during enzymatic digestion of the parent protein. Differential labeling of peptides is achieved by the action of proteases such as trypsin and endoprotease Glu-C, which catalyze the exchange of two ^{16}O for two ^{18}O atoms, thus causing a mass shift of 4 Da. Proteolytic labeling was first applied to MS-based comparative proteomics in 2001 by measuring the relative signal intensities of the labeled and unlabeled peak pairs from two serotypes of adenovirus *(96,97)*.

This technique confers significant advantages over metabolic and chemical labeling strategies; it is simple, inexpensive, free from side reactions, and applicable to virtually any type of sample. In spite of these advantages, proteolytic labeling suffers from variable incorporation of the isotopic label. Incomplete exchange of ^{18}O can occur, resulting in [^{16}O, ^{18}O] or both [^{18}O] oxygen atoms incorporated at the C-terminus. A complex isotopic distribution arising from the mixture of unlabeled and variably labeled peptides makes it difficult to correctly ascertain the $^{16}O/^{18}O$ ratios, thus complicating analysis *(98,99)*. Slow exchange of the second ^{18}O atom often leads to incomplete labeling, which can be circumvented by carrying out ^{18}O incorporation under acidic pH *(100)*. The isotopic label can be lost by acid-catalyzed nonenzymatic back exchange when the two protein digests are mixed together for analysis, though this is not observed under mildly acidic conditions typically employed for LC runs during mass spectrometry *(101)*. Protease-catalyzed back exchange can also lead to loss of the label *(102)*, but using a Lys-N enzyme for protein digestion, which introduces a single ^{18}O at the C-terminus and hence does not alter $^{16}O/^{18}O$ ratios, significantly addresses this issue *(103)*. Incorporation of a single ^{18}O atom also eliminates variable incorporation of the label, but the resultant mass shift of only 2 Da can go undetected if the mass spectrometer is not adequately high resolution.

A recently developed method for ^{18}O-based quantification of peptides uses catalysis with hydrochloric acid instead of standard proteases for labeling. Unlike proteolytic labeling, acid catalysis labels all carboxyl groups present in a peptide (Asp, Glu, carboxymethylated Cys, C-terminal carboxyl groups). Thus, most peptides labeled by this strategy have a mass difference >4 Da, making data analysis simpler and more reliable than conventional proteolytic labeling approaches *(104)*.

Proteolytic labeling has not yet been applied to peptidomic analysis, which is carried out in the absence of enzyme digestion. However, sample preparation under ^{18}O-enriched conditions, allowing for the endogenous peptide pool to incorporate the isotopic label during the general turnover process, can potentially answer critical questions regarding stability of the peptides predominant in the peptidome.

4 CONCLUDING REMARKS

A large variety of approaches have been developed for quantification of proteins, and some of these have been developed for analysis of peptides. Each of the methods has advantages and disadvantages, and as a result the choice of approach depends largely on the experimental goals and details. Remarkable progress has been made over the past decade in the development of reagents and mass spectrometry instruments. It is now possible to accurately measure the levels of peptides present in small amounts of tissue or cells, and to identify many of these peptides from MS/MS spectra. In looking back at techniques used several decades ago, what can now be done in a few days would have taken decades of work from large teams of scientists, first to purify each peptide and then to determine the sequence. Several Nobel prizes were awarded for discoveries of bioactive peptides (insulin, vasopressin, thyrotropin-releasing hormone, gonadotropin-releasing hormone, and others), based on the importance of the discovery as well as the many years of work involved with the isolation and sequencing of each of these peptides. With current methods, it would have taken days or weeks to identify the peptides. However, it still takes considerable time to learn the function of peptides found by peptidomic approaches. The use of quantitative methods to learn which peptides are regulated by particular treatments can help focus further studies to those peptides that are highly regulated, and therefore likely to play important biological roles. Thus, quantitative peptidomics is an advance over nonquantitative peptidomic methods.

REFERENCES

1. Clynen, E.; Baggerman, G.; Veelaert, D.; Cerstiaens, A.; Van der Horst, D.; Harthoorn, L.; Derua, R.; Waelkens, E.; De Loof, A.; Schoofs, L. *Eur. J. Biochem.* **2001**, *268*, 1929–1939.
2. Verhaert, P.; Uttenweiler-Joseph, S.; de Vries, M.; Loboda, A.; Ens, W.; Standing, K. G. *Proteomics* **2001**, *1*, 118–131.
3. Brockmann, A.; Annangudi, S. P.; Richmond, T. A.; Ament, S. A.; Xie, F.; Southey, B. R.; Rodriguez-Zas, S. R.; Robinson, G. E.; Sweedler, J. V. *Proc. Natl. Acad. Sci. U.S.A.* **2009**, *106*, 2383–2388.
4. Lee, J. E.; Zamdborg, L.; Southey, B. R.; Atkins, N., Jr.; Mitchell, J. W.; Li, M.; Gillette, M. U.; Kelleher, N. L.; Sweedler, J. V. *J. Proteome Res.* **2013**, *12*, 585–593.
5. Furukawa, Y.; Nakamaru, K.; Wakayama, H.; Fujisawa, Y.; Minakata, H.; Ohta, S.; Morishita, F.; Matsushima, O.; Li, L.; Romanova, E.; Sweedler, J. V.; Park, J. H.; Romero, A.; Cropper, E. C.; Dembrow, N. C.; Jing, J.; Weiss, K. R.; Vilim, F. S. *J. Neurosci.* **2001**, *21*, 8247–8261.

6. Sweedler, J. V.; Li, L.; Rubakhin, S. S.; Alexeeva, V.; Dembrow, N. C.; Dowling, O.; Jing, J.; Weiss, K. R.; Vilim, F. S. *J. Neurosci.* **2002**, *22*, 7797–7808.
7. Hui, L.; Zhang, Y.; Wang, J.; Cook, A.; Ye, H.; Nusbaum, M. P.; Li, L. *ACS Chem. Neurosci.* **2011**, *2*, 711–722.
8. Bernay, B.; Gaillard, M. C.; Guryca, V.; Emadali, A.; Kuhn, L.; Bertrand, A.; Detraz, I.; Carcenac, C.; Savasta, M.; Brouillet, E.; Garin, J.; Elalouf, J. M. *Mol. Cell. Proteomics* **2009**, *8*, 946–958.
9. Che, F. Y.; Eipper, B. A.; Mains, R. E.; Fricker, L. D. *Cell. Mol. Biol. (Noisy-le-grand)* **2003**, *49*, 713–722.
10. Svensson, M.; Skold, K.; Svenningsson, P.; Andren, P. E. *J. Proteome Res.* **2003**, *2*, 213–219.
11. Rubakhin, S. S.; Greenough, W. T.; Sweedler, J. V. *Anal. Chem.* **2003**, *75*, 5374–5380.
12. Che, F. Y.; Vathy, I.; Fricker, L. D. *J. Mol. Neurosci.* **2006**, *28*, 265–275.
13. Gelman, J. S.; Wardman, J.; Bhat, V. B.; Gozzo, F. C.; Fricker, L. D. *Methods Mol. Biol.* **2012**, *829*, 487–503.
14. Che, F. Y.; Fricker, L. D. *Anal. Chem.* **2002**, *74*, 3190–3198.
15. Lone, A. M.; Kim, Y. G.; Saghatelian, A. *Curr. Opin. Chem. Biol.* **2013**, *17*, 83–89.
16. Lyons, P. J.; Fricker, L. D. *J. Biol. Chem.* **2010**, *285*, 38234–38242.
17. Tinoco, A. D.; Saghatelian, A. *Biochemistry* **2011**, *50*, 7447–7461.
18. He, X.; Yang, S.; Wei, L.; Liu, R.; Lai, R.; Rong, M. *Amino Acids* **2013**, *44*, 481–487.
19. Osaki, T.; Sasaki, K.; Minamino, N. *J. Proteome Res.* **2011**, *10*, 1870–1880.
20. Bassani-Sternberg, M.; Barnea, E.; Beer, I.; Avivi, I.; Katz, T.; Admon, A. *Proc. Natl. Acad. Sci. U.S.A.* **2010**, *107*, 18769–18776.
21. Lu, K.; Knutson, C. G.; Wishnok, J. S.; Fox, J. G.; Tannenbaum, S. R. *J. Proteome Res.* **2012**, *11*, 4916–4926.
22. Xiang, Y.; Kurokawa, M. S.; Kanke, M.; Takakuwa, Y.; Kato, T. *Methods Mol. Biol.* **2010**, *615*, 259–271.
23. Clynen, E.; De Loof, A.; Schoofs, L. *Gen. Comp. Endocrinol.* **2003**, *132*, 1–9.
24. Fricker, L. D.; Lim, J.; Pan, H.; Che, F. Y. *Mass Spectrom. Rev.* **2006**, *25*, 327–344.
25. Menschaert, G.; Vandekerckhove, T. T.; Baggerman, G.; Schoofs, L.; Luyten, W.; Van Criekinge, W. *J. Proteome Res.* **2010**, *9*, 2051–2061.
26. Cunha, F. M.; Berti, D. A.; Ferreira, Z. S.; Klitzke, C. F.; Markus, R. P.; Ferro, E. S. *J. Biol. Chem.* **2008**, *283*, 24448–24459.
27. Ferro, E. S.; Hyslop, S.; Camargo, A. C. *J. Neurochem.* **2004**, *91*, 769–777.
28. Gelman, J. S.; Fricker, L. D. *AAPS J.* **2010**, *12*, 279–289.
29. Mitchell, J. W.; Atkins, N., Jr.; Sweedler, J. V.; Gillette, M. U. *Front. Neuroendocrinol.* **2011**, *32*, 377–386.
30. Che, F. Y.; Lim, J.; Pan, H.; Biswas, R.; Fricker, L. D. *Mol. Cell. Proteomics* **2005**, *4*, 1391–1405.
31. Che, F. Y.; Yan, L.; Li, H.; Mzhavia, N.; Devi, L. A.; Fricker, L. D. *Proc. Natl. Acad. Sci. U.S.A.* **2001**, *98*, 9971–9976.
32. Ciccimaro, E.; Blair, I. A. *Bioanalysis* **2010**, *2*, 311–341.
33. Picotti, P.; Aebersold, R. *Nat. Methods* **2012**, *9*, 555–566.
34. Gerber, S. A.; Rush, J.; Stemman, O.; Kirschner, M. W.; Gygi, S. P. *Proc. Natl. Acad. Sci. U.S.A.* **2003**, *100*, 6940–6945.
35. Kirkpatrick, D. S.; Gerber, S. A.; Gygi, S. P. *Methods* **2005**, *35*, 265–273.
36. Zhu, W.; Smith, J. W.; Huang, C. M. *J. Biomed. Biotechnol.* **2010**, *2010*, 840518.
37. Benk, A. S.; Roesli, C. *Anal. Bioanal. Chem.* **2012**, *404*, 1039–1056.

38. Old, W. M.; Meyer-Arendt, K.; Aveline-Wolf, L.; Pierce, K. G.; Mendoza, A.; Sevinsky, J. R.; Resing, K. A.; Ahn, N. G. *Mol. Cell. Proteomics* **2005**, *4*, 1487–1502.
39. Tagore, D. M.; Nolte, W. M.; Neveu, J. M.; Rangel, R.; Guzman-Rojas, L.; Pasqualini, R.; Arap, W.; Lane, W. S.; Saghatelian, A. *Nat. Chem. Biol.* **2009**, *5*, 23–25.
40. Lundgren, D. H.; Hwang, S. I.; Wu, L.; Han, D. K. *Expert Rev. Proteomics* **2010**, *7*, 39–53.
41. Tinoco, A. D.; Tagore, D. M.; Saghatelian, A. *J. Am. Chem. Soc.* **2010**, *132*, 3819–3830.
42. Bantscheff, M.; Schirle, M.; Sweetman, G.; Rick, J.; Kuster, B. *Anal. Bioanal. Chem.* **2007**, *389*, 1017–1031.
43. Oda, Y.; Huang, K.; Cross, F. R.; Cowburn, D.; Chait, B. T. *Proc. Natl. Acad. Sci. U.S.A.* **1999**, *96*, 6591–6596.
44. Mann, M. *Nat. Rev. Mol. Cell Biol.* **2006**, *7*, 952–958.
45. Ong, S. E.; Blagoev, B.; Kratchmarova, I.; Kristensen, D. B.; Steen, H.; Pandey, A.; Mann, M. *Mol. Cell. Proteomics* **2002**, *1*, 376–386.
46. Van Hoof, D.; Pinkse, M. W.; Oostwaard, D. W.; Mummery, C. L.; Heck, A. J.; Krijgsveld, J. *Nat. Methods* **2007**, *4*, 677–678.
47. Frohlich, F.; Christiano, R.; Walther, T. C. *Mol. Cell. Proteomics* **2013**, *12*, 1995–2005.
48. McClatchy, D. B.; Dong, M. Q.; Wu, C. C.; Venable, J. D.; Yates, J. R., 3rd. *J. Proteome Res.* **2007**, *6*, 2005–2010.
49. Price, J. C.; Guan, S.; Burlingame, A.; Prusiner, S. B.; Ghaemmaghami, S. *Proc. Natl. Acad. Sci. U.S.A.* **2010**, *107*, 14508–14513.
50. Che, F. Y.; Yuan, Q.; Kalinina, E.; Fricker, L. D. *J. Neurochem.* **2004**, *90*, 585–594.
51. Gygi, S. P.; Rist, B.; Gerber, S. A.; Turecek, F.; Gelb, M. H.; Aebersold, R. *Nat. Biotechnol.* **1999**, *17*, 994–999.
52. Hansen, K. C.; Schmitt-Ulms, G.; Chalkley, R. J.; Hirsch, J.; Baldwin, M. A.; Burlingame, A. L. *Mol. Cell. Proteomics* **2003**, *2*, 299–314.
53. Goshe, M. B.; Veenstra, T. D.; Panisko, E. A.; Conrads, T. P.; Angell, N. H.; Smith, R. D. *Anal. Chem.* **2002**, *74*, 607–616.
54. Kaji, H.; Saito, H.; Yamauchi, Y.; Shinkawa, T.; Taoka, M.; Hirabayashi, J.; Kasai, K.; Takahashi, N.; Isobe, T. *Nat. Biotechnol.* **2003**, *21*, 667–672.
55. Ahrends, R.; Pieper, S.; Kuhn, A.; Weisshoff, H.; Hamester, M.; Lindemann, T.; Scheler, C.; Lehmann, K.; Taubner, K.; Linscheid, M. W. *Mol. Cell. Proteomics* **2007**, *6*, 1907–1916.
56. Schwarz, G.; Beck, S.; Weller, M. G.; Linscheid, M. W. *Anal. Bioanal. Chem.* **2011**, *401*, 1203–1209.
57. Rivera-Monroy, Z.; Bonn, G. K.; Guttman, A. *Bioorg. Chem.* **2008**, *36*, 299–311.
58. Ross, P. L.; Huang, Y. N.; Marchese, J. N.; Williamson, B.; Parker, K.; Hattan, S.; Khainovski, N.; Pillai, S.; Dey, S.; Daniels, S.; Purkayastha, S.; Juhasz, P.; Martin, S.; Bartlet-Jones, M.; He, F.; Jacobson, A.; Pappin, D. J. *Mol. Cell. Proteomics* **2004**, *3*, 1154–1169.
59. Thompson, A.; Schafer, J.; Kuhn, K.; Kienle, S.; Schwarz, J.; Schmidt, G.; Neumann, T.; Johnstone, R.; Mohammed, A. K.; Hamon, C. *Anal. Chem.* **2003**, *75*, 1895–1904.
60. Choe, L.; D'Ascenzo, M.; Relkin, N. R.; Pappin, D.; Ross, P.; Williamson, B.; Guertin, S.; Pribil, P.; Lee, K. H. *Proteomics* **2007**, *7*, 3651–3660.
61. Dayon, L.; Hainard, A.; Licker, V.; Turck, N.; Kuhn, K.; Hochstrasser, D. F.; Burkhard, P. R.; Sanchez, J. C. *Anal. Chem.* **2008**, *80*, 2921–2931.
62. Werner, T.; Becher, I.; Sweetman, G.; Doce, C.; Savitski, M. M.; Bantscheff, M. *Anal. Chem.* **2012**, *84*, 7188–7194.
63. Pichler, P.; Kocher, T.; Holzmann, J.; Mazanek, M.; Taus, T.; Ammerer, G.; Mechtler, K. *Anal. Chem.* **2010**, *82*, 6549–6558.

64. Hardt, M.; Witkowska, H. E.; Webb, S.; Thomas, L. R.; Dixon, S. E.; Hall, S. C.; Fisher, S. J. *Anal. Chem.* **2005**, *77*, 4947–4954.
65. Lone, A. M.; Nolte, W. M.; Tinoco, A. D.; Saghatelian, A. *AAPS J.* **2010**, *12*, 483–491.
66. DeSouza, L. V.; Taylor, A. M.; Li, W.; Minkoff, M. S.; Romaschin, A. D.; Colgan, T. J.; Siu, K. W. *J. Proteome Res.* **2008**, *7*, 3525–3534.
67. DeSouza, L. V.; Krakovska, O.; Darfler, M. M.; Krizman, D. B.; Romaschin, A. D.; Colgan, T. J.; Siu, K. W. *Proteomics* **2010**, *10*, 3108–3116.
68. Kang, U. B.; Yeom, J.; Kim, H.; Lee, C. *J. Proteome Res.* **2010**, *9*, 3750–3758.
69. Suh, E. J.; Kabir, M. H.; Kang, U. B.; Lee, J. W.; Yu, J.; Noh, D. Y.; Lee, C. *Exp. Mol. Med.* **2012**, *44*, 36–44.
70. Yoon, J. Y.; Yeom, J.; Lee, H.; Kim, K.; Na, S.; Park, K.; Paek, E.; Lee, C. *BMC Bioinforma.* **2011**, *12*(Suppl 1), S46.
71. Potier, D. N.; Griffiths, J. R.; Unwin, R. D.; Walker, M. J.; Carrick, E.; Willamson, A. J.; Whetton, A. D. *Anal. Chem.* **2012**, *84*, 5604–5610.
72. Zhang, R.; Sioma, C. S.; Thompson, R. A.; Xiong, L.; Regnier, F. E. *Anal. Chem.* **2002**, *74*, 3662–3669.
73. Morano, C.; Zhang, X.; Fricker, L. D. *Anal. Chem.* **2008**, *80*, 9298–9309.
74. Che, F. Y.; Fricker, L. D. *J. Mass Spectrom.* **2005**, *40*, 238–249.
75. Che, F. Y.; Biswas, R.; Fricker, L. D. *J. Mass Spectrom.* **2005**, *40*, 227–237.
76. Lim, J.; Berezniuk, I.; Che, F. Y.; Parikh, R.; Biswas, R.; Pan, H.; Fricker, L. D. *J. Neurochem.* **2006**, *96*, 1169–1181.
77. Pan, H.; Nanno, D.; Che, F. Y.; Zhu, X.; Salton, S. R.; Steiner, D. F.; Fricker, L. D.; Devi, L. A. *Biochemistry* **2005**, *44*, 4939–4948.
78. Wardman, J. H.; Zhang, X.; Gagnon, S.; Castro, L. M.; Zhu, X.; Steiner, D. F.; Day, R.; Fricker, L. D. *J. Neurochem.* **2010**, *114*, 215–225.
79. Zhang, X.; Che, F. Y.; Berezniuk, I.; Sonmez, K.; Toll, L.; Fricker, L. D. *J. Neurochem.* **2008**, *107*, 1596–1613.
80. Che, F. Y.; Yuan, Q.; Kalinina, E.; Fricker, L. D. *J. Biol. Chem.* **2005**, *280*, 4451–4461.
81. Decaillot, F. M.; Che, F. Y.; Fricker, L. D.; Devi, L. A. *J. Mol. Neurosci.* **2006**, *28*, 277–284.
82. Gelman, J. S.; Sironi, J.; Castro, L. M.; Ferro, E. S.; Fricker, L. D. *J. Neurochem.* **2010**, *113*, 871–880.
83. Gomes, I.; Grushko, J. S.; Golebiewska, U.; Hoogendoorn, S.; Gupta, A.; Heimann, A. S.; Ferro, E. S.; Scarlata, S.; Fricker, L. D.; Devi, L. A. *FASEB J.* **2009**, *23*, 3020–3029.
84. Fricker, L. D.; Gelman, J. S.; Castro, L. M.; Gozzo, F. C.; Ferro, E. S. *J. Proteome Res.* **2012**, *11*, 1981–1990.
85. Gelman, J. S.; Sironi, J.; Berezniuk, I.; Dasgupta, S.; Castro, L. M.; Gozzo, F. C.; Ferro, E. S.; Fricker, L. D. *PLoS One* **2013**, *8*, e53263.
86. Gelman, J. S.; Sironi, J.; Castro, L. M.; Ferro, E. S.; Fricker, L. D. *J. Proteome Res.* **2011**, *10*, 1583–1592.
87. Ji, J.; Chakraborty, A.; Geng, M.; Zhang, X.; Amini, A.; Bina, M.; Regnier, F. *J. Chromatogr. B Biomed. Sci. Appl.* **2000**, *745*, 197–210.
88. Zhang, X.; Jin, Q. K.; Carr, S. A.; Annan, R. S. *Rapid Commun. Mass Spectrom.* **2002**, *16*, 2325–2332.
89. Wang, S.; Regnier, F. E. *J. Chromatogr. A* **2001**, *924*, 345–357.
90. Wang, S.; Zhang, X.; Regnier, F. E. *J. Chromatogr. A* **2002**, *949*, 153–162.
91. Zhang, R.; Sioma, C. S.; Wang, S.; Regnier, F. E. *Anal. Chem.* **2001**, *73*, 5142–5149.
92. Zhang, H.; Li, X. J.; Martin, D. B.; Aebersold, R. *Nat. Biotechnol.* **2003**, *21*, 660–666.
93. Hsu, J. L.; Huang, S. Y.; Chow, N. H.; Chen, S. H. *Anal. Chem.* **2003**, *75*, 6843–6852.

94. Ji, C.; Guo, N.; Li, L. *J. Proteome Res.* **2005**, *4*, 2099–2108.
95. Hsu, J. L.; Huang, S. Y.; Chen, S. H. *Electrophoresis* **2006**, *27*, 3652–3660.
96. Reynolds, K. J.; Yao, X.; Fenselau, C. *J. Proteome Res.* **2002**, *1*, 27–33.
97. Yao, X.; Freas, A.; Ramirez, J.; Demirev, P. A.; Fenselau, C. *Anal. Chem.* **2001**, *73*, 2836–2842.
98. Miyagi, M.; Rao, K. C. *Mass Spectrom. Rev.* **2007**, *26*, 121–136.
99. Ye, X.; Luke, B.; Andresson, T.; Blonder, J. *Brief Funct. Genomic. Proteomic.* **2009**, *8*, 136–144.
100. Hajkova, D.; Rao, K. C.; Miyagi, M. *J. Proteome Res.* **2006**, *5*, 1667–1673.
101. Schnolzer, M.; Jedrzejewski, P.; Lehmann, W. D. *Electrophoresis* **1996**, *17*, 945–953.
102. Storms, H. F.; van der Heijden, R.; Tjaden, U. R.; van der Greef, J. *Rapid Commun. Mass Spectrom.* **2006**, *20*, 3491–3497.
103. Rao, K. C.; Carruth, R. T.; Miyagi, M. *J. Proteome Res.* **2005**, *4*, 507–514.
104. Haaf, E.; Schlosser, A. *Anal. Chem.* **2012**, *84*, 304–311.

Chapter 14

Analysis of Deep Sequencing Data: Insights and Challenges

Jacob W. Malcom and John H. Malone
Department of Molecular and Cell Biology, University of Connecticut, Storrs, Connecticut, USA

Chapter Outline

1 INTRODUCTION

Sequencing the entire complement of hereditary information from any organism provides data that are fundamental to understanding evolution, human disease, and biological life. The large size of genomes (kilobase to hundreds of gigabases in size) creates data analysis issues and challenges that are common to contemporary "big data" problems but some that are unique to genomics. How can one find a small gene of interest embedded in billions of data points? How does one load and visualize the data when there are hundreds of millions of sequence reads? What are resources for data summary, exploratory, and confirmatory analyses? When one has so much data, how do you avoid finding spurious significant effects?

Comprehensive Analytical Chemistry, Vol. 63. http://dx.doi.org/10.1016/B978-0-444-62651-6.00015-5

Analysis of deep sequencing data is an extremely active area of research and there are now a large number of data analysis tools and software packages available both for desktop computers and for large distributed computing clusters. This chapter does not compare and contrast the advantages and disadvantages of software tools for all deep sequencing applications. Rather, the purpose of this chapter is to review and discuss first principles of data analysis and strategies for working with deep sequencing data, and to point out areas that should be given careful consideration. Most of the discussion is aimed at investigators that are beginning to use deep sequencing in their research, and it will be heavily biased to the Illumina platform.

Deep sequencing data consist of many millions, and now oftentimes billions, of short (50–200 base pairs (bp)) DNA sequences termed "reads" that are generated by a new class of automated DNA sequencers called "whole-genome" sequencers. Whole-genome sequencers analyze massively parallel sequencing reactions that produce many reads covering each position in a genome. The sheer number of reads from an experiment creates redundancy such that even large genomes can be sequenced by single laboratories in a reasonable time frame. The abundance of recent technological advances is a truly transformative change in obtaining a DNA sequence compared to just 10 years ago, when genomes required consortia of investigators and significant financial resources.

How to approach the analysis of billions of short reads of sequence data will vary depending on the question, experimental design, and application of the investigator. Common to any deep sequencing research, however, is the need to first understand the nature and quality of the data that are the basis of all downstream analyses and inferences.

2 FUNDAMENTALS

The basic methodological strategy in most deep sequencing experiments is to fractionate the genome or transcriptome into small fragments that can be subjected to amplification on a flow cell. The fragment size depends on the sequencing instrument and application, but a typical fragment size consists of 100–300 bp. Current whole-genome sequencers produce reads 100–200 bp long and can sequence one or both ends of the DNA fragment. For larger fragments, this means that reads will not overlap in sequence.

Whole-genome sequencers use cameras to image fluorescence signals of carefully controlled terminator reactions. The raw data consist of image files, and the imaging analysis combined with the massively parallel nature of the sequencing reactions creates a problem of data management and storage. In the early days of whole-genome sequencers, when the output was tens of millions of reads comprising hundreds of gigabytes of data, it was common practice to save the raw data; researchers quickly realized that saving the raw data would not scale with increasing sequencer output.

Furthermore, no data repository will accept raw image files for public distribution. The lack of access to raw data has created the uncomfortable situation that output of whole-genome sequencers exceeds the cost of data storage: *reanalyses of raw data are no longer realistic or straightforward because the cost of storing the data exceeds the cost to resequence the same sample.* Whole-genome sequencer manufacturers have proprietary data processing algorithms to transform the raw images into DNA sequence data, and the reads reported by a core facility or companies will usually include only the high-quality read set. At this writing, all reads can be obtained by parsing the export file from Illumina, but that standard may change as the field advances.

A standard data format useful for a variety of downstream analyses and applications is the FASTQ format (Figure 1A). A FASTQ record consists of four lines for each sequence. The first line contains a header that begins with an @ symbol followed by an identifier that represents the sequencer, lane number of the flow cell, positional information for the image and cluster on the flow cell, sample identifier, and read number. The second line is a string of DNA sequence. The third line begins with a +symbol that may be followed by the same information as line 1. The fourth line contains the ASCII-encoded quality score values for each nucleotide base. Phred scaling is a direct transformation of the probability that the base is not correctly called ($=-10 \log_{10} P$), for example, Phred=30 translates to a probability of 0.001. The total number of reads available for analysis is the number of lines in the FASTQ file divided by 4.

The scale of the base quality scores is an issue that can lead to confusion and incorrect analyses (Figure 1B). Different instruments and different versions of the same instrument have different scales for the quality score output; the most commonly reported scores are Phred +33 or Phred +64. When considering quality thresholds for particular bases, it is important to know which version of software was used to generate the sequence because incorrect translation of quality score can change inferences (see Section 2.1) or cause data processing software to fail. In the case of the Illumina, the scale of the quality score has changed frequently, and use of legacy data requires knowing what version of processing software was used to generate the data so that proper assessment of quality score thresholds can be considered (Figure 1B). Incorporating legacy data from other platforms, such as the Applied Biosystems SOLiD System, increases the complexity of the quality score issue but is manageable if recognized early in analysis.

2.1 Quality Control

Short reads from several genome sequencers exhibit a systematic decrease in quality from the 5′ to 3′ end (Figure 1C) *(1)*. The severity of this problem can

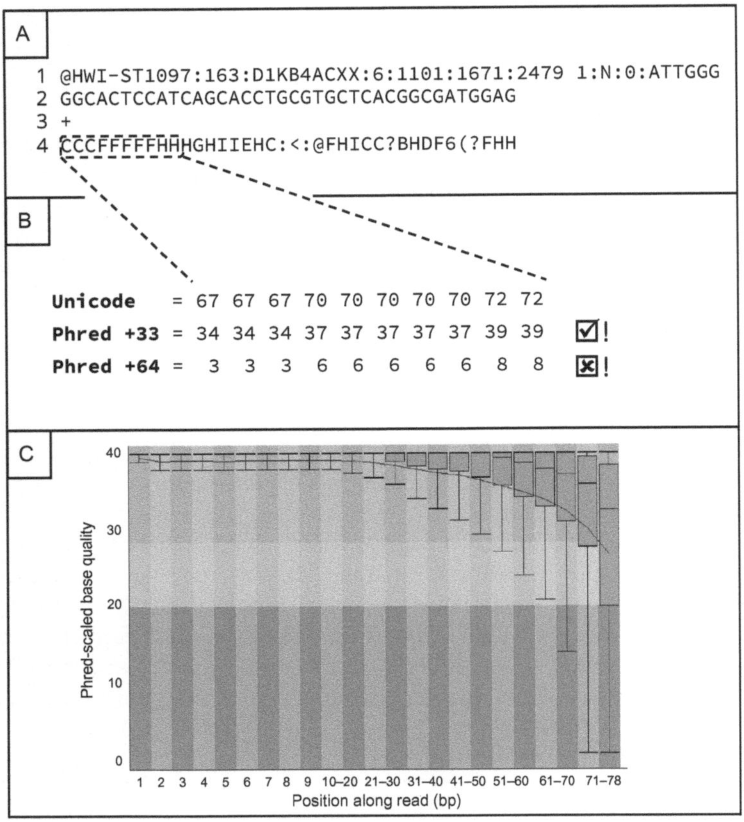

FIGURE 1 Basics of deep sequencing data. (A) FASTQ is the most common sequence format and, as represented here, consists of four lines: (1) the read information line; (2) the read sequence, in this case trimmed to 36 bp; (3) a null line with no additional information; and (4) the base-call quality scores in ASCII code points. (B) Translation of the first 10 quality scores of the read in (A). Quality scores, which are Phred-scaled p-values, range from 1 to 60 but are represented by a single ASCII character (top line). As discussed in the text, the researcher needs to know the quality offset used by the sequencing platform that generated the data (+33 in this case, middle line) because if the incorrect offset is used the quality scores are wildly incorrect (+64, bottom line). (C) A common characteristic of whole-genome sequencing reads is a systematic decline in quality score from 5′ to 3′ along the read. This panel was generated in FastQC, which is discussed in the text. (For color version of this figure, the reader is referred to the online version of this chapter.)

vary between experiments or between lanes within an experiment, and it is important to assess the extent of this problem and make decisions about how to manage this bias. Low-quality bases can affect many types of downstream analyses, particularly genome assembly, the detection of single-nucleotide polymorphisms (SNPs), distributing RNA-Seq reads among genes, and identifying allele-specific expression and cases of RNA editing.

One of the most basic analyses of sequencing data quality involves examining base-call quality scores along the length of sequence reads. All reads from a lane of sequencing are gathered and then a measure of central tendency (mean or median) of base quality scores is calculated for each position across the set of reads. Boxplots representing mean quality and variation around the mean are good for visualizing the problem (Figure 1C). Typical output shows a decrease in base quality toward the 3′ end, and the problem becomes more severe as read lengths increase. Several software packages exist to visualize this 5′ to 3′ bias problem. SolexaQA *(2)* is a Perl package that runs easily on *NIX (e.g., Linux or Mac) architectures and FastQC (http://www.bioinformatics.babraham.ac.uk/projects/fastqc/) is a Java tool that can run on desktop machines with any operating system running the Java Virtual Machine. These packages will generate plots of image tile quality, read quality scores, and base composition metrics that can help identify problems in sequencing data.

Quality assessment plots are critical to deciding whether the data, and which portion of reads, are useable for downstream analysis. A common practice for dealing with the 5′–3′ quality decline of reads is to trim them to a length that removes the lowest-quality bases. An alternative approach involves dynamically trimming reads, which produces reads with a variety of lengths; additional information can be garnered from reads with more high-quality bases while shorter reads are less informative but also less likely to propagate errors. Trimming all reads to a common length has some statistical advantages and is the favored approach because the effects of dynamic trimming on mapping probability estimates are unclear (see the next Section 2.2). The FASTX-Toolkit (http://hannonlab.cshl.edu/fastx_toolkit/index.html) performs a variety of critical assessments and ways to trim sequence. In addition to read-trimming to remove average low-quality base calls, several probabilistic methods exist to remove substitution errors before downstream analysis. The program Quake *(3)* provides a likelihood method for detecting and correcting sequencing errors in deep sequencing data.

2.2 Mapping

The general problem for any deep sequencing experiment is to determine from where in the genome each of millions or billions of short reads was generated (Figure 2A). This can be imagined as a "Where's Waldo?" problem (http://whereswaldo.com) in which millions of slightly different Waldos must be found among Earth's human population of >6 billion people. Furthermore, short read length, the 5′ to 3′ bias in sequence quality, and the presence of large repetitive sequences or small sequence duplications in the reference genome all create challenges for successful mapping. Fortunately, because the probability of a sequence match increases exponentially with the length of the sequence, relatively few matched bases can provide high confidence in mapping.

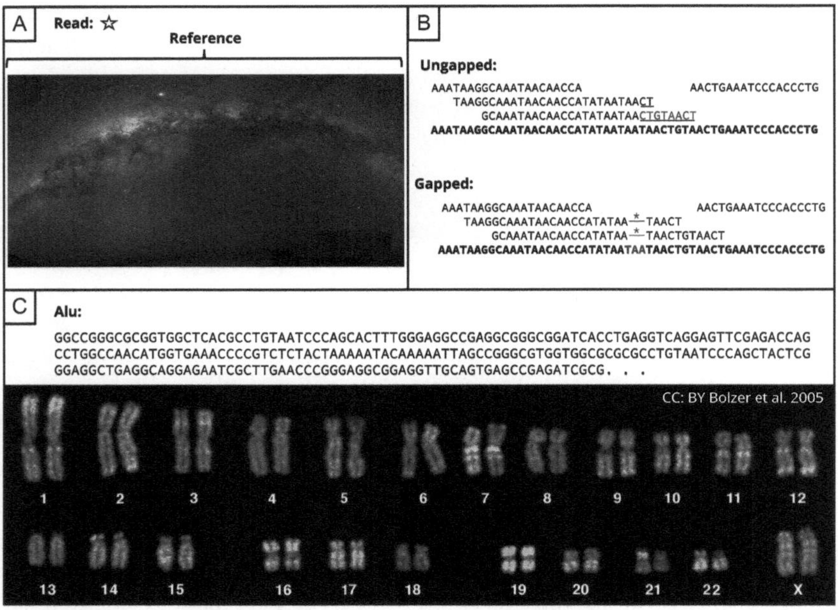

FIGURE 2 Conceptual representation of read mapping. (A) Mapping deep sequencing data to a genome is computationally challenging because it requires finding the origin of each of millions of short reads in a genome of hundreds of millions to several billion bases. This is analogous to finding the coordinates of millions of stars, based on just a handful of characteristics, from among the billions of stars in the Milky Way (image from the European Southern Observatory, CC-BY). (B) Mapping reads without accounting for insertions and deletions (i.e., ungapped alignment) can result in the appearance of several incorrect polymorphisms even with multiple read alignment. In this low-coverage example, ungapped, mapped reads stacked above the bold reference sequence suggest several polymorphisms (red, underlined) when the gapped mappings correctly identify a single three-base deletion (blue, asterisk). (C) Reads that map to multiple locations in the reference genome are "multimapped reads." The *Alu* retrotransposon (above, a portion of the canonical *Alu* sequence) comprises 10% of the human genome and is highlighted by green fluorescence in this fluorescent *in situ* hybridization (FISH) image (below; from Bolzer *et al.*). Determining from where in the genome a read with the *Alu* sequence was generated is an extreme example of the challenge posed by multimapped reads. (For interpretation of the references to color in this figure legend, the reader is referred to the online version of this chapter.)

If a genome sequence exists for the study species, then it can be used as a reference for mapping the short reads produced in a deep sequencing experiment. If no reference is available, then a *de novo* assembly may be required (see Section 3.1.1) or a reference-free analysis may be appropriate (see Section 3.1.2). Two particular genome features deserve consideration even with a high-quality reference. First is the degree of relatedness between the individual used to develop the reference genome, and the individual from which the short-read data was generated. Variation between individuals may include sequence substitutions or insertion and deletion polymorphisms (indels) that affect mapping; most read mapping software is relatively robust to minor variation, however. Second, and more important for analysis of sequence variation, is that the presence of small indels can introduce and propagate SNP errors (Figure 2B). If gaps and deletions are not properly handled by the mapping software, then polymorphism rate, and other statistics, will be incorrect *(4)*, and false-positive associations with phenotypic variation (e.g., with genome-wide association (GWA) studies *(5)*) will increase.

Many short-read aligners exist for working with deep sequencing data (Table 1), each with its strengths and weaknesses. For example, there is generally a speed versus sensitivity tradeoff for short-read mappers (see, e.g., Ref. *(6)*), and certain mapping algorithms tend to have better performance for shorter (<100 bp) or longer (>100 bp) reads *(7)*. Choosing a mapper requires weighing the balance between available computing resources, read length, sequencing depth (i.e., number of reads to be mapped), and the accuracy requirements of a project.

Two basic considerations of mapping are whether to produce an end-to-end (global) alignment or to align based on quality scores, and how many mismatches to allow between read and reference. Quality aware alignment is a favored approach because this considers the weight and accuracy of a particular base-pair call in the alignment *(8,9)*. The default mismatch rate is 2% of bases for many mappers, but recent work suggests that higher rates are acceptable for longer reads (up to 10% for 1 kb reads; http://bio-bwa.sourceforge.net). Most mappers allow users to choose what type of alignment is preferred and set the mismatch rate using command-line flags.

The mapping programs discussed above may be paired with other standalone software for the variety of deep sequencing analyses discussed in this chapter, but software suites are also available. The Tuxedo tool kit, freely available from the Steven Salzberg lab (http://ccb.jhu.edu/people/salzberg/Salzberg/Software.html), includes tools for tasks including basic read mapping (Bowtie2 *(10)*), splice-junction mapping (Tophat *(11)*), and RNA-Seq analysis (Cufflinks *(12)*); the site also contains links to well-integrated programs such as AMOScmp for genome assembly. The SOAP suite (http://soap.genomics.org.cn) includes a similar set of tools that build on the SOAPaligner algorithm (e.g., SOAPsnp, SOAPdenovo, SOAPindel). Use of these software suites may help researchers organize large and complex projects typical of deep sequencing technologies, but users should remain aware that specialized standalone programs may be better suited to their particular task.

Reads that map to a single genomic location are termed, "uniquely mapped"; a read mapped to two or more locations is termed as a "multi-mapped" read (Figure 2C). The uniqueness and quality of the read mappings are important to consider for downstream analyses *(13–15)*. Multimapped reads are problematic for RNA-Seq analyses because the number of reads mapping in a particular gene is used to estimate the expression level. Discarding multimapped reads will bias expression estimates of multicopy genes

TABLE 1 Representative Short-Read Mappers for Deep Sequencing Data

Mapping Program	Current Version	Specialty	Drawbacks	Open Source?	Websites
Bowtie	2.x	Fast alignment of short reads	Low sensitivity (precision/recall)	Y	http://bowtie-bio.sourceforge.net/index.shtml
BWA	0.7.5	Medium-speed, gap-aware alignments across a range of read lengths (with different algorithms for different lengths)	Designed for mapping reads with low sequence divergence from reference	Y	http://bio-bwa.sourceforge.net/
Cushaw	1.0.4	GPU-accelerated gapped alignments	Relatively low sensitivity (precision/recall)	Y	http://cushaw.sourceforge.net/homepage.htm#latest
LAST	323	Sensitive alignments and genome comparisons	Suitable for mapping reads from small genomes due to voluminous output	Y	http://last.cbrc.jp/
Novoalign	2.0.8	High-sensitivity alignments with indels	Very slow	N	http://www.novocraft.com/main/index.php
SOAP	2/3	Fast alignment of short reads with mismatches; SOAP3 with GPU acceleration	Lower sensitivity than some aligners	N	http://soap.genomics.org.cn/index.html
SHRiMP	2.2.3	High-sensitivity alignments with indels; correctly handles color space reads from SOLiD system	Slower than some aligners, but significant speed-up since v2.1.x	Y	http://compbio.cs.toronto.edu/shrimp/
Stampy	1.0.2	High-sensitivity mapping with indels and moderately diverged genome reference	Reportedly slower than other mappers, but speed improvements if BWA used for initial alignment	N	http://www.well.ox.ac.uk/project-stampy

downward relative to single-copy genes, so robust methods are needed to correctly distribute reads over candidate mapping sites (see Section 3.2.1). For *de novo* genome assembly or in DNA-Seq applications where the goal is to identify sequence polymorphisms, base errors can hamper the ability to assemble scaffolds and inflate sequence polymorphism estimates.

3 APPLICATIONS

Deep sequencing application and analysis consist of two basic stages. First is processing the reads, which can be divided into two groupings based on the nucleic acid under consideration: DNA-based and RNA-based (Table 2). Read

TABLE 2 Summary of DNA- and RNA-Based Deep Sequencing Applications

Nucleic Acid	Tasks	Requirements/Challenges
DNA	Genome assembly	DNA sequencing libraries with a variety of insert sizes; challenges posed by genome size, accurate, and efficient computational assembly
	Genetic variant calling (SNP and/or CNV)	Single- or paired-end DNA libraries; repeat-rich genomes and large sample sizes present significant computational challenges
	Regulatory sequence indent	ChIP-prepared DNA libraries; variation in sequencing depth peak profiles can introduce ambiguity
RNA	Gene finding/annotation	RNA[a] libraries, with variations for, for example, identifying UTRs; multimapped reads, novel gene models, and multiple isoforms present challenges
	Expression quantification	RNA[a] libraries (full transcript or 3′-tag); multimapped reads and multiple isoforms present challenges
	Identification of isoforms/splice sites	RNA[a] libraries and software for identifying spliced (junction) reads; rarity of junction reads is a challenge
	Identification of RNA editing sites, chimera transcripts, and RNA–protein interactions	RNA[a] libraries

[a] *"RNA" is interchangeable with "cDNA" libraries in most cases.*

quality and basic read mapping issues, as discussed above, are effectively the same for reads from both sources, but the questions each is able to address are very different. Second, after the analysis of the reads, we often want to associate DNA or RNA variation with a variety of phenotypes, which may range from transcript isoforms or abundance, to proteins and metabolites, to organismal or disease phenotypes.

3.1 DNA Applications

3.1.1 De Novo *Assembly*

Many deep sequencing projects require a reference genome against which the millions or billions of reads from different samples are aligned, but researchers are often interested in species for which no reference exists. Deep sequencing technologies permit the *de novo* assembly and annotation of a genome reference for almost any organism, except those with extremely large genomes for which targeted approaches (e.g., exome-only) are required. The success of *de novo* assembly depends on fragment lengths, read length, and computational power, and for most purposes these methods presently require extensive experimentation for success. The main strategy for *de novo* assembly using short-read sequence data has been to obtain different-sized fragment libraries for sequencing and to sequence the genome with high coverage. Specifically, sequencing paired ends of small-, medium-, and large-insert and mate-pair libraries provide useful information for draft genome assembly across a variety of scales.

There are numerous computational tools for *de novo* genome or transcriptome assembly. Some widely used open-source assemblers include Velvet *(16)*, SOAPdenovo *(17)* and SOAPdenovo2 *(18)*, Trinity *(19)*, and ALLPATHS-LG *(20)*. The ALLPATHS-LG assembler is considered the state-of-the-art, producing extremely accurate *de novo* assemblies from human and mouse genomes and Trinity is favored for assembly of transcriptomes. The Assemblathon project (www.assemblathon.org) has been examining the performance of a wide array of assemblers *(21,22)*; researchers interested in or requiring *de novo* assembly are encouraged to follow the project. Experimentally validating the draft assemblies from any particular analysis will continue to be a challenge: for even well-characterized genomes, new assemblies appear when new data are available, making the endpoint of genome sequencing hard to define. High-throughput validation of *de novo* assemblies needs major research attention.

3.1.2 DNA-Seq

Sequencing the genome in a cell, tissue, or individual organism provides the ability to identify SNPs, copy number variants, insertions and deletions, and other types of structural changes in a given genome. These data are useful for detecting associations between genomic variation and disease states and susceptibility *(23)*, evolutionary changes in protein-coding and regulatory regions *(24)*, and activation of transposable elements *(25)*. Given the continued decline in sequencing costs, resequencing the genomes of humans and other organisms will soon become routine.

The method of analysis for DNA-Seq data will depend on whether or not a reference genome exists. For organisms with a reference genome, short reads can be mapped back to a reference using a short-read mapper as discussed above. The position of reads and polymorphisms with respect to the reference are used to identify polymorphisms (Figure 3A). Accurately calling SNPs depends on several factors, including accurate mapping and sufficient coverage to conduct statistical analyses, correcting base-call quality scores, realigning reads around indels to reduce false-positive SNPs, and other factors. Generally, higher coverage provides more power to correct for biases inherent to deep sequencing projects and therefore more power to correctly identify

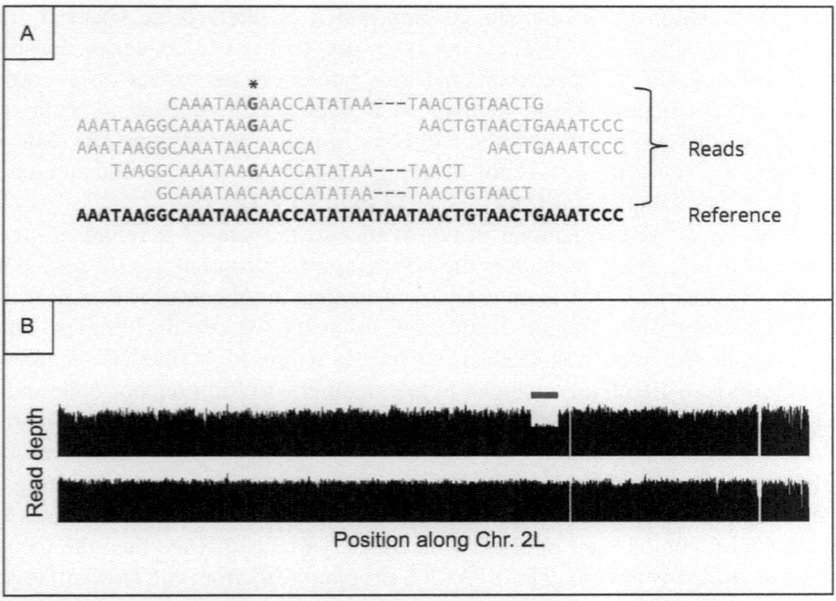

FIGURE 3 Read alignments in DNA-Seq analyses. (A) Many reads cover a set of loci in the genome permit identifying genetic variants such as SNPs (red, asterisk, heterozygous C/G) and indels (blue, a three-base deletion) relative to the reference (bold). The genotype at a locus is called using either likelihood or Bayesian models that assume binomial or multinomial error distributions. (B) Variation in average coverage depth permits identifying CNVs; here the red bar marks an 842-kbp deletion on Chr. 2L of *Drosophila melanogaster* (top), as evidenced by half-depth coverage of DNA-Seq alignments, relative to constant read depth when no deletion is present (bottom). *Data from Ref.* (26). (For interpretation of the references to color in this figure legend, the reader is referred to the online version of this chapter.)

SNPs. The Genome Analysis Tool Kit *(4,27)* in conjunction with Picard (www.picard.sourceforge.net) and SAMtools *(28)* constitute the state-of-the-art tools for variant calling when a reference genome is available and provide sophisticated models for handling necessary sequence corrections.

Beyond identifying individual variants from sequence data, it is often desirable to identify haplotypes to determine the parent-of-origin for a segment of DNA. In addition to increasing statistical power by reducing the number of tests required for association with, for example, disease phenotypes *(29,30)*, haplotype phasing permits imputing missing genotypes that may arise from inconsistent sequencing coverage between samples *(31)*. As with most of the field, computational tools for haplotype phasing and genotype imputing are readily available and under constant development. Widely used programs, each with different domains of use as well as strengths and weaknesses, include Beagle *(32)*, FastPhase *(33)*, Mach (http://www.sph.umich.edu/csg/abecasis/MACH/index.html), and IMPUTE *(34)*.

Another useful feature of DNA-Seq is the ability to detect copy number changes *(35)*. Recent analyses have revealed unprecedented levels of copy number variation in the human genome and it is likely these changes are important to disease *(36)*. These analyses use changes in sequence density to detect changes in DNA compared to a reference: we expect on average twice as many short reads to map to a duplicated sequence as map to sequences that are present in a single copy in a genome (Figure 3B). Many software packages now exist for detecting changes in DNA abundance and two popular examples are CNV-Seq *(37)* and rSW-Seq *(38)*.

When a reference genome is not available, the use of a reference-free approach with genome-reduction methods may be appropriate for the research question. Genomic DNA from samples of interest are digested with a restriction enzyme and the fragments are used to create sequencing libraries. The resultant deep sequencing reads from one or a few high-coverage samples are clustered—effectively mapped to one another—to form tens of thousands of small contigs. Consensus sequences of the contigs are defined, and reads from the entire population are mapped to the reduced genomic reference. SNPs can then be identified at sites across the thousands of small contigs, and allele frequency differences between groups (e.g., disease vs. nondisease phenotypes) can be used to identify candidate variants that are then the focus of in-depth investigation. The STACKS program *(39)* from the Cresko lab at the University of Oregon (http://creskolab.uoregon.edu/stacks/) provides a full pipeline for reference-free inference with deep sequencing data.

3.1.3 ChIP and Other Binding Seq Assays

Identifying the genomic sites that are responsible for gene expression, splicing, posttranscriptional, and other forms of regulation, in *cis* (i.e., proximal) and in *trans* (distant), is a critical challenge that deep sequencing is well suited to address. Chromatin immunoprecipitation (ChIP) provides a method for detecting the occupancy of a protein to a DNA sequence (e.g., the binding of transcription factors to gene promoter regions) *(40)*. The experimental procedure involves using an antibody that is specific to a transcription factor or other protein of interest then cross-linking proteins covalently to DNA with formaldehyde. The chromatin is sheared to 100–200 bp fragments using sonication or chemical digestion. After enrichment with protein-specific antibodies, the cross-links are reversed and the DNA fragments are purified. These fragments can be hybridized onto microarrays (ChIP-chip) or sequenced with deep sequencing (ChIP-Seq), then aligned to the reference using standard short-read mappers to profile the DNA sites occupied throughout the genome.

Biological replication and appropriate controls are essential to ChIP experiments to assess reproducibility because of the nonrandom shearability of DNA and off-target effects of enrichment. A minimum of two replicates are recommended for both experimental and control samples. Control samples consist of two types: a library consisting of formaldehyde-treated and -sheared DNA without antibody (input control) and a library where nonspecific, non-nuclear antibody is used for enrichment to the cross-linked DNA (mock library). Peaks in mapped reads from the enrichment library can be compared to peaks found from mapped reads control libraries to separate real and false-positive peaks (Figure 4A).

The ENCODE and modENCODE consortia have performed thousands of ChIP-Seq experiments and have produced standards that are actively updated *(41,42)*. They recognize three different classes of results based on the type of

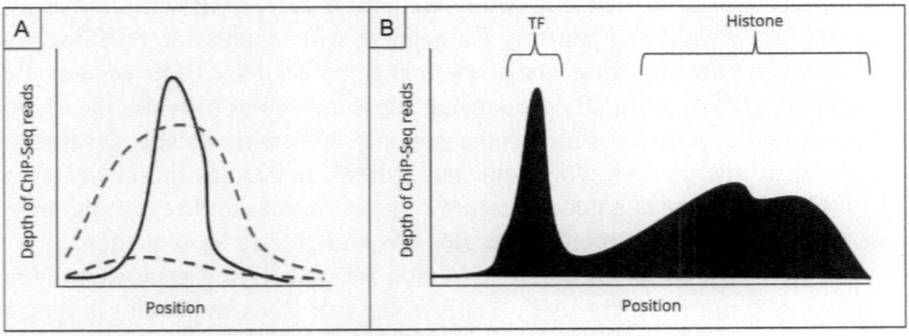

FIGURE 4 Peak variation in ChIP-Seq experiments. (A) Biological replication is required in ChIP-Seq to separate false-positive and true-positive binding sites. If the control (mock) read depth profile resembles the blue dashed line, then a true binding site is suggested by the experimental read depth profile (black line); by contrast, a false-positive is suggested if the control read depth profile resembles the red dashed line. (B) The breadth of the experimental read depth profile is indicative of the class of regulatory feature present in a genomic region, with tall, narrow peaks indicative of transcription factor binding sites and lower, broad plateaus indicative of chromatin marks. (For interpretation of the references to color in this figure legend, the reader is referred to the online version of this chapter.)

signal procedure (Figure 4B): point-source factors such as transcription factors produce very specific peaks; broad-source factors are factors such as chromatin marks and large proteins that have a broad signal; and mixed factors produce specific peaks in some regions of the genome and broad peaks in other regions. These different classes each have specific analysis issues that require different and specific approaches to data analysis. Because of the sensitivity of peak analysis and the fact that the number of peaks obtained in a ChIP-Seq experiment is roughly a function of read depth, ENCODE has defined methods to compare replicate samples for analysis. The irreproducible discovery rate provides a metric for replicate agreement and thresholds for peak detection that provides greater consistency between experiments, laboratories, and antibodies *(43)*. Statistical analysis of peak detection is based on a chi-square model to test whether there is significant signal compared to control sequences.

There are a variety of other enrichment-type assays that use a similar approach to ChIP-Seq for RNA. CLIP, HITS-CLIP, and RIP-Seq *(44,45)* are examples of methods that allow assays of binding by factors to RNA molecules to identify targets of regulation and degradation of RNA molecules. This new class of immunoprecipitation experiments promises a new way of looking at regulation of RNA molecules.

3.2 RNA Applications

3.2.1 RNA-Seq

Profiling the steady-state RNA levels provides a global view of gene expression for a particular experiment, cell type, disease state, or tissue. Microarrays were the first widespread platform for genome-wide expression profiling and for many experiments these tools are being replaced by deep sequencing. Sequencing cDNA or directly sequencing RNA molecules provides the ability to profile the expression state of the genome and carries some advantages compared to microarrays *(46)*. Sequencing RNA molecules or surrogates of RNA (cDNA) provides a good measure of gene expression because the more a gene is expressed, the more molecules are available to be sequenced.

There are two basic classes of RNA-Seq for quantifying gene expression. The first, and more widely used, employs the full RNA molecule. A typical experimental workflow is to extract total RNA followed by poly-A+ selection of mRNA. mRNA is fractioned with sonication or chemical hydrolysis into 200–300 bp fragments and the fragments sequenced with either single or paired-end sequencing. Paired-end sequencing is the better approach for *de novo* discovery of previously unannotated transcripts and for *de novo* transcriptome assembly. RNA-Seq reads are then mapped to a reference genome or transcriptome and the depth of reads for a particular genome feature (e.g., gene, ncRNA, isoform) represents expression level (Figure 5A).

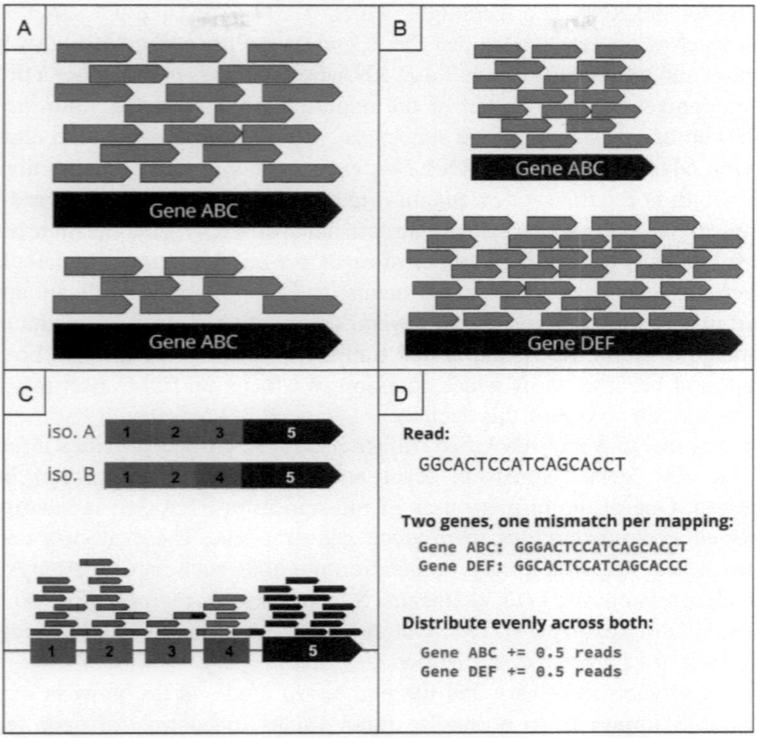

FIGURE 5 RNA-Seq coverage example. (A) The number of reads (gray arrows) mapping to a single-exon gene represents the gene's expression level, high in one sample (top) and low in a second sample (bottom). (B) Gene length confounds between-gene comparisons of expression level, such that two genes expressed at the same level can produce a very different number of reads if one gene is longer (bottom) than the other (top). The RPKM metric accounts for gene length differences. (C) The presence of multiple transcript isoforms for a single gene further complicates accurate estimation of gene expression levels. Here, a single gene (bottom) has two isoforms, with the blue (3) and purple (4) exons switched out between isoforms A and B. Counting both the number of reads (small arrows) mapping to the shared portions of the isoforms and the number of reads spanning exon–exon boundaries (bicolored small arrows, with each color representing the exon of origin for part of a read) permits estimating the quantity of each isoform. In this simplified model isoform B is expressed at a higher level than isoform A; also note that the reads mapping to the green (1), red (2), and black (5) exons are more numerous than either the purple (4) or blue (3) because these are shared exons. Multiple-isoform gene expression levels are calculated by the FPKM metric. (D) Distributing multimapped reads is necessary for accurate estimates of expression, which can be critical when comparing the expression of multicopy and single-copy genes. In this case the read possesses a single mismatch between subsequences of genes ABC and DEF; barring other variation (e.g., variation in base-call qualities, or the number of uniquely mapped reads to each gene), the count of aligned reads to each gene would be incremented by 0.5 for this read. (For interpretation of the references to color in this figure legend, the reader is referred to the online version of this chapter.)

The second RNA-Seq technique, which is less expensive per sample at a given level of coverage, uses just the 3′ end of the transcript for library construction and sequencing. Such 3′-tag RNA-Seq methods *(47–49)* use a primer that recognizes the poly-A tail of the mature mRNA such that only the last 50–200 bp of any transcript are sequenced, typically with single-end chemistry. One advantage of 3′-tag RNA-Seq experiments is that no correction for gene length is required when quantifying expression. Reads are mapped to a reference (or a *de novo* transcriptome), and for each gene the number of mapped reads is tallied; the number of reads per million mapped is calculated per gene to normalize across experiments; and multimapped reads are apportioned or distributed according to several quantitative metrics. A distinct disadvantage of 3′-tag RNA-Seq is that transcript isoforms are missed. For this reason, and because of its wider adoption, we focus on full-transcript RNA-Seq for the remainder of this section.

In contrast to 3′-tag RNA-Seq, full-transcript RNA-Seq provides information on both gene expression level and the transcript isoform(s) being expressed. One of the primary uses of full-transcript RNA-Seq is identifying expressed genomic regions to produce gene models. These models can be refined using modified library preparation methods such as Cap Analysis of Gene Expression *(50)* to delineate 5′-untranslated regions (UTRs) and 3P-Seq *(51)* to identify 3′-UTRs. Isoform quantification and splicing analyses are discussed in greater detail below.

Because gene sizes vary and the number of reads varies between experiments, it is important to normalize these values to control for gene length and experimental biases when quantifying gene expression levels. Larger genes will have more reads due to size, not because they are more highly expressed. This source of variation motivates the reads per kilobase per million mapped reads (RPKM) as a normalized metric for gene expression that can be used to estimate gene expression in a RNA-Seq experiment *(52)*. RPKM is conceptually very straightforward and easy to calculate for genes with no alternative splice products and no mapping bias. Given a genome annotation file, the number of reads that map within regions of interest (e.g., an open reading frame) are tallied, then divided by the size of the region and by the total number of mapped reads ($\times 1e^6$). This calculation makes RPKM a wonderful metric for gene-level expression and has been shown to be accurate with known inputs of RNA *(52,53)*.

Many genes encode multiple isoforms that arise from alternative splicing (Figure 5C); this introduces the challenge of assigning reads among the shared exons versus exons unique to the different isoforms. A model-based metric called the estimated fragments per kilobase per million mapped reads (FPKM; *(54)*) is at least a partial solution to this problem. This metric is produced from the output of Cufflinks after identifying reads that fall within or span two or more exons. A problem of FPKM is the inability to calculate the metric by hand and the lack of alternative methods to check that FPKM is biologically sensible. Unfortunately, much of this hinges on the reads that span exon boundaries ("junction reads," discussed further in Section 3.2.2), which, by their nature, are rare. Techniques that can enrich for junction reads in a sample may prove useful for estimating isoform levels of expression, which is more realistic biologically than gene-level metrics of expression.

We noted in Section 2.2 that multimapped reads are regular features of deep sequencing and affect RNA-Seq analysis. Specifically, multimapped reads need to be apportioned or distributed among candidate regions so that results are not biased with respect to gene family size and duplication (Figure 5D). BM-Map *(13)* is an example of a program that uses a Bayesian framework to account for differences in mapping quality, sequence differences between candidate regions detected in the RNA-Seq data, and the number of uniquely mapped reads to probabilistically distribute multimapped reads. Caution is required for low-expressed genes (see also below) when distributing multimapped reads because their expression tends to be inflated relative to mid- and high-expressed genes *(14)*.

The ENCODE project has produced standards for RNA-Seq. Two important considerations are worth mentioning. First, it is important to include controls for RNA-Seq experiments. One type of control are "spike-in" controls that have been produced by the External RNA Controls Consortium *(53)*. Spike-in controls consist of foreign or synthetic RNAs of known abundance that are put into the RNA cocktail before fractionation. These spike-ins provide an important control because the measurements produced from the RNA-Seq experiment can be compared to known inputs. While it is difficult to standardize target transcriptome measurements based on 96 synthetic RNAs that have different sequence characteristics, they are extremely useful as quality control checks for RNA-Seq experiments *(55)*.

Another important issue is the level of sequence coverage necessary to separate signal from noise. Some have argued that a single read mapping to the genome represents detectable expression, but when technical variation is greater than one read per gene model, mean and variance are confounded. This problem is exemplified by the level of expression for particular genes; those with high levels of expression require fewer reads to accurately measure compared to genes with low levels of expression. The ENCODE project recommends 100–200 million mapped 76 bp paired-end reads for mammalian genomes (http://encodeproject.org/ENCODE/protocols/datastandards/ENCODE_RNAseq_Standards_V1.0.pdf), and smaller genomes will require fewer reads. One approach is to sequence a sample library to great depth then compare estimates of gene expression for random selection for smaller numbers of reads to see how a decrease in read depth affects the repeatability of the expression metric. This will produce a receiver operator curve of reproducibility given varying read depths and provide a gauge for how many reads to acquire in order to obtain an accurate estimate of expression. Another method to define a threshold of detection is to compare the results of microarrays with RNA-Seq to provide a cutoff for detectable levels of expression. RNA-Seq and microarray experimental results can be used to cross-validate each approach, thereby providing a more conservative and accurate estimate of expression *(46)*.

While the best method to estimate coverage requirements is an active area of research, there is no replacement for careful thinking about the number of reads necessary for a particular question. For example, it is well known that gene expression is highly correlated because the set of major expression regulators—for example, transcription factors, metabolites, regulatory RNAs, and hormones—is relatively small compared to the full complement of genes. As a result, the information gained from sequencing and identifying genes with low expression may be very small compared to highly expressed genes that are estimable at lower RNA-Seq coverage *(54)*. If the question is to identify major patterns of response to particular experimental perturbations, whether they are developmental, genetic, or environmental, then lower RNA-Seq coverage may suffice. However, when the question involves identifying all, or the vast majority of, expressed genomic regions, then very deep coverage will be necessary to discover rare transcripts.

3.2.2 Splicing Analysis

In the preceding section, we discussed how alternative splicing complicates the analysis of quantifying expression from RNA-Seq data, but beyond the technical challenges splicing introduces, alternate isoforms can have important biological implications; for example, up to 60% of disease-causing mutations are thought to affect splicing *(56)*. Although the importance of particular gene isoforms is widely recognized, until the advent of RNA-Seq it has been very difficult to assess at genome-wide scales *(57)*. This is understandable because a single gene can sometimes generate a tremendous numbers of isoforms. For example, the gene *Dscam* in *Drosophila* is estimated to have 38,000+ isoforms and differences between the isoforms are very small *(58,59)*. The problem becomes increasingly difficult with larger numbers of isoforms, the extent of exon overlap, and the degree to which each isoform is expressed. Past approaches to identifying alternative splicing isoforms included tiling and junction microarrays, but these are somewhat limited in that prior information is needed to design maximally informative probes.

One of the main strengths of RNA-Seq is the ability to detect junctions between exons even in organisms for which gene models are poorly developed or where an isoform is unknown for a gene model. There is growing support for the high number of novel splice sites in gene models, the functions of which are not completely understood, but with RNA-Seq investigators can begin to quantify junction reads that are key to disentangling isoform variation. The number of junction reads can serve as a proxy for the expression level of the isoform because junction reads should be unique to isoforms, whereas reads that map within exons will be redundant across isoforms sharing an exon.

The detection of junction reads is performed using spliced aligners like TopHat *(11)*, SOAPsplice *(60)*, and others, which distinguish reads with splicing signals within the read compared to reads without splice signals. Junction reads will not map to genomic sequence (Figure 6A), so unmapped reads are examined for the presence of splice signals after the first round of mapping. Canonical splice donor/acceptor signals are particular sequences (GT-AG) found at most exon boundaries. RNA-Seq reads that do not map to the reference as a full read are split into smaller segments and these smaller segments are then mapped to the genome (Figure 6B). Mapping junction reads is challenging with short reads because of the need to split, but because of their length, longer paired-end reads (76 bp and longer) are relatively effective for mapping after each read is split. An ideal junction read would be in the middle of sequence read, such that a 100-bp read would give two, 50-bp sequences that are mapped.

One common approach to the splice-junction problem is sequencing libraries more deeply to obtain more reads that align to exon junctions. This introduces a problem of increased false-positives and, in fact, rarely identifies new junction reads because more of the most abundantly expressed junctions are sequenced repeatedly. The work of Sturgill and colleagues has produced a new splicing detection analysis called Spanki (Splicing Analysis Kit; http://www.cbcb.umd.edu/software/spanki/) that helps identify false-positives and provides a framework for junction-centric splicing analysis.

Junction reads from RNA-Seq will be key to quantifying the abundance of isoforms and their biological importance. As read length increases, the accuracy of junction reads should increase and create new opportunities for solving this important problem of RNA-Seq analysis.

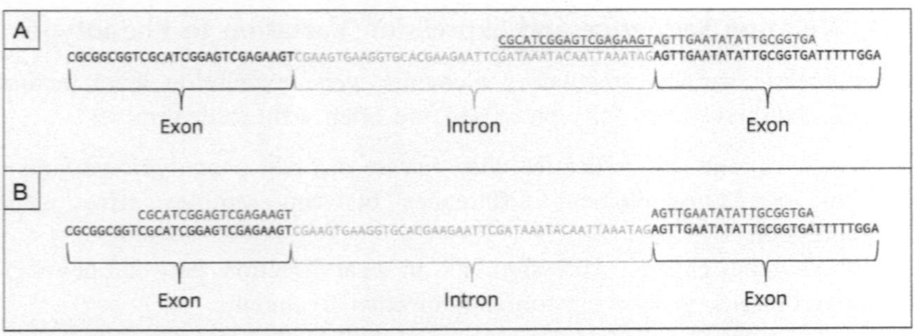

FIGURE 6 The challenge of splicing analysis. (A) A read generated from a transcript composed of more than one exon (top line) may not align to the reference genome (bottom line) because of mismatches over much of the read (red, underlined). (B) Reads that do not map as a single, unified read are split (top line, same sequence as A) and then aligned separately to identify exon boundaries, alternate isoforms, and chimeric transcripts. See also Figure 5C. (For interpretation of the references to color in this figure legend, the reader is referred to the online version of this chapter.)

3.2.3 Additional RNA-Seq-Based Analyses

Another strength of RNA-Seq analysis is the direct assay of sequence information, which enables detection of RNA editing events. RNA editing is a mechanism in which the RNA sequence is altered relative to the DNA template *(61)* and appears to occur frequently in brain and neuronal tissue *(62)*. To detect RNA editing events, reads are mapped back to the genome and/or exons and checked for sequence polymorphism relative to the reference. Between-individual genetic variation, such as between the reference and sample genomes, can result in false-positive estimates of RNA editing; it is often necessary to obtain genomic sequence from the sample individual to check for the presence of RNA editing, but genome annotations may include known editing sites.

Cancer and other genomic diseases often result in fusions of genomic regions through translocations, inversions, and rearrangements that create chimeric RNAs that are detectable with RNA-Seq. The importance of these fusion transcripts and the abundance of trans-splicing is widely debated *(63)*, but there are several software packages that will quantify the abundance of fusion transcripts (e.g., TopHat fusion, FusionHunter *(64)*). These programs rely on reference gene annotation, then search for novel rearrangements between exons of unrelated genes based on the presence of junction reads that bring together two different exons from different open reading frames. Longer reads improve the ability to detect fusion transcripts.

The ENCODE and modENCODE projects have uncovered evidence for extensive expression in unannotated regions, suggesting that nearly all the genome may be expressed at some level *(65,66)*. These novel expressed regions are detected by mapping reads back to a reference genome and assembling reads to create new transcript models. Software such as Cufflinks or Trinity toolkit will produce new transcript models that can be used in updated annotations.

3.3 Relating Sequence and Expression Variation to Phenotypes

Once genetic variants, regulatory elements, gene expression level, and/or splicing data have been fully processed, we often want to determine:

- Which variants are associated with disease or other phenotypic variants
- How regulatory element differences between samples affect gene expression
- Which genes change expression, and in what direction, between developmental stages, genetic experiments, or other treatments
- Which genes or isoforms are associated with phenotypic variation

GWA methods are commonly used to associate genetic variants with phenotypic variation, especially in whole-organism systems (e.g., humans) where experimentation is not possible *(29,67–69,5)*. The basic concept is that if a typed variant is causal—or is closely linked with the causal variant that may be untyped—then the phenotype of individuals possessing one allele will be significantly different than individuals possessing the alternate allele (Figure 7A). Researchers should be aware of three major points when considering GWA analyses. First, any phenotype can be the response variable in GWA models, from gene expression or splicing variants, to proteins and metabolites to organismal phenotypes, but as the number of response

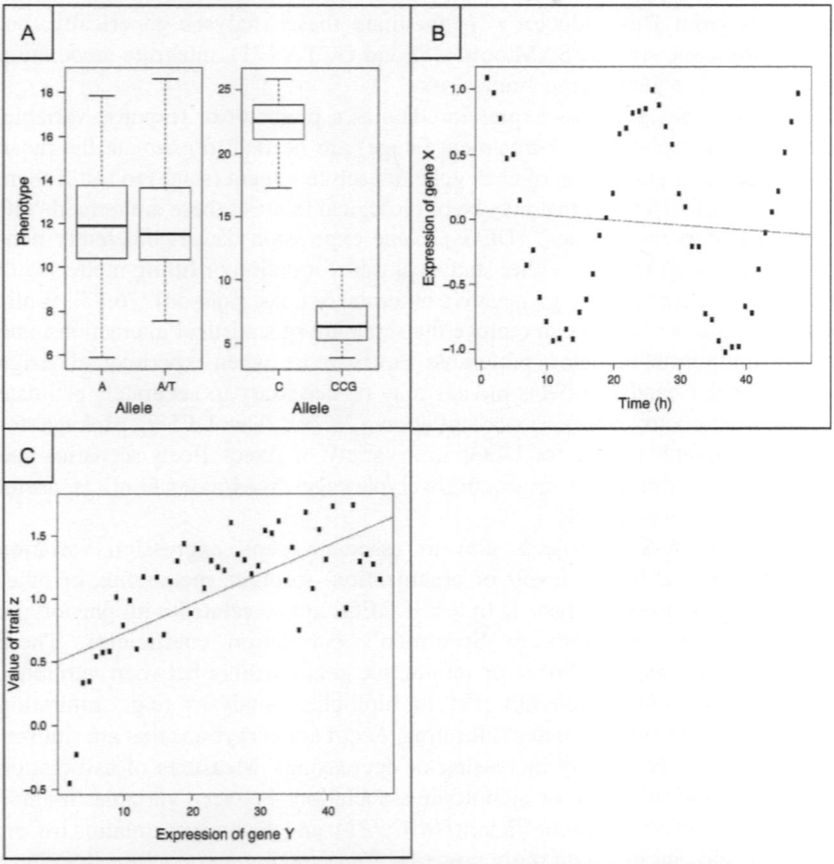

FIGURE 7 Relating deep sequencing data to phenotypic variation. (A) The goal of genome-wide association studies is to associate phenotypic variation with allelic variation at loci across the genome. Genetic variants at most loci will show no relationship to phenotype (left, $P=0.89$), but some will have alleles strongly associated with phenotype (right, $P<2.2\,e^{-16}$). (B) If an RNA-Seq study is focused on expression changes through time, such as across circadian periods, then Pearson's correlation indicates no statistical signal ($r=-0.07$) whereas MIC correctly detects a strong association (~1). (C) When a phenotype is a saturating function of gene expression, Pearson's and distance correlation return similar coefficients (0.67 and 0.68, respectively) but distance correlation returns a stronger *P*-value as would be expected given visual inspection ($P=2.1e^{-7}$ and $P<2.2e^{-16}$, respectively). (For color version of this figure, the reader is referred to the online version of this chapter.)

variables increases, power to detect associations decreases (see Section 3.4). Second, the relationship between minor allele frequency, sample size, and variant effect size are strongly interrelated such that larger and larger sample sizes are required to identify rarer variants or those with smaller effects *(5)*. Third, population structure confounds causal inference of SNP–phenotype associations *(5)*, but a variety of methods exist to correct for population structure in GWA studies *(70–73)*. Widely used statistical software such as R with packages from Bioconductor *(74)* facilitate these analyses generically, but other programs, such as SAMtools *(28)* and GCTA *(75)*, integrate association testing into the genotyping framework.

When an analysis uses expression data as a predictor or response variable, two or more replicates per treatment (stage) are needed to estimate the mean and variance of expression of each gene in each treatment (stage) to test if there is a statistical difference that may be of biological interest; these are termed "differentially expressed genes" (DEGs). Gene expression data is inherently non-normal, and thus requires either statistical transformation or fitting models with non-normal error terms (e.g., negative binomial or quasi-poisson) *(76,77)*. While simpler, transformations can remove the signature of statistical interactions, and fitting non-normal models is preferable. Furthermore, when experimental design is unbalanced, random-effects models may be necessary to accurately estimate means and control for variance inflation *(78,79)*. The DESeq package for R *(80)* facilitates testing for DEGs in a variety of fixed-effects scenarios and lme4 for R (http://cran.r-project.org/web/packages/lme4/index.html) is useful for random-effects models.

Many RNA-Seq projects aim to associate gene expression variation with variation at higher levels of organization—protein, metabolite, or other phenotypes. A basic analysis is to test if DEGs are correlated with phenotypic variation using Pearson's or Spearman's correlation coefficients. These statistical tests assume linear or monotonic relationships between variables, conditions that are often not met in biological analyses (e.g., saturating functions where stoichiometry is limiting, or circadian rhythms that are sinusoidal rather than constantly increasing or decreasing). Measures of association that do not assume linear or monotonic associations between variables include the maximal information coefficient (MIC; *(81)*) and distance correlation (dCor; *(82,83)*), and can be much more powerful for detecting associations than their linear counterparts. MIC requires large sample sizes (hundreds of samples) for reasonable statistical power but is very good at detecting sinusoidal relationships (Figure 7B); dCor (Figure 7C) can be used with much smaller samples but does not perform as well with sinusoidal relationships.

3.4 Application Pitfalls: Data Heterogeneity, Normalization, and False Discovery

Several types of heterogeneity exist in deep sequencing experiments, in particularly in RNA-Seq experiments. First, cDNA synthesis via random hexamer priming creates an uneven distribution of As, Ts, Cs, and Gs for the first 6–10 nucleotide bases. The distribution of the reads mapped to particular regions is also nonuniform and typically looks like smoothed peaks and mounds. This nonuniformity creates biases in assessing a metric for gene expression because, depending on the region chosen, fewer or additional reads may be available for the gene expression metric due to heterogeneity in mapping rather than changes in gene expression. Some biases are platform-specific, but different methods for preparing clusters on the Illumina instrument eliminate some of the sequencing bias in nucleotide distribution *(84)*. Additional sources of heterogeneity include different sequencing depths, or variation in library complexity, cluster density, and other technical factors can create heterogeneity between the biological replicates of interest. Additionally, the distribution of expression values for the measure genes can contribute to poor sampling.

To combat the various sources of heterogeneity, normalization techniques are used to minimize the effects of technical biases and to make meaningful comparisons between experiment and control. Normalization methods for RNA-Seq are still in development, but quantile normalization, similar to what is used in microarrays, is one approach implemented in Cufflinks to normalize biological replicate measurements. Quantile normalization rank makes the distribution of expression values equivalent between biological replicates by adjusting the distribution of expression values based on the data. Genes with the highest levels of expression will have the biggest impact on this distribution adjustment and quantile normalization can lead to large adjustments to the data.

Last, almost all deep sequencing analyses push the limits of statistical inference because the size of the data requires a tremendous number of statistical tests. Each additional test for a particular dataset entails a non-negligible increase in false significant results *(85)*. Consider SNP data consisting of tens or hundreds of thousands of variants tested against the disease state of each individual in the population: due to the finite size of the population and the large number of tests, some apparently "significant" results are expected by chance. To deal with this problem, the false discovery rate (FDR) has been developed for many scenarios *(85,86)* and several software packages exist to calculate and control for FDR (e.g., fdrtool in R; http://cran.r-project.org/web/packages/fdrtool/index.html).

4 THE COMPUTING SIDE OF DEEP SEQUENCING

The immense volume of data arising from deep sequencing research requires extensive and intensive computing resources. While a wide variety of computer programs have been discussed in the preceding sections, here we focus on some of the requirements and challenges that biologists and computer scientists (CSs) face with the large projects.

4.1 Data Management and Computational Power

Working with deep sequencing data requires substantial storage space and computational power. The bigger of these two problems is data storage. The processed data that are fundamental to most analyses discussed in the preceding sections can require huge amounts of disk space. Furthermore, it is frequently necessary to remap and explore variation in mapping parameters, which takes up even more space. Users of deep sequencing data should invest in multiple terabytes of storage space. One cheap strategy is to use external terabyte-size hard drives, which cost less than $100 U.S. and have the advantage of portability and ease of data transfer space. One problem is that these drives may crash and with it all data is lost. Redundancy by having two backups or employing mirrored RAID drives is an absolute necessity and can be achieved through products such as Drobo (http://www.drobo.com) or Synology (http://www.synology.com). Cloud computing backup services such as CrashPlan (http://www.crashplan.com) are another option, though access speeds can be limited.

Many of the popular short-read mappers can be distributed on high-throughput computing clusters, but work well enough on contemporary *NIX workstations (e.g., Linux or Mac). In contrast, *de novo* genome or transcript assembly may require substantial computational power. For these latter cases, a supercomputer is often necessary and may require several weeks on even a powerful workstation to complete an analysis. The global supercomputing infrastructure has rapidly developed over the past two decades, and publicly funded systems such as the National Science Foundation's XSEDE system (https://www.xsede.org) make such computational tasks feasible for deep sequencing researchers. With accounts on such systems, analysts can allocate thousands of processors and terabytes of random access memory for problems that require analyzing billions of reads from an arbitrary number of samples. An alternative that is increasingly attractive is use of "cloud" computing resources for data analysis. For computationally heavy but short-term analysis, cloud computing is a great option because there is no need to invest in hardware or infrastructure. Commercial products such as Amazon's EC2 (http://aws.amazon.com/ec2/), Microsoft's Azure (http://www.windowsazure.com/en-us/), Google's Compute Engine (http://cloud.google.com/products/compute-engine), and numerous other, typically smaller cloud computing and infrastructure services may be attractive options for short-term but intense computational analyses. While large computing clusters and cloud resources make large numbers of processors available for the abundance of required calculations, they also require specialized programming skills or languages to fully harness the power (see Section 4.4).

4.2 Visualization

The large size of genomic datasets means that visualizing even simple data representations can be challenging. There are several tools that can manage this problem. For genomes that have been sequenced, mapped reads can be visualized in a genome browser. Several genome browsers are available, and each has various perks and attributes. The University of California Santa Cruz Genome Browser (http://genome.ucsc.edu) is one of the most popular, and actively updates gene annotations and various Seq datasets in the browser. Other useful genome browsers include the Integrated Genome Viewer and JBrowse *(87,88)*. Users should install a copy of one or more of these browsers locally and then upload tracks of reads or read coverage for visualization.

Visualizing genome data in this fashion has both benefits and costs. On one hand, it is entirely satisfying to "see" the data, particularly for genes of interest that are the target of study. Browsing the data in this way is also dangerous, however, because one might "see" patterns that do not really exist or are not statistically supported, depending on browser settings. Herein lies both a strength and a drawback of genome-scale analyses: the approaches are powerful because of the vast amount of information generated, but the ability to visualize and work with the data is difficult and cannot be scaled across the entire genome.

4.3 Public Data

One of the most powerful aspects of Seq datasets is the ability to have other researchers analyze the same data, to add to the dataset by combining other datasets, and for performing meta-analyses at scales beyond what any single laboratory or working group can accomplish. Seq datasets should be deposited in publicly available databases. The Gene Expression Omnibus (GEO; http://www.ncbi.nlm.nih.gov/geo/) accepts RNA-Seq data and other Seq data associated with particular experiments to the Sequence Read Archive at the National Center for Biotechnology Information (http://www.ncbi.nlm.nih.gov/sra). One additional advantage of depositing the data in public repositories is the many mechanisms in place to back up data to guard against data loss that may not be available to small laboratories. Additionally, these centers have invested in high-performance hardware and software for data transfer to end users (e.g., the Aspera program used by SRA achieves 250–600 Mbps transfer rates, which is far faster than other transfer protocols).

Concomitant with the expansion of Seq database infrastructure, new models of data sharing and data deposition are emerging. Projects such as DRYAD (http://datadryad.org) and Gigadb (http://gigadb.org) are creating repositories that can be linked directly to manuscripts. This eliminates the difficulty with trying to find data and results that were used in the chapter.

4.4 Communicating Analyses: Galaxy and Code Repositories

Analyses of deep sequencing are growing more complex and ever harder to follow in manuscripts. Several groups have tried to repeat analyses based on descriptions found in published methods and were unable to reproduce the results *(89)*. These problems of data analysis are common to "big data" situations and to the large groups that are often necessary to corral these huge resources. The ensuing confusion has resulted in a variety of solutions that continue to be developed. On this topic, it is convenient, if not necessary, to identify two populations of researchers in deep sequencing research: biologists who are interested in and deeply understand the biology and biochemistry of the challenge, and the CSs whose specialized knowledge and skills often facilitate the computational analysis of these large datasets. These two groups are an ideal complement to analysis of deep sequencing data but have different requirements when it comes to communicating what they do.

One option to target the more biological, non-CS population is to develop robust, general, and high-level pipelines and workflows that can be easily shared between collaborators and investigators. The Galaxy project (http://galaxyproject.org) provides a format for reproducibility and shareability *(89,90)* and can be installed locally, in cloud computing environments, or run through the Internet. It is a graphical user interface portal to the common tools available for analysis of deep sequencing data and is highly regarded because it provides an interface that is comfortable to most biologists. The goal of Galaxy is to allow users to easily save and share an entire analytical workflow. This provides a lucid way for other scientists to quickly repeat the analysis. Galaxy also has a large community of users and a help forum that is actively used. This model for analysis of deep sequencing data is very attractive to many biologists because it provides immediate access to some of the most accepted and popular tools without the need to learn computer languages or high-level coding. For detailed analyses of the deep sequencing data, and the development of tools underlying interfaces such as Galaxy, there will continue to be a strong role for CS-oriented bioinformaticians.

A complement to systems such as Galaxy is openly distributing source code and scripts used in deep sequencing analyses, thus targeting the CS-oriented bioinformaticians who are an integral part of these projects. There have been several calls to increase the sharing of source code of biological analyses (see, e.g., Ref. *(91)*), and this pattern is slowly starting to take hold. Some journals permit Supplementary Information sections that could accommodate simple code, but high-performance programs often require thousands of lines of code that are better suited to dedicated code repositories. Sites such as GitHub (www.github.com), BitBucket (www.bitbucket.com), and many others offer free code repositories that can be shared among and contributed to by researchers across the globe. These are large-scale resources (millions of lines of code) whose presence is likely assured well into the future because of their scale and their reach beyond bioinformatics.

Open-source code is critical in the rapidly evolving field of bioinformatics, and can mean the difference between performing an analysis or not. For example, if a program's original author never needed to allocate more than

X kb of memory for a data point, but new technology or formats require X + 1 kb, then open-source code can be edited, and the program recompiled, to successfully run the analysis. The alternative—closed-source code—would require waiting for the program author to respond with what may be very simple updates that may never happen.

4.5 Resources

There are a variety of help forums and blogs devoted to analysis of deep sequencing data. SEQanswers (http://seqanswers.com) is perhaps the most widely used and developers of various Seq tools frequently monitor and answer questions, making it a wonderful resource to obtain feedback on problems with the analysis of Seq data. Seq tool developers also have mailing lists that users can subscribe to in order to find out the latest software updates. Software tools are frequently updated, sometimes multiple times per month. These frequent updates represent one of the frustrations with Seq analysis because analysis of data become obsolete quickly, requiring the reanalysis of entire datasets in order to keep pace with current software. For both the biologists and CS populations in deep sequencing projects, sites such as StackOverflow (http://stackoverflow.com) and StackExchange (http://stackexchange.com) offer excellent programming and computing community support, and CrossValidated (http://stats.stackexchange.com) is populated with highly proficient statisticians who can help address statistical challenges of the scale such projects entail.

5 SUMMARY AND FUTURE DIRECTIONS

The advent of high-throughput sequencing technologies and the attendant advances in computing resources required to manage and analyze the resultant "big data" offer both unique opportunities and challenges for researchers. Continued improvements of data quality and reduction of the 5′ to 3′ quality problem will improve our ability to accurately identify sequence variants that constitute the heritable basis of variation in phenotypes ranging from gene expression to disease. The development of longer reads will likely simplify some of the problems of read mapping and will improve *de novo* genome assembly. In these and other cases, the data going into computationally expensive analyses will likely improve both performance and accuracy as the field develops.

Analysis of deep sequencing data has become critical for addressing contemporary problems in biological and medical science. The sooner biologists from a variety of fields that touch on deep sequencing begin to understand the challenges of these datasets, the sooner communication between laboratories and integration across scientific disciplines can be harnessed to better address the future of biological research.

ACKNOWLEDGMENTS

We thank D. Sturgill, K. Hernandez, D. Goldhamer, T. Heider, and R. Kudra for their reviews, which improved the quality of this manuscript.

REFERENCES

1. Brockman, W.; Alvarez, P.; Young, S.; Garber, M.; Giannoukos, G.; Lee, W. L.; et al. *Genome Res.* **2008**, *18*, 763–770.
2. Cox, M. P.; Peterson, D. A.; Biggs, P. J. *BMC Bioinforma.* **2010**, *11*, 485.
3. Kelley, D. R.; Schatz, M. C.; Salzberg, S. L. *Genome Biol.* **2010**, *11*, R116.
4. DePristo, M. A.; Banks, E.; Poplin, R.; Garimella, K. V.; Maguire, J. R.; Hartl, C.; et al. *Nat. Genet.* **2011**, *43*, 491–498.
5. McCarthy, M.; Abecasis, G.; Cardon, L.; Goldstein, D.; Little, J.; Ioannidis, J.; et al. *Nat. Rev. Genet.* **2008**, *9*, 356–369.
6. Bao, S.; Jiang, R.; Kwan, W.; Wang, B.; Ma, X.; Song, Y.-Q. *J. Hum. Genet.* **2011**, *56*, 406–414.
7. Li, H. eprint arXiv:1303.3997, 2013.
8. Smith, A. D.; Xuan, Z.; Zhang, M. Q. *BMC Bioinforma.* **2008**, *9*, 128.
9. Li, H.; Durbin, R. *Bioinforma. Oxf. Engl.* **2009**, *25*, 1754–1760.
10. Langmead, B.; Salzberg, S. L. *Nat. Methods* **2012**, *9*, 357–359.
11. Trapnell, C.; Pachter, L.; Salzberg, S. L. *Bioinforma. Oxf. Engl.* **2009**, *25*, 1105–1111.
12. Trapnell, C.; Williams, B. A.; Pertea, G.; Mortazavi, A.; Kwan, G.; van Baren, M. J.; et al. *Nat. Biotechnol.* **2010**, *28*, 511–515.
13. Li, B.; Ruotti, V.; Stewart, R. M.; Thomson, J. A.; Dewey, C. N. *Bioinformatics* **2010**, *26*, 493–500.
14. Taub, M. A.; Corrada Bravo, H.; Irizarry, R. A. *Genome Med.* **2010**, *2*, 87.
15. Ji, Y.; Xu, Y.; Zhang, Q.; Tsui, K.-W.; Yuan, Y.; Norris, C., Jr.; et al. *Biometrics* **2011**, *67*, 1215–1224.
16. Zerbino, D. R.; Birney, E. *Genome Res.* **2008**, *18*, 821–829.
17. Li, R.; Zhu, H.; Ruan, J.; Qian, W.; Fang, X.; Shi, Z.; et al. *Genome Res.* **2010**, *20*, 265–272.
18. Luo, R.; Liu, B.; Xie, Y.; Li, Z.; Huang, W.; Yuan, J.; et al. *Gigascience* **2012**, *1*, 18.
19. Grabherr, M. G.; Haas, B. J.; Yassour, M.; Levin, J. Z.; Thompson, D. A.; Amit, I.; et al. *Nat. Biotechnol.* **2011**, *29*, 644–652.
20. Gnerre, S.; Maccallum, I.; Przybylski, D.; Ribeiro, F. J.; Burton, J. N.; Walker, B. J.; et al. *Proc. Natl. Acad. Sci. U.S.A.* **2011**, *108*, 1513–1518.
21. Earl, D.; Bradnam, K.; John, J. S.; Darling, A.; Lin, D.; Fass, J.; et al. *Genome Res.* **2011**, *21*, 2224–2241.
22. Bradnam, K. R.; Fass, J. N.; Alexandrov, A.; Baranay, P.; Bechner, M.; Birol, I.; et al. *Gigascience* **2013**, *2*, 10.
23. Chapman, S. J.; Hill, A. V. S. *Nat. Rev. Genet.* **2012**, *13*, 175–188.
24. Stern, D. L.; Orgogozo, V. *Evolution* **2008**, *62*, 2155–2177.
25. O'Donnell, K. A.; Burns, K. H. *Mob. DNA* **2010**, *1*, 21.
26. Malone, J. H.; Cho, D.-Y.; Mattiuzzo, N. R.; Artieri, C. G.; Jiang, L.; Dale, R. K.; et al. *Genome Biol.* **2012**, *13*, r28.
27. McKenna, A.; Hanna, M.; Banks, E.; Sivachenko, A.; Cibulskis, K.; Kernytsky, A.; et al. *Genome Res.* **2010**, *20*, 1297–1303.
28. Li, H.; Handsaker, B.; Wysoker, A.; Fennell, T.; Ruan, J.; Homer, N.; et al. *Bioinformatics* **2009**, *25*, 2078–2079.

29. Cardon, L. R.; Bell, J. I. *Nat. Rev. Genet.* **2001**, *2*, 91–99.
30. Consortium, I. H. *Nature* **2003**, *426*, 789–796.
31. Marchini, J.; Howie, B. *Nat. Rev. Genet.* **2010**, *11*, 499–511.
32. Browning, B. L.; Browning, S. R. *Am. J. Hum. Genet.* **2011**, *88*, 173–182.
33. Scheet, P.; Stephens, M. *Am. J. Hum. Genet.* **2006**, *78*, 629–644.
34. Howie, B. N.; Donnelly, P.; Marchini, J. *PLoS Genet.* **2009**, *5*, e1000529.
35. Alkan, C.; Coe, B. P.; Eichler, E. E. *Nat. Rev. Genet.* **2011**, *12*, 363–376.
36. Liu, P.; Carvalho, C. M. B.; Hastings, P. J.; Lupski, J. R. *Curr. Opin. Genet. Dev.* **2012**, *22*, 211–220.
37. Xie, C.; Tammi, M. T. *BMC Bioinforma.* **2009**, *10*, 80.
38. Kim, T.-M.; Luquette, L. J.; Xi, R.; Park, P. J. *BMC Bioinforma.* **2010**, *11*, 432.
39. Catchen, J. M.; Amores, A.; Hohenlohe, P.; Cresko, W.; Postlethwait, J. H. *G3 Genes Genomes Genet.* **2011**, *1*, 171–182.
40. Furey, T. S. *Nat. Rev. Genet.* **2012**, *13*, 840–852.
41. Park, P. J. *Nat. Rev. Genet.* **2009**, *10*, 669–680.
42. Landt, S. G.; Marinov, G. K.; Kundaje, A.; Kheradpour, P.; Pauli, F.; Batzoglou, S.; et al. *Genome Res.* **2012**, *22*, 1813–1831.
43. Li, Q.; Brown, J. B.; Huang, H.; Bickel, P. J. *Ann. Appl. Stat.* **2011**, *5*, 1752–1779.
44. Wang, Z.; Tollervey, J.; Briese, M.; Turner, D.; Ule, J. *Methods San Diego Calif.* **2009**, *48*, 287–293.
45. Zhao, J.; Ohsumi, T. K.; Kung, J. T.; Ogawa, Y.; Grau, D. J.; Sarma, K.; et al. *Mol. Cell* **2010**, *40*, 939–953.
46. Malone, J.; Oliver, B. *BMC Biol.* **2011**, *9*, 34.
47. Hong, L. Z.; Li, J.; Schmidt-Küntzel, A.; Warren, W. C.; Barsh, G. S. *Genome Res.* **2011**, *21*, 1905–1915.
48. Meyer, E.; Aglyamova, G. V.; Matz, M. V. *Mol. Ecol.* **2011**, *20*, 3599–3616.
49. Asmann, Y. W.; Klee, E. W.; Thompson, E. A.; Perez, E. A.; Middha, S.; Oberg, A. L.; et al. *BMC Genomics* **2009**, *10*, 531.
50. Jan, C. H.; Friedman, R. C.; Ruby, J. G.; Bartel, D. P. *Nature* **2011**, *469*, 97–101.
51. Shiraki, T.; Kondo, S.; Katayama, S.; Waki, K.; Kasukawa, T.; Kawaji, H.; et al. *Proc. Natl. Acad. Sci. U.S.A.* **2003**, *100*, 15776–15781.
52. Mortazavi, A.; Williams, B. A.; McCue, K.; Schaeffer, L.; Wold, B. *Nat. Methods* **2008**, *5*, 621–628.
53. Jiang, L.; Schlesinger, F.; Davis, C. A.; Zhang, Y.; Li, R.; Salit, M.; et al. *Genome Res.* **2011**, *21*, 1543–1551.
54. Kliebenstein, D. J. *Front. Plant Syst. Biol.* **2012**, *3*, 213.
55. Baker, S. C.; Bauer, S. R.; Beyer, R. P.; Brenton, J. D.; Bromley, B.; Burrill, J.; et al. *Nat. Methods* **2005**, *2*, 731–734.
56. López-Bigas, N.; Audit, B.; Ouzounis, C.; Parra, G.; Guigó, R. *FEBS Lett.* **2005**, *579*, 1900–1903.
57. Wang, G.-S.; Cooper, T. A. *Nat. Rev. Genet.* **2007**, *8*, 749–761.
58. Schmucker, D.; Clemens, J. C.; Shu, H.; Worby, C. A.; Xiao, J.; Muda, M.; et al. *Cell* **2000**, *101*, 671–684.
59. Neves, G.; Zucker, J.; Daly, M.; Chess, A. *Nat. Genet.* **2004**, *36*, 240–246.
60. Huang, S.; Zhang, J.; Peng, Z.; Yiu, S.-M. *Front. Genomic Assay Technol.* **2011**, *2*, 46.
61. Gott, J. M.; Emeson, R. B. *Annu. Rev. Genet.* **2000**, *34*, 499–531.
62. Park, E.; Williams, B.; Wold, B. J.; Mortazavi, A. *Genome Res.* **2012**, *22*, 1626–1633.
63. Costa, V.; Aprile, M.; Esposito, R.; Ciccodicola, A. *Eur. J. Hum. Genet.* **2013**, *21*, 134–142.

64. Li, Y.; Chien, J.; Smith, D. I.; Ma, J. *Bioinformatics* **2011**, *27*, 1708–1710.
65. Graveley, B. R.; Brooks, A. N.; Carlson, J. W.; Duff, M. O.; Landolin, J. M.; Yang, L.; et al. *Nature* **2011**, *471*, 473–479.
66. Djebali, S.; Davis, C. A.; Merkel, A.; Dobin, A.; Lassmann, T.; Mortazavi, A.; et al. *Nature* **2012**, *489*, 101–108.
67. Cordell, H. J. *Nat. Rev. Genet.* **2009**, *10*, 392–404.
68. Thomas, D. *Nat. Rev. Genet.* **2010**, *11*, 259–272.
69. Hirschhorn, J. N.; Daly, M. J. *Nat. Rev. Genet.* **2005**, *6*, 95–108.
70. Wang, W. Y.; Barratt, B. J.; Clayton, D. G.; Todd, J. A. *Nat. Rev. Genet.* **2005**, *6*, 109–118.
71. Pritchard, J. K.; Stephens, M.; Donnelly, P. *Genetics* **2000**, *155*, 945–959.
72. Kang, H. M.; Sul, J. H.; Service, S. K.; Zaitlen, N. A.; Kong, S.; Freimer, N. B.; et al. *Nat. Genet.* **2010**, *42*, 348–354, U110.
73. Price, A. L.; Patterson, N. J.; Plenge, R. M.; Weinblatt, M. E.; Shadick, N. A.; Reich, D. *Nat. Genet.* **2006**, *38*, 904–909.
74. R Development Core Team. *R: A Language and Environment for Statistical Computing*; R Foundation for Statistical Computing: Vienna, Austria, 2009.
75. Yang, J.; Lee, S. H.; Goddard, M. E.; Visscher, P. M. *Am. J. Hum. Genet.* **2011**, *88*, 76–82.
76. McCullagh, P.; Nelder, J. A. *Generalized Linear Models*. Chapman & Hall, CRC: Boca Raton, FL, 1999.
77. Di, Y.; Schafer, D. W.; Cumbie, J. S.; Chang, J. H. *Stat. Appl. Genet. Mol. Biol.* **2011**, *10*, 1–28.
78. Lynch, M.; Walsh, B. *Genetics and Analysis of Quantitative Traits*. Sinauer Associates: Sunderland, MA, 1998.
79. Cleophas, T. J.; Zwinderman, A. H. *Int. J. Clin. Pharmacol. Ther.* **2008**, *46*, 421–427.
80. Anders, S.; Huber, W. *Genome Biol.* **2010**, *11*, R106.
81. Reshef, D. N.; Reshef, Y. A.; Finucane, H. K.; Grossman, S. R.; McVean, G.; Turnbaugh, P. J.; et al. *Science* **2011**, *334*, 1518–1524.
82. Székely, G. J.; Rizzo, M. L.; Bakirov, N. K. *Ann. Stat.* **2007**, *35*, 2769–2794.
83. Székely, G. J.; Rizzo, M. L. *J. Multivar. Anal.* **2013**, *117*, 193–213.
84. Mamanova, L.; Andrews, R. M.; James, K. D.; Sheridan, E. M.; Ellis, P. D.; Langford, C. F.; et al. *Nat. Methods* **2010**, *7*, 130–132.
85. Ewens, W. J.; Grant, G. R. *Statistical Methods in Bioinformatics: An Introduction*. Springer: New York City, NY, 2010.
86. Strimmer, K. *BMC Bioinforma.* **2008**, *9*, 303.
87. Skinner, M. E.; Uzilov, A. V.; Stein, L. D.; Mungall, C. J.; Holmes, I. H. *Genome Res.* **2009**, *19*, 1630–1638.
88. Westesson, O.; Skinner, M.; Holmes, I. *Brief. Bioinform.* **2013**, *14*, 172–177.
89. Nekrutenko, A.; Taylor, J. *Nat. Rev. Genet.* **2012**, *13*, 667–672.
90. Goecks, J.; Nekrutenko, A.; Taylor, J. Galaxy Team. *Genome Biol.* **2010**, *11*, R86.
91. Barnes, N. *Nat. News* **2010**, *467*, 753.

Chapter 15

Gene Expression Analysis and Profiling of Microarrays Data and RNA-Sequencing Data

Javier De Las Rivas, Sara Aibar and Beatriz Roson
Bioinformatics and Functional Genomics Group, Cancer Research Center (CiC-IBMCC, CSIC/USAL), Salamanca, Spain

Chapter Outline

Comprehensive Analytical Chemistry, Vol. 63. http://dx.doi.org/10.1016/B978-0-444-62651-6.00016-7

1 HUMAN GENOME AND TRANSCRIPTOME: FROM GENE LOCI TO GENE PRODUCTS

1.1 Redefinition of Gene in the Context of Modern Transcriptomics

The human genome sequencing project *(1,2)* surprised us; the number of protein-coding genes found was smaller than expected. However, this work did not fundamentally change our perspective on what a gene is *(3)*. In contrast, the more recent human transcriptome project ENCODE *(4)* has uncovered complex patterns of regulation and extensive widespread transcription, together with nongenic conservation and an unexpected abundance of noncoding RNA (ncRNA) genes, challenging our current notion and definition of a gene *(5)*.

Looking at the etymology, the term "gene" derives from the Greek words *genesis* ("birth") and *genos* ("origin"). As described in a recent study of the history and updated definition of "gene" *(5)*, the word was first used by Wilhelm Johannsen in 1909, based on the previous concepts and experimental studies reported by Gregor Mendel in 1866. Johannsen called a gene the "special conditions, foundations and determiners which are present . . . in unique, separate and thereby independent ways by which many characteristics of the organism are specified". From this initial concept, the term "gene" evolved in biology from a "discrete unit of heredity" (1900s) to a "distinct *locus* or region in the genetic inherited material" (1930s). Later, the functional association of "one gene to one enzyme/polypeptide" was formulated (1940s). In the second part of the last century, the biochemical fundaments of the gene were discovered, moving its definition from "a physical DNA molecule of defined structure" (1950s) to "a molecule enclosing the genetic code that enables gene-to-protein translation" (1960s) and to "an open reading frame (ORF) with a specific sequence pattern" (1970s–1980s). Finally, the most recent definition provided in 2002 by the *Human Genome Nomenclature Organization* said that a gene is "a DNA segment that contributes to phenotype/function" *(6)*, and in 2006 the *Sequence Ontology Consortium* defined a gene as "a locatable region of genomic sequence, corresponding to a unit of inheritance, which is associated with regulatory regions, transcribed regions and/or other functional sequence regions" *(3)*.

Despite all these steps and efforts to achieve a precise definition of "gene," current advanced Omic technologies are providing detailed information about the transcriptomes, showing new features and complexity that blurs our current notion of a gene. This complexity is linked to several features: (i) the inclusion or not of the gene regulatory regions (promoters, enhancers, etc.) in the DNA sequence that defines a gene; (ii) the overlapping of coding sequences (CDS) between genes, sharing the same DNA sequence in a different reading frame or in the opposite strand; (iii) alternative splicing, where one genetic locus can code for multiple messenger RNA (mRNA) transcripts that may share a set of exons or may not have any exon in common; (iv) the phenomenon of trans-splicing or ligation of two separate mRNA molecules (i.e., transcripts coming from different loci) being joined before being spliced; and (v) the experimental observation that appreciable amounts of the supposedly noncoding DNA are detected as transcribed *(5)*.

All these complex features show the need for a different approach to define what a gene is in multicellular eukaryotic organisms. A new and original proposal was given by Gerstein and collaborators *(5)*, who defined a gene as a functional "subroutine" within a genomic operating system (OS). Given that counting genes in the genome is a large-scale computational endeavor and that genes fundamentally deal with information processing, the lexicon of computer science can be applied naturally to describe them. In particular, one metaphor that is increasingly popular is to think about genes in terms of "subroutines" inside a huge OS. The nucleotides of the genome are put together into a "code" that is executed through the process of transcription and translation. In this way, the genome can be thought of as an "OS for a living being". Genes are then individual subroutines in this overall system and are repetitively called in the process of transcription *(5)*. This approach to define a gene is open and dynamic, and it implies that the "gene sequence code" can be run in different ways depending on previous subroutines that may be present in one cell state or cell type but not in another. So, quoting the idea of Teilhard de Chardin that "everything is the sum of the past," previous subroutines may mark the evolving and dynamic nature of a gene in a given biological context. This interesting understanding explains why it is so difficult to impose static and pragmatic definitions where a gene is considered just a complementary DNA (cDNA) molecule synthesized from a given mRNA template, which includes one specific sequence that we want to clone and store in our laboratory refrigerators.

Despite this open discussion about the definition of gene in eukaryotic complex systems, for the sake of simplicity in the rest of this chapter we use the term "gene" following a definition similar to the one proposed by the *Sequence Ontology Consortium*: "a region of genomic functional sequence that is transcribed providing an active RNA signal".

1.2 Types of Genes and RNA Complexity: Emerging Interest in ncRNAs

Constant transcription of the genetic information in DNA is a central process for all living systems. The genetic information needs to be copied and transformed into what is called mRNA, which carries the information out into the protein-producing parts of the cells. It is the proteins that, in turn, actually construct the organism and its function. Therefore, protein-coding RNA (pcRNA) is a functional RNA molecule encoded by a genomic sequence that is transcribed into mRNA and translated into a protein. By contrast, ncRNA is a functional RNA molecule that is not translated into a protein. The DNA sequence from which an ncRNA is transcribed is often called an "RNA gene" and the DNA sequence from which a pcRNA is transcribed is called a protein-coding gene.

Around 98% of all transcriptional output in humans is ncRNA *(7,8)*. ncRNA genes include two well-known and highly abundant RNAs involved in the processes of translation and protein synthesis: transfer RNAs (tRNA) and ribosomal RNAs (rRNAs). rRNAs are the catalytic component of ribosomes. Eukaryotic ribosomes contain four different rRNA molecules: 18S, 5.8S, 28S, and 5S rRNA. Many other ncRNAs have been described in recent years *(9,10)*; most can be classified in three major groups: (i) RNA genes involved in posttranscriptional modifications (e.g., splicing or nucleotide modification of RNAs), such as small nuclear RNAs (snRNA) and small nucleolar RNAs (snoRNA); (ii) RNA genes involved in gene regulation (e.g., transcriptional attenuation, mRNA degradation, mRNA stabilization), such as antisense RNAs (aRNAs), small interfering RNAs (siRNA), microRNAs (miRNA), piwi proteins interacting RNAs (piRNA), extracellular RNAs (exRNA), and the long ncRNAs (lncRNA); and (iii) RNA genes that work in stable association with polypeptides forming ribonucleoproteins (RNP). Many of these RNAs are associated with nuclear proteins (i.e., inside the nucleus) such as enzyme telomerase, vault ribonucleoproteins, and ribonucleases (RNases), but new RNA–protein associations have been found outside the nucleus with very interesting regulatory roles (many of these RNA–protein associations include lncRNAs).

Considering all these types of RNA genes, it is worth to underlie the case of lncRNA because recently discovered functional RNA genes (e.g., *XIST* and *HOTAIR*) are encoded by lncRNA molecules that bear many features similar to mRNAs, including 5′ capping, splicing, and polyadenylation, but they do not have ORF *(11)*. Some lncRNAs are placed in intergenic DNA regions, called lincRNAs, and recent studies indicate that these RNA genes include up to 50,000 species. Therefore it seems that the fraction of the human and mouse euchromatic genomes encoding lincRNAs is more than two-fold greater than the fraction of protein-CDS. Although the sequences of most lincRNAs are much less conserved than protein sequences, the extent of orthology between them is unexpectedly high, with 60–70% of the lincRNA genes shared between human and mouse *(12)*.

In this complex and hazy scenario about the human or mouse transcriptome, another pragmatic approach for classification of RNA genes has been to divide them by size (i.e., by nucleotide length) in small ncRNAs

(<200nt: miRNA, small interfering RNA, etc.) and long ncRNAs (>1000 nt: lincRNA, etc.). At present these classifications are open to discussion, because more data and functional analysis are needed to unravel the biological roles that the ncRNAs play in the living systems. For the rest of this chapter we focus on transcriptomics and expression of protein-coding genes because these are the best studied with current advanced Omics technologies.

1.3 Protein-Coding Genes: cDNA Collections and ORFeomes

According to the last high coverage assembly (GRCh37.p10 from the *Genome Reference Consortium*, February 2009, and *Ensembl* version 71, November 2012) the human genome includes 20,387 known protein-coding gene loci and 81,626 protein-coding transcripts (http://www.gencodegenes.org/). The major evidence to identify the protein-coding genes and to build the consensus coding sequences (cCDS) come from UniProt curated human proteins (http://www.uniprot.org/) and from curated databases of human cDNA and expression sequence tag (EST) sequences obtained in experimental analyses of different human cell types and tissues. Probably the most comprehensive and integrated resource for protein-coding genes is the *Reference Sequence* (RefSeq) collection, which provides a nonredundant, well-annotated set of sequences, including explicitly linked nucleotide and protein sequences (http://www.ncbi.nlm.nih.gov/refseq/). This collection includes ongoing sieving by NCBI staff and collaborators to be updated and to reflect current knowledge of sequence data and biology.

A major scope of genome science is to provide comprehensive catalogs of all the genes that exist in a given organism. This aim seems obvious, but it remains quite difficult since the methods that have been used in the last 10 years do not yield even at least one complete ORF for every gene loci of the human genome *(13)*. Therefore, the common strategies—sequencing randomly selected cDNA clones, sequencing ESTs from different cell and tissue types, aligning protein sequences identified in several organisms, sequencing and comparing more genomes, and manual curation—have to be supplemented by large-scale amplification and sequencing of the predicted mRNAs that have to be proven. Thus, the use of high-throughput sequencing technologies (e.g., high-density microarrays and deep RNA sequencing) will help to complete the still imperfect strategies to define an ORF for each locus *(13)*.

In order to determine the gene structure of a given locus, there is a recommended preference order of evidences to be considered *(13)*: (1) alignments of full ORF cDNA sequences to the loci from which they were transcribed; (2) alignments of highly similar proteins (probably >95% identity), at loci where there is no full ORF cDNA alignment; (3) for a gene structure that is partially determined by an EST aligned to its native locus, extend it to a full ORF by *de novo* prediction; and (4) at loci where none of these are available, use pure *de novo* predictions using dual- or multigenome prediction algorithms.

The availability of the entire human genome sequences was expected to revolutionize the way in which biology and medicine were conducted. However, as described above, achieving this promise still requires significant effort in the research area of gene identification and also strong effort in the technological area of cloning and expression of thousands of protein-encoding genes still not available *(14)*. A really finished genome sequence should reveal the set of all the ORFs it encodes and this review reminds us that in the case of the human genome we still do not have a complete set. The name ORFeome defines the whole set of ORFs in a genome. The ORFs correspond to the CDS of the genes, and ORFs are usually cloned in expression vectors in order to be kept and used in particular experiments. Most of the published ORFeomes are partial collections of the real complete set of ORFs that exist for the corresponding organism. The ORFeome projects are gateways between genomics and many other Omics technologies *(14)*. The advent of systems biology necessitates the cloning of all known protein-encoding ORFs collected into ORFeomes to allow functional studies of the corresponding proteomes. The newest version available of the human ORFeome is hORFeome v8.1, containing 12,230 sequence-confirmed clonal isolates, which represent at least 11,149 human genes (http://horfdb.dfci.harvard.edu/). These numbers indicate that the coverage of the human ORFeome over the protein-coding human genome is still 54.7%, showing that it is still far from completeness.

1.4 Biological Databases to Identify and Explore Expression of Protein-Coding Genes: *Entrez Gene, ENSEMBL, GATExplorer,* and *ProteinAtlas*

There are many biological databases where a user can query for genes and gene IDs, but the original sequence databases that include the genes in their genomic context and provide the international IDs for any new discovered or annotated gene are *Entrez Gene* (at NCBI at http://www.ncbi.nlm.nih.gov/gene) *(15)* and *ENSEMBL Gene* (at EMBL-EBI/Sanger in Europe at http://www.ensembl.org/) *(16)*. These two primary resources include the CDS of the genes located in the genomic coordinates, and also integrate large amounts of available information about the genes. An example of valuable information to understand the role of the genes is the scientific literature information associated with each gene, which *Entrez Gene* includes in two sections: "Related Articles in PubMed" (PubMed links) and "Gene References Into Functions" (GeneRIFs). The retrieval of current information and knowledge about a query gene is facilitated by these Web sites, which provide online access to updated research publications linked to each gene.

Together with primary gene databases, the study of gene functions needs the integration of sequence data with expression data. There are multiple resources that provide this type of combined data. We mention and describe here two useful resources: *GATExplorer* and *ProteinAtlas*. *GATExplorer*

(Genomic and Transcriptomic Explorer; http://bioinfow.dep.usal.es/xgate/) *(17)* is a bioinformatic Web tool that includes a genomic browser for simple interactive exploration of gene loci, transcripts, and exons—for human, mouse, and rat genomes—integrated with a genomic map of probes corresponding to expression microarrays (i.e., high-density oligonucleotide arrays from *Affymetrix* platforms) and presentation of their respective expression levels in a broad set of normalized biological samples *(17)*.

Figure 1 presents an example view from *GATExplorer* corresponding to human gene *APOB* that is located in band p24 of chromosome 2 (2p24) and has two transcripts in a gene locus of 42.64 kb length. The currently known architecture of this locus shows two different transcripts that share the

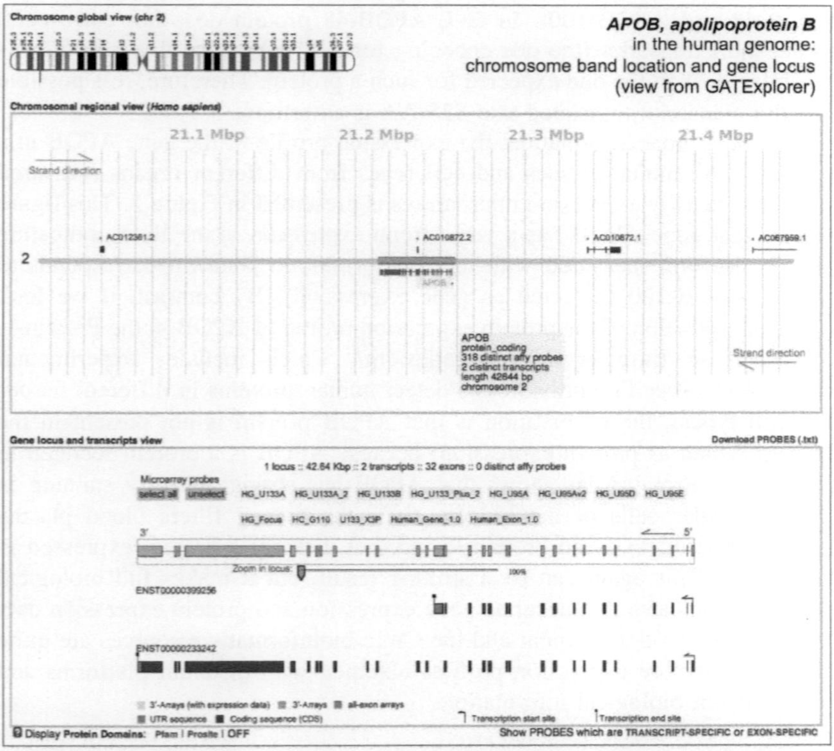

FIGURE 1 Graphical view taken from *GATExplorer* (http://bioinfow.dep.usal.es/xgate/) corresponding to the locus of human gene *APOB* that is located in band p24 of chromosome 2 (2p24) and has two transcripts in a locus of 42.64 kb length. The three panels show the chromosome 2 global view indicating the position of the gene, the chromosomal regional view just around gene *APOB* and the detailed view of *APOB* locus with the two known transcripts derived from it with their specific exon–intron architecture. The two transcripts share the 5′-end but provide two cDNA isoforms of very different lengths: 828 and 4563 AA (amino acids). (For color version of this figure, the reader is referred to the online version of this chapter.)

5′-end but provide two cDNA isoforms of very different lengths: 828 and 4563 AA (amino acids). The gene product derived from this locus is the apolipoprotein present in chylomicrons and low-density lipoproteins, which occurs in plasma as two protein isoforms: APOB-48 synthesized exclusively in the gut and APOB-100 synthesized only in the liver. It has been shown that both the intestinal and the hepatic forms of APOB are encoded by the long transcript, since the shorter APOB-48 protein is produced after mRNA editing of the APOB-100 transcript at amino acid residue 2180 (Q→Stop; CAA→UAA), resulting in the creation of a stop codon and early translation termination. All this indicates that despite APOB being a well-known human protein, there is possible disagreement and confusion between the two transcripts reported by the genome databases (cDNA isoforms of 828 and 4563 AA) and the two proteins reported by the proteome databases (proteins APOB-48 and APOB-100). In fact, APOB-48 protein does not correspond to the short transcript (the one encoding for 828 AA) since this transcript is much larger than the one expected for such a protein. Therefore, it is possible that this transcript translated into 828 AA is an artifact.

To show more information, the expression profile of the gene APOB in a dataset of 67 human tissues and cell types from different organs measured with high-density expression microarrays is presented in Figure 2. This figure reveals that gene APOB has a very strong expression in the liver, indicating that the isoform measured with this transcriptomic platform corresponds to protein APOB-100 (detected as gene expression). By contrast, if we look for direct detection of the protein expression profile of APOB in the ProteinAtlas database (http://www.proteinatlas.org/, which includes experimental results using specific antibodies to detect human proteins in different tissues and cell types), the observation is that APOB protein is not present in the liver (detected as protein expression) because APOB is a protein secreted to the plasma. ProteinAtlas shows that APOB has strong antibody staining in the glomerular cells of the kidney, since the kidney filters blood plasma in the glomeruli to avoid protein losing, but it is not a protein expressed in the kidney. This again can be a striking result, but it makes full biological sense and indicates that accurate gene expression and protein expression data should show good agreement and the Omic bioinformatic resources are quite helpful to explore expression profiles obtained with different platforms and find coherent biological information.

1.5 Gene Expression Regulation: Transcription Factors

To understand the properties of distinct cell types and cellular states we need to unravel the gene regulatory factors that control the expression of genomic information. Such regulation comes in a variety of elements, with transcription factors (TFs) and miRNAs representing the most numerous gene regulatory factors in multicellular genomes *(18)*. A TF is a protein that binds to

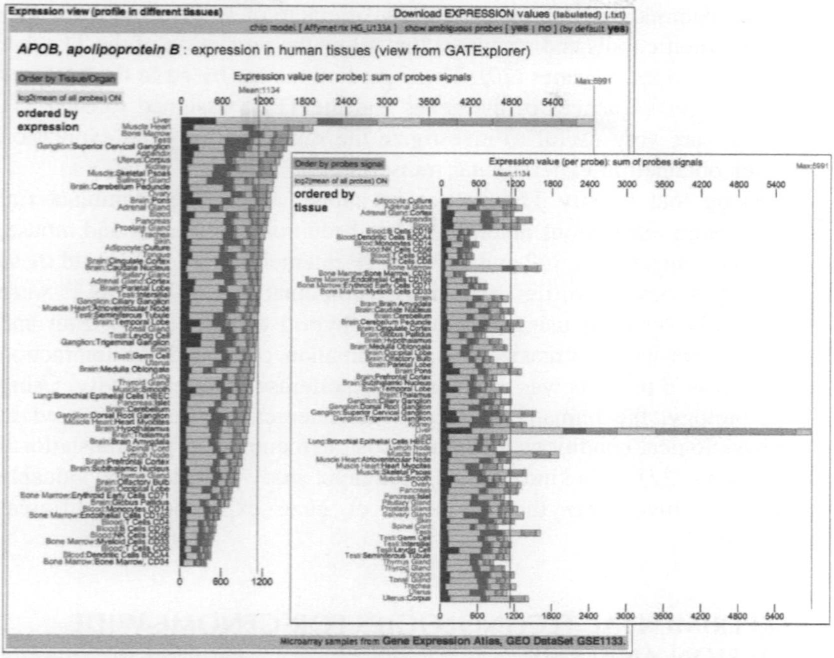

FIGURE 2 Expression profile of the gene *APOB* in a dataset of 67 human tissues and cell types from different organs measured with high-density expression microarrays taken from *GATExplorer* (http://bioinfow.dep.usal.es/xgate/). Each bar corresponds to the total expression of this gene in a given tissue, and the portions in different green colors in the bar correspond to the multiple probes that measure each gene in the microarrays used in these expression experiments. The figure shows that gene *APOB* has a strong expression in the liver. (For color version of this figure, the reader is referred to the online version of this chapter.)

specific DNA sequences to control the transcriptomic flow of genetic information from DNA to RNA. They are also called "sequence-specific DNA-binding factors" and the sequences where they bind are called "transcription factor binding sites" (TFBS), which define regulatory regions in the genome usually called "gene promoter regions". The TFs perform their function regulating the recruitment of RNA polymerase to trigger the expression of specific genes. Their functional binding to the DNA can be alone or with other TFs, but also with regulating proteins that do not bind DNA directly but form complexes interacting with the TFs (these are "activator proteins" if they promote the expression or "repressor proteins" if they block the expression).

Genome-wide studies in humans have identified the existence of a repertoire of 1391 TFs *(19)*. The study of the evolutionary profile of these TF genes and their orthologous across 24 eukaryotic genomes, ranging from yeast to chimpanzee, showed the existence of five groups of TFs with distinct patterns of conservation: the ones present only in primates and the ones predominantly present in mammals, vertebrates, metazoa, or eukaryotes *(19)*. There are several bioinformatic tools and resources that allow the prediction of the TFs that regulate a given set of genes *(20)*. These predictions are based in the analyses of the promoter sequences of the genes and the TFBS assigned to each TF. These tools are very useful to investigate the regulation of gene expression signatures obtained in experimental transcriptomic studies.

Knowing that usually TFs do not act alone, an atlas of combinatorial protein–protein interactions among TFs has been built for human and mouse, identifying, respectively, 762 and 877 TF–TF interactions *(21)*. To build these TF interaction binary profiles, all pairwise combinations of TF cDNAs were systematically screened using a mouse two-hybrid technique *(22)*. Bait and prey constructs were cotransfected in mammalian cells, and the interaction of the expressed proteins was monitored by luciferase reporter activity. Using this technology, the human and mouse TF interactions were measured in near-physiological conditions including possible mammalian posttranslational modifications *(22)*. This kind of high-throughput study provides very valuable resources to investigate the regulation of gene expression in different sample sets.

2 EXPERIMENTAL TECHNOLOGIES FOR GENOME-WIDE EXPRESSION ANALYSIS

2.1 Measuring Gene Expression: Real-Time qPCR

One of the most extended techniques in molecular biology to estimate the amount of transcript per gene that each cell is producing is the "quantitative polymerase chain reaction" (qPCR) or "real-time quantitative polymerase chain reaction" (RT-qPCR). It is based on the amplification properties of nucleic acids that can be simultaneously quantified during the reaction process. Following the polymerase chain reaction, each time the RNA (or its copy in DNA) is amplified, the light intensity of newly synthesized entities can be detected. Detection is performed by nonspecific fluorescent dyes that intercalate with any double-stranded DNA. Also fluorescent probes can be hybridized to a linked complementary sequence. RT-qPCR has become one of the most reliable and cost-effective techniques to quantify nucleic acids; this is why nowadays it is considered the gold-standard validation experiment of high-throughput gene expression results *(23)*. RT-qPCR is normally used for relative quantification, which measures the fold-change in abundance levels of cDNA with respect to a predefined internal reference. It does not require a calibration curve as the amount of the studied gene is compared to the amount of a control housekeeping gene. An extensively used device to perform quantitative standardized assays for gene expression analysis by real-time quantitative PCR is the TaqMan® Gene Expression Assay from *Applied Biosystems*.

2.2 Microarrays

High-performance genomic technologies developed mainly in the last two decades have automated and miniaturized experimental techniques of conventional molecular biology. Great scale production of data, in constantly shortened times, has empowered the study of cellular systems, from the simplistic view of a gene, to the global ("Omic") level of a whole genome. Since the 1990s when the first massive sequencer of DNA was created (Massively Parallel Signature Sequencing of Lynx Therapeutic), these technologies have evolved, increasing in potential and reducing in costs, turning into principal tools for current biological and biomedical research.

The term "Omics" refers to the global approach that these techniques provide. It can be applied to different biological compartments of study, such as the transcriptome, proteome, methylome, kinome, metabolome, and many other terms that have appeared along the growth of novel high-throughput platforms.

Into the transcriptomics area, high-density oligonucleotide microarrays have started a particular revolution in the biomedical field. Yet today, when sequencing of RNA has become a palpable reality, microarrays remain useful and utilized in plenty of studies. Their affordable cost and their great reproducibility have made them the platform of choice for most gene expression studies. Hundreds of publications and hundreds of accessible sample series available in different public repositories keep the interest of researchers in these devices *(24)*. The use of expression microarrays has been proven quite successful in Omic studies about the human gene coexpression landscape *(25)* and also about specific gene signatures discovered in complex human diseases *(26)*. In this section, we review the molecular basis of gene expression microarrays, focusing on high-density oligonucleotide microarrays, primarily manufactured by the American company *Affymetrix* (GeneChip® arrays, Santa Clara, CA) (http://www.affymetrix.com/). Each microarray technology has its specific technical properties and together with *Affymetrix* two other major manufacturers have launched genome-wide expression microarrays: *Applied Biosystems* (http://www.appliedbiosystems.com/) and *Agilent* (http://www.genomics.agilent.com/). We select in our studies *Affymetrix* microarrays because they outperform the other platforms. However, before going into more detail about these microarrays, the results of a public study comparing microarrays from different platforms are presented.

2.2.1 Public Comparison of Expression Microarrays from Different Platforms

During the first decade of this century, the use of microarray technology had a profound impact on gene expression research. The publication of some studies with dissimilar or contradictory results, obtained using different microarray platforms to analyze identical RNA samples, raised concerns about the repeatability, reproducibility and comparability of microarray technologies. To address these concerns the MicroArray Quality Control (MAQC) project was developed *(27)*, generating and analyzing expression data on four titration pools from two distinct reference RNA samples at multiple test laboratories using a variety of microarray-based and alternative technology platforms. In general, the results of this study show intraplatform consistency across test sites as well as a high level of interplatform concordance in terms of genes identified as differentially expressed. The study includes and compares three major human genome-wide expression one-color microarray platforms: one from *Affymetrix* called HG-U133 Plus 2.0 GeneChip (which includes 54,675 probesets); the one from *Agilent* called one-color Whole Human Genome Oligo Microarray G4112A (which includes 43,931 probesets); and one from *Applied Biosystems* called Human Genome Survey Microarray v2.0 (which includes 32,878 probesets). The results showed that *Affymetrix* microarrays have the best repeatability of expression signal within different test sites (i.e., different laboratories), measured by the coefficient of variation of the quantitative signal values between replicates *(27)*. The results of this study also showed that the *Affymetrix* microarrays displayed the highest correlation values ($r \geq 0.92$) with TaqMan assays, based on comparisons of ≈450–550 genes performed in three different sites. These correlations were calculated as the log-ratio differential expression (DE) values (using A vs. B replicates) from each microarray platform relative to values obtained by TaqMan quantitative standardized RT-qPCR assays *(27)*. Finally, as indicated above, the microarray that includes the largest number of probesets to test specific human genes is also the one from *Affymetrix*, with 54,675 probesets.

2.2.2 *Design and Use of* Affymetrix *Expression Microarrays*

Microarrays were designed to measure the quantity of mRNA for each gene that can be found into a biological sample at the moment of the extraction. They are physical devices that stand upon the hybridization principle, by which a single-strand molecule of RNA from the origin sample can bind specifically to a complementary single-strand oligonucleotide of DNA mounted on a solid surface. Those oligonucleotides, called "probes," have to be defined in advance based on the existing knowledge of the genes sequences in a reference genome. In the *Affymetrix* platform, the probes are 25-mer (bases) in length, and have been designed in groups of probes ("probesets") composed by 11–16 probes assigned to a gene entity. The physical view of a microarray or genechip is a small cartridge (≈7×4 cm) with a centered window of about 1 cm^2 where the array is, allocating thousands of micro-cells, each containing copies of a given sequence-specific probe.

Under the *Affymetrix* trademark, three hybridization assays can be found that generate three types of whole-genome arrays for expression analysis: the 3′-IVT array, the gene array, and the exon array. Figure 3 illustrates the

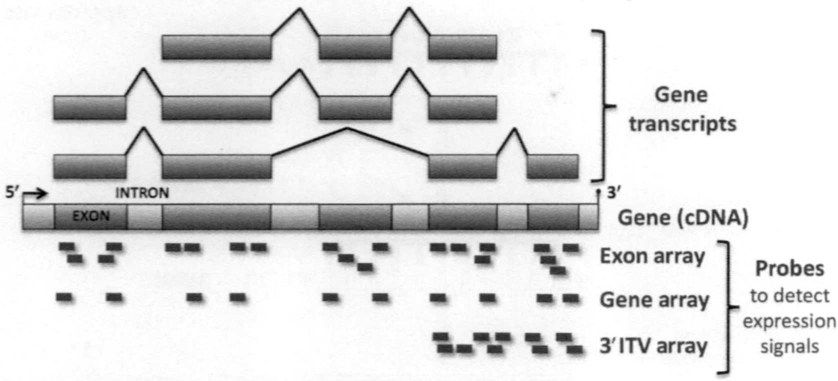

FIGURE 3 Example illustration of how the oligonucleotide probes of the expression microarrays from *Affymetrix* cover the exons and the transcripts in a given gene locus. The 3′-ITV arrays are designed to map the 3′-end of the transcripts. The gene and exon arrays are designed to map all the exons of the gene, called whole-transcript (WT) arrays. The figure shows that the probes in the WT models are disposed along the sequence covering the whole gene length to generate a more reliable expression signal. In the case of the exon arrays, they include multiple probes per exon, providing measures for the detection of splicing events. (For color version of this figure, the reader is referred to the online version of this chapter.)

probe coverage for each design type. The first one, and the oldest assay, was designed to map the 3′-end of the transcript, a model denoted as 3′-*in vitro transcribed*. The last two represent the whole transcript, and they are denoted as WT models. To avoid the bias produced by steps of PCR amplification, probes of 3′-IVT models were located in the 3′-neighborhood of the gene, which many times constitute the untranslated regions of the gene (i.e., 3′-UTRs). Probes in WT models are disposed along the sequence covering the whole gene length, generating a more reliable expression signal and even providing measures for the detection of different gene exons.

The 3′-IVT hybridization protocol (see Figure 4) begins with oligo-dT primer synthesis of copy-DNA (cDNA) using as a template the mRNA sample extracted from the tissue or cells submitted to study. After this step, double-strand cDNA samples can be stored long term until use because of the higher stability of DNA compared to RNA. When all the samples are ready to hybridize, the cDNA is *in vitro* transcribed to aRNA, amplified and labeled with biotin. This biotinylated aRNA is then fragmented in oligonucleotides of similar length to the array probes. The sample is inserted into the chip device for hybridization steps, then washed to remove the excess sample RNA, and finally incubated with a biotin staining system and scanned. The array is then laser-scanned, and the amount of fluorescence per micro-cell is detected, in other words, an array image is generated. Postscanner analysis of each array image allows the quantification of expression levels for each probe, identifying each probe's exact position by its *XY* coordinates into the

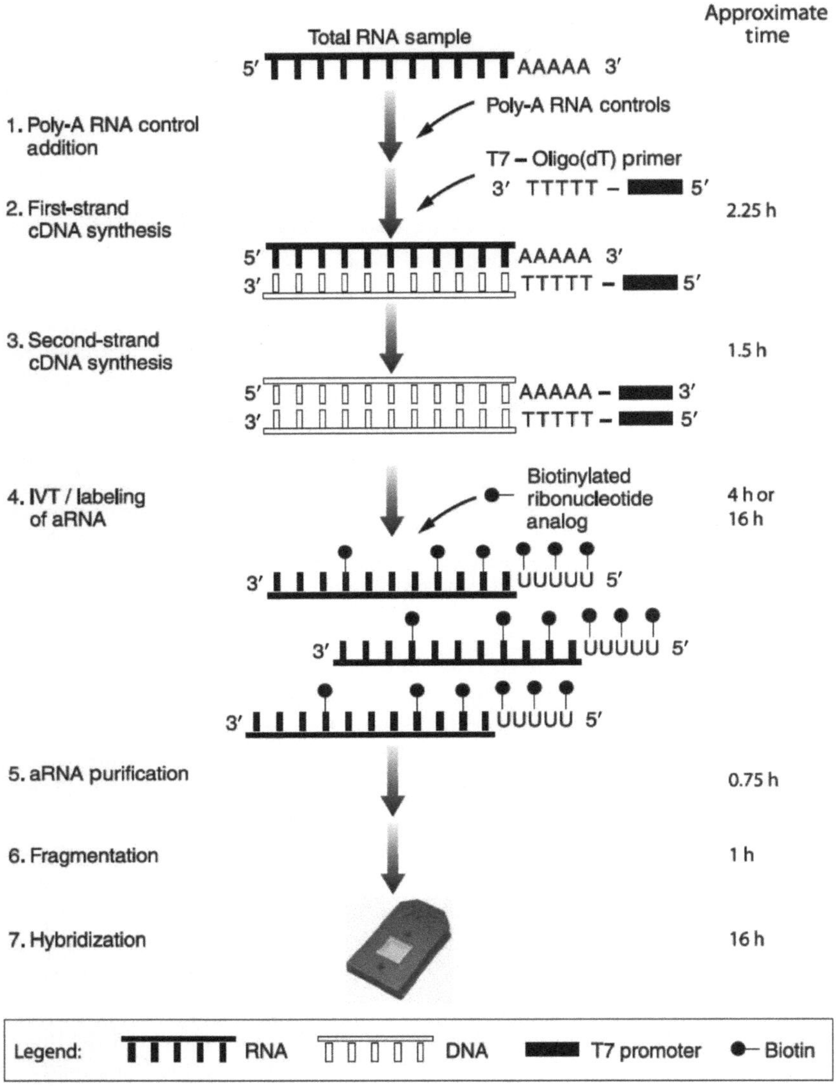

FIGURE 4 Hybridization protocol for 3′-IVT arrays. It starts with the oligo-dT primer synthesis of cDNA using as a template the mRNA sample extracted from the tissue or cells submitted to study. Following, the cDNA is *in vitro* transcribed to antisense RNA (aRNA), amplified and labeled with biotin. This biotinylated aRNA is then fragmented in oligonucleotides of similar length to the array probes. The sample is inserted into the chip device for hybridization, then washed to remove the excess of sample RNA, and finally incubated with a biotin staining system and scanned. The array is then laser-scanned, and the amount of fluorescence per each micro-cell is detected to generate an array high-resolution image.

array. The fluorescence intensities per probe-cell are written into an output file. These files, which have the.CEL extension, contain the so-called "raw data". The information of a raw data file is in reality a matrix with the numeric measures of the intensities for each probe in an individual microarray.

In the case of the WT models, the labeling during cRNA *in vitro* transcription is not executed. Instead, there is a second cycle of cDNA synthesis and random priming amplification from the cRNA strand, with a subsequent cRNA hydrolysis. In this case, the cDNA is the labeled molecule that will be hybridized against the surface attached probes. While using the same basic chemistry as its predecessors, it has been optimized to allow for lower sample input and faster processing times, while retaining comparable result quality and reproducibility.

The importance acquired by the microarray technology and its input to the research community can be illustrated by browsing at the best-known expression data repository, the Gene Expression Omnibus (GEO; http://www.ncbi.nlm.nih.gov/geo/) *(28)*. For example, in mid-2013 the array model "Human Genome U133 Plus 2.0" of *Affymetrix* (entry: GPL570) accounts for 82,767 hybridized samples, distributed by 3093 data series.

2.2.3 Hybridization Drawbacks: The Probe Effects

As explained before, the oligo probes (25 nucleotides in length) are designed to recognize specific sequences of the genes and measure the amount of complementary RNA bound. However, not all of them reflect the amount of RNA per gene with the same accuracy. Analyses of series of arrays hybridized with controlled growing concentrations of RNA (experiments denoted as "spiked-in" arrays) have revealed large variability in the intensity signals between probes measuring fragments of the same RNA *(29)*. Spiked-in experiments showed that against a known increase in the RNA amount, the signals of expression detected were not proportional in the same range *(29)*. This unscaled probe-dependent measurement of abundance is termed the "probe effect" and implies that the same oligo probe can be comparable between different samples, but different probes are not comparable because they do not react in the same way to the same amount of RNA in the sample. By extension this effect equally affects a group of probes (i.e., the probeset).

The causes of these differences lie on the physicochemical reactivity properties inherent to each probe sequence. For example, a broadly used parameter in molecular genetics for its essential role in primer affinity is the guanine–cytosine percentage (i.e., GC content) of the oligonucleotide. Another cause, in this case associated to the probe-mapping locus, is the distance to the 3′-end of the gene. This was primarily reported by the 3′-IVT designs of *Affymetrix* microarrays. The 3′-poly-A amplification performed before the hybridization in these devices can considerably bias the abundance measures detected, depending on how far one specific probe is located from the 3′-poly-A end. Another effect that has long been reported for microarray probes is called cross hybridization. Probes should bind the sample RNA fragments in a highly specific manner, but considering the design limitations to the existing knowledge of the genome, they might bind unspecifically RNA fragments, that come from regions or loci different to the desired locus. Nowadays, the researcher can find useful bioinformatic tools to mitigate those undesired effects (see Section 3).

2.3 RNA Sequencing

As explained before, microarrays have so far been the most used instrument in global gene expression profiling, but this hybridization-based technology has a "knowledge restriction" since it is capable to quantify only annotated genes or genes derived from known sequences of cDNA and EST libraries. By contrast, massively parallel sequencing of RNA (RNA-seq) has the potential to truly embrace the whole transcriptome, providing an excellent tool for the discovery of novel transcription entities. RNA-seq also offers a larger dynamic range to measure RNA abundance in a nonrelative quantitative mode and with high reproducibility *(30)*. These advantages, and the upcoming price reductions, make RNA-seq an increasingly attractive strategy, even for unknown genome organisms.

Since the beginning of the "next-generation sequencing" (NGS) revolution, several biotechnology companies entered into a marathon to produce the most reliable sequencing technology. In the case of applying sequencing to RNA molecules, there are three main high-throughput sequencing technologies associated with three commercial platforms: *Roche 454* sequencing (based in pyrosequencing chemistry, with 250–400 bp reads length, 100 Mb per run, and around 90 USD/Mb); *SOLiD* sequencing (ligation-based sequencing, with 35–75 bp reads length, 3000 Mb per run, and around 10 USD/Mb); *Illumina-Solexa* sequencing (polymerase-based sequence-by-synthesis, with 50–150 bp reads length, 300,000 Mb per run, and around 2 USD/Mb). These next-generation platforms are in continuous development to improve the sequencing accuracy, the reads length and to reduce costs. Moreover, other commercial platforms are launching new sequencing technologies to achieve such improvements in performance (as, e.g., *Ion Torrent* http://www.iontorrent.com/ or *Pacific Biosciences* http://www.pacificbiosciences.com/). Within the three most used NGS platforms described above, the preferred may be so far the *Illumina-Solexa* platform (*Illumina*, Inc., San Diego, CA) due to its low cost and large coverage when applied to genome-wide studies.

2.3.1 Design and Use of Illumina-Solexa *RNA-seq*

The first step in an RNA-seq experiment is the copy-DNA (cDNA) library preparation from sample RNA. At this step, sample RNA must be isolated and purified, defining previously which type of RNA wants to be analysed. As explained before, there are many types of RNA molecules based on their size and function. For example, if the aim of the study is to identify the protein-coding transcriptome of a given cell type, we can select the poly-A tail molecules (i.e., mainly mRNAs) and perform a depletion step of ribosomic RNA. Also, application of strand-specific protocols or pair-end sequencing should be decided before starting the work plan. Once the mRNA is isolated it will be fragmented, reverse-transcribed (to synthesize the double-stranded cDNA), and ends-repaired with ligation of adaptors for amplification steps (*Illumina* provides its own reagent kits and settled protocols). Once the cDNA library has been prepared, another step is cDNA concentration and quality assessment. Sample quality can be assessed using an electrophoretic chip (e.g., *Agilent DNA-1000* chip in a *2100 Bioanalyzer Instrument*; Agilent Technologies, Inc., Santa Clara, CA), which measures the fragment size and analyzes its distribution. In general, the better defined the fragment size peak, the highest the library quality will be. The sample cDNA library will next be loaded into a solid supporting unit where base-by-base sequencing will be performed per fragment, and the newly synthesized chains will be reported in an entity called "read". At present, the *Illumina HiSeq* machinery yields up to 3 billion reads per sequencing run (http://www.illumina.com/). The *Illumina HiSeq* platform has two sequencing units (flow cells), each providing eight separate sequencing reactions (lanes), which, depending on the total read length, takes between 1.5 and 11 days to complete. Different sequencing protocols are available, depending on the genome to sequence and the result expected by the researcher. The main parameters to have in mind are the "read length" desired (currently for *Illumina* from 50 to 150 bp) and the single-end versus pair-end option. An important concept to extract successful sequencing results is the number of total reads yielded per sample, also called "sequencing depth". A depth ranging from 30 to 50 million reads yields from a minimal to a near-saturated coverage of the transcriptome *(31)*. Any value between these, should report good abundance estimations of gene expression.

Previous to sequencing cycles, another amplification step is performed inside the sequencer, over the lane surface. This is denoted as *bridge-amplification* because of the morphology adopted by the fragments during priming steps of PCR. The generated fragments growth attached to the surface conforming clusters, each cluster coming from only one origin sequence. Once the clusters have been produced, they will be submitted to cycles of sequencing, and during each cycle each cluster incorporates one dideoxynucleotide (ddN) bound to a fluorescent label, which emits a light signal that is scanned and registered by the sequencing machine. Four different fluorophores are used, each one associated to a different ddN (ddA, ddT, ddC, ddG). The output is a collection of lectures, sequences, or reads with a quality assessment value per nucleotide written into a *fasta* file type. Usual extension formats are ".bam" or ".fastq" (the first is a more compressed format).

2.3.2 *Drawbacks of RNA-Seq Alignment*

Before any other functional analysis, the reads obtained from an RNA-seq experiment must be localized into the coordinates of the reference genome, a process known as alignment or "mapping of reads". Usually, reads have unique locations within the genome or transcriptome, although a significant proportion can map to more than one location, known as multiple-mapping reads. They might originate from gene duplications, repetitive regions, or overlapping genes. Thus, their assignment to specific genes becomes problematic and can lead to mistakes in expression measurements. The easiest way to handle the problem is removing these ambiguous reads from further analysis, but it implies the loss of information. If the coverage is high, the results should not change notably by removal of the ambiguous reads. Currently the scientific community is working toward the development of tools that deal with multi-reads in an optimized manner. In the case that the reference genome is unknown, the alignment is not possible and a more complicated scenario is presented since we will need to assemble the transcriptome *de novo*, building each cDNA. Such assembly requires much higher read length and sequencing depth to achieve maximum coverage and allow overlapping of consecutive reads. Upcoming third-generation sequencing of single-molecule (provided, for example, by *Pacific Biosciences*) is expected to have the potential to sequence complete transcripts at once. Moreover, if other steps can be avoided, like fragmentation of the cDNA or PCR amplification, these third-generation technologies would drastically reduce computing time and improve the identification of all types of RNAs. Finally, as in the case of microarrays, the RNA-seq analyses have to apply methods for accurate estimation of the abundance of genes, transcripts, or splice variants, and to test for DE between conditions or sample types. The development of successful bioinformatic methods for these purposes applied to RNA-seq is still a challenge.

2.4 Other High-Throughput Platforms to Investigate Regulation of Gene Expression

Control of the amount and time of transcripts production is essential for the cell to have functional gene products when and where needed. Furthermore, this control empowers the cell to adapt and respond to environmental stimuli. Three main transcriptional regulatory elements can be defined: (i) elements at gene locus level, that are regions defined by TFs or miRNAs interactions; (ii) elements at transcript level, that drive alternative splicing, trans-splicing, or RNA editing; and (iii) elements at epigenetic level, that are regions defined by DNA methylation or histone modification. Experimental data that measure changes along these regulation elements using large-scale technologies may be integrated with genome-wide expression data in order to study and understand the cellular systems that operate under determined behaviors, conditions, or diseases.

2.4.1 TFs Profiling: ChIP-on-Chip and ChIP-Seq Technologies

As previously indicated in this chapter, TFs are proteins that bind to specific DNA sequences and thereby control the expression flow of gene products. These factors can act alone, by pairs, or in complexes, and even more, over different genes at the same time, activating or blocking their transcription. To unravel the biological links established between different regulators and functional gene cassettes we need to freeze the interactions taking place between these proteins and DNA into an *in vivo* cellular context. ChIP-on-chip is a technique that combines chromatin immunoprecipitation ("ChIP") with microarray technology ("chip"). It seeks to evidence interactions between TFs and their DNA sequence binding sites (TFBS) on a genome-wide basis. ChIP sequencing, also known as ChIP-seq, is used with the same purpose, but combines ChIP with massively parallel DNA sequencing to identify the binding sites of DNA-associated proteins. The goal of these approaches is to identify functional regulatory elements in the genome that appear associated to a specific TF of interest. These analyses complement and enhance gene expression measurements because the finding of genome regulatory regions is not limited to classical "promoter" regions, but the Omic techniques also allow to identify "enhancers," "silencers," "activators," "repressors," or "insulators" that are usually at larger distances from the gene transcription initiation point *(32)*.

2.4.2 microRNA Expression Profiling: TaqMan® miRNA Arrays

As previously indicated in this chapter, miRNAs are a class of naturally occurring ncRNAs that play a key role in gene regulation. These transcripts are highly conserved single-stranded RNAs ($\approx$22 nucleotides) that are cleaved from larger hairpin precursor transcripts. They exert their regulatory effects through the RNA interference (RNAi) machinery by cleaving or, most often, repressing the translation of their mRNA targets, being predicted to regulate at least one-third of all human genes *(33)*. One of the Omic technologies best applied to detect miRNAs at large scale is the TaqMan® miRNA card (*Applied Biosystems*, Foster City, CA). This biochip runs several hundred qPCR assays on a micro fluidic system. It includes all entities of miRBase v14 (http://www.mirbase.org/) across a two-card set for a total of 754 specific human miRNAs. Each card contains four control assays: three endogenous controls and one negative control. This qPCR-based technology allows the detection of miRNAs since it is applicable to very small and usually low-expressed nucleic acids with high reliability, enabling the generation of a comprehensive expression profile using as little as 1 ng of input total RNA.

2.4.3 Alternative Splicing Detection: Exon Arrays

During alternative splicing a pre-mRNA is transformed into mature mRNAs through inclusion/exclusion of different exonic regions, generating different functional isoforms of a gene. Consequently, the same gene locus may produce a variable number of transcripts that considerably amplify the complexity of transcriptomics. The expression of different isoforms is usually associated with tissue and cell type specificity. Alternative splicing is considered a highly regulated process that appeared in the evolution of multicellular organisms *(34)*.

The *Affymetrix* Human Exon 1.0 ST array (2005) was designed to quantify expression at the individual exon level. It contains over 5 million individual probes distributed in 1.4 million probesets. Each probeset comprises four individual probes that usually correspond to a single exon, and longer exons have several probesets. Probesets targeting individual exons can be further grouped into "transcript clusters" and gene-level expression can be obtained by considering all probes comprising a transcript cluster. Taking apart RNA sequencing, this exon array has reached one of the highest rates of transcriptome coverage of high-throughput hybridization technologies *(17)*. The advantage of providing genome-wide expression data at exon level and its vast availability at public repositories clearly benefit the study of alternative splicing events. Several bioinformatic tools and algorithms have been developed since the appearance of exon arrays, most of them implemented in the R/Bioconductor framework.

2.4.4 Epigenetic Profiling: Methylation Arrays

Epigenetics is a relatively new subject dedicated to the study of nonsequence DNA changes and modifications that affect the chromatin structure and regulate gene expression *(35)*. DNA methylation is the most stable of all epigenetic modifications, and it is specified by tissue and cell type during development and differentiation, although under certain cellular signals or transformations it can be dynamically altered. It consists of the addition of methyl groups to DNA, mostly at CpG sites (two adjacent nucleotides of cytosine and guanine, respectively) to convert cytosine to 5-methyl-cytosine. Some areas of the genome are methylated more heavily than others, being less transcriptionally active. Some methylation patterns are also inherited from one of the parents, known as *genetic imprinting*.

For DNA methylation profiling across the whole genome, there are a couple of technologies that might also be combined to genome-wide expression data: Infinium® 450 K methylation BeadChip (*Illumina*, San Diego, CA) and bisulfite sequencing. Both assays treat sample DNA with sodium bisulfite ($NaHSO_3$ to uracils) that converts unmethylated cytosines into thymines, whereas methylated cytosines remain protected from conversion. The goal is to determine methylation at single CpG resolution, generating a quantitative signal of methylation pattern per CpG. The older version of Infinium® BeadChip (27 K) contained two types of 50 bp probes: one designed to specifically hybridize to the methylated CpG site (M) and the other to the unmethylated

CpG site (U). At each CpG site, single-base extension generates the same color signal for both the M and U probe. The probe design relies on the assumption that any CpG sites underlying the probe are methylated to the same extent as the target site. The current Infinium® Human Methylation 450 K BeadChip (≈450,000 probes) has added another assay for coverage purposes: the Infinium II probes, which use just one probe per CpG locus, and employ different dye colors (green and red) to differentiate between M and U signals, respectively. It includes degenerate (R) bases at CpG sites. However, the Infinium type II probes can only include a maximum of three R bases, so type I probes from previous versions are kept to assay regions of the DNA with a higher density of CpG sites (i.e., CpG islands associated with promoter regions).

3 BIOINFORMATIC ANALYSIS OF GENE EXPRESSION DATA

3.1 Computational Methods to Achieve Gene Expression Profiling and Find Gene Signatures

Gene expression is one of the lowest levels at which the phenotype arising from a genotype can be observed; however, gene activity differs not only depending on a cell's DNA, but also on the cell type (i.e., tissue) and on the conditions the cell has been exposed to. Since the combination of genes that are active in a cell type is what grants its unique properties, it is interesting to see which genes are turned on or off in a specific cell type or in a certain biological condition. This is what is called "gene expression profiling," the study of the activity of the genes in a cell in a given state. Therefore, studying the cell expression profile can provide insights into its inner cellular processes (e.g., metabolic processes, cell cycle phase, etc.) or the functions it is involved in. In addition, by comparing the expression profiles of cells in different situations, scientists can identify the processes that are affected by a given condition (e.g., processes altered in a disease or modified by drugs), select gene targets for further study, and, hopefully, understand the biomolecular mechanisms laying underneath.

When the combination of active genes is unique for a specific biological phenotype, this group of genes is known as "gene signature". Finding this group of genes, whose expression pattern is unique to the studied condition, is one of the primary objectives in most gene expression profiling studies. Gene signatures obtained through gene expression studies have already successfully been applied in commercial tests, for example, in the *Oncotype DX (36,37)*, which analyzes the expression of a set of genes to determine whether it is recommended to provide chemotherapy to breast cancer patients, or in the *MammaPrint (38)*, which evaluates the risk of recurrence. Gene signatures from published experiments are collected in databases such as GeneSigDB (http://compbio.dfci.harvard.edu/genesigdb/) *(39)*, in order to make them available to the research community. There are also databases that collect experimental data from raw gene expression studies: ArrayExpress (http://www.ebi.ac.uk/arrayexpress) *(40)* and GEO (http://www.ncbi.nlm.nih.gov/geo/) *(28)*.

Gene signatures can be found using three main computational approaches: DE, class assignment and prediction based on expression, and class discovery based on expression.

3.1.1 Differential Expression

This approach consists of a comparison of expression profiles between groups of samples to identify the genes that show a statistically significant difference in their level of expression. The analyzed samples can be either different cell types, which would allow identifying the transcriptomic profile that makes these cells different, or the same cell type in different conditions. A DE analysis is generally the first step in any expression profiling analysis and a required step in order to find a gene signature. However, keep in mind that a set of differentially expressed genes between compared conditions is not necessarily a gene signature of the studied cells or samples because the DE is always a relative and comparative signal. Therefore, a set of differentially expressed genes might be enough to distinguish the classes from each other, but not from other cell types or other conditions.

Genes found through DE will provide information regarding the processes affected in the studied states, and have many potential applications. These include, for example, finding genes associated with diseases and evaluating the response to treatments, either by exploring the expression changes that certain drugs produce or by monitoring the expression changes associated with certain treatments. The genes associated with diseases can be potential targets for drugs *(41)*, and drugs can be developed in order to control the expression of such key disease genes by blocking its regulators or inhibiting its protein products.

3.1.2 Class Assignment and Prediction (Classification) *Based on Expression*

Gene expression profiles can be used in order to differentiate several predefined classes or sample types in a dataset. An ideal situation would be to find a gene whose expression allows distinguishing between two classes. Unfortunately, the gene alone usually does not work and it will normally be required to find a combination of genes that, altogether, allow differentiating the states under study. These genes can then be provided with a classification system built using machine learning techniques, which will be able to determine the membership of a new sample whose class is not known. In order to be able to identify to which class a new sample is more likely to belong, the classifier first needs to be trained using a set of samples whose class is known.

The construction of classifiers also requires the selection of an optimal combination of genes that can discriminate between classes *(42)*. Once a classifier is trained, it can be provided with a new sample in order to determine to which class it is more likely to belong. Note that this technique is meant for classification of known states or subgroups. In order to successfully classify this new sample, it needs to belong to one of the biological states the classifier was provided as training. Potential applications for classifiers include tests for predisposition to a disease, diagnostic tests (i.e., identification of diseases and disease subtypes), and prognostic tests (i.e., prediction of a clinical outcome).

*3.1.3 Class Discovery (*Clustering*) Based on Expression*

In contrast to the previous applications, which rely on predefined classes or states, it is also possible to analyze the expression datasets in order to discover classes or subgroups. This type of analysis, also known as clustering, searches for subgroups of samples with common features within them and distinct from others. One application of these methods is to identify molecular classes and improve the understanding of the differences arising within apparently homogeneous conditions. In medicine, this method has been successfully applied to identify new subtypes of cancer *(43)*.

3.2 Algorithms to Analyze Microarray Expression Data

3.2.1 Expression Signal Calculation

Once expression microarrays have been hybridized and scanned, the fluorescence signal provided by each probe type within each cell of the array is transformed into a numeric value. In order to get an accurate expression value for each specific biological entity tested (gene, transcript, or exon), the raw signal values of the array probes need to be preprocessed. The first step is usually a background correction, which filters out the fluorescence noise obtained by nonspecific hybridization of mRNA. In addition, it is also required to compensate for the artificial variability on expression signal. This variability can be due to the amount of mRNA in different samples, hybridization efficiency, optical effects from scanning, location of the probe in the array, and so on. Since these are systematic differences between probes or arrays, they can be analyzed and normalized. This normalization step will get rid of the variability due to the technology or the measurement technique. During background correction and normalization, the numeric values are usually log-transformed (logarithmic transformation) in order to even out the distributions and better allow the use of parametrical statistical methods. Finally, in order to quantify the mRNA of a specific gene, it is also required to map and summarize the signal of the probes. The mapping of the array probes to the biological entities is determined by the *Chip Definition File* (CDF), which contains which probes correspond to each biological entity: gene, transcript, or exon.

By default, many chips map the probes to *probesets* (i.e., groups of probes) that are labeled with arbitrary IDs given by the manufacturer (e.g., 212983_at by *Affymetrix*), that are later assigned to genes. However, the groups of probes can be directly summarized into genes by changing the CDF file (i.e., providing direct remapping of probes to genes) *(17)*.

There are multiple methods to perform each of the steps described (background correction, normalization, and summarization). In addition, preprocessing steps are platform-dependent, which means they need to be adjusted to the specific array technology. Since the final expression value can vary greatly depending on the preprocessing algorithms used, there are multiple studies analyzing and comparing the different methods *(44,45)*. One of the best and most broadly used methods is the Robust Multiarray Average (RMA) algorithm *(29)*.

3.2.2 Differential Gene Expression

Once the expression signal has been calculated for each biological entity tested (gene, transcript, or exon), the most common approach is to proceed through comparative analyses of the expression signals in different conditions. Therefore, the most common analysis is the calculation of the difference or the ratio in expression between two conditions—studied and control—or between a series of conditions. There are multiple methods to calculate DE. Some of the most popular packages are Significance Analysis of Microarrays (SAM) *(46)* and Linear Models for Microarray Data Analysis (LIMMA) *(47)* (both are freely available in *Bioconductor*). SAM is a robust method to calculate differentially expressed genes that uses permutations to achieve better accuracy and provides p-values for each assigned gene and also a *false discovery rate* (FDR) to allow multiple testing corrections. LIMMA is a versatile package that includes a collection of models for DE and allows performing multiple contrasts, being quite suitable for complex comparisons.

3.3 Algorithms to Analyze RNA-Seq Data

3.3.1 Alignment of Reads

The first data outcome from an RNA-sequencing analysis constitute the raw "reads", which should be evaluated for quality control using a tool such as FastQC (http://www.bioinformatics.babraham.ac.uk/projects/fastqc/). This program reports a series of fragment quantifications that the researcher can use as quality control parameters. They can help to detect artifacts associated with library preparation or sequencing cycles. Once the quality of the reads is shown correct, the next step is the alignment (see Figure 5). During the alignment process, reads are mapped into the sequence of the reference genome. It should be considered that part of the reads will fall into points of splicing (known as "splice junctions"), which means that these reads will present part

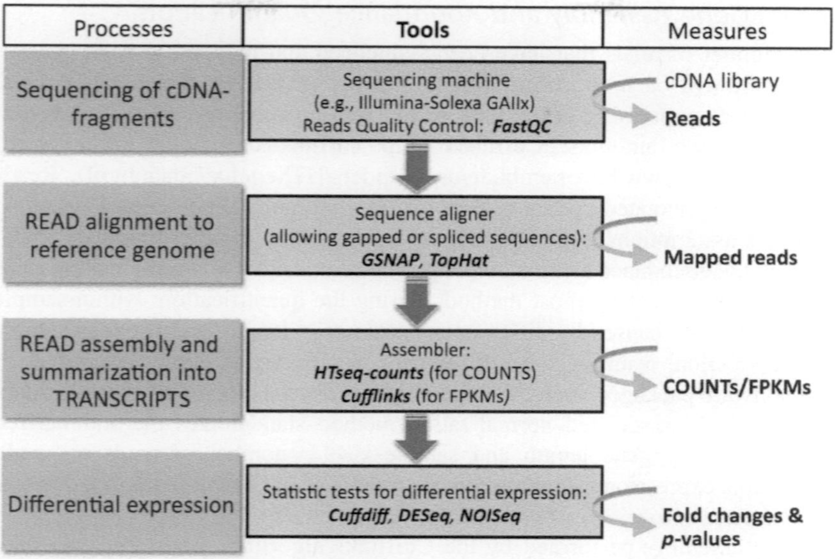

FIGURE 5 Schematic workflow showing the four main processing steps, tools, and measures for a RNA-seq data analysis. The first step corresponds to the data outcome from a RNA-seq experiment that provides the raw "reads", which can be evaluated for quality control using a tool such as *FastQC*. The second step is the "reads" alignment. During the alignment process, reads will be mapped into the sequence of the reference genome. Some of the tools for alignment are *TopHat* or *GSNAP*. The third step is the "reads" assembly and summarization into transcripts to transform the "reads" in COUNTs or FPKMs that are proportional to the quantity of mRNA per locus. Some tools applied in these calculations are *Cufflinks* or *Htseq-counts*. Finally, the fourth step is the application of differential expression analyses to detect the expression changes with respect to control samples. Some tools applied in these calculations are *Cuffdiff*, *DESeq*, or *NOISeq*. (For color version of this figure, the reader is referred to the online version of this chapter.)

of the sequence shared between two consecutive exons. The alignment tools usually implement methods to split apart such reads, called "gapped aligners," to allow a proper and more time-efficient mapping of splice junctions. Some of the tools for alignment are TopHat (http://tophat.cbcb.umd.edu/) or GSNAP *(48)*. Based on these alignment methods, important information about the different mRNA splice variants might be extracted later on the analysis workflow. Another approach to align is using the same annotated transcriptome as reference. In this way the aligner does not need to consider gaps into reads, and it is less computationally consuming. The output of the alignment can be divided into three types of reads: (i) reads with unique mapping (only one localization in the genome); (ii) reads with multiple mapping (more than one localization); and (iii) reads that do not map anywhere in the reference (not localized).

3.3.2 Gene Assembly and Abundance Quantification

The number of reads that have been sequenced and mapped to each gene or transcript will be proportional to the quantity of mRNA that the biological sample was expressing at the time of extraction. There are several tools that calculate such amounts: Cufflinks (http://cufflinks.cbcb.umd.edu) or htseq-counts (http://www.huber.embl.de/users/anders/HTSeq/doc/count.html). Reads assigned to annotated coding regions of the genome will be counted based on different assumptions, depending on the used method. In order to reduce deviations in the abundance estimation per gene or transcript, two normalization steps are usually applied by most methods during the quantification: within-sample bias correction, caused by differences in transcript length, and between-sample bias correction, resulting from differences in sequencing depth. Assembler or DE software packages often implement their preferred method of normalization. The Cufflinks-associated normalization method standardizes the summarized read counts by gene length and sample size (=number of reads mapped). The output expression values are denoted as RPKM or FPKM: *reads or fragments per kilobase of exon model per million mapped reads.*

The assembly performed by the Cufflinks algorithm gives the researcher three possible approaches: reference annotation assembly (adjusted to transcriptome knowledge), reference annotation-based assembly (transcriptome-driven but allows the discovery of new entities), and without reference annotation assembly (*de novo*). Estimating the expression levels of individual splice variants of a gene is rather complex, as they typically share a set of exons, and thus only a minor fraction of the reads will align uniquely to the distinct regions of a particular splice variant. Methods such as Cufflinks use a statistical model to estimate the proportion of reads that can be assigned to individual splice variants, generally known as the assembly step.

3.3.3 Differential Gene Expression

For DE testing, we will have to compare gene/transcript abundance values between samples. At this point only between-sample normalization would be necessary. For instance, the adjustment of counts to the total number of reads mapped in the sample would be enough. Although intuitive, this approach introduces inaccuracies if there is a minor fraction of genes that comprise the majority of the total read count. Consequently, normalization methods that are more robust to extreme values, such as the *median count ratio* and *trimmed mean of M-values* (TMM), have been developed. Another recently implemented approach to check for reproducibility among samples is called the *irreproducibility discovery rate* (IDR) *(49)*, which gives a probability value of the lack of concordance in gene abundances between samples.

To test for DE, distinct software tools have applied several statistical distributions previously, accommodating the discrete nature of RNA-seq data.

When the total number of reads per gene has been summarized, methods such as DESeq *(50)* evaluate DE using a negative binomial distribution. Comparing the relative abundance of different mRNA splice variants as meant before is a more challenging task. Methods such as cuffdiff (from Cufflinks) have been *ad-hoc* developed to handle testing for differences in alternative splicing variants between conditions.

3.4 Methods for Functional and Pathway Enrichment Analysis of Selected Gene Lists

Genome-wide expression analyses performed using the described high-throughput techniques (microarrays or RNA-seq) provide lists of "selected genes" that are associated with studies performed over specific sets of samples in definite biological contexts. One of the major challenges of current computational biology is to provide robust automatic methods for a meaningful functional annotation of the long lists of genes (or proteins, i.e., gene products) derived from such high-throughput studies. Functional enrichment analysis (EA) is at present the most popular methodology to derive functional implications of sets of cooperating genes. It uses statistical testing to find significant annotations in groups of genes. A recent review of enrichment tools categorizes them in three major classes: singular (SEA), modular (MEA), and gene-set (GSEA) *(51)*. Gene-set enrichment analysis (GSEA) is a popular tool to extract biological insight from complete ranked gene lists without the need for preselecting top genes *(52)*. Modular analysis can be considered an optimized second generation of functional enrichment since it uses "concurrent" annotation of genes in multiple annotation databases, improving the coverage and coherence of the results *(53,54)*. Another improved approach, that uses singular enrichment, is the functional annotation clustering method developed and published in 2007 by DAVID Bioinformatics Resources (http://david.abcc.ncifcrf.gov/) *(55)*. This method has a similar goal of finding functional modules (i.e., modules that include genes and terms), but, as indicated, it uses data derived from singular EA.

Functional EA, however, does not address several key problems associated with the biological annotations: (i) redundancy of the biological terms, which are repeated in many different annotation resources (e.g., *cell cycle* GO:0007049, *cell cycle* KEGG hsa04110, etc.) or are segregated in very similar terms with the same biological meaning (e.g., GO:0007049 *cell cycle* and GO:0022402 *cell cycle process*); (ii) bias in the annotation space due to highly frequent use of certain "promiscuous" terms that are unspecific (e.g., GO:0050789 *regulation of biological process*, includes more than 40% of all human genes annotated in GO-BP); (iii) inadequate functional annotation of many genes that are well known (e.g., NRAS human gene product P01111 is not annotated to GO:0043410 *positive regulation of MAPKKK cascade*, but the role of this gene in the MAPK signaling is well known, since it is paralogous to gene HRAS, which has a central role in such a pathway).

To overcome these limitations and challenges, a recent new computational method to find significant and coherent metagroups of genes and terms, eliminating redundant and noninformative data has been developed *(54)*. This method, called GeneTerm-Linker (http://gtlinker.cnb.csic.es/), takes the output of an EA and produces a simple result, including genes and co-annotations associated in metagroups of genes and terms. These metagroups are ranked by analysis of their statistical significance and coherence, in order to show the most relevant functions present in the query gene list. The innovation and genuine value of GeneTerm-Linker is to provide a single coherent solution to the problem of functional annotation of complex lists of genes *(54)*. To achieve this, it combines multiple nonorthogonal and nonhomogeneous biological annotation spaces going beyond EA approaches, since standard EA methods give as a result many separated lists of genes and annotations not integrated and redundant, and many times including low-information content terms. Therefore, GeneTerm-Linker helps to alleviate such difficulties, offering a step forward in the EA methods.

REFERENCES

1. International Human Genome Sequencing Consortium; Lander, E. S.; et al. *Nature* **2001**, *409*, 860–921.
2. Human Genome Celera Consortium; Venter, J. C.; et al. *Science* **2001**, *291*, 1304–1351.
3. Pearson, H. *Nature* **2006**, *441*, 398–401.
4. ENCODE Project Consortium; Birney, E.; et al. *Nature* **2007**, *447*, 799–816.
5. Gerstein, M. B.; Bruce, C.; Rozowsky, J. S.; Zheng, D.; Du, J.; Korbel, J. O.; Emanuelsson, O.; Zhang, Z. D.; Weissman, S.; Snyder, M. *Genome Res.* **2007**, *17*, 669–681.
6. Wain, H. M.; Bruford, E. A.; Lovering, R. C.; Lush, M. J.; Wright, M. W.; Povey, S. *Genomics* **2002**, *79*, 464–470.
7. Mattick, J. S. *EMBO Rep.* **2001**, *2*, 986–991.
8. Mattick, J. S. *PLoS Genet.* **2009**, *5*, e1000459.
9. Mattick, J. S.; Makunin, I. V. *Hum. Mol. Genet.* **2006**, *15*(Spec. No. 1), R17–R29.
10. Washietl, S.; Pedersen, J. S.; Korbel, J. O.; Stocsits, C.; Gruber, A. R.; Hackermüller, J.; Hertel, J.; Lindemeyer, M.; Reiche, K.; Tanzer, A.; Ucla, C.; Wyss, C.; Antonarakis, S. E.; Denoeud, F.; Lagarde, J.; Drenkow, J.; Kapranov, P.; Gingeras, T. R.; Guigó, R.; Snyder, M.; Gerstein, M. B.; Reymond, A.; Hofacker, I. L.; Stadler, P. F. *Genome Res.* **2007**, *17*, 852–864.
11. FANTOM Consortium; RIKEN Genome Exploration Research Group and Genome Science Group (Genome Network Project Core Group); Carninci, P.; et al. *Science* **2005**, *309*, 1559–1563.
12. Managadze, D.; Lobkovsky, A. E.; Wolf, Y. I.; Shabalina, S. A.; Rogozin, I. B.; Koonin, E. V. *PLoS Comput. Biol.* **2013**, *9*, e1002917.
13. Brent, M. R. *Genome Res.* **2005**, *15*, 1777–1786.
14. Rual, J. F.; Hill, D. E.; Vidal, M. *Curr. Opin. Chem. Biol.* **2004**, *8*, 20–25.
15. Maglott, D.; Ostell, J.; Pruitt, K. D.; Tatusova, T. *Nucleic Acids Res.* **2011**, *39*, D52–D57.
16. Flicek, P.; Ahmed, I.; Amode, M. R.; Barrell, D.; Beal, K.; Brent, S.; Carvalho-Silva, D.; Clapham, P.; Coates, G.; Fairley, S.; Fitzgerald, S.; Gil, L.; García-Girón, C.; Gordon, L.;

Hourlier, T.; Hunt, S.; Juettemann, T.; Kähäri, A. K.; Keenan, S.; Komorowska, M.; Kulesha, E.; Longden, I.; Maurel, T.; McLaren, W. M.; Muffato, M.; Nag, R.; Overduin, B.; Pignatelli, M.; Pritchard, B.; Pritchard, E.; Riat, H. S.; Ritchie, G. R.; Ruffier, M.; Schuster, M.; Sheppard, D.; Sobral, D.; Taylor, K.; Thormann, A.; Trevanion, S.; White, S.; Wilder, S. P.; Aken, B. L.; Birney, E.; Cunningham, F.; Dunham, I.; Harrow, J.; Herrero, J.; Hubbard, T. J.; Johnson, N.; Kinsella, R.; Parker, A.; Spudich, G.; Yates, A.; Zadissa, A.; Searle, S. M. *Nucleic Acids Res.* **2013**, *41*, D48–D55.

17. Risueño, A.; Fontanillo, C.; Dinger, M. E.; De Las Rivas, J. *BMC Bioinforma.* **2010**, *11*, 221.
18. Hobert, O. *Science* **2008**, *319*, 1785–1786.
19. Vaquerizas, J. M.; Kummerfeld, S. K.; Teichmann, S. A.; Luscombe, N. M. *Nat. Rev. Genet.* **2009**, *10*, 252–263.
20. Ho Sui, S. J.; Fulton, D. L.; Arenillas, D. J.; Kwon, A. T.; Wasserman, W. W. *Nucleic Acids Res.* **2007**, *35*, W245–W252.
21. FANTOM Consortium; Ravasi, T.; et al. *Cell* **2010**, *140*, 744–752.
22. Suzuki, H.; Fukunishi, Y.; Kagawa, I.; Saito, R.; Oda, H.; Endo, T.; Kondo, S.; Bono, H.; Okazaki, Y.; Hayashizaki, Y. *Genome Res.* **2001**, *11*, 1758–1765.
23. VanGuilder, H. D.; Vrana, K. E.; Freeman, W. M. *Biotechniques* **2008**, *44*, 619–626.
24. Rung, J.; Brazma, A. *Nat. Rev. Genet.* **2013**, *14*, 89–99.
25. Prieto, C.; Risueño, A.; Fontanillo, C.; De las Rivas, J. *PLoS One* **2008**, *3*, e3911.
26. Vicent, S.; Luis-Ravelo, D.; Antón, I.; García-Tuñón, I.; Borrás-Cuesta, F.; Dotor, J.; De Las Rivas, J.; Lecanda, F. *Cancer Res.* **2008**, *68*, 2275–2285.
27. MAQC Consortium; Shi, L.; et al. *Nat. Biotechnol.* **2006**, *24*, 1151–1161.
28. Barrett, T.; Wilhite, S. E.; Ledoux, P.; Evangelista, C.; Kim, I. F.; Tomashevsky, M.; Marshall, K. A.; Phillippy, K. H.; Sherman, P. M.; Holko, M.; Yefanov, A.; Lee, H.; Zhang, N.; Robertson, C. L.; Serova, N.; Davis, S.; Soboleva, A. *Nucleic Acids Res.* **2013**, *41*, D991–D995.
29. Irizarry, R. A.; Hobbs, B.; Collin, F.; Beazer-Barclay, Y. D.; Antonellis, K. J.; Scherf, U.; Speed, T. P. *Biostatistics* **2003**, *4*, 249–264.
30. Marioni, J. C.; Mason, C. E.; Mane, S. M.; Stephens, M.; Gilad, Y. *Genome Res.* **2008**, *18*, 1509–1517.
31. Tarazona, S.; García-Alcalde, F.; Dopazo, J.; Ferrer, A.; Conesa, A. *Genome Res.* **2011**, *21*, 2213–2223.
32. Hoheisel, J. D. *Nat. Rev. Genet.* **2006**, *7*, 200–210.
33. Zamore, P. D.; Haley, B. *Science* **2005**, *309*, 1519–1524.
34. Irimia, M.; Rukov, J. L.; Penny, D.; Roy, S. W. *BMC Evol. Biol.* **2007**, *7*, 188.
35. Bird, A. *Nature* **2007**, *447*, 396–398.
36. Paik, S. *Oncologist* **2007**, *12*, 631–635.
37. Cobleigh, M. A.; Tabesh, B.; Bitterman, P.; Baker, J.; Cronin, M.; Liu, M. L.; Borchik, R.; Mosquera, J. M.; Walker, M. G.; Shak, S. *Clin. Cancer Res.* **2005**, *11*, 8623–8631.
38. Drukker, C. A.; Bueno-de-Mesquita, J. M.; Retèl, V. P.; van Harten, W. H.; van Tinteren, H.; Wesseling, J.; Roumen, R. M. H.; Knauer, M.; van't Veer, L. J.; Sonke, G. S.; Rutgers, E. J. T.; van de Vijver, M. J.; Linn, S. C. *Int. J. Cancer* **2013**, *133*, 929–936.
39. Culhane, A. C.; Schwarzl, T.; Sultana, R.; Picard, K. C.; Picard, S. C.; Lu, T. H.; Franklin, K. R.; French, S. J.; Papenhausen, G.; Correll, M.; Quackenbush, J. *Nucleic Acids Res.* **2010**, *14*, D716–D725.
40. Rustici, G.; Kolesnikov, N.; Brandizi, M.; Burdett, T.; Dylag, M.; Emam, I.; Farne, A.; Hastings, E.; Ison, J.; Keays, M.; Kurbatova, N.; Malone, J.; Mani, R.; Mupo, A.; Pedro Pereira, R.; Pilicheva, E.; Rung, J.; Sharma, A.; Tang, Y. A.; Ternent, T.; Tikhonov, A.;

Welter, D.; Williams, E.; Brazma, A.; Parkinson, H.; Sarkans, U. *Nucleic Acids Res.* **2013**, *41*, D987–D990.

41. Chang, C. Y.; Kazmin, D.; Jasper, J. S.; Kunder, R.; Zuercher, W. J.; McDonnell, D. P. *Cancer Cell* **2011**, *20*, 500–510.
42. Bell, D. A.; Wang, H. *Mach. Learn.* **2000**, *41*, 175–195.
43. Golub, T. R.; Slonim, D. K.; Tamayo, P.; Huard, C.; Gaasenbeek, M.; Mesirov, J. P.; Coller, H.; Loh, M. L.; Downing, J. R.; Caligiuri, M. A.; Bloomfield, C. D.; Lander, E. S. *Science* **1999**, *286*, 531–537.
44. Millenaar, F. F.; Okyere, J.; May, S. T.; van Zanten, M.; Voesenek, L. A.; Peeters, A. J. *BMC Bioinformatics* **2006**, *7*, 137.
45. Qin, L. X.; Beyer, R. P.; Hudson, F. N.; Linford, N. J.; Morris, D. E.; Kerr, K. F. *BMC Bioinformatics* **2006**, *7*, 23.
46. Tusher, V. G.; Tibshirani, R.; Chu, G. *Proc. Natl. Acad. Sci. U. S. A.* **2001**, *98*, 5116–5121.
47. Smyth, G. K. *Stat. Appl. Genet. Mol. Biol.* **2004**, *3*, 1 (Article 3).
48. Wu, T. D.; Nacu, S. *Bioinformatics* **2010**, *26*, 873–881.
49. Djebali, S.; Davis, C. A.; Merkel, A.; Dobin, A.; Lassmann, T.; Mortazavi, A.; Tanzer, A.; Lagarde, J.; Lin, W.; Schlesinger, F.; Xue, C.; Marinov, G. K.; Khatun, J.; Williams, B. A.; Zaleski, C.; Rozowsky, J.; Röder, M.; Kokocinski, F.; Abdelhamid, R. F.; Alioto, T.; Antoshechkin, I.; Baer, M. T.; Bar, N. S.; Batut, P.; Bell, K.; Bell, I.; Chakrabortty, S.; Chen, X.; Chrast, J.; Curado, J.; Derrien, T.; Drenkow, J.; Dumais, E.; Dumais, J.; Duttagupta, R.; Falconnet, E.; Fastuca, M.; Fejes-Toth, K.; Ferreira, P.; Foissac, S.; Fullwood, M. J.; Gao, H.; Gonzalez, D.; Gordon, A.; Gunawardena, H.; Howald, C.; Jha, S.; Johnson, R.; Kapranov, P.; King, B.; Kingswood, C.; Luo, O. J.; Park, E.; Persaud, K.; Preall, J. B.; Ribeca, P.; Risk, B.; Robyr, D.; Sammeth, M.; Schaffer, L.; See, L. H.; Shahab, A.; Skancke, J.; Suzuki, A. M.; Takahashi, H.; Tilgner, H.; Trout, D.; Walters, N.; Wang, H.; Wrobel, J.; Yu, Y.; Ruan, X.; Hayashizaki, Y.; Harrow, J.; Gerstein, M.; Hubbard, T.; Reymond, A.; Antonarakis, S. E.; Hannon, G.; Giddings, M. C.; Ruan, Y.; Wold, B.; Carninci, P.; Guigó, R.; Gingeras, T. R. *Nature* **2012**, *489*, 101–108.
50. Anders, S.; Huber, W. *Genome Biol.* **2010**, *11*, R106.
51. Huang, D. W.; Sherman, B. T.; Lempicki, R. A. *Nucleic Acids Res.* **2009**, *37*, 1–13.
52. Subramanian, A.; Tamayo, P.; Mootha, V. K.; Mukherjee, S.; Ebert, B. L.; Gillette, M. A.; Paulovich, A.; Pomeroy, S. L.; Golub, T. R.; Lander, E. S.; Mesirov, J. P. *Proc. Natl. Acad. Sci. U. S. A.* **2005**, *102*, 15545–15550.
53. Nogales-Cadenas, R.; Carmona-Saez, P.; Vazquez, M.; Vicente, C.; Yang, X.; Tirado, F.; Carazo, J. M.; Pascual-Montano, A. *Nucleic Acids Res.* **2009**, *37*, W317–W322.
54. Fontanillo, C.; Nogales-Cadenas, R.; Pascual-Montano, A.; De Las Rivas, J. *PLoS One* **2011**, *6*, e24289.
55. Huang, D. W.; Sherman, B. T.; Tan, Q.; Collins, J. R.; Alvord, W. G.; Roayaei, J.; Stephens, R.; Baseler, M. W.; Lane, H. C.; Lempicki, R. A. *Genome Biol.* **2007**, *8*, R183.

Chapter 16

Bioinformatic Approaches to Increase Proteome Coverage

Francesco M. Mancuso[1], Salvatore Cappadona[1] and Eduard Sabidó
Proteomics Unit, Centre for Genomic Regulation (CRG) and Universitat Pompeu Fabra (UPF), Barcelona, Spain

Chapter Outline

1 INTRODUCTION

In many high-throughput proteomic experiments, protein mixtures are digested to peptides by a site-specific enzymatic protease. These peptides are then separated by liquid chromatography (LC), converted online to gas-phase ions, and analyzed by a mass spectrometer (MS), which scans the whole mass range and produces high-resolution MS spectra. From each spectrum, the acquisition software automatically selects a preset number of abundant peptides for fragmentation and further analysis by so-called tandem mass spectrometry (MS/MS). The resulting MS/MS spectra are then compared either to theoretical fragmentation spectra generated from a protein-sequence database or to spectral libraries to retrieve the corresponding peptide sequences. Protein inference is finally accomplished by assigning peptide sequences to proteins. This bottom-up approach, known as *shotgun*

1. F.M.M. and S.C. give an equal contribution in the redaction of the book chapter.

Comprehensive Analytical Chemistry, Vol. 63. http://dx.doi.org/10.1016/B978-0-444-62651-6.00017-9

proteomics, has become the method of choice in biological studies that demand deep *proteome coverage (1,2)*.

Recent studies investigated the current limits of shotgun proteomics. Michalski *et al. (3)* have estimated that in data-dependent acquisition (DDA) mode, of over 100,000 likely peptide features eluting in single shotgun runs, less than 20% are targeted for fragmentation; of those, typically less than 60% lead to peptide identification. In another recent study, Zubarev and coworkers *(4)* investigated how to improve shotgun performance using the advanced commercial LC–MS instrumentation available today. By means of multiparameter optimization and without any hardware or software modification, they showed that it is possible to identify more than 37,500 peptides and 4800 protein groups in a single ultra performance LC–MS run of the proteolytic digest from a human cancer cell line, with 3 h's gradient time and a very low amount of loaded material. These numbers were further increased by more than 50% and 10%, respectively, by running the sample in triplicates, thus achieving a coverage of over 50% of the expressed human cellular proteome.

Despite being the deepest protein coverage ever reported for a single-dimension experiment of such a short gradient, the study by Zubarev *et al. (4)* shows that there is ample room for further improvement as most of the sample information is not exploited, either because no peptide sequence data are available or because acquired spectra are not properly interpreted. Therefore, new advances in upstream analytical technologies such as the development of new acquisition methods and the refinement of protein/peptide separation techniques, as well as downstream bioinformatics data analysis, are required to improve proteome coverage. In particular, there are two ways to enhance proteome coverage: (i) by increasing the *fragmentation rate* (i.e., the number of peptide features targeted for tandem MS analysis), and (ii) by increasing the *identification rate* (i.e., by reducing the number of unassigned MS2 spectra).

The fragmentation rate can be enhanced by either increasing the scanning speed of the instruments or by reducing the complexity of peptide mixtures prior to MS detection. Scan rate has significantly evolved during recent years, and the current generation of MSs reaches scan rates around 100 Hz, which represents a qualitative improvement compared to older models. Similarly, sample preparation has also been refined, and multidimensional separation at the level of proteins, proteolytic peptides, or both have made it possible to achieve a comprehensive analysis of nearly the entire human proteome *(5)*. Unfortunately, this excellent performance is reached at the cost of huge sample consumption and long data acquisition time, which makes complex sample preparation inadequate for high-throughput biological and clinical research. Rapid advances in analytical techniques are increasingly narrowing the gap between monodimensional and multidimensional approaches. These improvements are likely to continue in the near future, and it is hoped that they will soon allow effective targeting of all currently detectable peptide features in standard shotgun analyses.

Despite the progress made in instrument acquisition methods and sample preparation techniques, the open issue that we highlight in this chapter is the significant number of fragmentation spectra that remain unassigned in high-throughput proteomics experiments. Indeed, in shotgun proteomic experiments frequently only a relatively small fraction of spectra are eventually assigned to a peptide sequence, with an identification success rate usually lying below 60%. Poor spectral matching can originate from two main causes: (i) low quality of the experimental spectra and/or (ii) absence of the corresponding theoretical spectra in the sequence database. Therefore, although there is an increase of fragmentation rates, there is still a need to reduce the number of *unidentified peptides* and increase the number of *identifiable peptides*.

The availability of high-resolution instrumentation has largely contributed to improving the number of assigned spectra by facilitating the correct detection of coeluting isobaric peptides and correct assignment of their charge state. Nevertheless, this issue is more strictly dependent on the bioinformatics tools used to infer peptide sequence from tandem mass spectra. Several initiatives have emerged to reduce the number of *unassigned spectra*, including the exploitation of different fragmentation techniques, the improvement of spectra quality, and the combination of search engines, among others. Similarly, information related to protein-splicing variants, amino acid peptide variants (polymorphisms), and posttranslational modifications (PTMs) have recently been introduced in data analysis pipelines to increase the rate of identifiable peptides.

In this chapter, we focus on the strategies that have been developed during the past few years to increase the proteome coverage in shotgun proteomics studies (Figure 1). The analytical techniques that might increase the fragmentation rate are outlined briefly in Section 2 (for a comprehensive review,

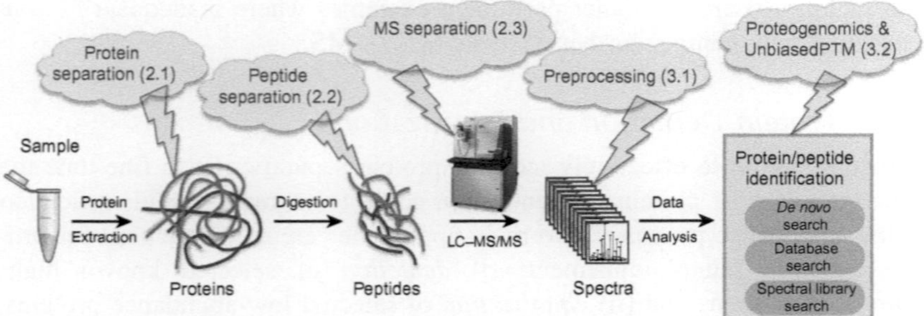

FIGURE 1 Workflow for a common analysis of MS-based proteomics. The flashes represent where it is possible to apply different methods (bioinformatics and not) in order to increase the number of identification. (For color version of this figure, the reader is referred to the online version of this chapter.)

see Zhang *et al.* *(2)*). The different bioinformatics strategies used to reduce the number of unassigned spectra and to improve the peptide identification rates are extensively detailed in Section 3.

2 INCREASING THE NUMBER OF FRAGMENTED FEATURES

No current analytical strategy is capable of fully resolving complex biological samples. For this reason, orthogonal separation techniques are often combined to maximize peptide separation before the mass spectrometric analysis. The aim is to reduce as much as possible the number of coeluting peptides introduced into the MS at any given time, so as to maximize peptide identification and proteome coverage.

In this section, we describe the complementary techniques that have been developed to increase the fragmentation rate in mass spectrometric protein analyses. Following the general shotgun proteomics workflow outlined in Figure 1, the first two paragraphs summarize the different strategies that are commonly used to maximize protein and peptide separation before mass spectrometric analysis, while the third paragraph outlines the strategies that allow for further separation after peptide ionization in the mass spectrometer.

2.1 Protein-Level Separation

2.1.1 Protein Isolation and Extraction

Proteins are part of a dynamic network of biomolecules that interact to regulate their localization and function within the cell. Disruption of this physical and chemical system of interactions has become the first paramount step in performing protein analysis by shotgun proteomics. *Protein isolation* techniques, for instance, have allowed understanding of the complex dynamics of proteins among cellular subcompartments, such as the nucleolus or the mitochondrion *(6)*. Membrane-embedded proteins *(7)* and DNA-binding transcription factors *(8)* are two other prominent examples where inadequate *protein extraction* may hamper further analysis by LC–MS.

2.1.2 Protein Depletion and Equalization

A second option to effectively increase protein separation is to fine-tune the dynamic range of protein concentration, so as to increase the identification of low-abundance proteins. Two main approaches are currently used for protein dynamic range adjustment: (i) *depletion* of selected known high-abundance proteins and (ii) *equalization* of selected low-abundance proteins. These two strategies have proved particularly useful for clinical samples *(9)*, where the dynamic range can reach up to nine orders of magnitude, which is much higher than the three to four orders of magnitude typically achievable by current instrumentation. Usually, only a few proteins are extremely abundant and can be selectively depleted before LC–MS analysis. These abundant proteins, such as serum albumin, can be either precipitated by chemical-based approaches or coupled to a specific antibody and depleted by immunoaffinity subtraction.

The equalization approach is based on the opposite strategy, in that only low-abundance proteins are collected and subjected to LC–MS analysis. The method relies on bead-bound combinatorial peptide ligand libraries *(10)* that enrich low-abundance proteins. This strategy is much cheaper than antibody precipitation, but it has the clear drawback that only proteins that have affinity to one of the millions of beads present in the library used can be equalized for further analysis.

2.1.3 Protein Separation and Fractionation

A third powerful option for the separation of complex protein mixtures is two-dimensional polyacrylamide gel electrophoresis (2D-PAGE) *(11)*. In this technique, proteins are first solubilized and denatured by a combination of detergents and chaotropes, such as SDS and urea, respectively; then they are separated in two orthogonal dimensions, based on their charge and their mass. The development of LC–MS-based proteomics strategies has caused a rapid decline of 2-D gel separation. Nevertheless, the two complementary techniques that allow protein separation on a 2-D gel are both still largely used independently for protein fractionation prior to LC–MS. 1D-SDS-PAGE remains a standard separation method for complex protein mixtures, which are separated based on molecular weight, cut into 10–20 slices, and digested in-gel. This technique is particularly appreciated for its ability to differentiate protein isoforms. The orthogonal technique, isoelectric focusing, separates proteins based on their isoelectric point (p*I*). This technique is particularly appreciated for its high resolution, separating proteins with a single charge difference into distinct gel bands.

2.2 Peptide-Level Separation

2.2.1 Proteolytic Digestion

Due to the better sensitivity of current MS platforms for low-molecular-weight molecules, shotgun proteomics has evolved into a somewhat counter-intuitive peptide-centric way, based on the assumption that proteins in the original sample can be identified by means of MS-mediated identification of their constituent proteolytic peptides. This reverse-engineering approach circumvents many of the challenges associated with intact protein separation because *proteolytic digestion* acts to normalize the biochemical heterogeneity of proteins. Furthermore, the presence of multiple peptides associated with the same protein increases the probability of identifying low-abundance proteins as well as PTMs.

Trypsin is by far the most popular enzyme used in proteomics *(12)* due to its high catalytic activity and specificity in cleaving proteins into peptides at the C-terminal of arginine and lysine residues. The generated peptides behave well under reversed-phase conditions and are very well suited for analysis by mass spectrometry, because of their average length (10–12 residues) and because of the presence of a C-terminal basic residue that enhances positive ionization. Furthermore, the balance in basicity between the free amine of the N-terminal of the peptide and the C-terminal arginine or lysine generally leads to good fragmentation spectra in MS/MS.

Other proteolytic enzymes commonly used in proteomics analysis include Lys-C, Lys-N, and chymotrypsin. In general, these enzymes differ by their specificity for cleaving the amide bond before or after a specific residue. For complex protein samples, a combination of highly selective proteases has proved to increase proteome coverage by creating complementary peptides *(13)*.

2.2.2 Peptide Separation and Fractionation

LC has proved to be an excellent separation technique for shotgun proteomics. It can be directly coupled to the electrospray source of the MS, permitting high resolving power, an important criterion to increase feature detection and identification. Nevertheless, as already mentioned, to date no separation method has proved able to fully resolve all components of a complex sample in a single analytical dimension. For this reason, in the last years, the combination of two or more orthogonal peptide separation procedures has been introduced to enhance proteome coverage. The rationale is that a 2-D separation system, resulting from the coupling of two orthogonal retention mechanisms, can achieve a higher resolving power than each single dimension *(14)*. Multidimensional separations can be coupled in either *online* or *offline* mode. Online methods involve a direct transfer of the eluant between the two columns, thus reducing manipulation, sample loss, and analysis time. Offline methods are based on the collection of fractions in the first dimension prior to analysis in the second. These methods are usually simpler in operation and design than other approaches and they offer greater flexibility in the choice of LC modes and their optimization.

The vast majority of 2D-LC methods implemented today are based on strong cation exchange (SCX) coupled to reverse-phase (RP) LC, in both online and offline mode. Yet, many more 2D-LC technologies have been proposed to tackle specific analytical challenges and they have been recently reviewed in *(15,16)*. In particular, HILIC is emerging as a valid alternative technique that has a higher orthogonality to RP than SCX, it can be used in both dimensions, and it is proving ideal for targeted analysis of PTMs, thanks to its suitability for the separation of highly polar molecules.

Novel combinations of separation techniques in three-dimensional configurations *(17)* represent a further step toward comprehensiveness and sensitivity.

2.3 MS-Level Separation

After entering the MS, eluting peptide cations have their mass-to-charge (*m/z*) values measured in the MS1 scan. The most abundant precursors are then selected for a series of sequential tandem MS events (MS2), usually ranging from 3 to 50, depending on the acquisition rate of the device. Following a second MS1 scan, a new group of targets is selected and the succession is cycled for the duration of the analysis. The process, called DDA, is at the very core of shotgun analysis and has been kept unchanged for many years. Despite significant improvements in scan rate, sensitivity, mass accuracy, and resolution that have constantly evolved with the MS hardware, the shotgun approach still suffers from two main limitations: (i) poor run-to-run reproducibility and (ii) inability to effectively target peptides of interest. Both issues are caused by the highly stochastic nature of choosing which *m/z* peaks to sample according to their abundance, which can preclude the identification of low-abundant peptides. Specifically, a recent interlaboratory study *(18)* determined that overlapping of peptide IDs between replicate experiments could be as low as 35%. This lack of reproducibility and completeness limits the quantity and quality of biological conclusions that can be drawn from a proteomic experiment. In fact, identifying statistically significant differences in protein or PTM abundance requires reproducibility.

In the following paragraphs, we introduce some instrument-level improvements and novel data-acquisition paradigms that have been recently exploited to increase the fragmentation rate.

2.3.1 Improvement of MS "Figures of Merit"

In recent years, the field of mass spectrometry has greatly evolved and MS instruments are experiencing significant progress in their main "figures of merit," such as sensitivity, scan rate, mass accuracy, and resolution. The main improvements are certainly coming from the combination of different mass analyzers, ion optics, and fragmentation sources into novel hybrid systems, capable of creating more informative and complementary peptide fragment ions through various fragmentation methods. Common mass analyzers that have been combined for peptide analysis from complex mixtures are linear ion traps, Orbitraps, Fourier Transform ion cyclotron resonance, and quadrupoles (Q).

2.3.2 Ion Mobility Configurations

An old technology that is regaining popularity for peptide cation separation is ion mobility *(19)*. This analyzer provides a complementary fractionation of the peptide ions as they are passed through a carrier buffer gas prior to mass analysis. Separation is achieved through the differential mobility of the ions due to their charge, size, and conformation. Application of ion mobility configurations at the inlet of MSs has proved very effective at separating peptides from background noise. It has also been shown to improve sensitivity, dynamic range, proteome coverage, and differential identification of isomeric peptides and phosphopeptides.

2.3.3 Gas-Phase Fractionation

In the former section, we have introduced online and offline fractionation as two techniques that reduce MS1 spectral complexity to increase the fragmentation rate. An alternative fractionation technique that can be used in the *m/z* dimension is gas-phase fractionation (GPF). In this technique, MS1 scans and precursor selection are performed over a narrow mass range (i.e., 100 *m/z*) instead of the full mass space, thus allowing detection of low-intensity features within that mass region. By repeating the acquisition in different sections of the MS1 range in sequential experiments, GPF achieves better proteome coverage as compared to DDA methods. A recent extension of this fractionation technique consists of scanning the whole mass range for precursor ion acquisition, while narrowing only the window from which a precursor can be selected for MS2 analysis. This technique has been termed *binning* or *tiling (20)*, depending on whether the full MS1 range is divided into continuous sequential segments or in several smaller segments distributed over the lower, middle, and upper mass regions. The main advantages of these techniques can be appreciated with label-free approaches, where sampling depth is increased without the requirement of additional instrument time or sample prefractionation.

2.3.4 Alternative Paradigms for Data Acquisition

Several DDA add-ons and alternatives have been proposed in order to tackle the issues of poor reproducibility and proteome coverage. These studies can be grouped into two categories: (i) hypothesis-driven acquisition (HDA), also called targeted acquisition and (ii) data-independent acquisition (DIA).

HDA approaches include inclusion/exclusion lists and selected reaction monitoring (SRM). Both inclusion and exclusion lists employ the basic framework of the DDA method. With inclusion lists, only MS1 signals falling into specific, preselected *m/z* ranges are interrogated, thus boosting sensitivity and improving reproducibility. The number of IDs, however, drops substantially because unlisted features are ignored. On the opposite side, with exclusion lists, features falling into specific, preselected *m/z* ranges are never interrogated. Hence, sampling depth is boosted by preventing selection of *m/z* values already identified in previous technical replicates. In the SRM approach, prior knowledge of the system allows the preselection of candidate proteins for detection and abundance measurement *(21)*. Exploiting the unique capabilities of triple quadrupole MS systems (QqQ), SRM selects predefined precursors and detects their product ions one at a time. In this way, the method can boost sensitivity for targeted peptides by orders of magnitude compared to the discovery-driven scanning techniques. SRM can also achieve the reproducibility that the shotgun approach lacks and it can be used to determine absolute abundance. These results have propelled the recent fervor in SRM methods for targeted proteomics. Nevertheless, the main obstacle hindering pervasive use of SRM is the technique's marked low throughput.

In DIA approaches *(22)*, no attention is paid to precursor abundance or even presence, and consecutive small (10–25 *m/z*) isolation windows are dissociated and mass-analyzed. Recent studies have demonstrated that DIA resolves the stochastic nature of the DDA approach and does indeed improve both sensitivity and reproducibility. However, the major disadvantage of the current data-independent implementation is a great loss in throughput, as the sample must be analyzed numerous times to cover the entire precursor mass range. Even worse, it also requires an enormous amount of sample, which is often not available. Recently, a novel technique called SWATH-MS *(23)* has been proposed that aims to complement a DIA method with an innovative data analysis approach based on targeted data extraction. Like in other DIA methods, the MS cycles through precursor acquisition windows designed to cover the whole range of 400–1200 *m/z*—in which most of the proteotypic peptide precursors of an organism fall—within 2–4 s. During each cycle, the MS fragments all the precursors from a given precursors window and it records a complete, high-accuracy fragment ion spectrum. The same range will be fragmented again in the next cycle, thus providing a time-resolved recording of fragment ions that elute on the chromatography. In this way, the SWATH-MS method provides highly multiplexed fragment ion spectra that are deterministically recorded over the complete chromatographic time. In principle, SWATH-MS allows a complete and permanent recording of all fragment ions of all peptide precursors in a biological sample and can thus combine the advantages of shotgun proteomics (high-throughput) with those of SRM (high reproducibility and sensitivity).

3 REDUCING THE NUMBER OF UNASSIGNED SPECTRA

Despite the rapid pace of analytical improvements, which are gradually increasing the fragmentation rate in high-throughput experiments, an open challenge of shotgun proteomics is the high proportion of peptide features that remain unassigned after peptide spectrum matching.

In this section, we introduce the strategies that have been devised to reduce the number of unassigned spectra. In the first section we explore the different methods that have been developed to improve spectra quality and reduce the number of *unidentified peptides*. In the second section, we highlight the main tools that can be included in data analysis pipelines in order to increase the rate of *identifiable peptides*.

3.1 Reducing Rates of Unidentified Peptides

3.1.1 Exploiting Different Peptide Fragmentation Methods

Besides the improvements in the performance of MS analyzers obtained by the novel hybrid instruments available, numerous alternative dissociation methods and scan types have recently been introduced, including high-energy collisional dissociation (HCD) and electron-transfer dissociation (ETD). Each of these techniques shows applicability and superlative performance for a subset of peptide precursors. The result is a set of techniques, scan types, and parameters that is not easily integrated into the current DDA paradigm.

To take advantage of these new dissociation techniques, Joshua Coon and colleagues have recently introduced a third acquisition paradigm, namely, the decision tree acquisition (DTA) approach, which has proven to significantly improve peptide identification rates *(24)*. DTA uses precursor mass, charge, and *m/z* to automatically determine, in real time, whether to employ CID or ETD during MS2. Later, this acquisition technique has been extended to also employ HCD *(25)*.

The most recent advance in DTA technology is the instant sequence confirmation (inSeq) algorithm *(26)*, also developed by Coon's laboratory. inSeq extends the former DTA method by adding several different decision nodes. These nodes enable automated functionalities including real-time identification, elution prediction, advanced quantification, PTM localization, large-scale targeted proteomics, and increased proteome coverage, among others. This technology provides a direct pathway to transform the current passive data collection paradigm to active acquisition modes. Specifically, knowing the identity of peptides that are eluting into the MS system in a certain moment permits an ensemble of advanced automated decision-making logic and allows for real-time targeted mass spectrometry experiments. This novel data-acquisition paradigm harnesses the mass accuracy and speed of new-generation high-resolution hybrid MS systems to achieve the sensitivity and reproducibility of SRM methods with the high-throughput capacities of the shotgun technique.

3.1.2 Improving Spectra Quality

Obtaining high-quality spectra is crucial to ensure confident peptide identifications. Several processing tools have been developed to alleviate the impact of bad-quality spectra *(27,28)*. These tools normally involve, among others, (i) spectrum quality assessment, (ii) spectrum/peak filtering, and (iii) spectrum clustering *(29,30)*.

To enable automated spectral quality assessment, several computational tools that can be used to detect unassigned high-quality spectra and mark them for subsequent reanalysis have been developed in recent years. A first approach, presented by Bern *et al.*, predicts spectral quality based on a manually curated set of features *(31)*. Other studies reported a quadratic discriminant function and a random forest classifier to separate good- and bad-quality spectra *(32,33)*. Na *et al.* suggested a cumulative intensity normalization method for quality assessment *(29)*. Flikka *et al.* proved that the instrument type significantly affects machine-learning classifier performances *(34)*. For postanalysis, Nesvizhskii *et al.* released *QualScore*, which produces accurate results to find unassigned good spectra after a database search *(35)*. More recently, Ma *et al.* developed *ScanRanker*, which evaluates spectral quality via sequence tagging *(36)*. Apart from reducing the computational time, spectral quality scores have also been used to validate peptide assignment in addition to the commonly used spectrum–peptide match scores. Moreover, they are useful in optimizing the mass spectrometry acquisition settings to maximize the number of high-quality spectra. Thus, they may help to select spectra for *de novo* sequencing or cross-linking analyses, and they can be used for quality control of datasets derived from large-scale proteomic studies. In spite of the importance of spectra quality assessment tools and the efforts performed in the last recent years for their development, they are not yet completely integrated into mass spectrometric pipelines. Only recently, *OpenMS* developers started a project to embed quality control tools to assess the quality of large datasets.

Despite efforts to assess and improve the quality of spectra, in current datasets there is still a high percentage of spectra that can be considered of bad quality, while others present a significant amount of redundancy, thus increasing the number of unassigned spectra and hampering protein and post-translational identification. MS/MS spectra usually contain many more peaks than the ones expected from the theoretical fragmentation scheme, including ions resulting from unknown fragmentation pathways, sample-specific or systematic chemical contaminations, and random noise produced by the electronic detection system. These additional peaks do not only complicate spectra interpretation, but they can also induce false identification of peptide sequences, and in particular their PTMs *(37–39)*. The aim of filtering is to increase data quality by reducing, at the same time, the number of peaks/spectra to be analyzed. The filtering methods can be divided into two main categories: (i) spectrum filtering and (ii) peak filtering. The first methods select low-quality spectra either by their mass-to-charge distribution or by their atypical intensities, and then exclude them from further analysis. The common strategy is to use peptide fragmentation rules to identify high-quality spectra *(34,35,40–42)*, similar to what is done in many *de novo* and sequence-tagging algorithms. On the other hand, peak filtering approaches discard unnecessary peaks from the spectra of a particular experiment either by their low intensities or because they do not follow fragmentation rules. In this kind of approach any peak known to be discarded by identification tools can be filtered out as "noise" *(43,44)*. Tools like *mzStar (45)* and *wiff2data (46)* include basic preprocessing methods in which spectra are selected by an intensity-based cutoff. On the contrary, *DTASuperCharge (47)*, *Mascot Distiller* (Matrix Science), and *MaxQuant (48)* select the top-most intense peaks within a user-defined intervals, besides identifying the charge state and correcting the monoisotopic masses of the precursors. The advantage of noise filtering methods is that they produce cleaner spectra that can then be further analyzed with any search engine. However, the signal-to-noise filters remove all low-intensity peaks, regardless of whether or not they follow the expected fragmentation rules. In order to address this problem, Reiz *et al.* recently published a peak-by-peak strategy designed to reduce the number of uninformative peaks by removing those that either do not form $b-y$ ($a-x$, $c-z$) pairs or are not separated by the mass of an amino acid. In this way, they succeeded in improving peptide annotation by 15–25% *(49)*.

As previously mentioned, MS/MS datasets could contain many redundancies, as a peptide can be selected for fragmentation multiple times, ending up with hundreds or even thousands of spectra from the same feature. Applying clustering methods to similar spectra allows to make one search per precursor, rather than repeating the process for each spectrum, thus speeding up the analysis *(50)*. Apart from reducing the number of false identifications due to low-quality spectra, clustering can also help to identify peptides with single amino acid polymorphisms as well as unexpected PTMs by restricting the search to only the unidentified clusters. In 2005, Tabb *et al.* released the *MS2Grouper* algorithm *(51)*, which was able to reduce the number of spectra that have to be searched, but with a decrease in the number of identified peptides. Beer *et al. (52)* developed the *Pep-Miner* clustering algorithm (IBM) that further reduced the runtime and generated additional peptide identifications. Ramakrishnan *et al. (53)* and Dutta and Chen *(54)* tried to apply metric space embedding, an adaptation of Internet and database clustering strategies, to MS/MS clustering algorithms. In 2006, Alm *et al.* presented *SPECLUST (55,56)*, which performs hierarchical clustering of MS spectra. In 2008, the Pevzner group published *MS-Cluster (57)*, a tool that combines all spectra from the same precursor and produces a consensus spectrum. By joining high-quality and low-quality spectra of the same peptide, it decreases the probability of false identification when searching low-quality spectra separately. Last year, two more algorithms were released: *BaMS*^{2} (Bayesian model selection for tandem mass spectra) that provides a score to all the spectra in a cluster against candidate peptides using Bayesian model selection *(58)*; and *CAMS* (clustering algorithm for mass spectra) that utilizes a graph theoretical framework based on a novel metric, called F-set, that allows accurate identification of the spectra that are similar *(59)*. Finally, the PRIDE consortium recently published *PRIDE Cluster (60)*, a tool that makes use of a refined version of *MS-Cluster* to recognize consistent peptide identifications in heterogeneous MS data present in the PRIDE database.

All these computational strategies improve quality assessment and are particularly useful not only to discard *a priori* low-quality spectra and to increase the identification rate but also to obtain more reliable genome validation and annotation, as also discussed in Section 3.2.1.

3.1.3 Combining Search Engines

Search engines can be classified in three main groups: (i) sequence search engines (reviewed in Ref. *(61)*), which try to match experimental spectra with theoretical *in silico* spectra; (ii) search engines based on spectral libraries (assessed in Ref. *(62)*), which attempt to match spectra with a library of previously observed and identified spectra; and (iii) *de novo* search engines (revised in Ref. *(63)*), which attempt to derive peptide identifications based on the MS/MS spectrum peak patterns.

Most search engines perform correctly on shotgun datasets, with the results having significant overlap between the algorithms, especially when multiple peptide sequences are identified with high confidence *(64–71)*. Due to slight differences in the search algorithms, different search engines often assign distinct subsets of spectra. Therefore, several initiatives have appeared to merge results from multiple engines and improve peptide identification. Nevertheless, each search method uses a different algorithm and makes a different use of the spectrum components. Thus, it is possible that the search results for one spectrum from various search engines may differ significantly, with peptides that have been identified in one search engine being below the chosen confidence threshold in another, or even with the same spectra assigned to two different peptides *(72)*.

Combining search results from different methods presents several remarkable obstacles such as the heterogeneity of output formats *(73)* and the combination of different score parameters used to indicate the quality of PSMs. The first problem was addressed by trying to convert raw files to a common identification format, such as *mzIdentML (74)* and *pepXML (75)*. A simplistic way to solve the difference in scoring is to just filter each list of PSMs based on specific search engine thresholds, merge them, and filter the final list according to a minimum threshold of congruency. Decoys can be maintained to estimate false discovery rates (FDR) for the output, but, in any case, spectra identified with high confidence by only one search engine are lost. To achieve the largest number of peptide assignments without inflating the FDR, more advanced algorithms have been developed. These tools are revised in Table 1.

PeptideProphet (76) is one of the most commonly used solutions for peptide identification. It integrates a statistical model that makes use of the iterative expectation-maximization (EM) algorithm in order to discriminate incorrect assignments by giving to every PSM a probability of being correct and by calculating global FDRs at several thresholds. If a dataset has been searched with multiple search engines, each of the separate *PeptideProphet* modeling results can be combined with the *iProphet* algorithm *(77)* that recomputes the PSM probabilities and calculates peptide-level probabilities

TABLE 1 Available Tools that Combine Different Search Engine Results

References	Tool	Availability
(76,77)	PeptideProphet–iProphet	Free
(78)	PepArML	Free
(79)	MSblender	Free
(80)	FDRAnalysis	Free
(81)	ConsensusID	Free
(82)	PeptideShaker	Free
(83)	Scaffold	Commercial
(84)	inChorus	Commercial

by taking into account supporting evidence of other PSMs. Other available solutions are *PepArML (78)*, an unsupervised, model-free, result-combining, machine-learning meta-search that performs multiple searches directly on a Web server; *MSblender (79)*, which uses a probabilistic approach to combine results from multiple search engines; *FDRAnalysis (80)*, which adjusts the distribution of *q*-values associated with PSMs from individual search engines, calculates a score called FDRScore, and counts decoy matches; *ConsensusID (81)*, integrated in the OpenMS suite *(85)*, which calculates a probabilistic consensus score for all peptides and search engines by using a sequence similarity scoring mechanism; and *PeptideShaker*, a recently released tool *(82,86)*, still under development, that offers the possibility to visualize peptide and protein identification results from multiple search engines. Two commercial solutions are also on the market: *Scaffold (83)*, which estimates the individual peptide probability for each search engine and then uses a Bayesian approach to calculate an agreement single-probability score; and *inChorus*, included in the Bioinformatics Solutions Inc. *PEAKS* suite *(84)*, which lets the user choose the final FDR and automatically adjusts the score threshold for each engine to meet the target FDR.

Multiple searches with a combiner that takes into consideration as many search engines as possible provide a large improvement in the confidence of identified PSMs, distinct peptide sequences, and proteins, with the benefit being significant for specific types of datasets. Combining results also maximizes the protein coverage by identifying peptides that otherwise would not be assigned with high confidence in a single search engine result. However, if after the statistical analysis different methods report different confident identifications for the same spectrum, it is necessary to apply *ad hoc* rules to choose the identification to be maintained. Furthermore, in many cases, it is required to assume *a priori* a score distribution for the true-positive and the false-positive identifications, thus reducing the strength of any statistical significance assignment obtained from these analyses.

3.2 Increasing Rates of Identifiable Peptides

As already mentioned, experimentally derived tandem mass spectra (MS/MS) are generally searched against a protein database of theoretically derived peptides, which provides identification of peptide sequences if the corresponding proteins are present in the databases *(87)*. In the previous section, we have seen that a certain percentage of the unassigned spectra comes from low-quality spectra and that these could be rescued by using different methods of filtering and clustering to increase the peptide identification rate. But what happens when protein databases are incomplete or even not available? Then, a high amount of spectra, regardless of their quality, cannot be annotated, and therefore, alternative approaches such as proteogenomics and the unbiased PTM methods need to be taken into consideration.

3.2.1 Exploiting Proteogenomics

Over the last 20 years, the process of completing a genome sequence has become relatively routine for prokaryotic and eukaryotic species. Next-generation sequencing technologies (NGS) enable characterization of DNA sequence variation, *de novo* sequencing of a number of species, genome detection of methylated regions, quantitation of transcript abundances, characterization of different isoforms of genes present in a given sample, and identification of the degree to which mRNA transcripts are being actively translated *(88,89)*. One of the main features of the NGS technologies is their high throughput at a modest cost, with hundreds of sequenced gigabases now being possibly obtained in a single run for a few thousand dollars *(90,91)*.

The extensive use of NGS technologies increases the availability of new raw genomic sequences, which needs to be correctly annotated. Genome annotation has mostly been achieved using *ab initio* gene prediction algorithms trained to recognize specific features of the open reading frames *(92,93)*. However, they often suffer from low accuracy, leading to a high number of false-positive gene predictions and wrongly predicted protein termini or exon–intron structure in eukaryotic genes *(94–99)*. Moreover, different prediction algorithms may produce contradictory results, which make it difficult to obtain an unambiguous gene annotation. Consequently, the eventual experimental validation of these putative genes relies on the detection of gene products at the transcript level, or at the protein level using proteomics data.

Proteogenomics refers to the integration of the proteomic data with the genomic and transcriptomic data with the goal of improving our understanding of complex living systems *(100,101)*. The main idea behind this approach is to identify peptides in a biological sample using mass spectrometry by searching the six-frame translation of the genome sequence, as opposed to searching in a protein database. Such approaches are, therefore, not biased toward existing gene annotations and enable identification of protein regions that are absent or incorrectly represented in current genome annotations *(102)*. The identification of peptides by proteomics techniques have confirmed correct translation *(103–105)*, discovered pseudogenes in presumed noncoding regions that are actually expressed *(106–108)*, established the frame of translation (even multiple overlapping frames), described sites of posttranslational processing such as signal peptides *(109)*, identified exact splicing boundaries and alternative splicing variants, and predicted completely novel genes by mapping the peptide coding region to an uncharacterized genomic location.

Proteogenomics can certainly provide many benefits, but also place substantial bioinformatics challenges *(110)*, like, for example, the need for different tools to perform the six-frame translation, peptide loci mapping, and FDR analysis *(111,112)*, as well as issues related to task segmentation to accommodate limited computer memory *(113)*. Translated six-frame databases normally are much larger (in orders of magnitude) than traditional protein-sequence databases employed in routine MS/MS peptide spectra analysis *(105,114)*, which reduces sensitivity and increases the number of false-positive identifications *(115)*. Therefore, the FDR estimation in peptide identification has become a critical step in proteogenomics studies. One way to overcome these difficulties consists in reducing the peptide searching space. For instance, Tanner *et al. (116)* and more recently Woo *et al. (117)* created an exon graph database, as a compact representation of all human putative exons, splice variants, and polymorphisms constructed from the expressed sequence tag (EST) database and from GENEID exon predictions. Similarly, Sevinsky *et al. (118)* developed a method that applies accurate peptide p*I* filtering and accurate mass to constrain the database size and complexity, whereas Kuster *et al. (119)*, McGowan *et al. (120)*, and Ferro *et al. (121)* restricted the spectrum interpretation to peptide sequence tags (PSTs), so that sequence tags of the PSTs were translated into a corresponding degenerated oligonucleotide sequence and matched to the genome. Recently, Risk *et al. (110,122)* published an integrated proteogenomic pipeline that also addresses the FDR problem by adopting the mass spectrum sequential subtraction technique described by Helmy *et al. (123)*, in which spectra were first searched in a proteomic database and then nonassigned spectra were searched again using a genomic database. Removing positively identified spectra with each database search did not only make searches faster, it also efficiently addressed the problem of false-positives in genome-only matches.

In addition to genomic sequences, the availability of high-throughput RNA sequence (RNA-Seq) technology *(124)* has triggered great interest for the integration of transcriptomic and proteomic analyses. Integration allows gaining a more complete understanding of the cellular systems at both the genomic and proteomic levels *(125–127)*. While protein abundance can only be partially explained by mRNA concentration *(128)*, mRNA expression is clearly a prerequisite for protein expression. RNA-Seq enables not only to obtain transcript levels, but also to identify RNA splicing events *(129)* and to pinpoint sequence variations *(130)*, which are two more recent hot topics in the field of proteogenomics.

3.2.1.1 Accounting for Splicing Variants

Alternative splicing is considered to be a key factor underlying increased cellular and functional complexity in higher eukaryotes as it allows a single gene to produce multiple protein isoforms, thereby playing a major role in the production of complex proteomes with a broad range of functional diversity. It is estimated that up to 70% of human genes have alternative splicing and that the proteins translated from alternatively spliced mRNAs contain differences in their amino acid sequence and, often, in their biological functions, cellular localization, binding proteins, stability, or enzymatic activity *(131–135)*. Alternative splicing is not only involved in normal molecular processes, but it is also associated with cancer and diverse disease, and thus, disease-specific splicing variants may be potentially used as diagnostic, prognostic, and predictive biomarkers *(136)*.

Comprehensive transcriptome profiling studies using RNA-Seq *(124)* have found evidence for many novel exons, and the number of identified alternative splicing forms is much larger than previously expected *(137)*. As alternatively spliced gene structures can be very different from the initial predicted gene structure, current *ab initio* and sequence similarity methods are not adequate to validate such predictions. Instead, mRNA sequences from RNA-Seq are translated to proteins and used as databases to identify the protein products of these novel exons, even though the final number of sequences that actually become functional proteins *in vivo* remains unknown *(138,139)*.

In the last few years many studies have been published employing mass spectrometry to increase the number of alternatively spliced proteins detected. The first approaches searched MS/MS spectra against databases containing splice variant sequences, confirming the existence of a spliced sequence by detecting unique peptides for that form *(106,140–148)*. Other methods made use of exon–exon databases in which exon sequences, obtained from databases such as Ensembl or from exon-prediction algorithms, are assembled into all theoretical exon–exon (and exon–intron) combinations, translated and used for MS-based searching *(149–152)*. Several research groups have decreased the search space by eliminating spurious sequences, and by including only those sequences confirmed by transcript expression data *(27,153)*. Other groups used the RNA-Seqs translation from EST contigs *(154–157)* or the entire genome as a template for peptide sequence alignment *(158,159)*.

Several of the above methods significantly increase the size of proteomic databases. This causes a decrease in peptide identification sensitivity *(111,130)*, complicates protein inference (because of many redundant sequences) *(160,161)*, and leads to longer search times. Although some efforts have been made to optimize search speeds *(162)*, these hurdles remain unsolved.

The possibility to identify and quantify protein-splicing variants by mass spectrometry has certainly attracted great interest from researchers in recent years, due to their variety of biological functions and their importance in many health- and disease-related processes. However, database searches are not yet optimized, and the ability to find a balance between the inclusion of all putative proteoform sequences *(163)* and the reduction of database size to control sequence redundancy and false-positives will definitely determine the success of this approach.

3.2.1.2 Mutation-Tolerant Search Tools

Single-nucleotide polymorphisms, insertions and deletions (INDELs), and copy number variations contribute to variations in the genomes from individuals of the same species, some of which can cause loss or gain of function of certain gene products *(164–166)*. Due to the significant interest of amino acid variants in fields such as cancer *(167)*, pharmacogenomics, hereditary diseases *(168)*, and population proteomics *(169)*, the development of algorithms to match variant peptide sequences to spectra has become a priority for computational proteomics. Indeed, reliable information about the protein-sequence variations is essential to have a deeper perception of disease genotype–phenotype correlations.

Annotation of amino acid mutations will not only increase the number of identified peptides but it will also allow uncovering of amino acid sequence polymorphisms of proteins across different proteomic samples. In principle, all standard database-searching algorithms can be used to search for a small number of amino acid mutations, as these changes can be defined as peptide modifications. However, because there are 19 alternative states for each amino acid, the number of possible candidates becomes large, which negatively affects processing time, sensitivity, and specificity. This issue can be partially solved by means of dedicated algorithms that enable the identification of peptide changes by employing different computational approaches. The pioneers of the computational analysis of mutation-modified peptides were Mann and Wilm in 1994 *(170)* and the Yates group in 1995 *(87)*. The former authors proposed a PST approach, while the latter group suggested the generation of an *in silico* database of all modified peptides for a small set of modifications and to match the spectrum against this virtual database. The mutation searching method assisted by a database has been refined and modified in a variety of tools *(171–175)*.

Other types of solutions have been presented during the years, such as the *PepNovo* algorithm published by Frank *et al. (172)*, *GutenTag (173)*, and many others *(174–176)*. In 2000, Pevzner *et al.* introduced a new notion of spectral similarity that allows the identification of related spectra even if the corresponding peptides have multiple modifications/mutations *(177)*. A similar approach was used by Hu *et al.* *(178)* by perturbing known spectra of similar sequences, such as those with single amino acid substitutions. In 2010 de Souza *et al.* presented *MSMSpdbb* (Multi-Strain Mass Spectrometry Prokaryotic DataBase Builder), in which protein databases from several sources are merged *(179)* and peptide sequences with polymorphisms are appended to the primary sequence in an MS-friendly manner using nonassigned characters *(180)*. Finally, *customProDB* is a dedicated R package to generate a customized database from RNA-Seq data, incorporating single-nucleotide variations and short INDELs into the protein database *(181)*.

All these tools are useful in the detection of PTMs and amino acid mutations, but often they were optimized for specific data formats or specific datasets, and many lack an easy-to-use graphical or Web user interface. In order to make mutation-tolerant searches easier, Dasari *et al.* developed *TagRecon* *(182)*, a new sequence-tagging algorithm to identify mutant peptides as part of an integrated bioinformatics pipeline that produces HTML reports of protein and peptide identifications. Commercial solutions, such as *Parago*n *(183)* integrated into the ABSciex *ProteinPilot* software and *SPIDER (184)* integrated into *PEAKS (84)*, also make the whole analysis process simpler but introduce direct economical costs.

3.2.2 Unbiased PTMs Analysis

PTMs are covalent processing events that change the properties of a protein by proteolytic cleavage or by addition of a modifying group to one or more amino acid *(185)*. PTMs are fundamental key regulators; they can change the intrinsic properties of a protein and can determine its activity state, localization, turnover, and interactions with other proteins. Therefore, identifying and understanding PTMs is critical in cell biology studies and disease treatment and prevention. More than 300 different types of PTMs have currently been identified and new ones are regularly discovered. A 2006 study of human proteome samples estimated that there are 8–12 modified versions for each unmodified tryptic peptide *(186)*.

Liquid chromatography coupled with tandem mass spectrometry (LC–MS/MS) is the central method for identifying proteins with PTMs *(187,188)* due to its high sensitivity, its ability to identify the site, and its ability to quantify the relative changes in PTM occupancy at distinct sites *(189)*, among others. However, PTM analysis presents a number of analytical challenges because most modifications are low-abundance and/or substoichiometric and some are labile during MS and MS/MS.

As for nonmodified peptides, a modified peptide is usually identified by comparing the observed spectral data to a protein database using database search engine programs. Identification by these algorithms is based on a restricted database search in which MS/MS spectra are aligned with protein sequences bearing a few specified PTMs attached to specific amino acids *(154)*. The expected modifications have to be set *a priori* as "variable": in other words, a researcher must guess in advance which PTMs are present in the sample. Thus the presence of unanticipated protein modifications is believed to be among the major reasons for the low spectral identification rate. However, when search engines consider all peptide sequences with and without variable modifications the search space expands exponentially with an increase in the number of PTMs. The so-called restrictive approach to modification identification presents two main drawbacks: (i) considering a large number of variable modifications simultaneously, database searches often take much longer than the time dedicated to collect the experimental dataset *(190)* and (ii) there is a rise in the rate of false-positives because of random matches *(87,100)*. To eliminate the latter a much higher score threshold for identification of peptides needs to be applied, which leads to a drop of sensitivity and thus to a greater number of false-negatives. For these reasons, conventional database search engines such as SEQUEST and Mascot allow no more than, respectively, 10 and 9 variable modification types in one search.

To bypass the problems of conventional database search methods, a number of strategies for unrestricted PTM identification have been developed, such as multistep database search, sequence tag by *de novo* sequencing, and spectral matching approaches (reviewed in Table 2).

In the multistep database search approach, a normal search with few user-defined modifications is first performed against the entire database of interest; the second step consists of a refinement search that considers an extensive list of variable modifications against a database with only those proteins identified in the first search *(190,191)*. Recently, Huang *et al.* published a new multistep tool named *ISPTM*, in which MS/MS spectra are refined instead of the database *(192)*, enabling the identification of PTMs from complex peptide mixtures without prior identification of the proteins in the sample. Using these strategies, the search speed and the number of allowed modifications increase considerably, but the user still needs to specify a list of modifications. Another method is the modification-tolerant database search, in which all peptides in a database, rather than only the ones matching the observed precursor mass, are compared with the input spectrum, and putative modifications are introduced to account for the offset of precursor masses. One of the first software of this type was MS-Alignment, based on spectral convolution, local or fitting spectral alignment, and branch-and-bound algorithms *(177,215–217)*. An adaption of the *Batch-Tag* algorithm integrated into *Protein Prospector* gave another example of modification-tolerant search *(193)*. In 2005, Hansen *et al.* published the *P-Mod* algorithm, which estimates mass shifts for every

TABLE 2 Available Tools for Unbiased PTM Analysis

Method	References	Tool	Availability
Iterative database search			
	(191)	—	—
	(190)	Mascot error-tolerant search	Commercial
	(192)	ISPTM	Free
Modification-tolerant database search			
	(177)	MS-Alignment	Free
	(193)	Protein Prospector	Free
	(194)	P-Mod	Free
	(195)	Interrogator	Free
	(196)	TwinPeaks	Free
	(197)	SeMoP	Free
	(198)	PTMap	Free
	(199)	PEAKS PTM	Commercial
Sequence tags by *de novo* sequencing			
	(172)	PepNovo	Free
	(174)	InspecT	Free
	(184)	SPIDER	Commercial
	(173)	GutenTag	Free for nonprofit
	(182)	TagRecon	Free
	(200)	OpenSea	Free (upon request)
	(201)	VEMS 3.0	Free
	(202)	MOD[i]	Free
	(203)	UStag	Free
	(176)	Point Process Model	Free
	(204)	Popitam	Free
	(40)	ByOnic	Free
	(205)	MODa	Free

Continued

TABLE 2 Available Tools for Unbiased PTM Analysis—Cont'd

Method	References	Tool	Availability
Spectral matching			
	(206)	Spectral-Networks	Free
	(207)	Bonanza	Free
	(208)	—	—
	(209)	pMatch	Free
	(210)	ModifiComb	Free (upon request)
	(211)	MDF	Free
	(212)	DeltAMT	Free
Other			
	(213)	SIMS	Free
	(214)	PILOT_PTM	Free

sequence-to-spectrum comparison with a mass difference between the peptide sequence and the observed precursor ion mass *(194)*. Similarly, *Interrogator (195)* is a database search engine that implements novel features that enable the rapid identification of peptides with a modification of arbitrary delta mass. *TwinPeaks* by Havilio *et al. (196)* is a variation of the SEQUEST algorithm that removes the restriction in variable modification and it is not limited by database size. *SeMoP* (search for modified peptides) is a three-step tool that mixes iterative and modification error-tolerant approaches. First, it does a basic database search to identify proteins in a sample; then, using only the identified proteins, it performs a comprehensive and unrestricted search to discover modified peptides within a specified mass range ($\pm$200 Da) with respect to the corresponding unmodified peptides; finally, it executes a targeted search using a standard database search for verifying selected modifications found in the second step *(197)*. *PTMap* is a sequence alignment algorithm that permits the identification of mass shifts associated to spectra with possible unspecified modifications. The main aim of this tool is to reduce the number of false-positives when identifying peptides that bear an unknown mass shift at an unknown amino acid residue, by adopting several strategies such as the selection of sequence-rich MS peaks, automation of mass-shift adjustment, precise localization of PTM sites, implementation of rules for manual verification, and development of a scoring system based on spectrum quality *(198)*. The only commercial solution present on the market is *PEAKS PTM (199)* integrated in the *PEAKS* suite.

Another set of tools to identify PTMs are those based on sequence tags recovered by *de novo* sequencing that are used to locate the full peptide sequence by mass-based alignment. One or more unanticipated modifications may be inferred according to the mass difference between the assigned sequence and the observed masses. Most of these tools, such as *PepNovo (172)*, *InspecT (174)*, *SPIDER (184)*, *GutenTag (173)*, *TagRecon (182)*, or *OpenSea (200)*, have already been mentioned in the previous section, as mutations are considered as variable modifications. Other examples of this approach include *VEMS 3.0 (201)*, *MOD^i (175)*, *UStag (203)*, *Point Process Model (176)*, *Popitam (204)*, *ByOnic (40)*, and *MODa (205)*. *VEMS 3.0* is an algorithm implemented into the *VEMS* (Virtual Expert Mass Spectrometrist) program *(218)* that takes advantage of the sequence tag approach of *VEMS* followed by a first iteration over all possible masses and a second iteration over all possible positions of modifications. It presented for the first time a scoring function, which considers diagnostic ions and neutral losses specific for certain modifications and an error model for fragment ion masses to efficiently validate the modified peptides. In 2006 the Paek group published the *MOD^i* algorithm, another sequence tag approach that differs from previous methods because it simultaneously uses multiple sequence tags derived from a spectrum to build a tag chain, a combination structure of multiple sequence tags. A tag chain offers an effective localization of modified regions within a spectrum and thus allows rapid identification of multiple PTMs in a peptide, avoiding search space explosion by inspecting PTMs only in the modified regions of a peptide. The tag chain algorithm is robust against *de novo* sequencing errors, whereas most tag-based approaches depend critically on good *de novo* interpretations *(202)*. In order to discover rare and unknown modifications, the same group developed a local alignment tool (*MODmap*) that refines *MOD^i* results *(219)*. Another *ab initio* approach to sequence tag selection has been adopted by Liu *et al.* into their novel tree decomposition-based algorithm, *Point Process Model (176)*. It can efficiently and effectively find all maximum weighted antisymmetric paths in a spectrum graph and use the notion of extended spectrum graph that contains additional edges to describe the relationships between pairs of complementary vertices. A similar approach was already proposed by Hernandez *et al. (204)*: a given number of peaks in the spectrum are selected according to their intensity, the MS/MS spectrum is transformed into a graph that is compared with theoretical peptides from a database, leading to a ranked list of scored candidate peptides. Finally, a very recent tool, *MODa*, enables multiblind unrestrictive PTM searches with a speed that is an order of magnitude faster than other existing approaches. It can also identify any number of modifications on a single peptide. As for *MOD^i*, *MODa* simultaneously uses multiple sequence tags from each MS/MS spectrum and a dynamic programming algorithm to identify modifications between sequence tags matched to a database peptide. Although these approaches are very promising, their applicability is limited because they rely too much on the accuracy of sequence tags, which in turn rely on the quality of spectra. Indeed, *de novo* sequencing still has limited capabilities and is not as practical as the conventional database search approach.

The third approach is spectral matching that includes, among others, spectral clustering and library search. The concept behind this type of method is that the modified and unmodified versions of the same peptide often coexist in a protein sample and that their fragmentation mass spectra are similar. The *Spectral-Networks* algorithm of Bandeira *(206)* uses spectral alignment *(177)* to combine uninterpreted tandem mass spectra from overlapping peptides (pairs produced from unmodified and modified versions of peptides) and only then determines consensus identifications of sequences and modifications for sets of aligned tandem mass spectra. Spectral pairs are constructed into networks of related spectra, from which more informative virtual spectra can be generated to facilitate *de novo* peptide sequencing. Similarly, *Bonanza (207)* clusters spectra regardless of precursor *m/z*, amino acid residue modifications, or whether the peptide sequence can be identified by a database search algorithm using an improved version of the common spectral-dot product similarity *(220)* with shifted mass matching window. Ahrne *et al. (208)* proposed to build a library of spectra confidently identified by a database search and subsequently search the remaining unidentified spectra against the spectral library with a very large precursor mass tolerance. Recently, Ye *et al. (209)* reported an open spectral library search tool, pMatch, which creates the library with spectra optimized by the full peptide sequence information to tolerate peptide fragmentation pattern variations. Moreover, they developed a new scoring system, which uses charge-dependent mass shifts for peak matching and combines a probability-based model with the general spectral-dot product. However, the above approaches present some drawbacks: (i) they are computationally expensive, (ii) their performance depends on the spectra quality, and (iii) they completely ignore retention-time information. The first tool that tried to address these problems was *ModifiComb (210)*, which pioneered the use of retention time as complementary information to detect modifications by using the mass and RT differences to build a two-dimensional map of the present PTMs. Another example is the *MDF* algorithm *(211)*, which is completely independent from both sequence databases and prior knowledge about modifications, since it uses only precursor mass information. Finally, in 2011, Fu *et al.* published *DeltAMT (212)*, which aims to find pairs of modification-related spectra even though fragmentation spectra are not matched with each other. Each pair is represented by a two-dimensional vector with information about the differences in precursor mass and retention time. Then, the complete set is scored according to bivariate Gaussian mixture models to discriminate modification-induced distributions from random ones.

In the last years, a set of new tools that cannot be classified in the previous approaches have also been developed. This is the case of *SIMS* (Sequential

Interval Motif Search) *(213)* and *PILOT_PTM (214)*. In *SIMS*, the search process takes into account mass intervals, corresponding to pairs of product ion peaks whose *m/z* difference is equivalent to amino acid residue masses (within a predefined mass tolerance). The PTM mapping problem is therefore formulated as a "stitching" procedure, in which the stringent selection of intervals that best optimizes the similarity between the sequence and the spectrum reveals both the site and mass of putative PTMs. In the case of *PILOT_PTM (214)*, the algorithm seeks to maximize the sum of intensity contributions from theoretical peak matches to the experimental spectrum given a set of logical constraints. It directly incorporates the intensity contribution from both sets of complementary ion peaks as well as all corresponding offsets in the objective function. Given a template sequence of amino acids, the model determines the optimal set of modifications among a selected "universal" list (built from Unimod, RESID *(221)*, and Delta Mass databases) based on the MS/MS spectral data assuming that all template positions can contain a PTM.

Blind PTM search engines are used because they can detect both known and novel PTMs, but PTM predictions still contain inaccurate modification masses and incorrect modification positions. Therefore, it is useful to incorporate PTM refinement as part of a PTM prediction pipeline to improve the quality of the PTM predictions. There are two types of PTM refinement: the first evaluates the quality of predicted modification sites from PTM search engines *(222)* and the second refines both observed modification masses and modification positions. Recently published algorithms that use this type of PTM refinement method are *PTMFinder (223)*, *PTMClust (224)*, and its update *iPTMClust (225)*. *PTMFinder* takes a peptide-level approach to PTM refinement, in which spectra mapping to the same modified sequence are grouped to produce, for each spectrum, a final peptide sequence with a modification mass and a modification position. Hence, refinement using this method is limited to modified peptides that occur multiple times in the same dataset. Instead, *PTMClust* models modifications at the PTM level. Its approach has the advantage of accounting for low-abundance modified peptides because other peptides with the same underlying PTM can help identify the correct but unknown modification mass and modified amino acid.

Despite myriad tools to reannotate spectra and identify PTMs, their general efficacy is still limited, and only a small part of the total spectra are identified with high confidence. Moreover, there is no consensus PTM analysis pipeline yet, and therefore efforts need to be placed in the development of new software and the combination of the existing solutions in order to increase both detection coverage and confidence of PTMs.

4 SUMMARY

Shotgun proteomics is the method of choice to obtain comprehensive proteome coverage from complex biological samples. Despite its high-throughput capabilities, recent studies evidenced the current limits of this technique and demonstrated suboptimal information retrieval from complex samples. Indeed, in high-throughput proteomics experiments, most features eluting from the chromatographic system remain either not fragmented or, if fragmented, not assigned to a peptide sequence after mass spectrometric analysis. Here, we have reviewed the efforts from the proteomics community to overcome these limitations by improving both fragmentation rates and the number of annotated spectra.

During the last years, the increase in scanning speed of mass spectrometric analyzers, together with the refinement of protein and peptide separation and fractionation techniques and the development of new acquisition modes, has significantly improved peptide fragmentation rates. Although being still far from targeting all eluting features for fragmentation, important progress has already been made toward this goal as presented in this chapter, and several examples show how acquisition optimization and orthogonal fractionation techniques can result in impressive proteome coverage. Further advances in upstream analytical techniques, such as faster MSs and new acquisition methods, as well as refined downstream data analysis are expected to further improve proteome coverage.

In parallel to the improvement of fragmentation rates, we have reviewed the efforts that are being done, especially by the bioinformatics community, to reduce the number of unassigned MS2 spectra, either by improving spectra annotation or by increasing the number of identifiable peptides. Combination of different fragmentation techniques, improvement of spectra quality, and combination of search engines have been described as solid examples of initiatives to improve spectra annotation. Indeed, the combination of the results from several search engines with finer statistical methods is providing better FDR evaluation and more peptide identifications. Moreover, the development of spectra quality assessment tools and their full integration in routine analysis pipelines will certainly contribute to the implementation of new real-time instrument operations and to the interpretation of the acquired data.

Similarly, we have described how information related to protein-splicing variants, amino acid peptide variants (polymorphisms), and PTMs is being introduced in data analysis pipelines in order to increase the rate of identifiable peptides. The introduction of genomics and transcriptomics information, often neglected in routine proteomics analyses, will enable the characterization of myriad splicing variants and amino acid polymorphisms, as well as the development of robust proteomics analysis pipelines, which will contribute to elucidate their role in health and disease states. These tools will make it possible to uncover a new layer of the proteome complexity and to expand the information available from cellular systems and the potential implication of these protein variants in different physiological processes. Finally, several bioinformatics tools that perform unbiased PTM analysis have also been reviewed. The tools that are currently being developed and refined tackle the characterization of unexpected PTMs and the crosstalk among signaling pathways.

Overall, all the efforts addressed to increase fragmentation rates and to reduce the number of unassigned MS2 spectra will enable maximizing the information retrieved from proteomics experiments, which will eventually provide a comprehensive quantitative overview of all cellular proteins and their PTMs in different states and conditions.

ACKNOWLEDGMENTS

The UPF/CRG Proteomics Unit is a member of ProteoRed-Instituto Nacional de Proteómica, funded by Instituto de Salud Carlos III. This work has been supported by the PRIME-XS project (Grant No. 262067), funded by the European Union 7th Framework Programme.

REFERENCES

1. Aebersold, R.; Mann, M. *Nature* **2003**, *422*, 198–207.
2. Zhang, Y.; Fonslow, B. R.; Shan, B.; Baek, M. C.; Yates, J. R., 3rd. *Chem. Rev.* **2013**, *113*, 2343–2394.
3. Michalski, A.; Cox, J.; Mann, M. *J. Proteome Res.* **2011**, *10*, 1785–1793.
4. Pirmoradian, M.; Budamgunta, H.; Chingin, K.; Zhang, B.; Astorga-Wells, J.; Zubarev, R. A. Rapid and Deep Human Proteome Analysis by Single-Dimension Shotgun Proteomics. *Mol. Cell. Proteomics* **2013**, *12*(11), 3330–3338.
5. Nagaraj, N.; Wisniewski, J. R.; Geiger, T.; Cox, J.; Kircher, M.; Kelso, J.; Paabo, S.; Mann, M. *Mol. Syst. Biol.* **2011**, *7*, 548.
6. Dreger, M. *Mass Spectrom. Rev.* **2003**, *22*, 27–56.
7. Bendz, M.; Moller, M. C.; Arrigoni, G.; Wahlander, A.; Stella, R.; Cappadona, S.; Levander, F.; Hederstedt, L.; James, P. *J. Proteome Res.* **2009**, *8*, 5666–5673.
8. Stergachis, A. B.; MacLean, B.; Lee, K.; Stamatoyannopoulos, J. A.; MacCoss, M. J. *Nat. Methods* **2011**, *8*, 1041–1043.
9. Pieper, R.; Su, Q.; Gatlin, C. L.; Huang, S. T.; Anderson, N. L.; Steiner, S. *Proteomics* **2003**, *3*, 422–432.
10. Righetti, P. G.; Castagna, A.; Antonioli, P.; Boschetti, E. *Electrophoresis* **2005**, *26*, 297–319.
11. Weiss, W.; Gorg, A. *Methods Mol. Biol.* **2009**, *564*, 13–32.
12. Bunkenborg, J.; Espadas, G.; Molina, H. *J. Proteome Res.* **2013**, *12*, 3631–3641.
13. Peng, M.; Taouatas, N.; Cappadona, S.; van Breukelen, B.; Mohammed, S.; Scholten, A.; Heck, A. J. *Nat. Methods* **2012**, *9*, 524–525.
14. Giddings, J. C. *Anal. Chem.* **1984**, *56*, 1258A–1260A, 1262A, 1264A passim.
15. Antberg, L.; Cifani, P.; Sandin, M.; Levander, F.; James, P. *J. Proteome Res.* **2012**, *11*, 2644–2652.
16. Di Palma, S.; Hennrich, M. L.; Heck, A. J.; Mohammed, S. *J. Proteomics* **2012**, *75*, 3791–3813.
17. Hennrich, M. L.; Groenewold, V.; Kops, G. J.; Heck, A. J.; Mohammed, S. *Anal. Chem.* **2011**, *83*, 7137–7143.
18. Tabb, D. L.; Vega-Montoto, L.; Rudnick, P. A.; Variyath, A. M.; Ham, A. J.; Bunk, D. M.; Kilpatrick, L. E.; Billheimer, D. D.; Blackman, R. K.; Cardasis, H. L.; Carr, S. A.; Clauser, K. R.; Jaffe, J. D.; Kowalski, K. A.; Neubert, T. A.; Regnier, F. E.;

Schilling, B.; Tegeler, T. J.; Wang, M.; Wang, P.; Whiteaker, J. R.; Zimmerman, L. J.; Fisher, S. J.; Gibson, B. W.; Kinsinger, C. R.; Mesri, M.; Rodriguez, H.; Stein, S. E.; Tempst, P.; Paulovich, A. G.; Liebler, D. C.; Spiegelman, C. *J. Proteome Res.* **2010**, *9*, 761–776.

19. Bohrer, B. C.; Merenbloom, S. I.; Koeniger, S. L.; Hilderbrand, A. E.; Clemmer, D. E. *Annu Rev Anal Chem (Palo Alto, Calif)* **2008**, *1*, 293–327.
20. Vincent, C. E.; Potts, G. K.; Ulbrich, A.; Westphall, M. S.; Atwood, J. A., 3rd.; Coon, J. J.; Weatherly, D. B. *Anal. Chem.* **2013**, *85*, 2825–2832.
21. Lange, V.; Picotti, P.; Domon, B.; Aebersold, R. *Mol. Syst. Biol.* **2008**, *4*, 222.
22. Venable, J. D.; Dong, M. Q.; Wohlschlegel, J.; Dillin, A.; Yates, J. R. *Nat. Methods* **2004**, *1*, 39–45.
23. Gillet, L. C.; Navarro, P.; Tate, S.; Rost, H.; Selevsek, N.; Reiter, L.; Bonner, R.; Aebersold, R. *Mol. Cell. Proteomics* **2012**, *11*, O111.016717.
24. Swaney, D. L.; McAlister, G. C.; Coon, J. J. *Nat. Methods* **2008**, *5*, 959–964.
25. Frese, C. K.; Altelaar, A. F.; Hennrich, M. L.; Nolting, D.; Zeller, M.; Griep-Raming, J.; Heck, A. J.; Mohammed, S. *J. Proteome Res.* **2011**, *10*, 2377–2388.
26. Bailey, D. J.; Rose, C. M.; McAlister, G. C.; Brumbaugh, J.; Yu, P.; Wenger, C. D.; Westphall, M. S.; Thomson, J. A.; Coon, J. J. *Proc. Natl. Acad. Sci. U.S.A.* **2012**, *109*, 8411–8416.
27. Ning, K.; Fermin, D.; Nesvizhskii, A. I. *Proteomics* **2010**, *10*, 2712–2718.
28. Mancuso, F.; Bunkenborg, J.; Wierer, M.; Molina, H. *J. Proteomics* **2012**, *75*, 5293–5303.
29. Na, S.; Paek, E. *J. Proteome Res.* **2006**, *5*, 3241–3248.
30. Gentzel, M.; Kocher, T.; Ponnusamy, S.; Wilm, M. *Proteomics* **2003**, *3*, 1597–1610.
31. Bern, M.; Goldberg, D.; McDonald, W. H.; Yates, J. R., 3rd. *Bioinformatics* **2004**, *20*(Suppl. 1), i49–i54.
32. Xu, M.; Geer, L. Y.; Bryant, S. H.; Roth, J. S.; Kowalak, J. A.; Maynard, D. M.; Markey, S. P. *J. Proteome Res.* **2005**, *4*, 300–305.
33. Salmi, J.; Moulder, R.; Filen, J. J.; Nevalainen, O. S.; Nyman, T. A.; Lahesmaa, R.; Aittokallio, T. *Bioinformatics* **2006**, *22*, 400–406.
34. Flikka, K.; Martens, L.; Vandekerckhove, J.; Gevaert, K.; Eidhammer, I. *Proteomics* **2006**, *6*, 2086–2094.
35. Nesvizhskii, A. I.; Roos, F. F.; Grossmann, J.; Vogelzang, M.; Eddes, J. S.; Gruissem, W.; Baginsky, S.; Aebersold, R. *Mol. Cell. Proteomics* **2006**, *5*, 652–670.
36. Ma, Z. Q.; Chambers, M. C.; Ham, A. J.; Cheek, K. L.; Whitwell, C. W.; Aerni, H. R.; Schilling, B.; Miller, A. W.; Caprioli, R. M.; Tabb, D. L. *J. Proteome Res.* **2011**, *10*, 2896–2904.
37. Mujezinovic, N.; Raidl, G.; Hutchins, J. R.; Peters, J. M.; Mechtler, K.; Eisenhaber, F. *Proteomics* **2006**, *6*, 5117–5131.
38. Salmi, J.; Nyman, T. A.; Nevalainen, O. S.; Aittokallio, T. *Proteomics* **2009**, *9*, 848–860.
39. Renard, B. Y.; Kirchner, M.; Monigatti, F.; Ivanov, A. R.; Rappsilber, J.; Winter, D.; Steen, J. A.; Hamprecht, F. A.; Steen, H. *Proteomics* **2009**, *9*, 4978–4984.
40. Bern, M.; Cai, Y.; Goldberg, D. *Anal. Chem.* **2007**, *79*, 1393–1400.
41. Hoopmann, M. R.; Finney, G. L.; MacCoss, M. J. *Anal. Chem.* **2007**, *79*, 5620–5632.
42. Geer, L. Y.; Markey, S. P.; Kowalak, J. A.; Wagner, L.; Xu, M.; Maynard, D. M.; Yang, X.; Shi, W.; Bryant, S. H. *J. Proteome Res.* **2004**, *3*, 958–964.
43. Xu, H.; Freitas, M. A. *BMC Bioinforma.* **2010**, *11*, 436.
44. Hansen, K. C.; Schmitt-Ulms, G.; Chalkley, R. J.; Hirsch, J.; Baldwin, M. A.; Burlingame, A. L. *Mol. Cell. Proteomics* **2003**, *2*, 299–314.

45. Pedrioli, P. G.; Eng, J. K.; Hubley, R.; Vogelzang, M.; Deutsch, E. W.; Raught, B.; Pratt, B.; Nilsson, E.; Angeletti, R. H.; Apweiler, R.; Cheung, K.; Costello, C. E.; Hermjakob, H.; Huang, S.; Julian, R. K.; Kapp, E.; McComb, M. E.; Oliver, S. G.; Omenn, G.; Paton, N. W.; Simpson, R.; Smith, R.; Taylor, C. F.; Zhu, W.; Aebersold, R. *Nat. Biotechnol.* **2004**, *22*, 1459–1466.
46. Boehm, A. M.; Galvin, R. P.; Sickmann, A. *BMC Bioinforma.* **2004**, *5*, 162.
47. Schulze, W. X.; Mann, M. *J. Biol. Chem.* **2004**, *279*, 10756–10764.
48. Cox, J.; Mann, M. *Nat. Biotechnol.* **2008**, *26*, 1367–1372.
49. Reiz, B.; Kertesz-Farkas, A.; Pongor, S.; Myers, M. P. *Bioinformatics* **2013**, *29*, 925–932.
50. Tabb, D. L.; MacCoss, M. J.; Wu, C. C.; Anderson, S. D.; Yates, J. R., 3rd. *Anal. Chem.* **2003**, *75*, 2470–2477.
51. Tabb, D. L.; Thompson, M. R.; Khalsa-Moyers, G.; VerBerkmoes, N. C.; McDonald, W. H. *J. Am. Soc. Mass Spectrom.* **2005**, *16*, 1250–1261.
52. Beer, I.; Barnea, E.; Ziv, T.; Admon, A. *Proteomics* **2004**, *4*, 950–960.
53. Ramakrishnan, S. R.; Mao, R.; Nakorchevskiy, A. A.; Prince, J. T.; Willard, W. S.; Xu, W.; Marcotte, E. M.; Miranker, D. P. *Bioinformatics* **2006**, *22*, 1524–1531.
54. Dutta, D.; Chen, T. *Bioinformatics* **2007**, *23*, 612–618.
55. Alm, R.; Johansson, P.; Hjerno, K.; Emanuelsson, C.; Ringner, M.; Hakkinen, J. *J. Proteome Res.* **2006**, *5*, 785–792.
56. Granlund, I.; Kieselbach, T.; Alm, R.; Schroder, W. P.; Emanuelsson, C. *J. Proteomics* **2011**, *74*, 1190–1200.
57. Frank, A. M.; Bandeira, N.; Shen, Z.; Tanner, S.; Briggs, S. P.; Smith, R. D.; Pevzner, P. A. *J. Proteome Res.* **2008**, *7*, 113–122.
58. Ryu, S.; Goodlett, D. R.; Noble, W. S.; Minin, V. N. A Statistical Approach to Peptide Identification from Clustered Tandem Mass Spectrometry Data. *Proceedings (IEEE Int. Conf. Bioinformatics Biomed.)* **2012** Oct 4, 648–653.
59. Saeed, F.; Pisitkun, T.; Knepper, M. A.; Hoffert, J. D. An Efficient Algorithm for Clustering of Large-Scale Mass Spectrometry Data. *Proceedings (IEEE Int. Conf. Bioinformatics Biomed.)* **2012** Oct 4, 1–4.
60. Griss, J.; Foster, J. M.; Hermjakob, H.; Vizcaino, J. A. *Nat. Methods* **2013**, *10*, 95–96.
61. Eng, J. K.; Searle, B. C.; Clauser, K. R.; Tabb, D. L. *Mol. Cell. Proteomics* **2011**, *10*, R111.009522.
62. Lam, H. *Mol. Cell. Proteomics* **2011**, *10*, R111.008565.
63. Ma, B.; Johnson, R. *Mol. Cell. Proteomics* **2012**, *11*, O111.014902.
64. Chamrad, D. C.; Korting, G.; Stuhler, K.; Meyer, H. E.; Klose, J.; Bluggel, M. *Proteomics* **2004**, *4*, 619–628.
65. Sadygov, R. G.; Cociorva, D.; Yates, J. R., 3rd. *Nat. Methods* **2004**, *1*, 195–202.
66. Resing, K. A.; Meyer-Arendt, K.; Mendoza, A. M.; Aveline-Wolf, L. D.; Jonscher, K. R.; Pierce, K. G.; Old, W. M.; Cheung, H. T.; Russell, S.; Wattawa, J. L.; Goehle, G. R.; Knight, R. D.; Ahn, N. G. *Anal. Chem.* **2004**, *76*, 3556–3568.
67. Kapp, E. A.; Schutz, F.; Connolly, L. M.; Chakel, J. A.; Meza, J. E.; Miller, C. A.; Fenyo, D.; Eng, J. K.; Adkins, J. N.; Omenn, G. S.; Simpson, R. J. *Proteomics* **2005**, *5*, 3475–3490.
68. Balgley, B. M.; Laudeman, T.; Yang, L.; Song, T.; Lee, C. S. *Mol. Cell. Proteomics* **2007**, *6*, 1599–1608.
69. Alves, G.; Wu, W. W.; Wang, G.; Shen, R. F.; Yu, Y. K. *J. Proteome Res.* **2008**, *7*, 3102–3113.
70. Kandasamy, K.; Pandey, A.; Molina, H. *Anal. Chem.* **2009**, *81*, 7170–7180.
71. Shteynberg, D.; Nesvizhskii, A. I.; Moritz, R. L.; Deutsch, E. W. *Mol. Cell. Proteomics* **2013**, *12*, 2383–2393.

72. Sultana, T.; Jordan, R.; Lyons-Weiler, J. *J. Proteomics Bioinform.* **2009**, *2*, 262–273.
73. Deutsch, E. W. *Mol. Cell. Proteomics* **2012**, *11*, 1612–1621.
74. Jones, A. R.; Eisenacher, M.; Mayer, G.; Kohlbacher, O.; Siepen, J.; Hubbard, S. J.; Selley, J. N.; Searle, B. C.; Shofstahl, J.; Seymour, S. L.; Julian, R.; Binz, P. A.; Deutsch, E. W.; Hermjakob, H.; Reisinger, F.; Griss, J.; Vizcaino, J. A.; Chambers, M.; Pizarro, A.; Creasy, D. *Mol. Cell. Proteomics* **2012**, *11*, M111.014381.
75. Keller, A.; Eng, J.; Zhang, N.; Li, X. J.; Aebersold, R. *Mol. Syst. Biol.* **2005**, *1*, 2005.0017.
76. Keller, A.; Nesvizhskii, A. I.; Kolker, E.; Aebersold, R. *Anal. Chem.* **2002**, *74*, 5383–5392.
77. Shteynberg, D.; Deutsch, E. W.; Lam, H.; Eng, J. K.; Sun, Z.; Tasman, N.; Mendoza, L.; Moritz, R. L.; Aebersold, R.; Nesvizhskii, A. I. *Mol. Cell. Proteomics* **2011**, *10*, M111.007690.
78. Edwards, N.; Wu, X.; Tseng, C. W. *Clin. Proteomics* **2009**, *5*, 23–36.
79. Kwon, T.; Choi, H.; Vogel, C.; Nesvizhskii, A. I.; Marcotte, E. M. *J. Proteome Res.* **2011**, *10*, 2949–2958.
80. Wedge, D. C.; Krishna, R.; Blackhurst, P.; Siepen, J. A.; Jones, A. R.; Hubbard, S. J. *J. Proteome Res.* **2011**, *10*, 2088–2094.
81. Nahnsen, S.; Bertsch, A.; Rahnenfuhrer, J.; Nordheim, A.; Kohlbacher, O. *J. Proteome Res.* **2011**, *10*, 3332–3343.
82. Vaudel, M.; Barsnes, H.; Berven, F. S.; Sickmann, A.; Martens, L. *Proteomics* **2011**, *11*, 996–999.
83. Searle, B. C.; Turner, M.; Nesvizhskii, A. I. *J. Proteome Res.* **2008**, *7*, 245–253.
84. Ma, B.; Zhang, K.; Hendrie, C.; Liang, C.; Li, M.; Doherty-Kirby, A.; Lajoie, G. *Rapid Commun. Mass Spectrom.* **2003**, *17*, 2337–2342.
85. Sturm, M.; Bertsch, A.; Gropl, C.; Hildebrandt, A.; Hussong, R.; Lange, E.; Pfeifer, N.; Schulz-Trieglaff, O.; Zerck, A.; Reinert, K.; Kohlbacher, O. *BMC Bioinforma.* **2008**, *9*, 163.
86. Barsnes, H.; Vaudel, M.; Colaert, N.; Helsens, K.; Sickmann, A.; Berven, F. S.; Martens, L. *BMC Bioinforma.* **2011**, *12*, 70.
87. Yates, J. R., 3rd.; Eng, J. K.; McCormack, A. L.; Schieltz, D. *Anal. Chem.* **1995**, *67*, 1426–1436.
88. Shendure, J.; Ji, H. *Nat. Biotechnol.* **2008**, *26*, 1135–1145.
89. Schadt, E. E.; Turner, S.; Kasarskis, A. *Hum. Mol. Genet.* **2010**, *19*, R227–R240.
90. Eid, J.; Fehr, A.; Gray, J.; Luong, K.; Lyle, J.; Otto, G.; Peluso, P.; Rank, D.; Baybayan, P.; Bettman, B.; Bibillo, A.; Bjornson, K.; Chaudhuri, B.; Christians, F.; Cicero, R.; Clark, S.; Dalal, R.; Dewinter, A.; Dixon, J.; Foquet, M.; Gaertner, A.; Hardenbol, P.; Heiner, C.; Hester, K.; Holden, D.; Kearns, G.; Kong, X.; Kuse, R.; Lacroix, Y.; Lin, S.; Lundquist, P.; Ma, C.; Marks, P.; Maxham, M.; Murphy, D.; Park, I.; Pham, T.; Phillips, M.; Roy, J.; Sebra, R.; Shen, G.; Sorenson, J.; Tomaney, A.; Travers, K.; Trulson, M.; Vieceli, J.; Wegener, J.; Wu, D.; Yang, A.; Zaccarin, D.; Zhao, P.; Zhong, F.; Korlach, J.; Turner, S. *Science* **2009**, *323*, 133–138.
91. Metzker, M. L. *Nat. Rev. Genet.* **2010**, *11*, 31–46.
92. Stein, L. *Nat. Rev. Genet.* **2001**, *2*, 493–503.
93. Harrow, J.; Nagy, A.; Reymond, A.; Alioto, T.; Patthy, L.; Antonarakis, S. E.; Guigo, R. *Genome Biol.* **2009**, *10*, 201.
94. Starmer, J.; Stomp, A.; Vouk, M.; Bitzer, D. *PLoS Comput. Biol.* **2006**, *2*, e57.
95. Devos, D.; Valencia, A. *Trends Genet.* **2001**, *17*, 429–431.
96. Lamontagne, J.; Beland, M.; Forest, A.; Cote-Martin, A.; Nassif, N.; Tomaki, F.; Moriyon, I.; Moreno, E.; Paramithiotis, E. *BMC Genomics* **2010**, *11*, 300.

97. Parra, G.; Agarwal, P.; Abril, J. F.; Wiehe, T.; Fickett, J. W.; Guigo, R. *Genome Res.* **2003**, *13*, 108–117.
98. Reboul, J.; Vaglio, P.; Rual, J. F.; Lamesch, P.; Martinez, M.; Armstrong, C. M.; Li, S.; Jacotot, L.; Bertin, N.; Janky, R.; Moore, T.; Hudson, J. R., Jr.; Hartley, J. L.; Brasch, M. A.; Vandenhaute, J.; Boulton, S.; Endress, G. A.; Jenna, S.; Chevet, E.; Papasotiropoulos, V.; Tolias, P. P.; Ptacek, J.; Snyder, M.; Huang, R.; Chance, M. R.; Lee, H.; Doucette-Stamm, L.; Hill, D. E.; Vidal, M. *Nat. Genet.* **2003**, *34*, 35–41.
99. Flicek, P.; Keibler, E.; Hu, P.; Korf, I.; Brent, M. R. *Genome Res.* **2003**, *13*, 46–54.
100. Jaffe, J. D.; Berg, H. C.; Church, G. M. *Proteomics* **2004**, *4*, 59–77.
101. Pandey, A.; Lewitter, F. *Trends Biochem. Sci.* **1999**, *24*, 276–280.
102. Renuse, S.; Chaerkady, R.; Pandey, A. *Proteomics* **2011**, *11*, 620–630.
103. Ansong, C.; Yoon, H.; Norbeck, A. D.; Gustin, J. K.; McDermott, J. E.; Mottaz, H. M.; Rue, J.; Adkins, J. N.; Heffron, F.; Smith, R. D. *J. Proteome Res.* **2008**, *7*, 546–557.
104. Lipton, M. S.; Pasa-Tolic, L.; Anderson, G. A.; Anderson, D. J.; Auberry, D. L.; Battista, J. R.; Daly, M. J.; Fredrickson, J.; Hixson, K. K.; Kostandarithes, H.; Masselon, C.; Markillie, L. M.; Moore, R. J.; Romine, M. F.; Shen, Y.; Stritmatter, E.; Tolic, N.; Udseth, H. R.; Venkateswaran, A.; Wong, K. K.; Zhao, R.; Smith, R. D. *Proc. Natl. Acad. Sci. U.S.A.* **2002**, *99*, 11049–11054.
105. Molina, H.; Bunkenborg, J.; Reddy, G. H.; Muthusamy, B.; Scheel, P. J.; Pandey, A. *Mol. Cell. Proteomics* **2005**, *4*, 637–650.
106. Brosch, M.; Saunders, G. I.; Frankish, A.; Collins, M. O.; Yu, L.; Wright, J.; Verstraten, R.; Adams, D. J.; Harrow, J.; Choudhary, J. S.; Hubbard, T. *Genome Res.* **2011**, *21*, 756–767.
107. Rusk, N. *Nat. Methods* **2011**, *8*, 448–449.
108. Lewis, B. P.; Green, R. E.; Brenner, S. E. *Proc. Natl. Acad. Sci. U.S.A.* **2003**, *100*, 189–192.
109. Nielsen, P.; Krogh, A. *Bioinformatics* **2005**, *21*, 4322–4329.
110. Krug, K.; Nahnsen, S.; Macek, B. *Mol. Biosyst.* **2011**, *7*, 284–291.
111. Castellana, N.; Bafna, V. *J. Proteomics* **2010**, *73*, 2124–2135.
112. Blakeley, P.; Overton, I. M.; Hubbard, S. J. *J. Proteome Res.* **2012**, *11*, 5221–5234.
113. Risk, B. A.; Spitzer, W. J.; Giddings, M. C. *J. Proteome Res.* **2013**, *12*, 3019–3025.
114. UniProt Consortium. *Nucleic Acids Res.* **2010**, *38*, D142–D148.
115. Wang, G.; Wu, W. W.; Zhang, Z.; Masilamani, S.; Shen, R. F. *Anal. Chem.* **2009**, *81*, 146–159.
116. Tanner, S.; Shen, Z.; Ng, J.; Florea, L.; Guigo, R.; Briggs, S. P.; Bafna, V. *Genome Res.* **2007**, *17*, 231–239.
117. Woo, S.; Cha, S. W.; Merrihew, G.; He, Y.; Castellana, N.; Guest, C.; Maccoss, M.; Bafna, V. *J. Proteome Res.* **2013**, *13*(1), 21–28.
118. Sevinsky, J. R.; Cargile, B. J.; Bunger, M. K.; Meng, F.; Yates, N. A.; Hendrickson, R. C.; Stephenson, J. L., Jr. *J. Proteome Res.* **2008**, *7*, 80–88.
119. Kuster, B.; Mortensen, P.; Andersen, J. S.; Mann, M. *Proteomics* **2001**, *1*, 641–650.
120. McGowan, S. J.; Terrett, J.; Brown, C. G.; Adam, P. J.; Aldridge, L.; Allen, J. C.; Amess, B.; Andrews, K. A.; Barnes, M.; Barnwell, D. E.; Berry, J.; Bird, H.; Boyd, R. S.; Broughton, M. J.; Brown, A.; Bruce, J. A.; Brusten, L. M. J.; Draper, N. J.; Elsmore, B. M.; Freeman, C. D.; Giles, D. M.; Gong, H.; Gormley, D.; Griffiths, M. R.; Hawkes, T. D. R.; Haynes, P. S.; Heesom, K. J.; Herath, A.; Hollis, K.; Hudsen, L.; Inman, J.; Jacobs, M.; Jarman, D.; Kibria, I.; Kilgour, J.; Kinuthia, S. K.; Lane, K. E.; Lees, M. L.; Loader, J.; Longmore, A.; McEwan, M.; Middleton, A.; Moore, S.; Murray, C.; Murray, H. M.; Myatt, C. P.; Ng, S. S.; O'Neil, A.; Parekh, R. B.; Patel, A.;

Patel, K. B.; Patel, S.; Patel, T. P.; Philp, R. J.; Platt, A. E.; Poyser, H.; Prendergast, C.; Prime, S.; Redpath, N.; Reeves, M.; Robinson, A. W.; Rohlff, C. R.; Rosenbaum, J. M.; Schenker, M.; Scrivener, E.; Shipston, N.; Siddiq, S.; Southan, C.; Spencer, D. I. R.; Stamps, A.; Steffens, M. A.; Stevenson, D.; Sweetman, G. M. A.; Taylor, S.; Townsend, R.; Ventom, A. M.; Waller, M. N. H.; Weresch, C.; Williams, A. M.; Woolliscroft, R. J.; Yu, X.; Lyall, A. *Curr. Proteomics* **2004**, *1*, 41–48.

121. Ferro, M.; Tardif, M.; Reguer, E.; Cahuzac, R.; Bruley, C.; Vermat, T.; Nugues, E.; Vigouroux, M.; Vandenbrouck, Y.; Garin, J.; Viari, A. *J. Proteome Res.* **2008**, *7*, 1873–1883.
122. Risk, B. A.; Edwards, N. J.; Giddings, M. C. *J. Proteome Res.* **2013**, *12*, 4240–4247.
123. Helmy, M.; Sugiyama, N.; Tomita, M.; Ishihama, Y. *Genes Cells* **2012**, *17*, 633–644.
124. Wang, Z.; Gerstein, M.; Snyder, M. *Nat. Rev. Genet.* **2009**, *10*, 57–63.
125. Desgagne-Penix, I.; Khan, M. F.; Schriemer, D. C.; Cram, D.; Nowak, J.; Facchini, P. J. *BMC Plant Biol.* **2010**, *10*, 252.
126. Lundberg, E.; Fagerberg, L.; Klevebring, D.; Matic, I.; Geiger, T.; Cox, J.; Algenas, C.; Lundeberg, J.; Mann, M.; Uhlen, M. *Mol. Syst. Biol.* **2010**, *6*, 450.
127. Fanayan, S.; Smith, J. T.; Lee, L. Y.; Yan, F.; Snyder, M.; Hancock, W. S.; Nice, E. Proteogenomic Analysis of Human Colon Carcinoma Cell Lines LIM1215, LIM1899, and LIM2405. *J. Proteome Res.* **2013**, Mar 13. [Epub ahead of print] PubMed PMID: 23458625.
128. Vogel, C.; Abreu Rde, S.; Ko, D.; Le, S. Y.; Shapiro, B. A.; Burns, S. C.; Sandhu, D.; Boutz, D. R.; Marcotte, E. M.; Penalva, L. O. *Mol. Syst. Biol.* **2010**, *6*, 400.
129. Sheynkman, G. M.; Shortreed, M. R.; Frey, B. L.; Smith, L. M. *Mol. Cell. Proteomics* **2013**, *12*, 2341–2353.
130. Wang, X.; Slebos, R. J.; Wang, D.; Halvey, P. J.; Tabb, D. L.; Liebler, D. C.; Zhang, B. *J. Proteome Res.* **2012**, *11*, 1009–1017.
131. Mironov, A. A.; Fickett, J. W.; Gelfand, M. S. *Genome Res.* **1999**, *9*, 1288–1293.
132. Blencowe, B. J. *Cell* **2006**, *126*, 37–47.
133. Matlin, A. J.; Clark, F.; Smith, C. W. *Nat. Rev. Mol. Cell Biol.* **2005**, *6*, 386–398.
134. Ben-Dov, C.; Hartmann, B.; Lundgren, J.; Valcarcel, J. *J. Biol. Chem.* **2008**, *283*, 1229–1233.
135. Stamm, S.; Ben-Ari, S.; Rafalska, I.; Tang, Y.; Zhang, Z.; Toiber, D.; Thanaraj, T. A.; Soreq, H. *Gene* **2005**, *344*, 1–20.
136. Hatakeyama, K.; Ohshima, K.; Fukuda, Y.; Ogura, S.; Terashima, M.; Yamaguchi, K.; Mochizuki, T. *Proteomics* **2011**, *11*, 2275–2282.
137. Wang, E. T.; Sandberg, R.; Luo, S.; Khrebtukova, I.; Zhang, L.; Mayr, C.; Kingsmore, S. F.; Schroth, G. P.; Burge, C. B. *Nature* **2008**, *456*, 470–476.
138. Birzele, F.; Csaba, G.; Zimmer, R. *Nucleic Acids Res.* **2008**, *36*, 550–558.
139. Melamud, E.; Moult, J. *Nucleic Acids Res.* **2009**, *37*, 4873–4886.
140. Tress, M. L.; Martelli, P. L.; Frankish, A.; Reeves, G. A.; Wesselink, J. J.; Yeats, C.; Olason, P. I.; Albrecht, M.; Hegyi, H.; Giorgetti, A.; Raimondo, D.; Lagarde, J.; Laskowski, R. A.; Lopez, G.; Sadowski, M. I.; Watson, J. D.; Fariselli, P.; Rossi, I.; Nagy, A.; Kai, W.; Storling, Z.; Orsini, M.; Assenov, Y.; Blankenburg, H.; Huthmacher, C.; Ramirez, F.; Schlicker, A.; Denoeud, F.; Jones, P.; Kerrien, S.; Orchard, S.; Antonarakis, S. E.; Reymond, A.; Birney, E.; Brunak, S.; Casadio, R.; Guigo, R.; Harrow, J.; Hermjakob, H.; Jones, D. T.; Lengauer, T.; Orengo, C. A.; Patthy, L.; Thornton, J. M.; Tramontano, A.; Valencia, A. *Proc. Natl. Acad. Sci. U.S.A.* **2007**, *104*, 5495–5500.
141. Tress, M. L.; Bodenmiller, B.; Aebersold, R.; Valencia, A. *Genome Biol.* **2008**, *9*, R162.

142. Menon, R.; Zhang, Q.; Zhang, Y.; Fermin, D.; Bardeesy, N.; DePinho, R. A.; Lu, C.; Hanash, S. M.; Omenn, G. S.; States, D. J. *Cancer Res.* **2009**, *69*, 300–309.
143. Bitton, D. A.; Smith, D. L.; Connolly, Y.; Scutt, P. J.; Miller, C. J. *PLoS One* **2010**, *5*, e8949.
144. Blakeley, P.; Siepen, J. A.; Lawless, C.; Hubbard, S. J. *Proteomics* **2010**, *10*, 1127–1140.
145. Menon, R.; Omenn, G. S. *Cancer Res.* **2010**, *70*, 3440–3449.
146. Severing, E. I.; van Dijk, A. D.; van Ham, R. C. *BMC Plant Biol.* **2011**, *11*, 82.
147. Leoni, G.; Le Pera, L.; Ferre, F.; Raimondo, D.; Tramontano, A. *Genome Biol.* **2011**, *12*, R9.
148. Ezkurdia, I.; del Pozo, A.; Frankish, A.; Rodriguez, J. M.; Harrow, J.; Ashman, K.; Valencia, A.; Tress, M. L. *Mol. Biol. Evol.* **2012**, *29*, 2265–2283.
149. Mo, F.; Hong, X.; Gao, F.; Du, L.; Wang, J.; Omenn, G. S.; Lin, B. *BMC Bioinforma.* **2008**, *9*, 537.
150. Chang, K. Y.; Georgianna, D. R.; Heber, S.; Payne, G. A.; Muddiman, D. C. *J. Proteome Res.* **2010**, *9*, 1209–1217.
151. Zhou, A.; Zhang, F.; Chen, J. Y. *BMC Bioinforma.* **2010**, *11*(Suppl. 6), S7.
152. Xing, X. B.; Li, Q. R.; Sun, H.; Fu, X.; Zhan, F.; Huang, X.; Li, J.; Chen, C. L.; Shyr, Y.; Zeng, R.; Li, Y. X.; Xie, L. *Genomics* **2011**, *98*, 343–351.
153. Lopez-Casado, G.; Covey, P. A.; Bedinger, P. A.; Mueller, L. A.; Thannhauser, T. W.; Zhang, S.; Fei, Z.; Giovannoni, J. J.; Rose, J. K. *Proteomics* **2012**, *12*, 761–774.
154. Yates, J. R., 3rd.; Eng, J. K.; McCormack, A. L. *Anal. Chem.* **1995**, *67*, 3202–3210.
155. Chen, J. H.; Shi, J. S.; Tian, D. G.; Yang, L. M.; Luo, Y. M.; Yin, D. H.; Hu, X. Y. *Plant Omics* **2011**, *4*, 257–263.
156. Power, K. A.; McRedmond, J. P.; de Stefani, A.; Gallagher, W. M.; Gaora, P. O. *PLoS One* **2009**, *4*, e5001.
157. Edwards, N. J. *Mol. Syst. Biol.* **2007**, *3*, 102.
158. Castellana, N. E.; Payne, S. H.; Shen, Z.; Stanke, M.; Bafna, V.; Briggs, S. P. *Proc. Natl. Acad. Sci. U.S.A.* **2008**, *105*, 21034–21038.
159. Castellana, N. E.; Pham, V.; Arnott, D.; Lill, J. R.; Bafna, V. *Mol. Cell. Proteomics* **2010**, *9*, 1260–1270.
160. Nesvizhskii, A. I.; Aebersold, R. *Mol. Cell. Proteomics* **2005**, *4*, 1419–1440.
161. Meyer-Arendt, K.; Old, W. M.; Houel, S.; Renganathan, K.; Eichelberger, B.; Resing, K. A.; Ahn, N. G. *J. Proteome Res.* **2011**, *10*, 3060–3075.
162. Roos, F. F.; Jacob, R.; Grossmann, J.; Fischer, B.; Buhmann, J. M.; Gruissem, W.; Baginsky, S.; Widmayer, P. *Bioinformatics* **2007**, *23*, 3016–3023.
163. Smith, L. M.; Kelleher, N. L. *Nat. Methods* **2013**, *10*, 186–187.
164. Sachidanandam, R.; Weissman, D.; Schmidt, S. C.; Kakol, J. M.; Stein, L. D.; Marth, G.; Sherry, S.; Mullikin, J. C.; Mortimore, B. J.; Willey, D. L.; Hunt, S. E.; Cole, C. G.; Coggill, P. C.; Rice, C. M.; Ning, Z.; Rogers, J.; Bentley, D. R.; Kwok, P. Y.; Mardis, E. R.; Yeh, R. T.; Schultz, B.; Cook, L.; Davenport, R.; Dante, M.; Fulton, L.; Hillier, L.; Waterston, R. H.; McPherson, J. D.; Gilman, B.; Schaffner, S.; Van Etten, W. J.; Reich, D.; Higgins, J.; Daly, M. J.; Blumenstiel, B.; Baldwin, J.; Stange-Thomann, N.; Zody, M. C.; Linton, L.; Lander, E. S.; Altshuler, D. *Nature* **2001**, *409*, 928–933.
165. Jorde, L. B.; Wooding, S. P. *Nat. Genet.* **2004**, *36*, S28–S33.
166. Levy, S.; Sutton, G.; Ng, P. C.; Feuk, L.; Halpern, A. L.; Walenz, B. P.; Axelrod, N.; Huang, J.; Kirkness, E. F.; Denisov, G.; Lin, Y.; MacDonald, J. R.; Pang, A. W.; Shago, M.; Stockwell, T. B.; Tsiamouri, A.; Bafna, V.; Bansal, V.; Kravitz, S. A.; Busam, D. A.; Beeson, K. Y.; McIntosh, T. C.; Remington, K. A.; Abril, J. F.; Gill, J.; Borman, J.; Rogers, Y. H.; Frazier, M. E.; Scherer, S. W.; Strausberg, R. L.; Venter, J. C. *PLoS Biol.* **2007**, *5*, e254.

167. Bacolod, M. D.; Schemmann, G. S.; Giardina, S. F.; Paty, P.; Notterman, D. A.; Barany, F. *Cancer Res.* **2009**, *69*, 723–727.
168. Shi, M. M. *Clin. Chem.* **2001**, *47*, 164–172.
169. Nedelkov, D. *Expert Rev. Proteomics* **2005**, *2*, 315–324.
170. Mann, M.; Wilm, M. *Anal. Chem.* **1994**, *66*, 4390–4399.
171. Gatlin, C. L.; Eng, J. K.; Cross, S. T.; Detter, J. C.; Yates, J. R., 3rd. *Anal. Chem.* **2000**, *72*, 757–763.
172. Frank, A.; Pevzner, P. *Anal. Chem.* **2005**, *77*, 964–973.
173. Tabb, D. L.; Saraf, A.; Yates, J. R., 3rd. *Anal. Chem.* **2003**, *75*, 6415–6421.
174. Tanner, S.; Shu, H.; Frank, A.; Wang, L. C.; Zandi, E.; Mumby, M.; Pevzner, P. A.; Bafna, V. *Anal. Chem.* **2005**, *77*, 4626–4639.
175. Na, S.; Jeong, J.; Park, H.; Lee, K. J.; Paek, E. *Mol. Cell. Proteomics* **2008**, *7*, 2452–2463.
176. Liu, C.; Yan, B.; Song, Y.; Xu, Y.; Cai, L. *Bioinformatics* **2006**, *22*, e307–e313.
177. Pevzner, P. A.; Dancik, V.; Tang, C. L. *J. Comput. Biol.* **2000**, *7*, 777–787.
178. Hu, Y.; Li, Y.; Lam, H. *Proteomics* **2011**, *11*, 4702–4711.
179. de Souza, G. A.; Arntzen, M. O.; Fortuin, S.; Schurch, A. C.; Malen, H.; McEvoy, C. R.; van Soolingen, D.; Thiede, B.; Warren, R. M.; Wiker, H. G. *Mol. Cell. Proteomics* **2011**, *10*, M110.002527.
180. Schandorff, S.; Olsen, J. V.; Bunkenborg, J.; Blagoev, B.; Zhang, Y.; Andersen, J. S.; Mann, M. *Nat. Methods* **2007**, *4*, 465–466.
181. Wang, X.; Zhang, B. *Bioinformatics* **2013**, *29*, 3235–3237.
182. Dasari, S.; Chambers, M. C.; Slebos, R. J.; Zimmerman, L. J.; Ham, A. J.; Tabb, D. L. *J. Proteome Res.* **2010**, *9*, 1716–1726.
183. Shilov, I. V.; Seymour, S. L.; Patel, A. A.; Loboda, A.; Tang, W. H.; Keating, S. P.; Hunter, C. L.; Nuwaysir, L. M.; Schaeffer, D. A. *Mol. Cell. Proteomics* **2007**, *6*, 1638–1655.
184. Han, Y.; Ma, B.; Zhang, K. *J. Bioinform. Comput. Biol.* **2005**, *3*, 697–716.
185. Mann, M.; Jensen, O. N. *Nat. Biotechnol.* **2003**, *21*, 255–261.
186. Nielsen, M. L.; Savitski, M. M.; Zubarev, R. A. *Mol. Cell. Proteomics* **2006**, *5*, 2384–2391.
187. Witze, E. S.; Old, W. M.; Resing, K. A.; Ahn, N. G. *Nat. Methods* **2007**, *4*, 798–806.
188. Young, N. L.; Plazas-Mayorca, M. D.; Garcia, B. A. *Expert Rev. Proteomics* **2010**, *7*, 79–92.
189. Larsen, M. R.; Trelle, M. B.; Thingholm, T. E.; Jensen, O. N. *Biotechniques* **2006**, *40*, 790–798.
190. Creasy, D. M.; Cottrell, J. S. *Proteomics* **2002**, *2*, 1426–1434.
191. Craig, R.; Beavis, R. C. *Rapid Commun. Mass Spectrom.* **2003**, *17*, 2310–2316.
192. Huang, X.; Huang, L.; Peng, H.; Guru, A.; Xue, W.; Hong, S. Y.; Liu, M.; Sharma, S.; Fu, K.; Caprez, A. P.; Swanson, D. R.; Zhang, Z.; Ding, S. J. *J. Proteome Res.* **2013**, *12*, 3831–3842.
193. Chalkley, R. J.; Baker, P. R.; Medzihradszky, K. F.; Lynn, A. J.; Burlingame, A. L. *Mol. Cell. Proteomics* **2008**, *7*, 2386–2398.
194. Hansen, B. T.; Davey, S. W.; Ham, A. J.; Liebler, D. C. *J. Proteome Res.* **2005**, *4*, 358–368.
195. Tang, W. H.; Halpern, B. R.; Shilov, I. V.; Seymour, S. L.; Keating, S. P.; Loboda, A.; Patel, A. A.; Schaeffer, D. A.; Nuwaysir, L. M. *Anal. Chem.* **2005**, *77*, 3931–3946.
196. Havilio, M.; Wool, A. *Anal. Chem.* **2007**, *79*, 1362–1368.
197. Baumgartner, C.; Rejtar, T.; Kullolli, M.; Akella, L. M.; Karger, B. L. *J. Proteome Res.* **2008**, *7*, 4199–4208.
198. Chen, Y.; Chen, W.; Cobb, M. H.; Zhao, Y. *Proc. Natl. Acad. Sci. U.S.A.* **2009**, *106*, 761–766.
199. Han, X.; He, L.; Xin, L.; Shan, B.; Ma, B. *J. Proteome Res.* **2011**, *10*, 2930–2936.

200. Searle, B. C.; Dasari, S.; Wilmarth, P. A.; Turner, M.; Reddy, A. P.; David, L. L.; Nagalla, S. R. *J. Proteome Res.* **2005**, *4*, 546–554.
201. Matthiesen, R.; Trelle, M. B.; Hojrup, P.; Bunkenborg, J.; Jensen, O. N. *J. Proteome Res.* **2005**, *4*, 2338–2347.
202. Kim, S.; Na, S.; Sim, J. W.; Park, H.; Jeong, J.; Kim, H.; Seo, Y.; Seo, J.; Lee, K. J.; Paek, E. *Nucleic Acids Res.* **2006**, *34*, W258–W263.
203. Shen, Y.; Tolic, N.; Hixson, K. K.; Purvine, S. O.; Anderson, G. A.; Smith, R. D. *Anal. Chem.* **2008**, *80*, 7742–7754.
204. Hernandez, P.; Gras, R.; Frey, J.; Appel, R. D. *Proteomics* **2003**, *3*, 870–878.
205. Na, S.; Bandeira, N.; Paek, E. *Mol. Cell. Proteomics* **2012**, *11*, M111.010199.
206. Bandeira, N. *Biotechniques* **2007**, *42*, 687, 689, 691 passim.
207. Falkner, J. A.; Falkner, J. W.; Yocum, A. K.; Andrews, P. C. *J. Proteome Res.* **2008**, *7*, 4614–4622.
208. Ahrne, E.; Masselot, A.; Binz, P. A.; Muller, M.; Lisacek, F. *Proteomics* **2009**, *9*, 1731–1736.
209. Ye, D.; Fu, Y.; Sun, R. X.; Wang, H. P.; Yuan, Z. F.; Chi, H.; He, S. M. *Bioinformatics* **2010**, *26*, i399–i406.
210. Savitski, M. M.; Nielsen, M. L.; Zubarev, R. A. *Mol. Cell. Proteomics* **2006**, *5*, 935–948.
211. Potthast, F.; Gerrits, B.; Hakkinen, J.; Rutishauser, D.; Ahrens, C. H.; Roschitzki, B.; Baerenfaller, K.; Munton, R. P.; Walther, P.; Gehrig, P.; Seif, P.; Seeberger, P. H.; Schlapbach, R. *J. Chromatogr. B Analyt. Technol. Biomed. Life Sci.* **2007**, *854*, 173–182.
212. Fu, Y.; Xiu, L. Y.; Jia, W.; Ye, D.; Sun, R. X.; Qian, X. H.; He, S. M. *Mol. Cell. Proteomics* **2011**, *10*, M110.000455.
213. Liu, J.; Erassov, A.; Halina, P.; Canete, M.; Nguyen, D. V.; Chung, C.; Cagney, G.; Ignatchenko, A.; Fong, V.; Emili, A. *Anal. Chem.* **2008**, *80*, 7846–7854.
214. Baliban, R. C.; DiMaggio, P. A.; Plazas-Mayorca, M. D.; Young, N. L.; Garcia, B. A.; Floudas, C. A. *Mol. Cell. Proteomics* **2010**, *9*, 764–779.
215. Pevzner, P. A.; Mulyukov, Z.; Dancik, V.; Tang, C. L. *Genome Res.* **2001**, *11*, 290–299.
216. Tsur, D.; Tanner, S.; Zandi, E.; Bafna, V.; Pevzner, P. A. *Nat. Biotechnol.* **2005**, *23*, 1562–1567.
217. Tanner, S.; Pevzner, P. A.; Bafna, V. *Nat. Protoc.* **2006**, *1*, 67–72.
218. Matthiesen, R.; Bunkenborg, J.; Stensballe, A.; Jensen, O. N.; Welinder, K. G.; Bauw, G. *Proteomics* **2004**, *4*, 2583–2593.
219. Na, S.; Paek, E. *J. Proteome Res.* **2009**, *8*, 4418–4427.
220. Stein, S. E.; Scott, D. R. *J. Am. Soc. Mass Spectrom.* **1994**, *5*, 859–866.
221. Garavelli, J. S. *Proteomics* **2004**, *4*, 1527–1533.
222. Chalkley, R. J.; Clauser, K. R. *Mol. Cell. Proteomics* **2012**, *11*, 3–14.
223. Tanner, S.; Payne, S. H.; Dasari, S.; Shen, Z.; Wilmarth, P. A.; David, L. L.; Loomis, W. F.; Briggs, S. P.; Bafna, V. *J. Proteome Res.* **2008**, *7*, 170–181.
224. Chung, C.; Liu, J.; Emili, A.; Frey, B. J. *Bioinformatics* **2011**, *27*, 797–806.
225. Chung, C.; Emili, A.; Frey, B. J. *Bioinformatics* **2013**, *29*, 821–829.

Chapter 17

Transcriptome and Metabolome Data Integration—Technical Perquisites for Successful Data Fusion and Visualization

Michael Witting* and Philippe Schmitt-Kopplin*,†

*Research Unit Analytical BioGeoChemistry, Helmholtz Zentrum München, German Research Center for Environmental Health, Neuherberg, Germany

†Chair of Analytical Food Chemistry, Technische Universität München, Freising-Weihenstephan, Germany

Chapter Outline

1 INTRODUCTION

Understanding complex living systems or supersystems on a molecular level needs comprehensive data analysis, collection, and mining strategies. Today's Omics technologies, such as transcriptomics, proteomics, and metabolomics,

Comprehensive Analytical Chemistry, Vol. 63. http://dx.doi.org/10.1016/B978-0-444-62651-6.00018-0

enable collection of such data on a routine basis, even in a high-throughput manner. In classic biology view data flows from genes via transcripts and proteins to metabolites, but scientists recognize that these network layers are more tightly connected than previously assumed. In several cases, collection of single-level data will never tell the whole story. A metabolite may show significantly higher concentrations under a certain physiological state because it is produced in higher amounts by the downstream pathway or it is less metabolized by upstream reactions. If no further information on surrounding metabolites or data from other methods is available, it is difficult to decipher causality. Similarly, transcripts can tell only if a gene is transcribed, not if the protein is functional and active. Therefore, more and more scientists integrate different Omics approaches in one experimental setup to confine the studied systems. Different combinations of Omics technologies are possible, whereas transcriptomics and metabolomics are recently often preferred.

Transcriptomics provides information about the presence and relative abundance of transcripts. Different methods for gene expression profiling at the genome level exist. Gene array technology is widely used and versatile enough for high-throughput studies. It is based on the principle of Watson-Crick base pairs and hybridization of two complementary DNA or RNA strains. Next-generation sequencing of transcripts, often referred as RNA-Seq, uses the latest generation sequencing machines to complete RNA content of a cell or system, including mRNA, transfer RNA, small interfering RNA, and so on. Although readily applied, this technology is still under development and data analysis is not as straightforward as with gene expression arrays. The major advantage of RNA-Seq is the coverage of the transcriptome because sequencing is not based on predefined oligonucleotides for hybridization. Results can include previously unknown genes, splicing variants, or novel open reading frames (ORFs) *(1)*. Therefore, with RNA-Seq more novel discoveries are most likely possible *(2)*.

Metabolomics is defined as the systematic study of metabolites in a biological system. It is the youngest approach in the Omics family. The metabolome is defined as the sum of all metabolites given in a biological system under particular physiological conditions. It can be divided into the exometabolome (metabolites outside the cell) and the endometabolome (intracellular metabolites). The term *metabonomics* was defined by Jeremy Nicholson *(3)* in 1999 as "the quantitative measurement of the dynamic multiparametric metabolic response of living systems to pathophysiological stimuli or genetic modification." First measurements of metabolites in body fluids have already been carried out earlier without calling it metabolomics (e.g., the pioneering work of Linus Pauling) *(4)*. In contrast to transcripts and proteins, metabolites are not encoded in the genome and are highly dependent on the surrounding environment, making it complicated to give definite numbers of metabolites present in an organism. Additionally, chemical diversity of metabolites is far bigger than that of DNA, RNA, and proteins, which are based on 4 nucleotides or 20 amino acids. The terms *metabolomics* and *metabonomics* are often used synonymously in the literature. To avoid any confusion, the expressions nontargeted and targeted metabolomics are widely used. *Nontargeted metabolomics* refers to the hypothesis-free elucidation of metabolism. The aim in nontargeted metabolomics is to detect as many metabolic features from different metabolite classes as possible and correlate them with a certain physiological state. Nontargeted metabolomics approaches are data- and work-intensive, mostly generating hypothesis during analysis, which has to be proved afterward. The counterpart is *targeted metabolomics*, which is analyzing and quantifying only a subset of metabolites (e.g., compounds belonging to energy metabolism). Several analytical techniques are used in both approaches. Nontargeted metabolomics is mostly based on discovery type and high- or ultrahigh-resolution mass spectrometric methods, such as Q-ToF-MS, ICR-FT/MS, or nuclear magnetic resonance (NMR). Targeted metabolomics prefers routine very sensitive and fast techniques, such as triple quadrupole mass spectrometry (MS). However, the line between both is blurring.

Figure 1 shows a timeline of a PubMed search for papers sharing the terms "metabolomics" and "transcriptomics" from 2000 until 2012. The amount has increased steadily and may increase further because both technologies can be applied today on a routine basis.

This chapter aims to review different combinations of data fusion. Genomics and transcriptomics currently reach total coverage for a sequenced organism. Not all gene function might be known, but at least predicted ORFs can be tested for expression. Currently, metabolomics is far away for complete metabolome coverage. Today numbers between 5% and 30% of known metabolites in a typical MS-based dataset are the normal case. The remaining

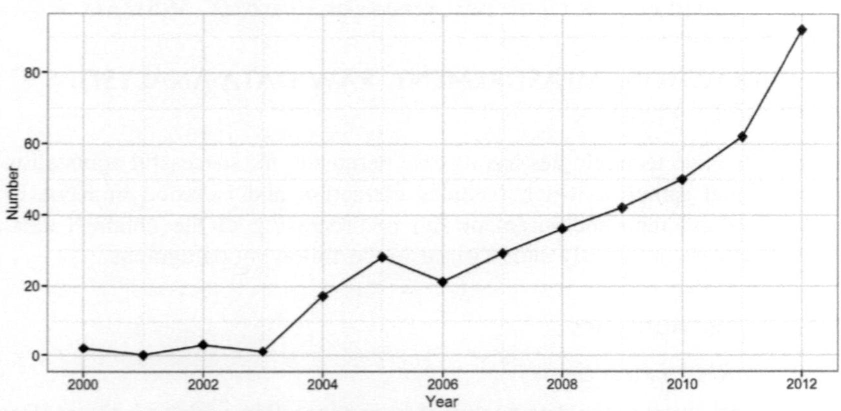

FIGURE 1 Timeline for PubMed search using the keywords *metabolomics* and *transcriptomics*. Amount of publications using a combination of both technologies was steadily increasing since 2000. Both types of analysis can be used in routine analysis; therefore, in the near future a vast amount of combined efforts can be expected.

70–95% may correspond to previously unknown and novel metabolites. The combined analysis of both levels together gives room for novel insights and discoveries, including gene function annotation and metabolite identification.

Different associations between transcripts and metabolites can be found depending on the nature of analysis. For instance, a gene of unknown function is correlating with well-known metabolites. In this case, the possible gene represents a novel enzyme of a specific pathway. In another case, a gene encodes for an enzyme catalyzing reaction for a specific class of metabolites. Together with known metabolites, several unknowns are correlated with a specific gene. This may imply that both known and unknown metabolites have similar structures or at least share a structural scaffold.

Another possible goal for combining metabolomics and transcriptomics is to substantiate results from both on known pathways. Because data integration of transcriptomics and metabolomics aims to provide better understanding of processes on pathway levels, we focus here on the gene expression array technology. These arrays also include genes of unknown function, and room for novel findings on gene functions is still possible.

We focus on the combination of transcriptomics and metabolomics and more specifically on microarray data, which is currently the most used method for gene expression profiling and is used on a routine basis. In the first paragraphs, we briefly revise the extraction of mRNA or metabolites, their measurement, quality control of data, and analysis methods. Afterward two different types of data fusion and recent tools and publications are reviewed, followed by visualization methods for obtained data. Lastly, the metabolite annotation Web server *MassTRIX* is presented. This Web server allows combined analysis of transcriptomic and metabolomic data in the context of metabolic pathways. We compared the metabolomics part against similar tools and give a short outlook on the next version of *MassTRIX*, *MassTRIX 4*.

2 EXTRACTION, MEASUREMENT, RAW DATA ANALYSIS, AND DATA FUSION

Each of the two technologies has its own perquisite for successful application to biological samples, which includes extraction and isolation of RNA or metabolite and their measurement and preprocessing of the obtained data. These methods are briefly summarized in the following paragraphs.

2.1 Transcriptomics

2.1.1 mRNA Extraction

A major influence on the data quality of gene expression analysis is the quality of the RNA preparation. RNA derived from the sample should be free or at least contain a minimal amount as possible of genomic DNA, hemoglobin, fat, glycogen, Ca^{2+}, and others *(5,6)*. Contamination with genomic DNA may lead to false-positive observations in RT-PCR or microarray analysis. The most common extraction method, often also referred as TRIzol extraction, is based on phenol and chloroform and available as a kit from many different companies. Using acidic pH RNA can specifically recovered from the aqueous phase and finally purified by 2-propanol or ethanol precipitation. This method uses hazardous chemicals and a fume hood is needed to carry out this protocol. An appropriate alternative is represented by column-based extraction kits. Purification is based on interaction of nucleic acids with a solid phase, either silica or others. This interaction is dependent on pH and salt concentration. High salt buffers and low pH are used for binding and elution can be performed with water or a suitable buffer *(7)*. Purity of RNA is determined using the A260/A280 ratio, where a value $\sim$2 is considered pure. The most utilized system for quality assessment is the Agilent 2100 Bioanalyzer and delivers an RNA integrity number as a measure of quality. If DNA could not be completely removed, usage of DNAse is a suitable follow-up procedure. Typical amounts of RNA needed for microarray analysis range around 300–500 ng, including material for QC.

2.1.2 Microarray Technology

Several companies offer microarrays for gene expression profiling, whereas the most popular are the systems of Agilent Technologies and Affymetrix (Figure 2). Agilent microarrays are based on a glass slide with the approximate size of a specimen holder. The oligonucleotide is a 60mer and is directly synthesized on the glass slide. The Agilent design is based on the competitive binding of two samples differentially labeled with two different fluorescence tags, a red and a green one. In contrast, the Affymetrix arrays are based on silicon wafer chips and 25 oligomers and only one sample is hybridized per array. Both companies offer several microarrays for humans, but also for different model organisms such as a mouse, barley, tomato, Arabidopsis, *Caenorhabditis elegans*, *Escherichia coli*, and others. If the studied organism is not available, both companies offer the possibility to create custom arrays. However, most genetics laboratories produce their own microarrays based on cDNA. Examples for other microarray vendors are HySeq, Nanogen, or PerkinElmer.

2.1.3 Quality Control and Calculation of Gene Expression

At the beginning of gene expression data analysis, visual quality control is important to find artifacts, scratches, air bubbles, and unspecific signal background. Other parameters for QC include the brightness, uniformity, and morphology of the spot. An important measure of quality is the ratio of average foreground to background signal, which should be bigger than 3. Dependent on the type of error, either normalization of the system or an error model for random noise should be applied to ensure data integrity for downstream analysis.

A step with big influence on further downstream data analysis of microarray data is the method of calculating gene expression. In 2006, Millenaar

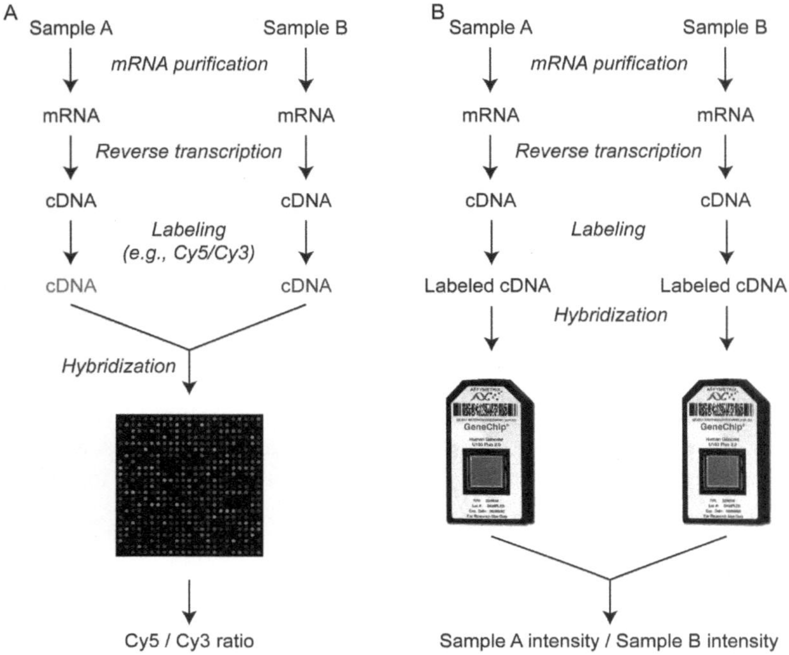

FIGURE 2 (A) Typical workflow for transcriptomics experiments using Agilent microarrays. After purification of mRNA, it is reverse-transcribed to cDNA, which is labeled with two different fluorescent tags. Equal amounts of cDNA preparation are mixed and applied to the microarray for competitive binding. (B) Most steps for Affymetrix microarrays are similar. For labeling biotin is used and readout of expression values is based on a streptavidin tag binding to biotin. Compared to Agilent, only one sample is hybridized per array. (For color version of this figure, the reader is referred to the online version of this chapter.)

presented a work comparing six different algorithms with the same data derived from the Arabidopsis ATH1-121501 genome array of Affymetrix. These algorithms included the Microarray Suite and its successor GeneChip Operating Software from Affymetrix; robust multiarray average (RMA) and the derived GeneChip RMA (GCRMA), which additionally uses probe affinity information; Model-based expression indexes implemented in dChip; and the positional-dependent-nearest-neighbor model. These algorithms were compared on their reproducibility and correlation with RT-PCR. Out of the six, RMA performed best. The authors claim that the combination of results from several of these algorithms lead to better results if the aim is to find candidate genes *(8)*.

After these preprocessing steps, raw data can be used for biological interpretation. In a typical experiment comparing two conditions, raw gene expression data is converted to log 2 fold-changes. Significances are calculated with different statistical tests.

2.2 Metabolomics

2.2.1 Metabolite Extraction and Sample Preparation

Like for all other analytical chemistry applications, in metabolomics sample preparation and cleanup is the essential first step. In comparison to RNA extraction methods, no generally accepted extraction protocol for metabolites exists. Most metabolomics applications are starting from cell exudates, cell pellets, or tissue. Protocols for sample preparation are dependent on the cell structure of the biological sample and always involve a quenching step to immobilize the metabolite content of the extract. Plant cell walls need a different treatment than a mammalian cell culture sample. One fact is similar for all protocols: After sample preparation, the sample should be highly enriched and pure in the targeted analytes. For metabolomics, especially nontargeted metabolomics, this is a tough task, as different classes of molecules will be analyzed simultaneously and each shows, in contrast to biopolymers such as DNA, RNA, or proteins, different physicochemical parameters and concentrations, which range over several orders of magnitude from pico- to millimolar. Thus it should be mentioned that a truly nontargeted extraction does not exist, as an extraction system is always more discriminant for some metabolite classes than for others. Lastly, an optimal sample preparation method should include only the minimal steps that are needed to achieve this goal to keep the samples as close as possible to its native state. Major tasks in metabolomics sample preparation are quenching of metabolism, lysis of cells, extraction of metabolites, and removal of interfering substances (e.g., proteins and salts).

Metabolome analysis can be compared to a more or less resolved photography, which is a static picture of a dynamic environment. Metabolism is highly variable and time- and condition-dependent. Metabolites from the primary metabolism are produced and consumed with high turnover rates (e.g., 1.5 mM/s for ATP) *(9,10)*, whereas secondary metabolites accumulate in the cell or are excreted. To obtain a snapshot of system physiology, metabolism has to be stopped quickly, meaning inactivation of metabolic enzymes. This can be carried out by using organic solvent (hot or cold), extreme pH conditions ($pH < 3$ or $pH > 11$), or snap-freezing in liquid nitrogen. A problem is the separation of intra- and extracellular matrix. Methods used in microbial metabolomics, such as spraying culture into cold methanol, lead to a mixture of both matrices. This is especially complicated if rich culture media with unknown exact formulation are used. These can contaminate the sample too much, which makes a reliable data interpretation impossible. Separation by centrifugation or filtration may alter the cellular metabolome. A tradeoff has to be found between both possibilities.

After the metabolism has been stopped, cells have to be lysed to access the intracellular metabolites. Different protocols are available for this task. Roughly separated mechanical and nonmechanical methods exist, including ultrasonic, grinding, enzymatic, or chemical lysis or osmotic shock. During the whole process attention to temperature in the sample should be given. Most mechanical methods heat the samples up, so lysis has to be carried on dry ice or ice. In the case of chemical lysis, some metabolites classes will be degraded or converted to other forms *(10)*. Together with the lysis, often metabolites are directly extracted into an appropriate solvent. For metabolites from primary metabolism and polar to midpolar secondary metabolites, solvent mixtures of water and a miscible organic solvent are used with concentrations $\geq$50% organic *(11)*. Commonly methanol, ethanol, acetonitrile, or isopropanol are used. Organic solvents precipitate proteins, a major interference in metabolomics studies. If more nonpolar substances such as lipids need to be extracted, isopropanol, chloroform, or methyl-tert-butyl ether (MTBE) can be used. A method to obtain a total lipid extract from solid material was described by Folch using a chloroform/methanol (2/1) mixture *(12)*, while the Bligh and Dyer method is used for lipid extraction from aqueous samples *(13)*. Both methods are using chloroform for extraction, meaning the lipid-rich layer after centrifugation will be the lower phase, making it hard to automate this procedure. An alternative was described by Matyash using MTBE. In this method, the organic solvent forms the upper layer, which is useful for automation on a robotic system. Extraction yields are comparable to the other methods mentioned *(14)*.

If more than one "Omics" approach needs to be used for analysis, often subsamples of the same biological sample are used. This is not applicable if only low amounts of sample are available. Therefore, combined extraction regimes have to be developed. Proteins for proteomic analysis can be recovered from this precipitate and used for further downstream proteomic analysis *(15)*. In 2004, a sequential extraction of metabolites, proteins, and RNA from the same biological sample was described. Metabolites were extracted with cold one-phasic methanol/chloroform/water. The supernatant contained both hydrophilic and hydrophobic compounds and was further subfractionated into these classes. Both were analyzed with GC-ToF-MS. From the remaining, pellet proteins and RNA were extracted. Two-dimensional LC–MS was used for protein analysis. Interestingly the amount of extracted RNA was higher than with a conventional RNA extraction kit *(16)*. A more recent work used a similar approach based on chromatographic spin columns to avoid hazardous chemicals. Simultaneous extraction of genomic DNA, large and small RNA, proteins, and metabolites was optimized for different microbial ecosystems (e.g., wastewater sludge, river water, or human feces). Quality of the respective fractions was compared to single dedicated extraction methods. Similar to the above-mentioned work, the authors found that combined methods yield similar or even better-quality material *(17)*.

After metabolite extraction, enrichment of target metabolites or other polishing steps such as desalting or solvent exchange may be necessary. One possibility for such a cleanup is solid-phase extraction (SPE). The principle is similar to chromatography and is based on the distribution of analytes between a solid and a mobile phase. Analytes of interest are trapped on a suitable solid phase and interfering substances are washed away. Afterward, compounds are eluted with a suitable organic solvent. Several materials for SPE exist, including reverse-phase materials, ion exchanger and mixed mode. As base material mostly silica gel is used, but use of polymer-based material is becoming popular. SPE is not only useful for removal of interfering salts but also for targeted analysis and cleanup of the targeted compounds and their concentration by using a smaller elution volume compared to original applied sample volume. Other methods for concentration of a sample can be used if necessary (e.g., lyophilization, gentle streams of nitrogen, or vacuum centrifuges to remove solvents). Attention has to be drawn to this step because some analytes might get lost during this procedure. With this method also starting conditions of a sample can be optimized for a specific analytical method (e.g., the change to deuterated solvents for NMR). Also, sensitivity is increased if the final volume is smaller than the previous sample volume.

2.2.2 Metabolomics Technologies

Different analytical chemistry methods are used for analysis of the metabolome (Figure 3). Direct-infusion mass spectrometry (DI-MS) on both, low and (ultra)high resolution MS, infuses raw metabolite extracts in the mass spectrometer without prior chromatography or electrophoretic separation and often uses high-resolution mass spectrometers. This method offers fast analysis with low duty cycles; however, isomeric and isobaric substances cannot be

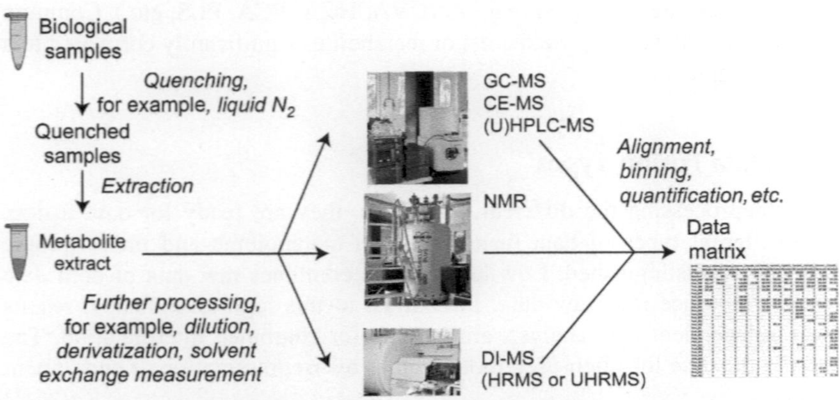

FIGURE 3 Typical metabolomics workflow. Biological samples are quenched, for example, with liquid nitrogen to stop enzymatic reactions. Afterward, they are extracted with a suitable solvent. For different analytical methods, further processing steps like SPE or solvent exchange to deuterated solvents for NMR are needed. After measurement, different data processing steps are needed to yield a suitable data matrix for downstream analysis. (For color version of this figure, the reader is referred to the online version of this chapter.)

resolved. Utilization of gas chromatography (GC), liquid chromatography (LC), or capillary electrophoresis (CE) can overcome this drawback. NMR lacks sensitivity compared to MS but provides qualitative and quantitative information in one experiment.

2.2.3 Quality Control and Data Preprocessing

Data pretreatment of metabolomics data is dependent on the employed method. In DI-MS- and GC/LC/CE-MS-based nontargeted metabolomics, peak lists have to be aligned in *m/z* or retention time direction in the latter case across different samples to yield a suitable data matrix. Several open-source and commercial software for this task are available *(18–20)*. In NMR-based methodologies, two different approaches exist: binning of the NMR spectra in defined bins or metabolites identified from resonances. In targeted metabolomics, analysis peaks of interest are integrated and compared against standards with known concentration to reveal absolute concentration of metabolites.

Metabolomics quality control uses different methods. In most cases for GC/LC/CE-MS-based nontargeted metabolomics, a pooled sample from the study serves as a quality-control sample and is injected prior to real samples to condition the chromatographic system and between samples for control of performance. Using this QC samples, retention time drifts or other alteration in performance can be monitored and possibly corrected. In targeted metabolomics retention time shifts, LOD and LOQ values and recovery rates of known materials are used similar to classic analytical chemistry.

For statistical analysis of metabolomics data, different uni- and multivariate techniques are employed (e.g., ANOVA, HCA, PCA, PLS, etc.). Common to all, after analysis they yield a list of metabolites significantly correlated to a certain sample state.

2.3 Data Fusion Types

After preprocessing the different data types, they are ready for data fusion. Two different types of data fusion between metabolome and transcriptome data can be distinguished. Low-level fusion combines raw data of both data types to produce new raw data. In contrast to this high-level fusion, results from independent data analysis are merged for combined interpretation. The latter is the case for often-used tools such as overrepresentation or enrichment analysis.

2.3.1 Low-Level Fusion

Low-level fusion is particularly interesting for nontargeted metabolomics. This type of data fusion can help with unknown structural elucidation and linkage to known metabolic pathways and reactions. However, important points have to be considered if correlation analysis is used. Care of different scales of metabolome and transcriptome data has to be taken. Dependent on the metabolomic method, targeted or nontargeted, either absolute concentrations or relative measures (e.g., peak intensities or peak areas) are derived. Both can change over several orders of magnitude.

If correlation analysis is used with the raw data, ideally both data types should have similar ranges and distributions. If data is directly linearly correlated, this can be neglected, but is rarely the case for metabolome and transcriptome data. Changes in gene expression may not alter metabolite pools significantly. Therefore, data have to be normalized in an appropriate way and correlation methods other than linear correlation have to be used (e.g., Spearman's rank-order correlation or Kendall rank correlation should be preferred over Pearson correlation).

Usage of correlation analysis is especially suited for combination with nontargeted analysis. Different combinations offer possibilities for identification of gene function or unknown metabolites. Possibly unknown metabolites correlate with genes of known function, which allows elucidation of functional groups or structural scaffolds of the unknown. These can possibly speed up identification of unknown metabolites and their chemical structure.

Interesting problems and challenges arise from clustering of genes with unknown functions and unknown metabolites. Coclustering with other genes and metabolites may help in understanding their biological roles. Self-organizing maps are an interesting approach to reveal clusters of similar functionality.

Several interesting papers on low-level fusion of transcriptomic and metabolomic data can be found in the plant research field. One of the first works published was conducted for potato tuber. A custom microarray on nylon filters was used and metabolome analysis was based on GC–MS. Spearman rank-order correlation with a significance threshold at $p=0.01$ was used. The authors explicitly stated that they used this method because mRNA and metabolites were correlated in a nonlinear manner. From 26,616 possible pairs, only 571 showed significant correlation. The approach was validated on known relationships (e.g., negative correlation between sucrose and sucrose transporter expression). Several transcripts correlated with more than one metabolite. A major point discussed is that no direct causality can be derived from this correlation analysis and further experiments are needed to elucidate underlying mechanisms *(21)*. Another example for low-level data fusion from the same group can be found in Hannah *et al.*, which profiled *Arabidopsis thaliana* challenged by different extreme environments. Metabolome data was collected using GC– and LC–MS. Transcriptomics was carried out using the Arabidopsis Affymetrix ATH1 array. Normalization of all arrays was performed using RMA. The complete dataset consisted of 562 analytes and approximately 12,500 transcripts. Spearman rank correlation was used together with Bonferroni correction to reveal significant correlations. Special attention was drawn to identify novel gene-regulating metabolites. The study also showed a possible mediating role for leucine *(22)*. Redestig *et al.* described a method for detection of metabolite–transcript coresponses using Pearson and lagged Pearson correlation in conjunction with hidden Markov model (HMM)-based similarity in time-series experiments. Their methodology was validated using Arabidopsis stress response from different available datasets. In all cases, HMM outperformed Pearson and lagged Pearson correlations. Authors claimed if enough known associations are present in the dataset, *de novo* associations could be found *(23)*.

These are just three examples of low-level data integration showing the capabilities of this approach. In the above-mentioned articles, previously known metabolite-transcripts were found along novel ones. In most cases, metabolites are known, but such analysis can be even further developed for the biological interpretation of unknown molecules. The major advantage of low-level data fusion is that *a priori* no knowledge about the studied system is needed, although for method validation known associations are needed.

2.3.2 High-Level Fusion

In contrast to low-level data fusion, high-level fusion relies on previous knowledge from databases and metabolic pathways. After statistical analysis obtaining possible markers, biological analysis is the next step. This can be carried out by enrichment analysis of coordinately changed metabolites or genes. This method is originally derived from gene expression analysis. It is known that genes belonging to the same pathway are altered in a coordinated manner. This led to the development of gene set enrichment analysis, which searches for enrichment of significantly different expressed genes on specific pathways or functions. Similar methods have been proposed for metabolomics (e.g., described by Xia and Wishart for human metabolism *(24)*). Their methodology included three different types of enrichment analysis, overrepresentation analysis (ORA), single sample profiling (SSP), and quantitative enrichment analysis (QEA). ORA compares a list of metabolites against a random generated list and searches for significantly enriched pathways. p-Values, Bonferroni corrected p-values, and FDR are reported as measures of significance. SSP uses normal concentration ranges of metabolites in blood, urine, or CSF, which are compared against the measured values from a single sample. Enrichment analysis is carried out on metabolites that are below or above the reported normal concentration analysis. QEA is calculating enrichment directly from raw metabolite concentrations without previous statistical investigations for a complete matrix of samples *(24)*. All methods are implemented in the metabolomic data analysis server MetaboAnalyst *(25)*. Another server that allows the direct analysis of MST from GC–MS data was described by Kankainen *et al.* *(26)*.

For combined analysis of transcriptomics or proteomics and metabolomics data, the IMPaLA Web server was designed. This Web server allows either ORA or Wilcoxon enrichment analysis (WEA). For ORA, similar to the above-described genes/proteins and metabolites that are significantly different are uploaded. Additionally this server allows the upload of a background list, which contains all measured genes/proteins and/or metabolites, to avoid potential bias. WEA directly compares two different conditions and identifiers together with either average expression/concentrations or fold-changes. p- and q-values are reported, according to Benjamini and Hochberg *(27)*. If metabolites and gene/proteins are uploaded, a combined p-value is calculated *(28)*.

3 VISUALIZATION

After identification of gene–metabolite associations or enrichment analysis, visualization is a second key point. Results from low-level data fusion often yields pairwise correlations, which can be visualized using networks. In the case of high-level data fusion, a combined visualization is used on metabolic pathways (e.g., the well-known pathway maps from KEGG are preferred). We discuss some technical resources for both visualization types. However, much more tools for different kinds of visualization exist and are reviewed elsewhere *(29)*.

3.1 Visualization on KEGG Pathways

The newest version of the KEGG database supplies different possibilities accessing different pathways, from simple pathway descriptions to custom-colored pathways. In this version, the API is changed from SOAP to a REST-based Web service. This Web service uses URI-based links for data retrieval (e.g., http://rest.kegg.jp/list/hsa returns a list of all human genes stored in the KEGG database). In a similar manner, colored pathways can be retrieved. Two different possibilities exist for transfer of data, the GET and POST methods: the POST method is preferred for longer datasets to color on a pathway. Because the URI always looks the same, easy implementation in routines in different programming languages is possible. HTML output is returned by the Web service, which can be used in your own implementations or on a Web server (Figure 4). Unfortunately, no functionality retrieving only the *.png* file, which was available in the old deprecated SOAP API, is available until now.

Both the GET and POST methods accept KEGG identifiers as input and colors in hexadecimal code (hex code) format for fore- and background color. The usage of hex code for colors allows use of color gradients for mapping of differential gene expression or different metabolite concentrations.

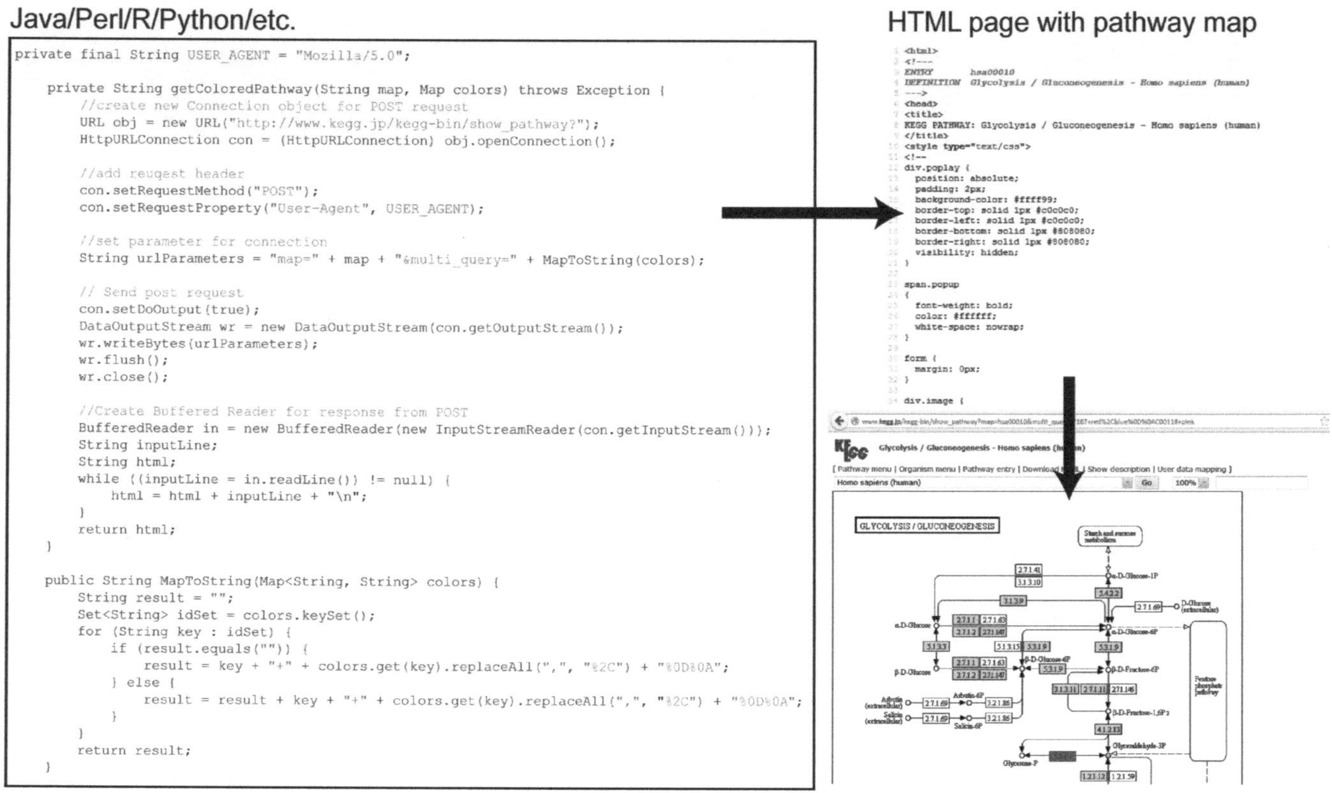

FIGURE 4 Different programming languages can access KEGG API REST Web services to retrieve colored pathways. The API returns a HTML page with respective pathway and metabolites marked. (For color version of this figure, the reader is referred to the online version of this chapter.)

3.2 Visualization on MetaCyc Pathways

The MetaCyc database collection allows mapping and visualization of different Omics data on the metabolic pathway present in this database. It is accessible via a webpage (http://biocyc.org/overviewsWeb/celOv.shtml). Basic pathway images can be retrieved via a REST-based Web service (http://biocyc.org/web-services.shtml). The major advantage compared to KEGG pathway mapping is the multiplexing for visualization of complex data (e.g., time series or different samples). Data can be uploaded as a simple tab-delimited file containing the respective identifiers and a numeric value, corresponding to expression or metabolite level.

3.3 Network Visualization and Analysis

Correlation analysis yields data matrix, which are hardly human-readable. Therefore, different methods for visualization are employed. A correlation matrix can be visualized as a heat map together with clustering analysis to reveal a subcluster of similar correlated transcripts or metabolites. However, correlation networks are often the preferred tool for visualizing this complex data. In such networks, metabolites and transcripts are represented as nodes connected through edges representing correlations. Only significant correlations can be visualized to reduce complexity of networks. Using network analysis and graph theory hubs, strongly connected and therefore probably important metabolites and transcripts can be identified. In most cases, using data-dependent layouts for network representation is also important so that a subcluster of biological closely related functions can be revealed. Many software tools for analysis of biological networks exist; Cytoscape and VANTED represent the two most employed.

VANTED, short for visualization and analysis of networks with related experimental data, uses networks produced by the software tool itself or derived from the KEGG database. It allows representation of transcript, enzyme, and/or metabolite data on the networks (e.g., for time-series data). A standardized Excel sheet serves as input for the application. It offers advanced data analysis methods, such as correlation analysis or self-organizing maps *(30,31)*.

Cytoscape is an open-source software framework for analysis of networks. It offers several plug-ins for customization of its functionalities and new plug-ins are released on a regular basis. It is not limited to biological data and can be used for visualization of very large networks (e.g., protein interaction maps) *(32)*.

Lastly, visualization of results in networks also allows mapping of additional data. For example, high-resolution metabolomics data can be analyzed with mass difference networks *(33)* and results can be used together with correlation analysis to find novel metabolic reactions.

4 *MassTRIX* RELOADED—COMBINED ANALYSIS AND VISUALIZATION OF METABOLOME AND TRASCRIPTOME DATA

MassTRIX is Web server for metabolite annotation using exact mass. It was originally developed by Suhre and Schmitt-Kopplin in 2008 and an updated version has been published, which allows additional analysis of transcriptome data supplied as Affymetrix .cel files *(34,35)*. The basic functionality is briefly reviewed here, together with a look to the future.

4.1 Annotation of Mass Spectrometric Data

The core functionality uses a given mass list and compares it against theoretical masses of adducts of metabolites from a chosen database within a certain error range. Table 1 shows the mass spectrometric adducts that are covered by *MassTRIX*.

Metabolites from different databases are used by *MassTRIX* for the annotation process. The monoisotopic masses were recalculated based on exact atomic masses using the molecular formulas stored in the respective database *(36)*. At the moment databases supported by *MassTRIX* are KEGG, HMDB, Lipidmaps, and MetaCyc in different combinations *(37–40)*. Because most likely not all metabolites of interest are present in this database, the new version of *MassTRIX* includes the possibility to upload a list of own molecules as

TABLE 1 All Possible Adduct Masses Are Calculated Based on Exact Atomic Masses

Scan mode	Adduct	Calculation
Negative	$[M-H]^-$	$M-1.007825037-e^-$
	$[M+Br]^-$	$M+78.9183361+e^-$ (^{79}Br, 50.69%)
		$M+80.91629+e^-$ (^{80}Br, 49.31%)
	$[M+Cl]^-$	$M+34.96885273+e^-$ (^{35}Cl, 75.77%)
		$M+36.96590262+e^-$ (^{37}Cl, 24.23%)
Neutral	[M]	M
Positive	$[M+H]^+$	$M+1.007825037-e^-$
	$[M+Na]^+$	$M+22.9897697-e^-$
	$[M+K]^+$	$M+38.9637079-e^-$ (^{39}K, 93.26%)
		$M+40.9618254-e^-$ (^{41}K, 6.73%)

For atoms with significantly abundant isotopes, all isotopes were included ($e^-=5.48579\times10^{-4}$ u).

precalculated adducts, which will be included in the annotation process. If KEGG IDs are supplied with this list, pathway mapping of these compounds is possible. Moreover, with this function adducts not covered by *MassTRIX* (e.g., $[M+H-H_2O]^+$ or $[M+2H]^{2+}$) can be included. Uploaded masses are matched against the theoretical adduct mass of metabolites from the chosen database within a certain error range, usually expressed in ppm. A maximum error up to 3 ppm is possible; for instruments with lower resolution, an absolute error range has been added in the new version. Several elements of these adducts have isotopes with significant natural abundances. To avoid false-positive annotations, adducts are filtered according to isotopes. Bromine, for example, has two different isotopes (^{79}Br and ^{81}Br) with a natural abundance of about 50%. Peaks identified as $[M+Br]^-$ adduct are only kept if both isotopes were found. Isotopic filtering is also applied to ^{13}C, ^{15}N, and ^{34}S species in molecules, meaning an isotope peak is considered true if the corresponding monoisotopic peaks are also found. Figure 5 shows the main workflow of *MassTRIX*. As alternative a list of KEGG compound IDs can be submitted, bypassing the whole annotation procedure.

4.2 Analysis of Transcriptomic Data

Transcriptomic data can be submitted to *MassTRIX* in two different formats, either a self-annotated file or .cel files for Affymetrix gene chips. The first one contains KEGG IDs, KEGG KO numbers, EC number or gene identifiers, and a fold-change or UP and DOWN as keywords. The submitted values are used for coloring of the respective enzyme on metabolic pathway maps together with annotated compounds. This format allows the use of non-Affymetrix gene expression chips or other techniques such as serial analysis of gene expression or next-generation sequencing of transcripts. In the second variant, two .cel files as output of Affymetrix gene expression chips are submitted. One serves as a reference file and the other is specific for the sample state. The data is analyzed with the gene chip robust multiarray averaging (GCRMA) package in R. GCRMA is an improved version of the RMA method of normalization and summarization. GCRMA uses sequence-specific probe affinities of gene chip probes for more accurate gene expression values. Results from this analysis are fully downloadable for further investigation.

4.3 Comparison Against Other Existing Resources

Besides *MassTRIX*, several other solutions for annotation of mass spectrometric data exist. Two examples are the *Pathos* Web server (http://motif.gla.ac.uk/Pathos/pathos.html) *(41)* and *Paintomics* (www.paintomics.org) *(42)*. *Pathos* principally is based on the same functionality as *MassTRIX* and is written in Java and uses an underlying MySQL database. It annotates possible metabolites within an error range to experimental masses. Additionally, for this

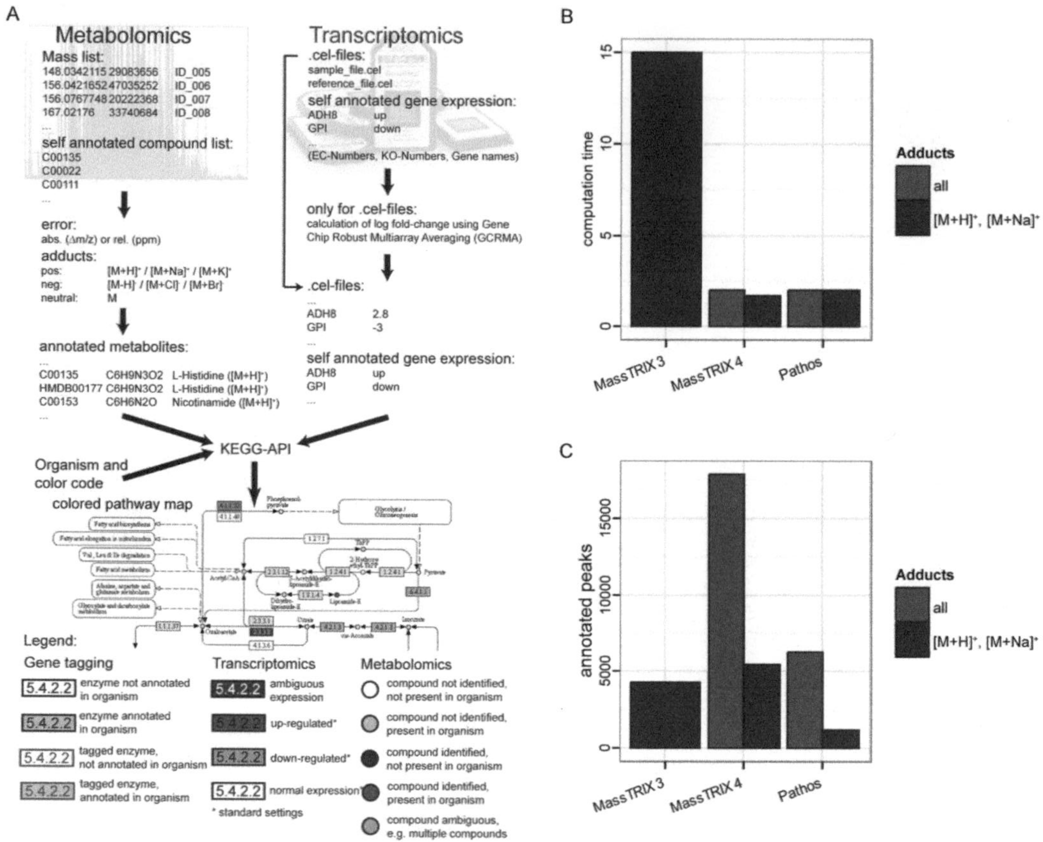

FIGURE 5 (A) Workflow for metabolomics and transcriptomic data. Results from both data types are mapped together on metabolic pathways obtained from KEGG. (B) Computation time for annotation of 25,644 MS peaks derived from *C. elegans* measured on a 12T Bruker solariX FT-MS. (C) Number of peaks with annotation. (For color version of this figure, the reader is referred to the online version of this chapter.)

comparison the annotation module of the newly programmed *MassTRIX 4* was included. Two major transitions were made in *MassTRIX 4* compared to version 3. First, the programming language was changed to Java for better maintenance of large a project, and second, the database was changed from flat files to MySQL. We used different comparisons to evaluate performance of each tool. *Pathos*, *MassTRIX 3*, and *MassTRIX 4* were compared by only annotating possible $[M+H]^+$ and $[M+Na]^+$ adducts. *Pathos* and *MassTRIX 4* were compared for all possible adducts. Data from a *C. elegans* metabolome extract measured on a Bruker solariX ICR-FT/MS containing 25,644 masses were subjected to the different tools. If only $[M+H]^+$ and $[M+Na]^+$ adducts are allowed for annotation, *Pathos* yielded 1223 annotated peaks in 2 min. Colored pathway maps are created on demand after the annotation process. However, the pathway maps are not cross-linked with other result pages as in *MassTRIX*. Additionally, submitted jobs are not stored on the server and have to be recalculated every time from the beginning. A basic comparison between different sample states is possible, mapping masses from different samples with different colors on the pathways. With *Pathos* no joined analysis and visualization of metabolomics and transcriptomics data is possible. *MassTRIX 3* needed for the same calculation finished in 15 min and yielded 4312 annotated masses. *MassTRIX 4* finished in 1.7 min with 5490 annotated masses. Because *Pathos* is only using masses occurring on metabolic pathways, it is limited to a certain subset of KEGG. *MassTRIX 3* uses a flat-file database, which slows down performance compared to *Pathos*. To obtain additional colored pathways in *MassTRIX 3*, 2–3 min more per pathway are needed, due to connection via the KEGG API to the KEGG database. Additional transcriptome data will just need several minutes more for calculation. Using all possible adducts for positive ionization, *Pathos* annotated 6251 peaks with possible metabolites and *MassTRIX 4* annotated 17,826 peaks. Both needed 2 min for the whole annotation process. On average, *MassTRIX 3* needed 0.34 s and *Pathos* and *MassTRIX 4* 0.05 s for processing of one peak.

The last Web server, *Paintomics*, only allows joint visualization of preanalyzed and identified metabolites and genes, making it different from the two previous Web servers. Representations based on the KEGG pathways are completely rewritten with XML and SVG technology to allow a multiplexed data representation. This is the big advantage of *Paintomics*.

The main functionality of *MassTRIX* is the direct annotation of mass spectrometric data to putative metabolites and direct mapping of these results to metabolic pathways. For more complex data visualization in networks, we recommend using *MassTRIX* annotation together with raw data in VANTED.

4.4 Future Directions for *MassTRIX*

MassTRIX is currently completely redesigned using Java instead of Perl, which offers more flexibility for the design of more complex data analysis steps. The underlying database is changed to MySQL, which offers more flexibility in maintenance and search speed, as shown above. Furthermore, all possible adducts mentioned in Huang *et al. (43)* will be available in the next release. A major focus in the next release will be the analysis of LC–MS metabolomics data. The only support for this data type in the current *MassTRIX* version is ability to use a bigger maximum error for instruments with smaller mass accuracy than ICR-FT/MS. Correct annotation in LC–MS data is much more complex due to chromatographic separation. It overcomes the major drawback of direct-infusion MS separation and overlap of isomeric and isobaric molecules, but produces a multiplicity of peaks with the same mass at different time points. The question that rises is which peak is belonging to which metabolite. We include improved analysis tools (e.g., correlation analysis of peaks) to find major adducts and fragments that derive from single metabolites. Furthermore, implementation of quantitative structure retention relationships will help in filtering false-positive annotations.

From the transcriptomic data side, more arrays will be added for increased usability (e.g., support for Agilent microarrays). Additional support for more than one file for reference and sample state will be included for improved statistics. Lastly, combined interpretation of metabolites and transcripts using enrichment and overrepresentation methods will be implemented.

5 CONCLUSIONS

Combination of "Omics" technologies in one biological experimental setup holds great opportunities for novel insights in systems regulation, metabolism, and overall homeostasis. Genomes can now be sequenced within days; the current bottleneck is the functional annotation. Therefore, the functional genomics tools, transcriptomics, proteomics, and metabolomics, together with all their subdisciplines evolved.

Increasing number of papers using combinations of different Omics approaches are published, whereas transcriptomics and metabolomics are often preferred. Combined analysis of both can be carried in different ways, as shown above. Different software tools have been developed for analysis of each single technology, but solutions for combined analysis are emerging. This is especially true for overrepresentation or enrichment analysis in high-level data fusion. With more powerful computer infrastructure available even low-level data fusion (e.g., correlation analysis) will be conducted. Here computational power will be needed because calculation time increases not linearly with data size but rather quadratic or higher.

One last interesting point should be drawn to the combination of transcriptomics, proteomics, and metabolomics. Proteomics and metabolomics are both based on similar chemical analysis techniques: LC–MS. It might be possible that in the future new work based on the combination of both or all three will be published. Virtually, high-resolution instruments such as the latest

Orbitrap or Q-ToF generations can be used for both. Integrating proteomics can help to overcome the major gap between gene expression and observed phenotype because altered expression of an enzyme may not change metabolite pools, but additional posttranslational modification does. Furthermore, from the metabolomics side increased metabolome coverage can improve combined data analysis. Currently, no methods that can cover all metabolites are available, but a combination of different analytical approaches (e.g., RP and HILIC separation) can improve the detected metabolite space.

Although only paper focusing on plant systems are mentioned here mainly, several other publications using metabolomics/transcriptomics exist (e.g., in the field of cancer research *(44)* or allergy *(45)*).

In summary, true systems biology seems to not be far away from the current point of view, although for successful application of more work on standardization of data exchange, annotation of biological entities has to be carried out.

REFERENCES

1. Fellner, L.; et al. Phenotype of htgA (mbiA), A Recently Evolved Orphan Gene of Escherichia Coli and Shigella, Completely Overlapping in Antisense to yaaW. *FEMS Microbiol. Lett.* **2014**, *350*, 57–64. http://dx.doi.org/10.1111/1574-6968.12288.
2. Wang, Z.; Gerstein, M.; Snyder, M. *Nat. Rev. Genet.* **2009**, *10*, 57–63.
3. Nicholson, J. K.; Lindon, J. C.; Holmes, E. *Xenobiotica* **1999**, *29*, 1181–1189.
4. Pauling, L.; Robinson, A. B.; Teranishi, R.; Cary, P. *Proc. Natl. Acad. Sci. U.S.A.* **1971**, *68*, 2374–2376.
5. Wilson, I. G. *Appl. Environ. Microbiol.* **1997**, *63*, 3741–3751.
6. Rossen, L.; Nørskov, P.; Holmstrøm, K.; Rasmussen, O. F. *Int. J. Food Microbiol.* **1992**, *17*, 37–45.
7. Boom, R.; Sol, C. J.; Salimans, M. M.; Jansen, C. L.; Wertheim-van Dillen, P. M.; van der Noordaa, J. *J. Clin. Microbiol.* **1990**, *28*, 495–503.
8. Millenaar, F. F.; Okyere, J.; May, S. T.; van Zanten, M.; Voesenek, L. A.; Peeters, A. J. *BMC Bioinforma.* **2006**, *7*, 137.
9. Rizzi, M.; Baltes, M.; Theobald, U.; Reuss, M. *Biotechnol. Bioeng.* **1997**, *55*, 592–608.
10. Villas-Bôas, S. G.; Roessner, U.; Hansen, M. A. E.; Smedsgaard, J.; Nielsen, J. *Metabolome Analysis—An Introduction*; 1st ed.; John Wiley & Sons Inc.: New Jersey, USA, 2007.
11. Rabinowitz, J. D.; Kimball, E. *Anal. Chem.* **2007**, *79*, 6167–6173.
12. Folch, J.; Lees, M.; Stanley, G. H. S. *J. Biol. Chem.* **1957**, *226*, 497–509.
13. Bligh, E. G.; Dyer, W. J. *Can. J. Biochem. Physiol.* **1959**, *37*, 911–917.
14. Matyash, V.; Liebisch, G.; Kurzchalia, T. V.; Shevchenko, A.; Schwudke, D. *J. Lipid Res.* **2008**, *49*, 1137–1146.
15. Schmidt, S. A.; Jacob, S. S.; Ahn, S. B.; Rupasinghe, T.; Krömer, J. O.; Khan, A.; Varela, C. *Metabolomics* **2013**, *9*, 173–188.
16. Weckwerth, W.; Wenzel, K.; Fiehn, O. *Proteomics* **2004**, *4*, 78–83.
17. Roume, H.; Muller, E. E.; Cordes, T.; Renaut, J.; Hiller, K.; Wilmes, P. *ISME J.* **2013**, *7*, 110–121.
18. Benton, H. P.; Wong, D. M.; Trauger, S. A.; Siuzdak, G. *Anal. Chem.* **2008**, *80*, 6382–6389.
19. Pluskal, T.; Castillo, S.; Villar-Briones, A.; Oresic, M. *BMC Bioinforma.* **2010**, *11*, 395.

20. Lommen, A.; Kools, H. J. *Metabolomics* **2012**, *8*, 719–726.
21. Urbanczyk-Wochniak, E.; Luedemann, A.; Kopka, J.; Selbig, J.; Roessner-Tunali, U.; Willmitzer, L.; Fernie, A. R. *EMBO Rep.* **2003**, *4*, 989–993.
22. Hannah, M. A.; Caldana, C.; Steinhauser, D.; Balbo, I.; Fernie, A. R.; Willmitzer, L. *Plant Physiol.* **2010**, *152*, 2120–2129.
23. Redestig, H.; Costa, I. G. *Bioinformatics* **2011**, *27*, i357–i365.
24. Xia, J.; Wishart, D. S. *Nucleic Acids Res.* **2007**, *38*, W71–W77.
25. Xia, J.; Mandal, R.; Sinelnikov, I. V.; Broadhurst, D.; Wishart, D. S. *Nucleic Acids Res.* **2012**, *40*, W127–W133.
26. Kankainen, M.; Gopalacharyulu, P.; Holm, L.; Oresic, M. *Bioinformatics* **2011**, *27*, 1878–1879.
27. Benjamini, Y.; Hochberg, Y. *J. R. Statist. Soc. B* **1995**, *57*, 289–300.
28. Kamburov, A.; Cavill, R.; Ebbels, T. M.; Herwig, R.; Keun, H. C. *Bioinformatics* **2011**, *27*, 2917–2918.
29. Chagoyen, M.; Pazos, F. *Brief. Bioinform.* **2013**, *14*, 737–744. http://dx.doi.org/10.1093/bib/bbs055.
30. Junker, B. H.; Klukas, C.; Schreiber, F. *BMC Bioinforma.* **2006**, *7*, 109.
31. Rohn, H.; Junker, A.; Hartmann, A.; Grafahrend-Belau, E.; Treutler, H.; Klapperstück, M.; Czauderna, T.; Klukas, C.; Schreiber, F. *BMC Syst. Biol.* **2012**, *6*, 139.
32. Saito, R.; Smoot, M. E.; Ono, K.; Ruscheinski, J.; Wang, P. L.; Lotia, S.; Pico, A. R.; Bader, G. D.; Ideker, T. *Nat. Methods* **2012**, *9*, 1069–1076.
33. Tziotis, D.; Hertkorn, N.; Schmitt-Kopplin, P. *Eur. J. Mass Spectrom.* **2011**, *17*, 415–421.
34. Suhre, K.; Schmitt-Kopplin, P. *Nucleic Acids Res.* **2008**, *36*, W481–W484.
35. Wägele, B.; Witting, M.; Schmitt-Kopplin, P.; Suhre, K. *PLoS One* **2012**, *7*, e39860.
36. Wapstra, A. H.; Audi, G.; Thibault, C. *Nucl. Phys. A* **2003**, *729*, 129–336.
37. Kanehisa, M.; Goto, S. *Nucleic Acids Res.* **2000**, *28*, 27–30.
38. Wishart, D. S.; Knox, C.; Guo, A. C.; Eisner, R.; Young, N.; Gautam, B.; Hau, D. D.; Psychogios, N.; Dong, E.; Bouatra, S.; Mandal, R.; Sinelnikov, I.; Xia, J.; Jia, L.; Cruz, J. A.; Lim, E.; Sobsey, C. A.; Shrivastava, S.; Huang, P.; Liu, P.; Fang, L.; Peng, J.; Fradette, R.; Cheng, D.; Tzur, D.; Clements, M.; Lewis, A.; De Souza, A.; Zuniga, A.; Dawe, M.; Xiong, Y.; Clive, D.; Greiner, R.; Nazyrova, A.; Shaykhutdinov, R.; Li, L.; Vogel, H. J.; Forsythe, I. *Nucleic Acids Res.* **2009**, *37*, D603–D610.
39. Caspi, R.; Foerster, H.; Fulcher, C. A.; Kaipa, P.; Krummenacker, M.; Latendresse, M.; Paley, S.; Rhee, S. Y.; Shearer, A. G.; Tissier, C.; Walk, T. C.; Zhang, P.; Karp, P. D. *Nucleic Acids Res.* **2008**, *36*, D623–D631.
40. Sud, M.; Fahy, E.; Cotter, D.; Brown, A.; Dennis, E. A.; Glass, C. K.; Merrill, A. H., Jr.; Murphy, R. C.; Raetz, C. R.; Russell, D. W.; Subramaniam, S. *Nucleic Acids Res.* **2007**, *35*, D527–D532.
41. Leader, D. P.; Burgess, K.; Creek, D.; Barrett, M. P. *Rapid Commun. Mass Spectrom.* **2011**, *25*, 3422–3426.
42. García-Alcalde, F.; García-López, F.; Dopazo, J.; Conesa, A. *Bioinformatics* **2011**, *27*, 137–139.
43. Huang, N.; Siegel, M. M.; Kruppa, G. H.; Laukien, F. H. *J. Am. Soc. Mass Spectrom.* **1999**, *10*, 1166–1173.
44. Zhang, G.; He, P.; Tan, H.; Budhu, A.; Gaedcke, J.; Ghadimi, B. M.; Ried, T.; Yfantis, H. G.; Lee, D. H.; Maitra, A.; Hanna, N.; Alexander, H. R.; Hussain, S. P. *Clin. Cancer Res.* **2013**, *19*, 4983–4993.
45. Singh, A.; Yamamoto, M.; Kam, S. H.; Ruan, J.; Gauvreau, G. M.; O'Byrne, P. M.; FitzGerald, J. M.; Schellenberg, R.; Boulet, L. P.; Wojewodka, G.; Kanagaratham, C.; De Sanctis, J. B.; Radzioch, D.; Tebbutt, S. J. *PLoS One* **2013**, *8*, e67907.

Chapter 18

Computational Approaches for Visualization and Integration of Omics Data

Vasudha Sehgal, Tyler J. Moss and Prahlad T. Ram
Department of Systems Biology, UT MD Anderson Cancer Center, Houston, Texas, USA

Chapter Outline

1 INTRODUCTION

The rapid advancement of research technology has made possible the greater than exponential growth of biological data over the past decade. The generation of large-scale Omics data provides the opportunity for a greater biological understanding, but one can get lost in the wash of data without proper biological context. Functional genomics, proteomics, transcriptomics, and so on, play an important role in biology to identify the component parts of systems and to help understand how the system components work in coordination in the proper functioning of cells and organisms. Genome-wide datasets are increasingly viewed as foundations for discovering pathways and networks relevant to phenotypes *(1)*.

New computational approaches are needed to contextualize, visualize, and understand these large, complex, and often heterogeneous data. In this

Comprehensive Analytical Chemistry, Vol. 63. http://dx.doi.org/10.1016/B978-0-444-62651-6.00019-2

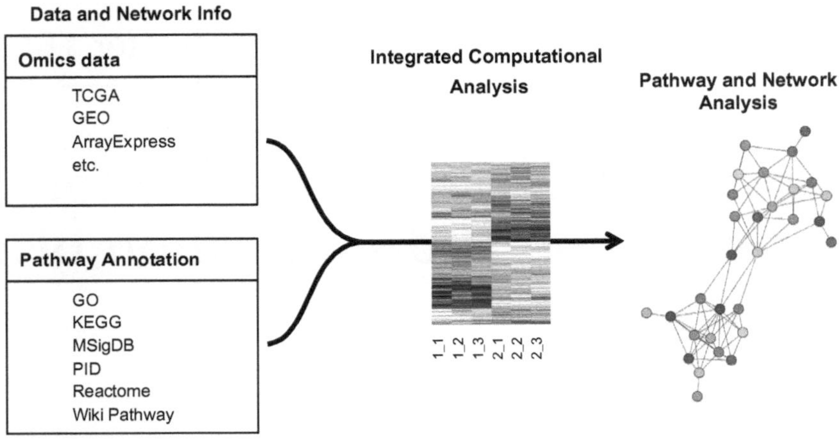

FIGURE 1 Scheme of integration and visualization of Omics data. (For color version of this figure, the reader is referred to the online version of this chapter.)

chapter, we discuss some of the approaches to functionally annotate, integrate, and visualize Omics data (Figure 1).

2 DATA OVERVIEW

With ever-increasing data from high-throughput technologies, the need for integration and visualization to understand their biological complexity is needed. Herein, we discuss the types of data that are being generated and their available sources as well as interaction databases and approaches to integrate and visualize data to come to a biological understanding of data in the proper context.

2.1 Data Types

The Omics data is a measurement of the totality of all molecules, interactions, and functions within given levels of cellular processes. Exploiting the information from Omics data is a key factor to develop new therapeutic insights for several diseases such as cancer. Genetic instability and abnormal levels and alterations in genomics are highly characteristic of cancer.

2.1.1 Genomics

Genomics is concerned with DNA-level information within the cell: genetic sequences, mutation, and so on, and all their functions; in other words, the complete genetic makeup of the cell. The first human genome was sequenced over a decade ago and since then, with the advent of next-generation sequencing, the cost per base-pair read has decreased by ~100,000-fold, with reads

~50,000 times faster *(2)*. Whole-genome sequencing at this scale makes possible the identification of mutations across thousands of tumors. Yet with all this genomic data now being produced, it is still difficult to identify mutations that are drivers of cancer. The genomic data must be integrated with other data to complete the picture. Moreover, further biological insights are required to understand the functional consequences of the genome alterations to fully understand the genomic data *(3)*. In addition to sequence data, the DNA copy number variations (CNV) and modification are also measured at the high-throughput scale. A subset of genomics, epigenomics is concerned with DNA modification such as methylation.

2.1.2 Transcriptomics

The totality of RNA produced from DNA transcription makes up the transcriptome. *Transcriptomics* are consemed *(4,5)* with all RNA transcribed: mRNA, as well as noncoding RNA such as miRNA and lincRNA. RNA levels were first measured on a high-throughput scale using oligonucleotide microarray technology. Recently, high-throughput techniques, by Illumina for instance, are used to generate high-throughput deep sequencing of the RNA and enables reads per million for isoforms of the mRNA. RNA deep sequencing shows the presence of RNA and measures the levels of RNA in cells.

2.1.3 Proteomics

Proteomics, protein abundance, function, and interactions within the cell, can be measured by liquid chromatography *(6)* and mass spectrometry or by reverse-phase protein arrays. The interactions are done by yeast-two-hybrid, cochromatography, and other methods.

2.1.4 Metabolomics

Metabolomics data are produced by mass spectrometry (MS)–based techniques and nuclear magnetic resonance to measure the totality of metabolites *(7)* in the cell and their levels. These types of analyses frequently utilize known metabolite data *(8)* to accurately quantify and annotate metabolites measured from the samples. Over the past decade, these technologies have developed to measure these different types of alterations, which are helpful for analyses of cancer Omics.

2.2 Data Sources

When microarray and other high-throughput data were in frequent use, the need arose to make the data accessible to other researchers as part of good practice. Today most journals require high-throughput data to be made publicly available and reach minimal information about a microarray experiment standards *(9)*.

Following publication, large-scale data can be uploaded to search data repositories. The two most popular are arrayExpress and the Gene Expression Omnibus (GEO). Community-generated data stored in ArrayExpress and GEO, maintained by the European Bioinformatics Institute and the National Center for Biotechnology Institute, respectively, are searchable and readily available for download and analysis. Verification analyses can be performed using these previously published datasets.

Several initiatives have resulted in the generation of public Omics data for use by the scientific community to tackle complex problems such as cancer. Although data from large-scale cancer genome projects have been made publicly available, accessing and using this data for analyses still remains a challenge for most investigators without prior experience in bioinformatics or systems biology. In addition, crowd-sourcing to analyze and visualize Omics data has also been very successful with the launch of the DREAM Project (http://www.the-dream-project.org/). There is greater emphasis now to openly share these large datasets and use what we have learned from social media to facilitate crowd-sourcing-type methods to help in the analysis. In this section, we provide a brief understanding of the data available from these projects along with the sources to access these datasets.

The Cancer Genome Atlas (TCGA) was begun by the National Cancer Institute to generate data from many institutes and integrate the data from different platforms to capture a wider picture of cancer genomics. With the TCGA data portal (http://cancergenome.nih.gov/), researchers can access the data for multiple types of cancers on a reasonably large number of samples on methylation, CNV, RNA sequencing, and so on. Two other major large-scale projects, The Cancer Genome Project (CGP), data available through the Welcome Trust Sanger Institute (http://www.sanger.ac.uk/genetics/CGP), and the International Cancer Genome Consortium (ICGC), data available through (http://dcc.icgc.org), are publically published and recognized *(10)*. Although these large-scale projects have made their data available publicly, the privacy of the patients is preserved since public access is only open to higher levels of this data with certain normalization already performed. Apart from these large-scale projects, several other sources are available where one can browse, search, and view the copy number and expression data. The UCSC Cancer Genome Browser (https://genome-cancer.soe.ucsc.edu) is user-friendly and a popular copy number and expression data viewer. The user can perform searches for their gene of interest (mRNA and miRNA) and view the alterations in copy number, other genes present, and information on the entire chromosome or zoom in to the specific chromosome location of interest. One other data portal is cBio Cancer Genomics (http://www.cbioportal.org/public-portal/), another searchable and available viewer for alterations in copy number and expression data. The cBio portal integrates data from the TCGA datasets and enables researchers to perform some quick analyses of CNV and expressions for the genes (mRNA and miRNA) of interest.

Oncomine (http://www.oncomine.org) is a commonly used site for querying analyzed cancer gene expression data. Oncomine allows comparison of any two datasets from different cancers to determine the genes that are specifically expressed in the dataset of interest *(11)*.

Nucleotide substitutions and small insertions and deletions also hamper the normal functioning of the cancer genomics. One useful database developed under the Sanger Institute, the Catalog of Somatic Mutations in Cancer (COSMIC) available at http://www.sanger.ac.uk/genetics/CGP/cosmic is an open database that contains data for somatic mutations and copy number alterations by gene, amino acid position, tumor type, and literature references. Table 1 displays the summary of various databases for cancer genomics that are publicly available.

TABLE 1 Databases for Cancer Genomic Data

Database	Link	Data Types
Databases Generating Data		
TCGA	http://cancergenome.nih.gov/dataportal	Methylation, copy number, mRNA and miRNA expression, and mutation sequencing
ICGC	http://dcc.icgc.org/	Copy number, expression, and mutation sequencing
CGP	http://www.sanger.ac.uk/genetics/CGP/Archive	SNP genotype profile and first-generation trace archive
CCLE	http://www.broadinstitute.org/ccle/home	Genomic data, analysis, and visualization for about 1000 cell lines
Databases Using the Patient/Known Data		
UCSC Cancer Genome Browser	https://genome-cancer.soe.ucsc.edu	Copy number and expression viewer
cBio Cancer Genomics Portal	http://www.cbioportal.org/public-portal/	Viewer for copy number and expression data; integrates data from TCGA, etc.
GEO	http://ncbi.nlm.nih.goc/geo	Gene expression
Oncomine	http://www.oncomine.org	Gene expression and copy number
COSMIC	http://www.sanger.ac.uk/genetics/CGP/cosmic	Somatic mutations and copy number alterations

3 DATA PROCESSING AND ANALYZING TOOLS

The sources mentioned above are useful for accessing publicly available data on several Omics data types including gene expression, copy number alterations, proteomics, and so on, as well as connecting these data to important health-related questions such as patient outcome. The data for most of these portals has gone through the initial normalization. Other major aspects in making sense of cancer genomics data is to be able to understand the biological functions responsible for causing alterations in cancer genomics. A lot of tools have been developed to do so efficiently. Some of these include the above-mentioned data portals themselves: the cBio Cancer Genomics Portal and the UCSC Cancer Genome Browser. Measuring the allelic gains and losses is an important aspect of analyzing copy number alterations. GISTIC is a major algorithm developed in this aspect *(12)*. One can also generate the GISTIC scores by using the cBio data portal. Bioconductor (http://www.bioconductor.org) in programming language R also has some packages available (cghMCR and CNTools) for analyzing copy number alterations.

The gene lists or signatures for the differentially expressed genes can be integrated with known pathway interactions information. Gene set enrichment analysis (GSEA) (http://www.broadinstitute.org/gsea/index.jsp) is an algorithm *(13)* that allows one to check whether the differentially expressed gene set is enriched for a particular gene set. Another important aspect of understanding functional genomics is to be able to find the pathways that are enriched for particular gene sets. Table 2 lists some of the open source analytical tools for analyzing genomics data.

4 NETWORK AND PATHWAY DATABASES

Molecular data generated from different high-throughput platforms inform us of the presence of SNPs or mutations in a gene, the degree of expression of a particular gene, or the abundance of a protein or metabolite. These data do not, however, tell us about the functions of the molecules, its interaction with

TABLE 2 Open-Source Analytical Tools for Cancer Genomics Data

Tools	Link
Bioconductor	http://www.bioconductor.org
Gene Pattern	http://www.broadinstitute.org/genepattern
Gene Ontology	http://geneontology.org/GO.tools.microarray.shtml
UCSC Cancer Genome Browser	https://genome-cancer.soe.ucsc.edu
Integrative Genomics Viewer (IGV)	http://www.broadinstitute.org/igv

other molecules, or the impact that dysregulation may have upon cells, tissues, or organisms. For a complete picture, Omic data must be integrated with functional, network, and pathway data.

4.1 Protein Interaction Databases

Genes and proteins do not operate in a vacuum. Macromolecules perform their function in the context of thousands of other genes and proteins through physical and genetic interaction. Publicly available databases have curated these interactions. Table 3 displays some of these databases.

4.2 Pathway Commons

Pathway information captures knowledge of biological processes at the molecular level and can be an important tool for interpreting the growing amount of biological data from genomic studies. Pathway and network information can also be usefully combined with high-throughput genomic data and

TABLE 3 Publically Available Databases with Curated Protein–Protein Interactions

Database	Description	URL
BioGrid	Supports major model organisms	http://thebiogrid.org/
CORUM	Database of mammalian protein complex information	http://mips.helmholtz-muenchen.de/genre/proj/corum/index.html
DIP	Database for experimentally determined interaction for proteins	http://dip.doe-mbi.ucla.edu/dip/Main.cgi
IntAct	Molecular interaction data	http://www.ebi.ac.uk/intact/
HPRD	Integrates architecture, posttranslational modifications, interaction networks, and disease association for each protein in the human proteome	http://www.hprd.org/
MINT	Experimentally verified protein–protein interactions	http://mint.bio.uniroma2.it/mint/Welcome.do
MIPS	Mammalian database	http://mips.helmholtz-muenchen.de/proj/ppi/
I2D	Database for five model organisms and human	http://ophid.utoronto.ca/ophidv2.204/

clinical phenotype data to investigate the properties of specific disease types *(14)* and to build classifiers for disease subtypes *(15)*.

Pathway Commons (http://www.pathwaycommons.org) is a freely available database that collects, normalizes, and integrates publicly available biological pathway and molecular interaction data about cellular processes. It is a collection of publicly available pathway data from multiple organisms. Pathway Commons provides a Web-based interface that enables biologists to browse and search a comprehensive collection of pathways from multiple sources *(16)*. Users can ask questions such as "What proteins interact with a protein under study?", "What pathways involve a given protein?", "Is a specific protein involved in transport events or biochemical reactions?", or "What enzymes use a specific metabolite as a substrate?".

5 VISUALIZATION OF OMICS DATA

The visual representation of data is vital in the analysis, understanding, and communication of these data. Clear, concise, and accurate visualization becomes more important when working with Omics data due its scale and inherent complexity. With large-scale data, new visualization techniques are required to analyze and interpret Omic data. A common method of visualizing transcriptome data is overlaying data on interaction data in the form of networks. Networks representations exhibit the broad omic data in the framework of biological networks, focusing the data down to a comprehensible scale while preserving the systems-level information. One of the challenges of working with large-scale data is focusing on the data on a comprehensible scale while preserving the systems-level information and representing the data visually. A common visual representation of Omic data is in the form of networks that frame the data in the context of biological interactions and focus the data to their biological relevance. As the volume of Omic data has expanded, so too have tools that interpret and represent the data. Here we focus on tools to integrate transcriptomic and proteomic data with interaction data in network form, and discuss some of the tools available for network creation, visualization, and analysis of the proteomic data (Table 4).

5.1 Clustering and Heatmaps

In gene expression data, it is important to be able to visualize the data for different genes and samples. The most common form of visualizing this data is in the form of heatmaps. A heatmap is a graphical representation of data where the individual values of the matrix are represented in color. Most commonly a red-green heatmap is displayed, with high intensity represented by red and low intensity represented by green. This is useful to be able to visually identify the subset of genes or samples exhibiting the same or consistent pattern. A key step in the analyses of gene expression data is the identification

TABLE 4 Top Downloaded Plug-ins Available for Cytoscape 3.0+

Cytoscape app	Description	Downloads
ClueGo	Creates and visualizes a network of grouped functional and pathway terms (GO terms)	4966
GeneMania	Imports and creates interacting networks from public databases based on input gene sets	3334
MCODE	Topology-driven network clustering	2758
jActiveModules	Expression-driven network clustering	2747
CluePedia	ClueGo plug-in search tool to identify new nodes associated with pathway terms	2552
DynNetwork	Animated visualization of dynamic networks	2301
AgilentLiteratureSearch	Network creation based on literature mining	2197

of groups of genes exhibiting similar expression patterns. Clustering gene expression data into homogeneous groups was shown to be useful in functional annotation, motif identification, and so on *(17)*. There are various clustering algorithms and software available, such as the tool for analyzing and visualizing the results of complex microarray experimental data from the Eisen lab (http://rana.lbl.gov/EisenSoftware.htm). It performs various types of clustering analysis and the other preprocessing required for large datasets, some of which include hierarchical clustering and k-means clustering, which are described in Eisen *et al. (18)*.

5.2 Tools for Network Creation, Visualization, and Analysis

5.2.1 Cytoscape

Cytoscape (www.cytoscape.org) is a popular network tool developed and maintained by the National Institute of General Medical Sciences. Cytoscape is an open-source software platform for visualizing complex networks and integrating these with any type of attribute data. Cytoscape's software Core provides basic functionality to layout and queries the network; visually integrates the network with expression profiles, phenotypes, and other molecular states; and links the network to databases of functional annotations. An array of community-developed plug-ins is available on the cytoscape website to expand the functionality of this bioinformatic tool *(19–24)*.

5.2.2 Circos

Circos is a software package for visualizing data and information (http://circos.ca) in a circular layout. This new and attractive method of visualizing large data has been developed in the form of circos plots. Circos plots are often chromosomes laid out in a circle with genomic, epigenetic, and transcriptomic data aligned and layered about the chromosomes in concentric rings *(25)*. Edges can also be drawn between genetic interactions within the circle. Visualization of data in this circular layout format enables exploration of relationships between different objects in the data. Circos plots are useful to visualize multidimensional data of different types and sources where relationships between elements are too complex to visualize in a normal two-dimensional table or matrix. One of the caveats of using circos plots currently is the requirement of the users to be familiar with the PERL programming language and interface.

5.2.3 Netwalker

NetWalker is a desktop application (Figure 2) developed in our lab for functional analyses of large-scale genomics datasets within the context of molecular networks (https://netwalkersuite.org). It is based on network-based analyses of data that integrate experimental gene expression data with prior knowledge interactions for the retrieval of most relevant biomolecular networks. NetWalker architecture is designed to enable network analyses based on global (no cutoff) integration of experimental data with *a priori* networks and to allow extensive interoperability between analysis components and external applications. NetWalker features random walk-based analysis methods for prioritization of network interactions and functional processes,

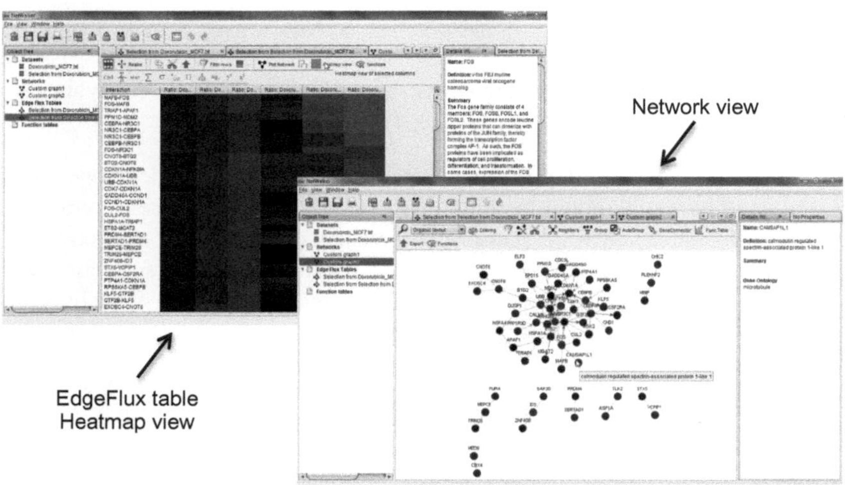

FIGURE 2 Netwalker is a user-friendly desktop application for analyzing gene expression data in the context of known biological interactions. (For color version of this figure, the reader is referred to the online version of this chapter.)

respectively, based on assessment of local network connectivity in conjunction with experimental data. This serves as a good platform for querying, analysis, and visualization of networks of interest *(26)*.

6 CONCLUSION

Understanding and being able to analyze the large-scale omics data produced on a daily basis is key to answering many important biological questions. Through the combined efforts of the scientific community at large a wealth of multilevel omic data has been generated and made publicly available. To understand these numerous and complex data focus has shifted towards developing diverse methodologies to analyze and visualise omics data to understand diseases and develop therapeutics. In this chapter, we have focused on presenting the different types of omics data available to the public and the various publicly available platforms and sources to visualize and integrate the data together with known knowledge bases for a contextual understanding of biological pathways. One popular way of understanding omics data is analyzing the data in the context of cellular pathways. Molecular omics data in the context of protein–protein and interaction networks help identify pathways that are dysregulated in diseases. Software such as Cytoscape and Netwalker make use of the interaction networks and provide a framework within which we can analyze and visualize the omics data.

REFERENCES

1. Hirschhorn, J. N. *N. Engl. J. Med.* **2009**, *360*, 1699–1701.
2. Lander, E. S. *Nature* **2011**, *470*; 187–197.
3. Chin, L.; Hahn, W. C.; Getz, G.; Meyerson, M. *Genes Dev.* **2011**, *25*, 534–555.
4. Rinck, A.; Preusse, M.; Laggerbauer, B.; Lickert, H.; Engelhardt, S.; Theis, F. J. *RNA Biol.* **2013**, *10*, 1125–1135.
5. Forrest, A. R.; Carninci, P. *RNA Biol.* **2009**, *6*, 107–112.
6. Nika, H.; Nieves, E.; Hawke, D. H.; Angeletti, R. H. *J. Biomol. Tech.* **2013**, *24*, 154–177.
7. Zheng, X.; Qiu, Y.; Zhong, W.; Baxter, S.; Su, M.; Li, Q.; Xie, G.; Ore, B. M.; Qiao, S.; Spencer, M. D.; Zeisel, S. H.; Zhou, Z.; Zhao, A.; Jia, W. *Metabolomics* **2013**, *9*, 818–827.
8. De Livera, A. M.; Dias, D. A.; De Souza, D.; Rupashinghe, T.; Pyke, J.; Tull, D.; Toessner, U.; McConveille, M.; Speed, T. P. *Anal. Chem.* **2012**, *84*, 10768–10776.
9. Brazma, A.; Hingamp, P.; Quackenbush, J.; Sherlock, G.; Spellman, P.; Stoeckert, C.; Aach, J.; Ansorge, W.; Ball, C. A.; Causton, H. C.; Gaasterland, T.; Glenisson, P.; Hosltege, F. C.; Kim, I. F.; Markowitz, V.; Matese, J. C.; Parkinson, H.; Robinson, A.; Sarkans, U.; Schulze-Kremer, S.; Stewart, J.; Taylor, R.; Vilo, J.; Vingron, M. *Nat. Genet.* **2001**, *29*, 365–371.
10. The International Cancer Genome Consortium. *Nature* **2010**, *464*, 993–998.
11. Rhodes, D. R.; Yu, J.; Shanker, K.; Deshpande, N.; Varambally, R.; Ghosh, D.; Barrete, T.; Pandey, A.; Chinnaiyan, A. M. *Neoplasia* **2004**, *6*, 1–6.
12. Beroukhim, R.; Getz, G.; Nghiemphy, L.; Barretina, J.; Hsueh, T.; Linhart, D.; Vivanco, I.; Lee, J. C.; Huang, J. H.; Alexander, S.; Du, J.; Kau, T.; Thomas, R. K.; Shah, K.; Soto, H.;

Perner, S.; Prensner, J.; Debiasi, R. M.; Demichelis, F.; Hatton, C.; Rubin, M. A.; Garraway, L. A.; Nelson, S. F.; Liau, L.; Mishcel, P. S.; Cloughesy, T. F.; Meyerson, M.; Golub, T. A.; Lander, E. S.; Mellinghoff, I. K.; Seller, W. R. *Proc. Natl. Acad. Sci. U.S.A.* **2007**, *104*, 20007–20012.

13. Subramanian, A.; Tamayo, P.; Mootha, V. K.; Mukherjee, S.; Ebert, B. L.; Gillette, M. A.; Paulovich, A.; Pomeroy, S. L.; Golub, T. R.; Lander, E. S.; Mesirov, J. P. *Proc. Natl. Acad. Sci. U.S.A.* **2005**, *102*, 15545–15550.
14. Cerami, E. G.; Demir, E.; Shcultz, N.; Taylor, B. S.; Sander, C. *PLoS One* **2010**, *5*, e8918.
15. Chuang, H. Y.; Lee, E.; Liu, Y. T.; Lee, D.; Ideker, T. *Mol. Syst. Biol.* **2007**, *3*, 140.
16. Cerami, E. G.; Gross, B. E.; Demir, E.; Rodchenkov, K.; Babur, O.; Anwar, N.; Schultz, N.; Bader, G. D.; Sander, C. *Nucleic Acids Res.* **2011**, *39*, D685–D690.
17. Sharan, R.; Elkon, R.; Shamir, R. *Ernst Schering Res. Found Workshop* **2002**, *38*, 83–108.
18. Eisen, M. B.; Spellman, P. T.; Brown, P. O.; Botstein, D. *Proc. Natl. Acad. Sci. U.S.A.* **1998**, *95*, 14863–14868.
19. Camilo, E.; Bovolenta, L. A.; Acencio, M. L.; Rubarczyk-Filho, J. L.; Castro, M. A.; Moreira, J. C.; Lemke, N. *Bioinformatics* **2013**, *29*, 2505–2506.
20. Ligtenberg, W. P.; Bosnacki, D.; Hilbers, P. A. *J. Bioinform. Comput. Biol.* **2013**, *11*, 1350004.
21. O'Brien, K. T.; Haslam, N. J.; Shields, D. C. *BMC Bioinforma.* **2013**, *14*, 224.
22. Shannon, P. T.; Grimes, M.; Kutlu, B.; Bot, J. J.; Galas, D. J. *BMC Bioinforma.* **2013**, *14*, 217.
23. Zhang, C.; Wang, J.; Hanspers, K.; Xu, D.; Chen, L.; Pico, A. R. *Bioinformatics* **2013**, *29*, 2066–2067.
24. Shannon, P.; Markiel, A.; Ozier, O.; Baliga, N. S.; Wang, J. T.; Ramage, D.; Amin, N.; Schwikowski, B.; Ideker, T. *Genome Res.* **2003**, *13*, 2498–2504.
25. Krzywinski, M.; Schein, J.; Briol, I.; Connors, J.; Gascoyne, R.; Horsman, D.; Jones, S. J.; Marra, J. A. *Genome Res.* **2009**, *19*, 1639–1645.
26. Komurov, K.; Durson, S.; Erdin, S.; Ram, P. T. *BMC Genomics* **2012**, *13*, 282.

Index

Note: Page numbers followed by "*f*" indicate figures and "*t*" indicate tables.

D

E

F

G

H

I

N

O

P

Q

R